Biogeochemical Investigations at Watershed, Landscape, and Regional Scales

Refereed papers from **BIOGEOMON**, *The Third International Symposium on Ecosystem Behavior; Co-Sponsored by Villanova University and the Czech Geological Survey; held at Villanova University, Villanova Pennsylvania, USA, June 21–25, 1997*

Edited by

R. KELMAN WIEDER
Villanova University
Villanova, Pennsylvania, USA

MARTIN NOVÁK and JIŘÍ ČERNÝ
Czech Geological Survey
Prague, Czech Republic

Reprinted from *Water, Air, and Soil Pollution* 105(1–2), 1998

Kluwer Academic Publishers
Dordrecht / Boston / London

A C.I.P. Catalogue record for this book is available from the Library of Congress.

ISBN 0-7923-5167-3

Published by Kluwer Academic Publishers
P.O. Box 17, 3300 AA Dordrecht, The Netherlands

Sold and distributed in North, Central and Latin America
by Kluwer Academic Publishers
101 Philip Drive, Norwell, MA 02061, U.S.A.

In all other countries, sold and distributed
by Kluwer Academic Publishers
P.O. Box 322, 3300 AH Dordrecht, The Netherlands

Printed on acid-free paper

Printed in the Netherlands

TABLE OF CONTENTS

PREFACE

This Special Issue of *Water, Air, and Soil Pollution* offers original contributions from ***BIOGEOMON****, The Third International Symposium on Ecosystem Behavior*, which was held on the campus of Villanova University from June 21-25, 1997. Previous meetings were held in Prague in 1987 and again in 1993. The ***BIOGEOMON*** series was initiated in 1987 when a group of researchers from the Czech Geological Survey organized a conference called ***GEOMON****, Geochemical Monitoring in Representative Basins.* ***GEOMON*** was fairly narrowly focused on monitoring of element pools and fluxes on a small watershed scale. As signalled by the change in name to ***BIOGEOMON***, the second conference explicitly recognized that assessment of anthropogenic effects on ecosystem processes requires a combination of geochemical monitoring with other approaches, including watershed-level manipulations, use of radioactive and stable isotopic tracers, and both empirical and process modeling. The 1997 ***BIOGEOMON*** conference was the largest, with over 240 participants from 28 countries on five continents in attendance, and broadest in scope. The conference featured a plenary speaker, six keynote speakers, 35 invited speakers, over 60 oral contributed presentations, and over 75 poster presentations. Presentations were organized into nine thematic sessions: acidification and its reversal; nitrogen cycling, limitation, and saturation in terrestrial ecosystems; natural abundance isotopes in biogeochemistry; nutrient cycling; interfacing biogeochemistry with geographic information systems; watershed monitoring; watershed manipulation; watershed modeling; and carbon gas flux across the land/atmosphere boundary. We are pleased that all of these themes are represented in the collection of papers published in this Special Issue - a volume that is 13% larger that the one resulting from the 1993 conference.

The ***BIOGEOMON*** series has continued to provide a forum for the dissemination and discussion of recent research findings, to explore future directions for biogeochemical research, and to foster interdisciplinary collaboration between researchers of all ages at regional, national, and international scales. More broadly, the philosophy of ***BIOGEOMON*** is stated in the Preface to the special issue of *Water, Air and Soil Pollution* (vol. 79) that resulted from the 1993 conference:

> *With international openness comes the increasing realization that every industrialized nation has its own legacy of environmental degradation. Anthropogenic impacts differ in severity and scale; air and water transport of pollutants transform local impacts into regional and global ones, ignoring political boundaries and economic differences. Environmental consequences of anthropogenic activities often are detectable at the ecosystem level. Thus, the challenge of ecosystem science, and to those who practice it, is to develop a comprehensive understanding of ecosystem function in the past and at present, and to apply such understanding toward minimizing future insults to the local, regional, and global environment.*

The future of biogeochemistry is bright and promising. As the world's human population continues to grow into the 21st century, human activities will increasingly alter the natural cycles of elements and flows of energy. The maturing disciplines of biogeochemistry and ecosystem science are uniquely suited for assessing past and present anthropogenic disruptions and disturbances to the earth's ecosystems at local, regional, and global scales. Biogeochemists will need to interact closely with ecosystem modelers to strive to make better predictions of known and projected human activities on the structure and function of the earth's ecosystems, again at scales ranging from local to global.

The future of the ***BIOGEOMON*** series remains, as it always has been, uncertain. It is our hope that through our continuing efforts, and through increasing involvement of the international biogeochemistry community, that ***BIOGEOMON*** will continue as a regular convening of researchers. If successful, through regular meetings and through dissemination of research findings in published abstracts and papers, ***BIOGEOMON*** can contribute meaningfully to the ongoing development of biogeochemistry and ecosystem science.

R. Kelman Wieder
Martin Novák
Jiří Černý

ACKNOWLEDGMENTS

This Special Issue would not have come into being without the dedication and enthusiasm of many members of the Villanova community and of colleagues in the Czech Geological Survey. Because of the efforts of these people, ***BIOGEOMON*** was a smoothly run and successful conference. We especially thank the members of the ***BIOGEOMON*** Local Organizing Committee, Merritt Turetsky, Scott Starr, and Lee Walton, along with the numerous Villanova University staff and students (Brian Coupe, Kathy Daumer, Tim Dietzler, Brian Hermann, John Jendro, Dan Lammey, Kira McCarthy, Hiren Patel, Leah Urbaniak, Melanie Vile, Maryann Welsch, Anne Webster, Chris Williams), as well as Eva Pačesová and Jaroslavá Zusková of the Czech Geological Survey. Prior to the conference, Melanie Vile, Jiří Černý, and Martin Novák worked tirelessly as the Editorial Committee for the ***BIOGEOMON*** abstracts, which were published in the *Journal of Conference Abstracts* (vol. 2, no. 2, June 1997; Cambridge Publications, UK; 334 pp.) and were distributed in hard copy to all ***BIOGEOMON*** attendees. All ***BIOGEOMON*** presenters had the option of submitting a manuscript for consideration for inclusion in this Special Issue. Of the 77 manuscripts that were submitted, 45 appear here. Each manuscript was reviewed by a minimum of two researchers. We are very grateful to those who gave of their time and effort to review submitted manuscripts thoroughly, carefully, and critically. The names and institutional affiliations of these reviewers are listed at the end of the Special Issue. Also, we thank the staff of Kluwer Academic Publishers, especially Mariette de Jong and Astrid Zandee for guidance and for facilitating our efforts to produce this Special Issue in a timely and efficient manner. Our respective home institutions, Villanova University and the Czech Geological Survey, provided generous support for all of our activities related to ***BIOGEOMON***, including the preparation of this Special Issue. Finally, none of this could have been possible without considerable external financial assistance. In this regard, we acknowledge the financial support for ***BIOGEOMON*** that was obtained from the National Science Foundation, the Electric Power Research Institute, and the National Oceanic and Atmospheric Administration, with additional support provided by the Valley Forge Brewing Company.

R. Kelman Wieder
Martin Novák
Jiří Černý
Tomáš Pačes

INTEGRATED ASSESSMENT OF SOIL CHEMICAL STATUS. 1. INTEGRATION OF EXISTING MODELS AND DERIVATION OF A REGIONAL DATABASE FOR SWITZERLAND

M. ALVETEG[1], H. SVERDRUP[1] and D. KURZ[2]

[1]*Department of Chemical Engineering II, University of Lund, S-221 00 Lund, Sweden*
[2]*EKG Geo-Science, Bern, Switzerland*

Abstract. A regional soil acidification model was developed by integration and adaptation of existing models. The regional model consists of the dynamic multi-layer soil chemistry model SAFE, its steady-state version INITSAFE, the atmospheric deposition and nutrient uptake reconstruction model MAKEDEP, and a routine with empirical relations concerning depth-dependent parameters. A scheme for the extraction of input to the regional model from available information of different geographical detail also was developed. Basic data sources considered were: 1) national surveys such as the National Forest Inventory, covering site specific information, 2) available point measurements of parameter values, and 3) literature sources. Not all parameters were available on a regional scale with sufficient resolution. Input required for the model calculations therefore was derived from the available data sources by means of transfer algorithms including spatial interpolation. Interpolation was done allocating parameter values determined at reference sites to conventionally mapped entities such as geological units, soil type, and other kinds of geographical information. The exercise resulted in a data base of the required 68 site-specific parameter values covering climatic, deposition and land use parameters, as well as stand characteristics and soil properties.

Keyword: dynamic modeling, INITSAFE, MAKEDEP, recovery, SAFE, soil acidification

1. Introduction

Over the last decade steady-state models, from the simple mass balance (SMB) model (Posch *et al.*, 1997) to more complex models such as PROFILE (Sverdrup and Warfvinge, 1992; Warfvinge and Sverdrup, 1992a) have been applied regionally throughout Europe. The main purpose with the regional exercises often has been to prepare critical loads on the EMEP150 grid resolution (Posch *et al.*, 1997). The critical loads have been used in turn by the RAINS model (Alcamo *et al.*, 1990) to assess cost optimal emission reductions strategies for Europe.

Steady-state models do not, however, address time aspects. Steady-state models answer questions regarding the state toward which a system is drifting, but do not give any answers on how long the transition will take, or how the present state relates to the steady-state. The critical load concept is based on a steady-state approach, the critical load being the load at which a chemical criterion eventually reaches a critical limit if the external conditions are held constant forever. For forest ecosystems, the chemical criterion is usually the base cation to aluminum (Bc/Al) molar ratio and the critical limit usually is set to unity. Only Ca, Mg, and K are included in the base cation (Bc) concentration, as Na is not regarded as an important nutrient. The present exceedance of the critical load does not give us any information about the current status of the chemical criterion. If the differences between sites in exceedance or soil resistance to acidification is large enough, we are likely to find a relation between exceedance and effects at different sites. The absence of such a relation, however, does not falsify the critical load concept as we

Water, Air, and Soil Pollution **105:** 1–9, 1998.

compare the ever-changing state variables of dynamic systems at different points in time.

To be able to compare present conditions of a dynamic system with computer model results in a consistent manner, it is necessary to turn to dynamic models. Dynamic models may answer such questions as the timing of effects and the time delays involved in recovery. Dynamic models, however, suffer a severe drawback as compared to steady-state models because dynamic models are more data demanding. In this study, we present a method to extract data on the regional level from available sources of information and derive sufficient information to apply dynamic soil chemistry models, specifically the SAFE model (Warfvinge and Sverdrup; 1992; Warfvinge *et al.* 1993).

2. Method

The application of the dynamic soil chemistry models on a regional level can be divided into five steps and includes three existing models (Figure 1): 1) data acquisition (Table I); 2) empirical relations (Table II); 3) the MAKEDEP reconstruction model (reconstructs atmospheric deposition, nutrient uptake and nutrient cycling; Alveteg *et al.*, 1997); 4) the INITSAFE model (steady-state version of the SAFE model, similar to the PROFILE model); and 5) the SAFE model (adynamic multi-layer soil chemistry model).

Of the previously existing models, MAKEDEP was revised and adapted to central European conditions, whereas only technical changes where made to the INITSAFE and SAFE models. The MAKEDEP model, the INITSAFE model and the empirical relations were combined into one computer program, REGPREP, which reads regionalized input and produces input to the SAFE model and diagnostic output. For a comprehensive description of the regionalized model, see FOEFL (1997).

2.1. The SAFE Model

The SAFE model is a dynamic multi-layer soil chemistry model. The SAFE model includes process-oriented descriptions of cation exchange reactions, chemical weathering of minerals, solution equilibrium reactions involving carbon dioxide, organic acids and Al species, and finally leaching and accumulation of dissolved chemical components. The Al concentration in soil solution is simulated using layer-specific apparent gibbsite coefficients. SAFE needs time series of input regarding atmospheric deposition, net uptake of nutrients, litterfall, canopy exchange, net mineralization and the precipitation flux, as well as input regarding physical and chemical soil parameters (see Table II).

The change in soil solution chemistry and the subsequent change in the distribution of elements on the exchange matrix is calculated by means of conservation equations. The cations Ca^{2+}, Mg^{2+} and K^{+} are grouped together on an equivalent basis and are regarded to react as divalent component base cations (Bc). Sodium Na^{+} is neglected as an exchangeable species as well as a nutrient, and is assumed to behave as a conservative element. Hydrogen ion H^{+} is treated as dependent on acid neutralizing capacity (ANC). Solution ANC is defined as the sum of concentrations in equivalents of OH^{-}, carbonate and organic anions minus that of H^{+} and Al cations.

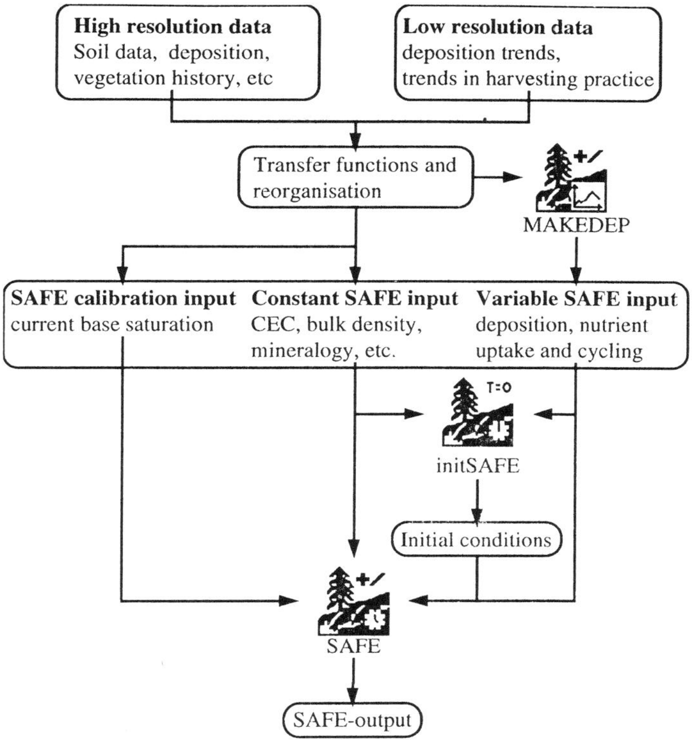

Fig. 1. Schematic view of the integrated regional model. Assessing the high and low resolution data is, in this study, described as data acquisition.

2.1.1. Cation Exchange and Weathering

Cation exchange is simulated according to the Gapon exchange equation. It is assumed that the selectivity coefficient, as well as the cation exchange capacity (CEC), is constant over time. This is a simplification because CEC usually varies with pH and because the amount of soil organic matter may vary. The rate of exchange is assumed to be controlled by mass transport.

Weathering rates are calculated using four weathering rate reactions Sverdrup and Warfvinge, 1995). The simulated weathering rate is dependent on mineralogy, water content, temperature and soil solution chemistry.

2.1.2. Initial Conditions and Calibration

There are at least two reasons why SAFE should be started from pristine conditions. First, starting from a steady-state assures that the model output is determined by changes in the external and internal load, rather than by transients due to unstable initial conditions. To be able to assume steady-state starting conditions, it is necessary to start the simulations before acidification sets in. Seond, measurement, e.g. of deposition and soil solution chemistry, is usually only available at a single point in time. It is therefore usually not possible to validate the dynamic behavior in applications of soil chemistry models quantitatively. Consequently, if the model can not reconstruct reasonable dynamics for the past, we have no reason to believe model forecasts of future dynamics. Starting from pristine conditions thus increases the transparency and reliability of the model application.

TABLE I
Overview of the data acquisition strategy used to derive input parameters to REGPREP.

Parameter	Comment
Precipitation	Hydrological Atlas of Switzerland (NHG, 1992). Simplified corrections for interception and surface runoff
Soil Temperature	Interpolated from a limited set of measurements (FOEFL, 1994). Long-term annual means were used.
Wet Deposition	From altitude-dependent rain water concentration (FOEFL, 1994) of S and N and precipitation rates.
Dry Deposition	Concentration of the pollutant in air multiplied by component-specific deposition velocities (FOEFL, 1994).
Throughfall	Interpolated from a limited set of measurements (FOEFL, 1997).
Bc Deposition	From open-field bulk deposition using filtering factors for dry deposition as proposed by Hettelingh *et al.* (1991).
Planting Year	Maps from 1898 and observed average stand ages (Brassel and Kaufmann, 1996).
Biomass	Estimated from the 1982-1986 NFI (Brassel and Kaufmann, 1996).
Nutrient Content	Single site measurements and various literature sources.
Annual Harvest	Drawn from the NFI 1982-1986 (Brassel and Kaufmann, 1996).
Thickness of Layers	Layer thicknesses for nine elevation classes were assessed using measurements from 192 soil profiles of the 1993-1995 NFI (Zimmerman, 1995).
Surface Area	From particle size distributions.
Mineralogy	Total elemental analysis, normative mineral compositions and back-calculation methods (FOEFL, 1994).
Soil Water Content	From water retention capacity maps, evapotranspitation, and soil depth (see layer thickness).
CEC and Base Saturation	Measurements for a key layer for all sites. Extrapolated versus depth by means of transfer functions derived from reference profiles (FOEFL, 1997).

The initial base saturation usually is calibrated, because historic measurements of base saturation so far only have been found for Rothamsted, UK (Sverdrup *et al.*, 1994). The target for the calibration is to fit the simulated present base saturation with measurements. Because the initial soil solution chemistry is determined by the INITSAFE model, calibrating an initial base saturation for each layer is equivalent with calibrating a Gapon exchange coefficient. Soil solution chemistry and runoff chemistry are not used for calibration.

TABLE II

Default values and transfer functions for deriving input to SAFE for all soil layers. P is the precipitation flux, Q is the percolation flux, both in m yr^{-1}, z is the layer thickness in m, index j indicates layer number, and index i the current layer. An 'x' in the type column denotes site input (transfer function or multiplication factor in the layer column), while a '-' denotes default value.

Parameter	Unit	Type	Layer			
			O	A/E	B	C
Percolation	m yr^{-1}	x	$Q_i = P(P - Q_4)\ \Sigma^i_{j=1} z_j / \Sigma_j z_j$			
Bc uptake	fraction	x	$z_j / \Sigma_j z_j$			
N uptake	fraction	x	$(\Sigma^4_{j=1} z_j - \Sigma^4_{j=i+1} z_j) / \Sigma_j z_j$			
Density	kg m^{-3}	-	360	900	1,350	1,520
Water content	m^{-3} m^{-3}	x	constant			
Layer thickness		x	see Table I			
CEC	kmol$_c$ kg^{-3}	x	see Table I			
Base saturation	%	x	see Table I			
Surface area	m^2 m^{-3}	x	0.4	0.8	0.9	1.0
DOC	mg L^{-1}	-	20.0	5.0	2.0	0.5
PCO_2	times ambient	-	3	5	10.0	20.0
log K_G	-	-	6.5	7.5	8.5	9.2
Mineralogy						
K-Feldspar	fraction	x	1.0	1.0	1.0	1.0
Plagioclase	fraction	x	0.7	0.8	1.0	1.0
Albite	fraction	x	1.0	1.0	1.0	1.0
Hornblende	fraction	x	0.0	0.4	1.0	1.0
Pyroxene	fraction	x	0.0	0.4	1.0	1.0
Epidote	fraction	x	0.0	0.4	1.0	1.0
Carbonate	fraction	x	0.1	0.3	1.0	1.0
Biotite	fraction	x	0.0	0.4	1.0	1.0
Muscovite	fraction	x	0.6	0.8	1.0	1.0
Chlorite	fraction	x	0.1	0.5	1.0	1.0
Vermiculite	fraction	x	0.4	0.6	1.0	1.0
Apatite	fraction	x	1.0	1.0	1.0	1.0

2.2. THE INITSAFE MODEL

The INITSAFE and PROFILE models are steady-state versions of the SAFE model. Whereas PROFILE adds some processes not included in the SAFE model, INITSAFE contains exactly the same processes and process descriptions as the SAFE model. INITSAFE is tailored to read deposition and nutrient uptake for the starting year of SAFE and to produce output, steady-state Bc, Na^+ and H^+ ion concentrations for each layer, in a format that can be read directly by the SAFE model.

The INITSAFE model also checks the N balance, to make sure that the specified net uptake is less or equal to the specified input of N to the soil, i.e. the sum of N deposition, net mineralization and canopy exchange. The diagnostic output from the INITSAFE model consists of ANC and corresponding pH values for various input fluxes, such as wet deposition, throughfall, throughfall plus litterfall, etc., assuming the water flux to be the same as the precipitation. The historic evolution of ANC and pH in throughfall should compare with available measurements from the literature. The diagnostic output from INITSAFE also can be used to gain a better understanding of the output from the SAFE model and to check if the relative importance of the biological interactions is reasonable.

2.3. The MAKEDEP Reconstruction Model

The MAKEDEP model reconstructs atmospheric deposition and nutrient uptake from current deposition rates, standing biomass, and nutrient concentrations in standing biomass. The nutrient uptake is reconstructed using a growth function

$$w = w_{max} \frac{Y^n}{Y^n + k^n}$$

where w is the current biomass, w_{max} is the maximum possible biomass at the site, n is the order of the growth function, and Y is the stand age. The time derivative of the logistic growth function is used to obtain a relation between biomass and growth rate, a logistic growth rate function, which is calibrated so that the reconstructed biomass is in agreement with measurements of current biomass:

$$\int_{t=0}^{t=t} (\mathrm{NU}_N - \mathrm{H}_N)dt = \mathrm{N}_{tree}^{t=t} - \mathrm{N}_{tree}^{t=0}$$

where NU is net uptake of N, H is harvest in terms of N, and N_{tree} is the N contained in tree biomass at a given time. It is assumed that the growth is N-limited. The tree is divided into two compartments, stem and canopy.

Deposition is separated into five different categories: wet marine, dry marine, wet non-marine, dry non-marine, and canopy exchange - assuming all Na^+ to be of marine origin and Bc to be deposited by the same mechanisms as Na^+. Dry deposi tion and canopy exchange is assumed to be linearly dependent on canopy biomass, and non-marine deposition is scaled according to general deposition trends.

2.1.1. Adaptation of MAKEDEP

A number of changes were made to adapt MAKEDEP to central European conditions. A third tree compartment, bark and branches, was introduced (Figure 2). Clearcut harvesting was removed in favor of small plot harvesting. which was assumed to affect the forest by decreasing total biomass, thereby increasing the potential growth rate. Harvesting of litterfall and time-dependent harvesting practice also was included. Three new standard curves were introduced to describe changes in harvesting (see Figure 2): 1) stem harvest rate; 2) harvesting practice, i.e. the fraction of twigs and branches removed at harvesting for different years; and 3) litterfall harvesting, i.e. the fraction of litterfall removed from the forest.

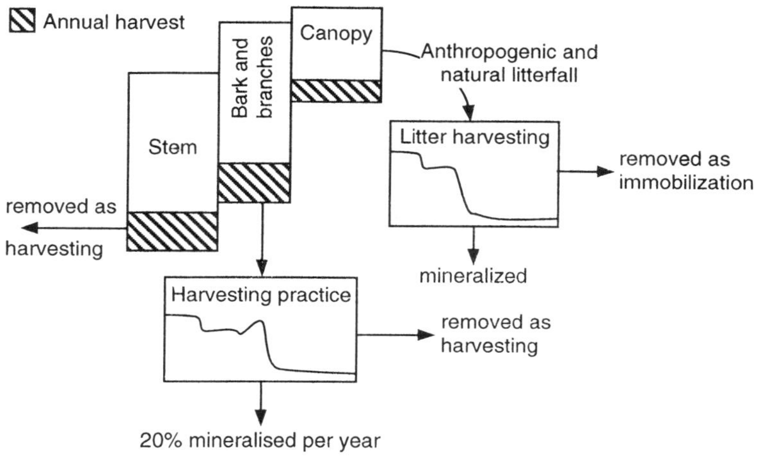

Fig. 2. Schematic view of the tree subsystem and its connection to the soil subsystem. Technically, the canopy harvesting is subtracted from the net mineralization rate, but does not influence the nutrient pool. No changes are therefore necessary in the SAFE model and the output from MAKEDEP can be used directly as input to the SAFE model. The fraction, same for all three pools, harvested each year is determined by the current harvest rate and standard curve for harvesting.

Harvesting of litterfall breaks the nutrient cycling and may create situations where there are not enough nutrients to rebuild the canopy. It can be shown (FOEFL, 1997) that if the canopy biomass only, and not the biomass of the rest of the tree, decreases, then net uptake is constrained by

$$-LF - CE \leq NU < 0$$

where NU is net uptake, LF is litterfall and CE is canopy exchange (all in mol m^{-2} yr^{-1}).

Whereas the standard version of MAKEDEP estimates dry deposition from information on throughfall and wet deposition, the adapted MAKEDEP may use data on throughfall, wet deposition and dry deposition, where input on dry deposition is optional. No division into marine and non-marine deposition is made. Canopy exchange is, if throughfall is specified, assumed to be zero Cl^-, Na^+, and SO_4^{2-} and is determined by

$$CE = TF - WD - DD$$

for the other ions, where TF is throughfall, WD is wet deposition and DD is dry deposition. In the adapted version of MAKEDEP, canopy exchange and dry deposition are assumed to be linearly dependent on canopy biomass. Dry deposition and wet deposition are scaled according to general Swiss trends, according to which deposition of Bc and Na^+ have been constant.

2.4. Empirical Relations and Data Acquisition

As not all soil data needed to run SAFE are commonly available on the regional scale, default values and transfer functions had to be used for certain parameters (see Table II). It was assumed that all sites had the same depth-dependent concentrations of dissolved organic carbon (DOC), partial carbon dioxide pressure (P_{CO_2}), and the same apparent gibbsite solubility constants (log K_G). Fix ratios were used to assign surface area and mineralogy to all layers from input on one layer only. The input parameters for the REGPREP model were derived from various sources of information with different resolution (see Table I).

3. Results and Discussion

The regional approach and data acquisition strategy described above was built in a modular fashion to facilitate verification, validation and improvement of the different submodules. When exchanging subsystems, it is important to keep the regional model internally consistent, so that conflicting assumptions are not used in different submodules. The data acquisition strategy (Table I) was used to derive a database with input parameters for 622 forested sites in Switzerland and the regional model was applied to all 622 sites (Kurz *et al.*, 1998). It is our experience that model transparency is of vital importance in this kind of exercise. Different studies may have been done with different objectives and with different underlying assumptions that may introduce inconsistencies when these studies are combined. A transparent modeling approach facilitates consistency checking and plausibility checking for both the modeller and external field experts.

Acknowledgments

The study received appreciative support from B. Acherman and R. Volz (FOEFL), P. Blaser, P. Brassel and E. Kaufmann (Swiss Fed. Inst. Forest, Snow and Landscape Res.), S. Braun (Inst. Appl. Plant. Biol.), U. Eggenberger (EKG Geo-Science) and B. Rihm (Meteotest) and was funded by the Swiss Federal Office of Environment, Forest and Landscape.

References

Alcamo, J., Shaw, R. and Hordijk, L.: 1990, *The RAINS Model of Acidification Science and Strategies in Europe*, Kluwer Academic Publishers, Dordrecht, the Netherlands.

Alveteg, M., Walse, C. and Warfvinge, P.: 1997, *Water Air, Soil Pollut.*, in press.

Brassel, P. and Kaufmann, E.: 1996, *Sweizerisches Landesforstinventar (LFI) 1982-86: Datenbankauszug vom 19, März 1996., Technical Report*, Eidgen. Forschungsanstalt für Walds, Schnee und Landschaft (WSL), Birmensdorf, Switzerland.

FOEFL (ed.): 1994, *Environmental Series Air, vol. 238*, Federal Office of Environment, Forests and Landscape (FOEFL), Berne, Switzerland.

FOEFL (ed.): 1997, *Environmental Documentation Air/Forest*, Federal Office of Environment, Forests and Landscape (FOEFL), Berne, Switzerland.

Hettelingh, J., Downing, R. and de Smet, P.: 1991, *Mapping Critical Loads for Europe, CCE Technical Report No. 1, RIVM Report No. 259101001*, Coordination Center for Effects, RIVM, Bilthoven.

Kurz, D., Alveteg, M. and Sverdrup, H.: 1998, *Water, Air, Soil Pollut.*, this volume.

NHG (ed.): 1992, *Hydrological Atlas of Switzerland*, National Hydrological and Geological Survey (NHG), Berne, Switzerland.

Posch, M., Hettelingh, J.-P., de Smet, P. and Downing, R.J. (eds): 1997, Calculation and Mapping of Critical Thresholds in Europe, Status Report 1997, Coordination Center for Effects, RIVM, Bilthoven.

Sverdrup, H. and Warfvinge, P.: 1992, *Appl. Geochem.* **8**, 273.

Sverdrup, H. and Warfvinge, P.: 1995, 'Estimating Field Weathering Rates using Laboratory Kinetics', in A.F. White and S.L. Brantley (eds), *Chemical Weathering Rates of Silicate Minerals, Reviews in Mineralogy, vol. 31*, Mineralogical Society of America, Washington D.C., pp. 485-541.

Sverdrup, H., Warfvinge, P., Blake, L. and Goulding, K.: 1994, *Agric., Ecosystems Environ.* **53**, 161.

Warfvinge, P., Falkengren-Grerup, U. and Sverdrup, H.: 1993, *Environ. Pollut.* **80**, 209.

Warfvinge, P. and Sverdrup, H.: 1992a, *Water, Air, Soil Pollut.* **63**, 119.

Warfvinge, P. and Sverdrup, H.: 1992b, 'Hydrochemical Modeling', in P. Sandén and P. Warfvinge (eds), *Modeling Acidification of Groundwater*, Swedish Meteorological and Hydrological Institute (SMHI), Norrköping, pp. 79-114.

Zimmerman, S.: 1995, pers. commun., Swiss Federal Institute for Forest Snow and Landscape Research (WSL), Birmensdorf.

INTEGRATED ASSESSMENT OF SOIL CHEMICAL STATUS. 2. APPLICATION OF A REGIONALIZED MODEL TO 622 FORESTED SITES IN SWITZERLAND

D. KURZ[1], M. ALVETEG[2] and H. SVERDRUP[2]

[1]*EKG Geo-Science, Ralligweg 10, CH-3012 Bern, Switzerland*
[2]*Lund University, Department of Chemical Engineering II, P.O. Box 124, S-221 00 Lund, Sweden*

Abstract. A regionalized version of the dynamic, process-oriented, multi-layer soil chemistry model SAFE was applied to 622 forest sites in Switzerland to assess effects of acidifying atmospheric deposition on the soil chemistry between 1850 and 2050. Simulation indicates that the present day chemical status of Swiss soils is a result of the last 50 years of acid deposition. Indicative soil parameters such as soil solution pH, acid neutralizing capacity, total Al concentration, base cation to total Al molar ratio and base saturation consistently deteriorate since the beginning of the 1950s, when acid loads start to increase. The simulated adverse evolution of soil chemistry persists until some time between 1975 and 2010, and apart from being site-dependent, the turning-point is also dependent on which parameter is considered at which soil depth.

Keywords: acidification, recovery, forest soils, regional dynamic modeling, Switzerland

1. Introduction

The dynamic, process-oriented, multi-layer soil chemistry model SAFE was applied to 622 forest sites in Switzerland. SAFE reconstructs acidification of terrestrial ecosystems by calculating the values of different chemical state variables as a function of time. The dynamic perspective of acidification completes and extends static modeling approaches used so far for the evaluation of emission abatement scenarios in Europe (e.g., Hettelingh *et al.*, 1991; Downing *et al.*, 1993). Static calculations have been used to evaluate critical loads, i.e., loads that can be tolerated by sensitive ecosystems without development of adverse changes in ecosystem structure and function in a long-term perspective. Recently established emission reductions will, however, not lead to acidifying pollutant deposition below critical loads of acidity neither everywhere in Europe, nor in Switzerland (Posch *et al.*, 1995). Continued exceedances of critical acid loads already have led or will lead to detrimental effects on forest ecosystems in sensitive areas. Additionally, temporary exceedances of critical loads of acidity have affected in the past or are still affecting forest ecosystems. The dynamics of the system, however, determine whether and when a violation of the critical limit used to calculate the critical loads will occur. To investigate temporal aspects of the adverse evolution and to evaluate damage and recovery time-lags, a dynamic approach is needed. Dynamic soil chemistry models allow quantitative assessment of the past, present and future status of key parts of the ecosystem and the temporal aspects of recovery from acidification under decreasing deposition loads. With the ability to estimate the time at which predicted changes will occur, dynamic model applications may give useful information for designing adequate step by step air pollution abatement policies toward non-exceedance of critical loads.

Water, Air, and Soil Pollution **105:** 11–20, 1998.

The modeling was preceded by a comprehensive revision of the input data generation necessary for a national application (Alveteg *et al.*, this volume). The objective of this exercise was to clarify whether background data and model development allow a national dynamic assessment of forest soil acidification. The present contribution analyzes the development of selected soil chemistry variables in time, essentially at a national scale, thereby illustrating the capability of the SAFE model and the associated input derivation to capture the most important features of soil chemistry evolution on a regional basis.

2. Materials and Methods

2.1. THE SAFE MODEL

SAFE calculates the evolution of key chemical soil parameters from measured or extrapolated input. Calculations are started from 'pristine' steady-state conditions determined by means of a variant of PROFILE. Principles of the models are discussed in Alveteg *et al.* (this volume). The SAFE model includes process-oriented descriptions of cation exchange reactions, chemical weathering of minerals, leaching and accumulation of dissolved chemical components and finally solution equilibrium reactions involving carbonate, organic acid and Al-species. SAFE also includes schematic descriptions of deposition, nutrient circulation of base cations (Ca^{2+}, Mg^{2+} and K^{+}) and nitrogen (N) as litterfall and canopy exchange, uptake of base cations and N from soil solution, and net mineralization and/or immobilization of base cations and N, the contributions of these processes being specified as time-series by the model user.

The temporal resolution is one year, therefore input and output parameters are expressed as annual values. The magnitude of capacity factors such as mineral abundance and cation exchange capacity is constant in time, sulfur (S) reactions are neglected, and base cations are lumped into a divalent component (Bc) in the present version of the SAFE model. All deposited reduced N is assumed to be either taken up or nitrified in the topmost soil layer. Dissolved organic matter is specified as input data, not generated by mineralization. Each soil compartment is further assumed to be chemically isotropic and the soil solution to be perfectly mixed. In this study, the soil profile is restricted to the major rooting zone (0.4 to 0.6 m deep) and is divided into four layers.

2.2. INPUT DATA

In general, the data acquisition strategy consisted of deriving input data from either national or regional survey information or single spot measurements, using transformation functions or computerized submodels. Time-invariant input such as climate, hydrology data and soil physico-chemical properties were drawn from the national critical loads data base (FOEFL, 1994, 1997). Particular focus was put on reconstructing consistent time-variant input comprising deposition, nutrient uptake and cycling. National deposition trends of S, oxidized N, and reduced N were inferred from historic and expected national (FOEFL, 1995) and European (e.g., Mylona, 1993) emission trends. Emissions were transformed to deposition considering EMEP import/export budgets (Barrett and Seland, 1995). Present wet, dry and throughfall deposition chemistry were extrapolated from survey and field measurements, using individual deposition models for wet and dry deposition (FOEFL, 1994) and a statistical approach for throughfall deposition (FOEFL, 1997). The site specific deposition as a function of time was then calculated by scaling the site's current deposition with the national deposition trend. Deposition is influenced by the canopy, increasing as the forest grows and decreasing as the forest is harvested. This was accounted for by linking reconstruction of deposition with forest growth implemented with the MAKEDEP routine (Alveteg *et al.*, 1997), which was adapted for central European conditions. Forest growth was modeled using the stands current biomass (Brassel and Kaufmann, 1996) and nutrient content, a forest growth function and information on historic land use. Forest exploitation in historic times and contemporary harvesting were assessed from historic sources including maps, from statistical annuals and the National Forest Inventory (NFI) 1982-86.

3. Results and Discussion

Trends in soil acidification are analyzed by means of trends in the indicative soil parameters pH, acid neutralizing capacity (ANC), the concentration of the physiologically relevant component Bc, total Al concentration (Al_{tot}), the Bc to Al_{tot} molar ratio (Bc/Al), and base saturation. The national scale was chosen for this initial analysis and trends are presented by selected percentile time-series. Percentiles, i.e. quantiles scaled to 100, are derived from linearly interpolated cumulative distribution functions (Posch *et al.*, 1995) determined by the respective parameter values of each model year.

3.1. Solution pH

Historic solution pH values calculated for the 622 sites generally decrease smoothly until the precipitation chemistry starts to change dramatically after 1930. In 1875, upper soil pH ranges from 3.88 to 6.83 in the O-layers (Figure 1A) and from 4.54 to 7.23 in the A/E-layers (Figure 1B). Model B- and C-layer pH (Figures 1C-D) similarly ranges from 4.51 and 4.81 to 8.05 and 7.96. In the 55 year time span up to 1930, modeled pH decreases on average by 0.24 units in the O-layers, and by 0.57, 0.47 and 0.35 units in the respective deeper consecutive layers. In the time period between 1945 and 1980, when acid loads peak, pH decreases by an average of 0.62 units in the O-layers and by an average of 1.06, 0.80 and 0.63 units in the A/E-, B- and C-layers, respectively. This decrease in solution pH is accompanied by a narrowing of the scatter, particularly for the lower 50 % pH values in all layers, indicating a decreasing ability of the respective soil systems to buffer the acidity load. The range of the lower 50 % pH values predicted for 1980 is on average only 40 % of the value range of 1945. The decline of acid deposition, having started in the late 1970s, and the assumed deposition scenario for the period after 2000 do not substantially affect the given pH pattern on the national scale. Particularly in the lower B- and C-layers, 70 % of the calculated pH values remain more or less below 4.7 and 5.0, respectively. Topsoil layer solution pH, on the other hand, more influenced by the deposition chemistry, slightly recovers from the low values predicted for the 1980s.

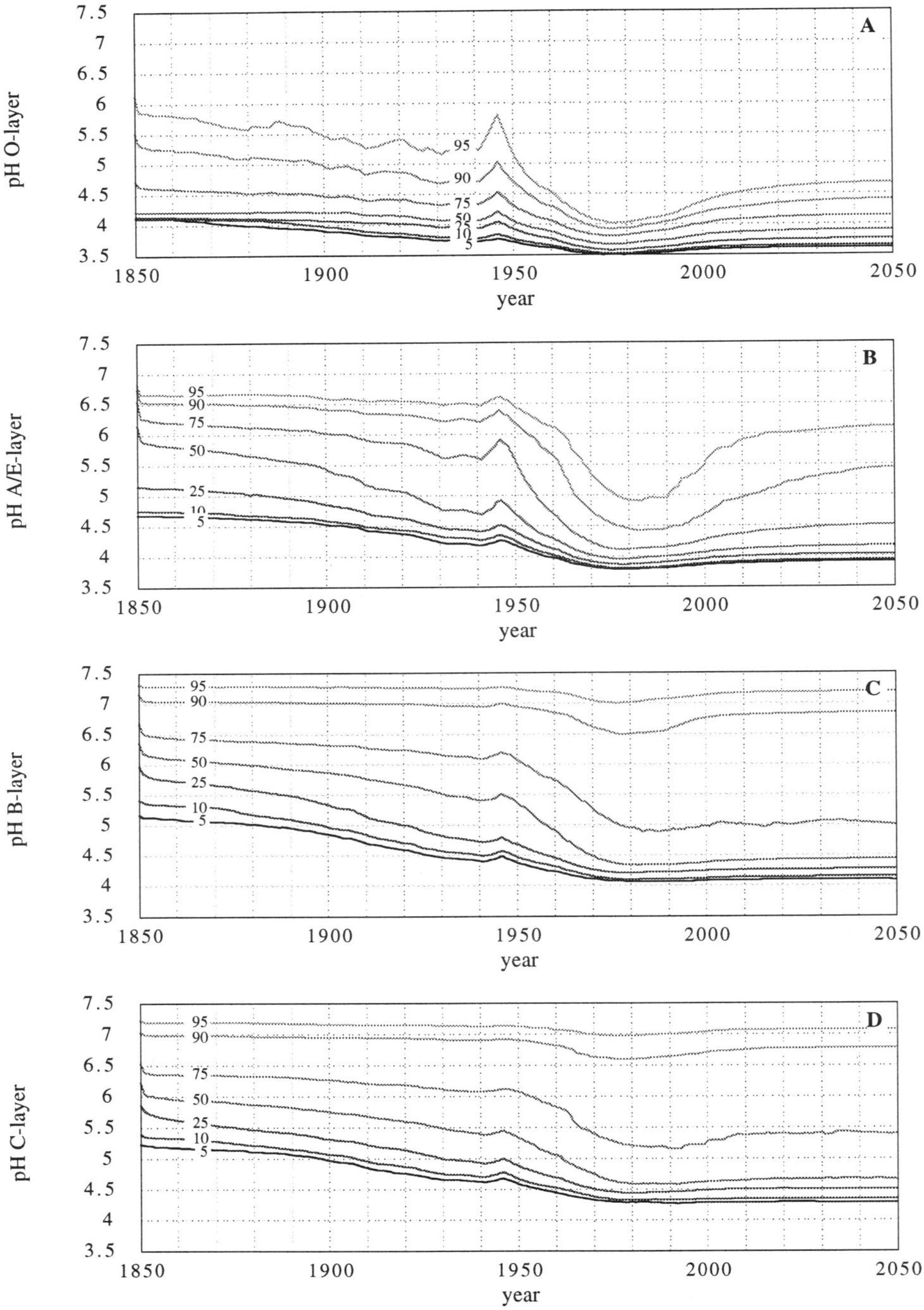

Fig. 1. Predictions of the solution pH evolution in the four model layers. National scale, 622 sites considered, small numbers in the plots refer to percentiles calculated from cumulative distribution functions of annual pH.

3.2. Solution Acid Neutralization Capacity

O- and A/E-layer ANC ranges from -80 to 300 μeq L^{-1} and from -10 to 700 μeq L^{-1}, respectively, by 1875. By the late 1970s, the minimum ANC falls to -1,000 μeq L^{-1} in the O-layers and to -770 μeq L^{-1} in the A/E layers. B- and C-layer ANC ranges from -1,500 to 7,500 μeq L^{-1} and from -1,900 to 12,500 μeq L^{-1}, respectively, by this time. In general, ANC is governed by the carbonate and the organic matter buffer systems at positive values for ANC, while the Al buffering system has its effect below this level. Only about 2 % of the sites have positive ANC in all soil layers throughout the simulation period, while approximately 20 % of the lower soil layers (those buffered by soil mineralogy) give solution ANC above 0 even at peak acid load. For 1875, on the other hand, the model predicts positive solution ANC for nearly 75 % of the soil profiles and for over 95 % of the lower soil layers. This demonstrates the modeled general decrease in ANC primarily induced by the input of acidity due to atmospheric deposition. Additionally, there is an increasing number of sites (layers) with increasingly lower ANC, expressed by the increasing differences in the lower percentile values of the 1875 and 1985 ANC distributions, for example. While the median 1875 and 1985 ANC differs by merely 215, 249, 240 and 261 μeq L^{-1} in the four model soil layers, the pentile ANC differs by 469, 540, 768 and 815 μeq L^{-1} from top to bottom of the rooting zone. ANC shows a certain recovery potential under the given deposition scenario, generally increasing to concentration levels calculated for the 1960s by the end of the simulation period.

3.3. Solution Base Cation Concentration

The concentration of Bc in the soil solution in the topsoil is dominated by the input from deposition, canopy exchange and litterfall, whereas soil internal processes such as cation exchange and weathering exert more influence at depth. Eighty percent of solution Bc concentrations of all four soil layers fall below 1,000 μeq L^{-1} throughout the simulation period. SAFE predicts a rather smooth increase of Bc concentrations on the national scale until 1940, when the increasing acidic deposition starts to induce, particularly in the lower soil layers, peak concentrations, which are finally reached by 1970 to 1975. By this time O- and A/E-layer solution Bc concentrations range from 30 to 1,100 μeq L^{-1}, while in the B- and C-layers the range covers nearly 4 orders of magnitude, from 2 to 13,500 μeq L^{-1}. On a site by site basis, the average increase of Bc concentrations between 1875 and 1970 is, however, only 152 μeq L^{-1}; 176 and 179 μeq L^{-1} in the A/E-, B-, and C-layers, respectively. Reduction of the acidity input may lead to decreasing solution Bc concentrations, mainly as a result of the decline of Bc production from exchange reactions. Solution Bc concentrations in the O-layer persist on an elevated level and cover a range of 80 to 700 μeq L^{-1} by the end of the simulation period in 2050. In the lower layers, the decline in acid deposition leads to a rapid decrease in Bc concentration, which levels off between 2000 and 2010. About 77 % of the sites have both B- and C- layer Bc concentrations by 2050 deviating not more than ±100 μeq L^{-1} from the respective values in 1875. The Bc concentration decrease in the A/E-layer, on the other hand, comes to a halt midway; Bc concentrations in 1875 on average are only 58 % of the 2010 values.

3.4. Solution Total Al Concentrations

SAFE generally predicts low historic soil solution Al_{tot} concentrations in the four model layers. Minima are always less than 0.01 μeq L^{-1} and 95 % of the calculated values fall below 10 μeq L^{-1} in 1875. The pronounced change in precipitation chemistry after 1945 induces a prominent increase of Al_{tot} concentrations in all four model layers, except for around 20 % of lower soil layers, which evidently are buffered by soil mineralogy. Peak concentrations in the O- and A/E-layers (several 100 μeq L^{-1}) are identified for the late 1970s, in the lower soil layers (up to 1850 μeq L^{-1}) not before the 1980s. Al_{tot} concentrations in the upper soils decreases, when the acid input to the soil decreases, and the distributions of the O- and A/E-layer Al_{tot} concentrations by 2050 approach the distributions of the early 1960s. B- and C-layer Al_{tot} concentrations show a substantially less distinct reaction and the median concentration remains at elevated levels, being 53 % of the peak value in the B-layer and 62 % in the C-layer, respectively. With a multi-layer model it is generally agreed that the maximum of the Al_{tot} concentrations calculated for the individual soil layers within the rooting zone is used as the relevant value for the soil profile. Historic Al_{tot} concentration maxima increase until 1940, 95 % of the values falling between 0.1 and 14 μeq L^{-1} in 1875 and between 0.2 and 181 μeq L^{-1} in 1940 (Figure 2A). Peak concentrations are generally reached in the late 1970s and early 1980s, when 95 % of the population values cover a range of approximately 10 to 1,000 μeq L^{-1}. By this time 40 % of the Al_{tot} concentration maxima are found in the upper soils' O- and A/E-layers, 10 % in the B- and 50 % in the C-layers (Figure 2B). Before 1950, on the other hand, maxima of the Al_{tot} concentrations are met in roughly equal shares in the O- and the C-layers. Ninety-five percent of the maximum Al_{tot} concentrations still range between approximately 1.0 and 950 μeq L^{-1} by the end of the modeling period, the lower soil B- and C-layers contributing nearly 75 % of the values.

3.5. Solution Base Cation to Al Molar Ratio

The soil solution Bc/Al molar ratio has become the most widely used chemical criterion in the assessment of critical loads of acidity for forest soils in Europe, as it correlates well with the health status of forests (Sverdrup and Warfvinge, 1993). The Bc/Al molar ratio integrates the evolution trends of solution Bc and Al_{tot} concentrations, respectively.

The assumed deposition trends led to a decrease of Bc/Al molar ratio by up to several orders of magnitude compared to historic levels. By 1977, O-layer Bc/Al molar ratios cover a narrow range, 90 % of the population scattered between 1.7 and 35.8, the respective values being 6.7 and 80.500 for 1875. A/E-layer Bc/Al molar ratios show similar evolution trends, although generally starting from higher values and consequently suffering from an even more distinct decline. Whereas upper soil layers' Bc/Al molar ratios reach their minima in the late 1970s, the decrease in B- and C-layer Bc/Al molar ratio levels out sometimes between 1980 and 2010. In 2050, 50 % of the sites have Bc/Al molar ratios in the lower soil layers comparable to values modeled for the time after the peak of acid deposition in the late 1970s and 20 % reach their all-time low in 2050. To get a single Bc/Al value for a soil profile, it is generally agreed to use the minimum calculated within the rooting zone independent of depth. In historic times, up to 60 % of the minima are supplied by the O-layers and the remaining 40 % are essentially obtained from the C-layers

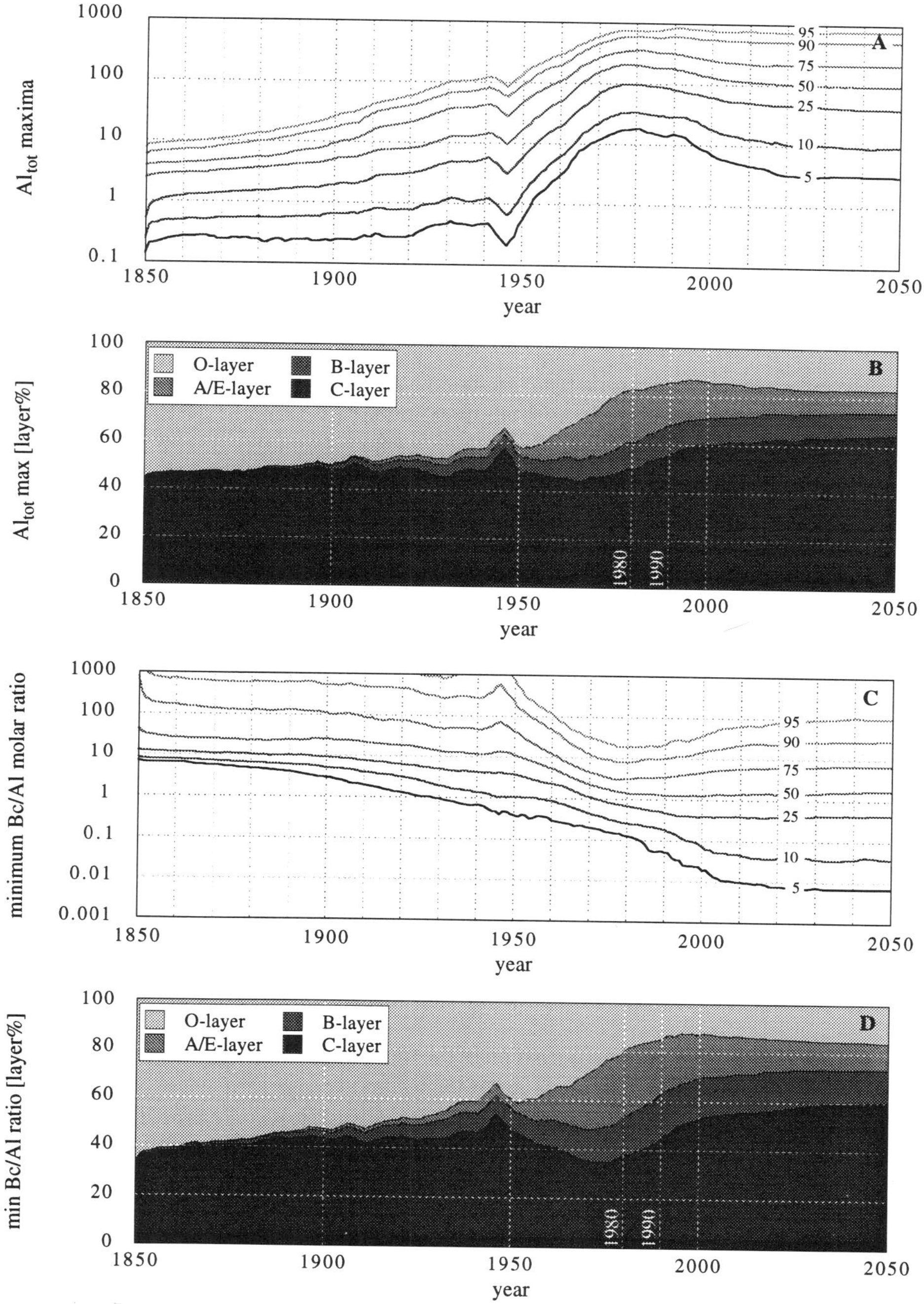

Fig. 2. Time-series of Al_{tot} maxima in $\mu eq\ L^{-1}$ (**A**) and Bc/Al minima (**C**) and of the percentage of layers which provide Al_{tot} maxima (**B**) and Bc/Al minima (**D**). National scale, 622 sites considered, small numbers in the plots refer to percentiles.

(Figure 2D). Consequently, the higher percentiles reproduce Bc/Al molar ratio evolution trends of the O-layers, whereas the lower percentiles reflect the trends in the C-layers. Ninety percent of the Bc/Al molar ratio minima fall between 5.1 and 2,535 in 1875 and between only 0.03 and 29.2 in 1993 (Figure 2C). Minima are simulated for the O-layer at 19 %, for the A/E-, B- and C-layers at 27 %, 16 % and 38 % of the sites, respectively, by this time. No substantial improvement is achieved with the assumed deposition reductions, as the lower soil layers deteriorate further until the end of the simulation period. By 2050, 90 % of the Bc/Al molar ratio minima remain scattered between 0.007 and 105; still nearly 40 % of the sites have minimum Bc/Al molar ratio below the critical limit of 1.

3.6. Soil Base Saturation

Fifty percent of Swiss forest soils have modeled base saturation in all layers above 90 % throughout the simulation period. As SAFE is calibrated by means of the present day base saturation, the simulated time-series are influenced by present day distribution patterns, indicating still exceptionally high base saturation. For O- and A/E-layer lower percentiles (Figures 3A-B), SAFE predicts only modest depletion of the total exchangeable bases for the times before 1950, even a slight increase of the lowest 15 % of the values in the O-layer. The increasing acidity input starting after 1950, on the other hand, appears to deplete the exchangeable bases, the lower 50 % of the estimates in the O-layers on average 22 % by 1980, and in the A/E-layers on average 40 % by 1990. For the lower 50 % of the B- and C-layer base saturations (Figures 3C-D), SAFE predicts a steady decrease leveling out not until 2000 to 2010. By 2000, the lower 50 % of base saturations only average 43 % of the estimates for 1875 in the B-layer, and 47 % in the C-layer, respectively. SAFE predicts that base saturation will continue to decrease in the lower soil layers, which are not buffered by soil mineralogy. This refers in particular to roughly 50 % of the C-layer base saturations which are lowest by 2050, on average 90 % of the estimates in 2000. The lower 50 % of the base saturations in the B-layers remain more or less on the low levels reached around 2010, while A/E-layer and in particular O-layer base saturations moderately improve toward the end of the simulation period.

4. Conclusions

Dominant causes of forest soil acidification in the time period after the Second World War are found in the deposition patterns, as forest exploitation ceased and harvesting practices changed. This suggests that the adverse effects on Swiss forest soil chemistry in the past 50 years are essentially a result of the increasing acidifying deposition. The impact has to be considered substantial with average solution pH dropping by up to 1 pH unit, an up to forty-fold increase of average solution Al_{tot} concentration and a decrease of solution Bc/Al molar ratios by up to several orders of magnitude. To properly evaluate damage and recovery time-lags it is essential to apply a multi-layer soil chemistry model. Simulations suggest that lower soil layers react sluggishly to increasing as well as decreasing acid input. The persistent violation of the critical Bc/Al molar ratio limit of 1 in the lower soil layers leads to the conclusion that current acidifying emission reduction plans are not sufficient to attain a long-term sustainable forest ecosystem.

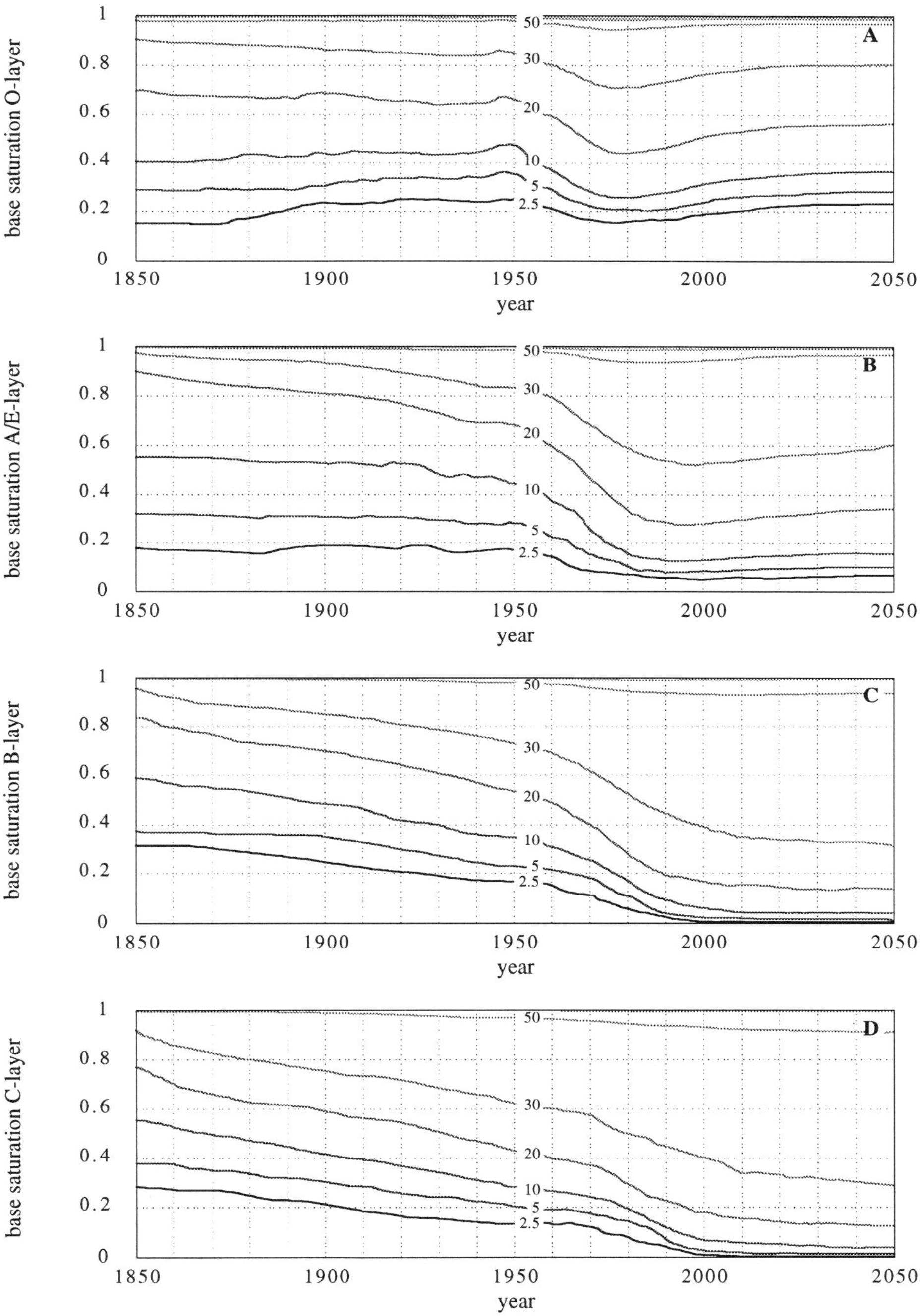

Fig. 3 Predicted evolution of base saturation (as fraction) in the four model layers National scale, 622 sites considered, small numbers in the plots refer to percentiles.

Acknowledgments

The study received appreciative support from B. Achermann and R. Volz (FOEFL), P. Blaser, P. Brassel and E. Kaufmann (Swiss Fed. Inst. Forest, Snow and Landscape Res.), S. Braun (Inst. Appl. Plant Biol.), U. Eggenberger (EKG Geo-Science) and B. Rihm (Meteotest). Funding is due to the Swiss Federal Office of Environment, Forest and Landscape. The contents of this paper do not necessarily reflect the views of the FOEFL and no official endorsement should be inferred. The manuscript profited from elaborate comments of K. Wieder and two anonymous reviewers.

References

Alveteg, M., Walse, C. and Warfvinge, P.: 1998, *Water, Air, Soil Pollut.*, in press.

Barrett, K. and Seland, Ø. (eds): 1995, *EMEP MSC-W Report 1/95*, Norwegian Meteorological Institute Oslo, Norway.

Brassel, P. and Kaufmann, E.: 1996, *NFI 1982-86 Data Base Extract*, Swiss Fed. Inst. Forest, Snow and Landscape Res. Birmensdorf, Switzerland.

Downing, R.J., Hettelingh, J.-P. and de Smet, P.A.M. (eds): 1993, *RIVM Report 259101003*, Coordination Center for Effects, Natl. Inst. Publ. Health and Environ. Prot. Bilthoven, Netherlands.

FOEFL (ed): 1994, *Environmental Series Air 238*, Federal Office of Environment, Forests and Landscape (FOEFL) Berne, Switzerland.

FOEFL (ed): 1995, *Environmental Series Air 256*, Federal Office of Environment, Forests and Landscape (FOEFL) Berne, Switzerland.

FOEFL (ed): 1997, *Environmental Documentation Air/Forest*, Federal Office of Environment, Forests and Landscape (FOEFL) Berne, Switzerland.

Hettelingh, J.-P., Downing, R.J. and de Smet, P.A.M. (eds): 1991, *RIVM Report 259101001*, Coordination Center for Effects, Natl. Inst. Publ. Health and Environ. Prot. Bilthoven, Netherlands.

Mylona, S.: 1993, *EMEP MSC-W Report 2/93*, Norwegian Meteorological Institute Oslo, Norway.

Posch, M., de Smet, P.A.M., Hettelingh, J-P. and Downing, R.J. (eds): 1995, *RIVM Report 259101004*, Coordination Center for Effects, Natl. Inst. Publ. Health and Environ. Prot. Bilthoven, Netherlands.

Sverdrup, H. and Warfvinge, P.: 1993, *Reports in Ecology and Environmental Engineering 1993:2*, Lund University Lund, Sweden.

MAGIC, SAFE AND SMART MODEL APPLICATIONS AT INTEGRATED MONITORING SITES: EFFECTS OF EMISSION REDUCTION SCENARIOS

M. FORSIUS[1], M. ALVETEG[2], A. JENKINS[3], M. JOHANSSON[1], S. KLEEMOLA[1], A. LÜKEWILLE[4], M. POSCH[5], H. SVERDRUP[2], and C. WALSE[2]

[1]*Finnish Environment Institute, P.O.Box 140, FIN-00251 Helsinki, Finland*
[2]*Lund University, Chemical Engineering II, P.O.Box 124, S-22 100 Lund, Sweden*
[3]*Institute of Hydrology, Wallingford, OXON, Oxfordshire OX10 8BB, United Kingdom*
[4]*Norwegian Institute for Air Research, P.O.Box 100, N-2007 Kjeller, Norway*
[5]*RIVM, P.O.Box 1, NL-3720 BA Bilthoven, The Netherlands*

Abstract. Three well-known dynamic acidification models (MAGIC, SAFE, SMART) were applied to data sets from five Integrated Monitoring sites in Europe. The calibrated models were used in a policy-oriented framework to predict the long-term soil acidification of these background forest sites, given different scenarios of future deposition of S and N. Emphasis was put on deriving realistic site-specific scenarios for the model applications. The deposition was calculated with EMEP transfer matrices and official emissions for the target years 2000, 2005 and 2010. The alternatives for S deposition were current reduction plans and maximum feasible reductions. For N, the NO_x and NH_y depositions were frozen at the present level. For NO_x, a reduction scenario of flat 30 % reduction from present deposition also was utilized to demonstrate the possible effects of such a measure. The three models yielded generally consistent results. The 'Best prediction'-scenario (including the effects of the second UN/ECE protocol for reductions of SO_2 emissions and present level for NO_x-emissions), resulted in many cases in a stabilization of soil acidification, although significant improvements were not always shown. With the exception of one site, the 'Maximum Feasible Reductions' scenario always resulted in significant improvements. Dynamic models are needed as a complement to steady-state techniques for estimating critical loads and assessing emission reduction policies, where adequate data are available.

Keywords: acidification, emission, MAGIC, model, nitrogen, SAFE, SMART, sulfur

1. Introduction

Both steady-state and dynamic models have been developed to predict the acidification of soils, lakes, streams and groundwater (e.g., Cosby *et al.,* 1985; De Vries *et al.*, 1989; Warfvinge and Sverdrup, 1992). Steady-state models for mapping of so-called critical loads and their exceedances presently are extensively used under the framework of the Convention on Long-Range Transboundary Air Pollution (LRTAP) of the United Nations Economic Commission for Europe (UN/ECE). In these models, all time-dependent processes and finite pools are neglected. Therefore, the models can be applied with a limited amount of information and are suitable for national to continental scale mapping purposes. National critical load data are collected and processed at the Coordination Center for Effects (CCE/RIVM) in the Netherlands, and are used in the international negotiations on protocols for emission reductions carried out under the framework of the UN/ECE (Bull, 1995).

Dynamic models are used to predict the gradual chemical response of a receptor to changing depositions by including the various buffer and adsorption/desorption mechanisms. The time development of acidification is important, e.g., for determining the timing of necessary measures for emission control and for assessing the dynamic response of possible critical load exceedances. However, even though dynamic modeling has been carried out on a regional basis (e.g., Jenkins *et al.*, 1990; De Vries *et al.*, 1994), there are

Water, Air, and Soil Pollution **105:** 21–30, 1998.

uncertainties involved in deriving the large amount of input data required by these regionalized models (by the use of transfer functions and/or statistical distributions). Hence, site-specific dynamic model applications, based on detailed, measured high resolution data, are useful for the development and assessment of emission reduction policies as a complement to regional/national mapping and modeling approaches.

The present study describes the application of three well-known dynamic acidification models (MAGIC, SAFE, SMART) to data sets from Integrated Monitoring sites in Europe. The Integrated Monitoring Programme (ICP IM) is one of the monitoring activities established under the UN/ECE LRTAP Convention for the assessment and monitoring of air pollution effects (Kleemola and Forsius, 1996). The calibrated models are used to predict long-term soil acidification of these background forest sites, given different scenarios of future deposition of S and N. A large emphasis was put on deriving realistic site-specific deposition scenarios, partly based on agreed measures for emission reductions.

2. Methods

2.1. Site Descriptions

According to data availability in the IM data base at the Finnish Environment Institute (Kleemola and Forsius, 1996), five sites were selected for the model applications: Afon Hafren (Wales), Birkenes (southern Norway), Forellenbach (southern Germany), Gårdsjön (southwestern Sweden) and Hietajärvi (eastern Finland). The sites are located in protected regions and are covered mainly with coniferous forests. The sites receive varying deposition loads and have different catchment characteristics. Therefore, these sites represent a wide range of possible future responses to the atmospheric loadings. Some main features of the sites are listed in Table I.

2.2. Model Descriptions: MAGIC, SAFE and SMART

The three models are all process-oriented dynamic models that attempt to describe the long-term impact of atmospheric deposition, net uptake by vegetation, weathering and cation exchange on the chemical composition of soil and the outflow water. Solution chemistry is governed by charge and mass balance principles, using lumped process descriptions. Detailed descriptions of the models have been published elsewhere, and therefore only some main features are briefly summarized here.

MAGIC (Cosby *et al.*, 1985) was used as a one-box model. It calculates a separate Gaines-Thomas equilibrium for the exchange of each of the cations, Al^{3+}, Ca^{2+}, Mg^{2+}, K^+ and Na^+. Sulfate adsorption is described by a Langmuir isotherm. The model keeps track of mass budgets and chemical equilibria of all major ions including organic acids. SAFE (Sverdrup *et al.*, 1995) is a multi-layer model, which calculates weathering rates from measurements of soil texture, mineralogy and moisture. The model uses a mass transfer equation for the ion exchange of base cations, which converges upon an equilibrium described by the Gapon equation. SMART (De Vries *et al.*, 1989) was used as a one-box model. SMART uses Gaines-Thomas equilibrium equations for the various exchange reactions. Nitrogen immobilization is modeled as a function of the C:N-ratio. A simple lake module, describing retention of SO_4^{2-}, NO_3^- and NH_4^+, as well as inorganic carbon equilibria, recently has been added. All models have been used in numerous studies on both catchment and regional scales (e.g., Jenkins *et al.*, 1990; Kämäri *et al.*, 1994; Sverdrup *et al.*, 1995).

TABLE I
Catchment and soil characteristics of the modeled sites.

Site	Total area (ha)	Dominant vegetation	Dominant soil type	Mean soil thickness (m)	Base saturation (%)
Afron Hafren	358.0	Sitka and Norway spruce	Peaty podzol	0.88	11.6
Birkenes	41.6	Norway spruce	Thin humus	0.22	29.1
Forellenbach	68.8	Spruce and birch	Dystric cambisol	1.00	4.6
Gårdsjön	3.7	Norway spruce	Orthic podzol	0.63	15.8
Hietjärvi	464.0	Scots pine	Fibric histosol	0.85	47.3

2.3. Derivation of Deposition and Nutrient Uptake Scenarios

The models require deposition and uptake by forest growth as external inputs. The derivation method used for generation of this input originates from the work of the SAFE modeling group (Alveteg *et al.*, 1998). The derivation method of historical development and future scenarios for deposition and uptake is explained in more detail in Johansson *et al.* (1996).

The deposition of each compound was divided into different components using bulk and throughfall deposition measurements, originating from sea or anthropogenic sources, and these both in dry and wet fractions. Historical and future development and forest filtering effects (i.e., the ability of the tree canopy to increase the dry deposition fraction compared to that in bulk deposition) were then allocated to proper components.

The total dry component for S, Cl^- and Na^+ was assumed to be the difference between throughfall and bulk deposition, as these are mobile ions in the tree canopy. For base cations other than Na^+, the minimum dry deposition fraction of Na^+ or Cl^- was used as a conservative estimate. For N compounds, the dry component fraction was estimated to be the minimum of dry deposition fraction of S, Na^+ or Cl^-, since the canopy exchange effect could not be determined.

The wet marine components for all compounds except N were estimated with sea salt correction factors from the Na^+ wet deposition; for the Afon Hafren site, Cl^- was used due to inconsistency between bulk and throughfall values measured at a distance from each other. The wet anthropogenic components were calculated by subtracting the wet marine component from total wet deposition. The same was applied to the dry anthropogenic component. The wet anthropogenic component for both N compounds was the measured bulk deposition.

The wet marine component was assumed to remain constant over time. Other wet and dry anthropogenic components were dependent on deposition histories, calibrated to the present measurements of one year. The S deposition history for 1880-1991 was extracted from data by Mylona (1993) for each 150 km × 150 km grid cell (the so-called EMEP grid) separately. The earlier scaling was extended back to 1800 using an average estimate for Europe by Sverdrup *et al.* (1995). Average N deposition histories for the whole of Europe were estimated on the basis of work by Sverdrup *et al.* (1995) and Asman and Drukker (1988). The non-marine dry deposition component of base cations was assumed to be partly connected to anthropogenic activities reflected by measurement data during a period of large emission reductions (Hedin *et al.,* 1994), and was here assumed to follow the historical S deposition curve. All dry deposition components are affected by forest filtering, which was assumed to depend linearly on the tree needle volume represented by the needle mass in internal calculation. The present filtering factor was calibrated to current canopy volume and dry deposition estimates.

In the scenarios, base cation and Cl^- depositions were kept at the present level. For acidifying deposition, there were three future scenarios for the model application (Table II). The deposition was calculated with transfer matrices from the UN/ECE EMEP/MSC-W center and reported official UN/ECE emissions for the target years 2000, 2005 and 2010. The alternatives for S deposition were current reduction plans (according to second S protocol, SP2) and maximum feasible reductions, MFR (Cofala and Schöpp, 1995). For N, the NO_x and NH_y depositions were frozen at the present level. For NO_x, a scenario of a flat 30 % reduction from present deposition was utilized to demonstrate the possible effects of such a measure.

TABLE II
Three scenarios employed in the acidification model runs, roughly representing the range of future acidifying deposition.

Scenario	SO_2	NO_x	NH_y
A) Best prediction	SP2	Present level	Present level
B) Lower limit	MFR	-30 % from present	Present level
C) Upper limit	Present level	Present level	Present level

The uptake by forest growth was based on the biomass density and element content together with the annual increment of each tree compartment: stem over bark, branches, and needles. Potential annual growth and standing volume were calculated from a logistic forest volume growth curve, representing the whole forest. The branch growth was assumed to be one fifth of stem over bark growth (Mälkönen, 1975). The needle growth was described with a growth curve similar to that of stem over bark. All growth curve parameters were calibrated to current volume and growth.

The forest growth was assumed to be N limited. The growth of all aboveground compartments is reduced if the N via atmosphere and mineralization are not sufficient to satisfy the estimated potential growth. The growth estimate is most proper for a temporally homogeneous forest with one major tree species, or for a well-known tree species distribution each of known age, after clearcut or forest fire.

2.4. Model Calibrations: MAGIC, SAFE and SMART

The models were applied by the same groups that have participated in the actual model development: MAGIC applications were carried out at the Institute of Hydrology and the Norwegian Institute for Water Research; SAFE was run at Lund University; SMART applications were carried out at the RIVM and the Finnish Environment Institute.

Application of the models at the catchment scale requires averaging of physical and chemical properties of the soil profile into one or several homogenous layers characteristic of the entire catchment. Yearly time-steps were used for these applications. The models were calibrated to produce present-day (1990-94) soil and streamwater chemistry. All models require an estimate of the weathering rate of base cations. In SAFE, these are obtained from information on soil texture and mineralogy, and in the other models by calibration.

The deposition and uptake patterns were used to drive the models and to produce the chemical changes in soil and streamwater over time. The models all used a common data set for each site, but were calibrated independently of each other. Standard model calibration procedures were used (Cosby *et al.,* 1985; Kämäri *et al.*, 1994; Sverdrup *et al.,* 1995).

3. Results and Discussion

3.1. Deposition and Nutrient Uptake Scenarios

The average S and N depositions for all IM sites for scenario A are shown as an example of the model outputs (Figure 1). The effect of large decreases in S emissions from the peak values in the late 1970s can be seen clearly. The Finnish Hietajärvi catchment has the lowest deposition of all the sites.

The N deposition histories are estimated averages for the whole of Europe. The difference between actual and derived history is likely to be largest at sites where the short-range transport of pollutants is significant, as for ammonia. However, this source of uncertainty is assumed to be relatively small for these background IM-sites. The absolute values of modeled historical and current deposition are calibrated with measured values at the site; therefore, the magnitude of historical deposition has less uncertainty when measurement data from several years are available.

Forest growth and nutrient uptake estimates could be refined by using locally estimated growth curves for the region and tree species concerned. A more complete description would, however, call for data presently not available for all sites in the IM program (Kleemola and Forsius, 1996). The growth curve calibration to present volume and growth only may lead to incorrect predictions in the early and late phases of the rotation period. Various biomass functions (e.g., Marklund, 1988) could be used to describe the growth of all tree compartments in more detail. However, the data available from IM sites restrict this approach.

To estimate uptake, a more straightforward approach would have been to estimate a modeled open land deposition with a constant filtering factor and historical scaling and a simple growth curve, not restricted by nutrient N available. Instead, a more realistic but rather complex description of hypothesized deposition breakdown and of an N-limited tree

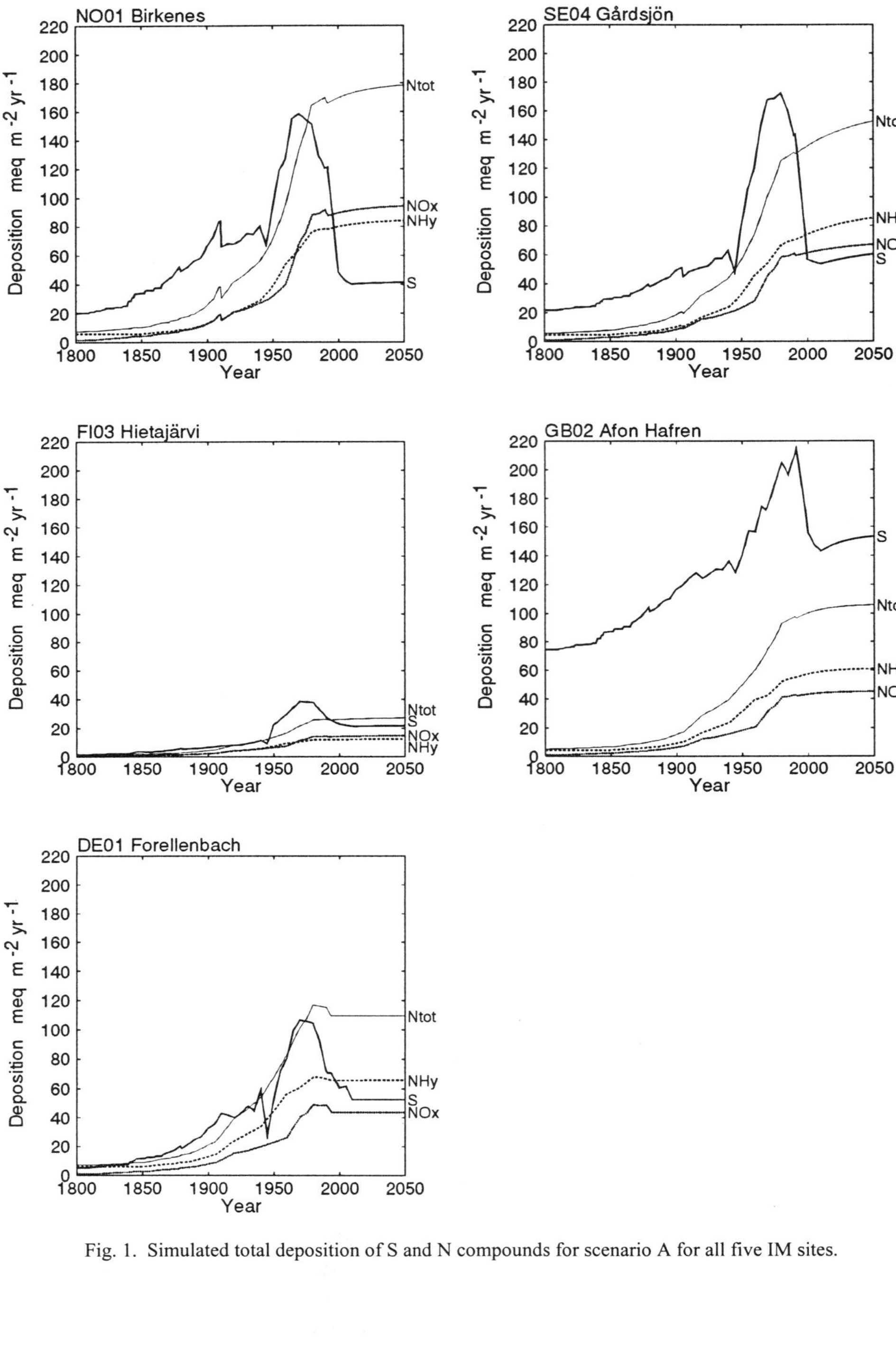

Fig. 1. Simulated total deposition of S and N compounds for scenario A for all five IM sites.

growth with a dynamic filtering capacity of dry deposition, was made. These results retain uncertainty due to, e.g., too few calibration values. The chosen scheme demonstrates one possible method of compiling fairly detailed scenarios at a level sufficient to dynamic acidification models. The advantage of the current approach is the rather modest requirement of input data.

3.2. MAGIC, SAFE AND SMART RESULTS

Of the large number of available model results at the different sites, soil base saturation was here chosen to demonstrate the model predictions. Base saturation is a key measure of soil acidification. This variable also is calculated in a comparable way by the three models, thus allowing a comparison of model results and their uncertainties. Base saturation was considered to be a more suitable variable than the base cation to aluminum ratio (BC/Al), which is used in the calculations of critical loads. The BC/Al ratios are highly variable, and the whole concept recently has been criticized (Løkke *et al.*, 1996). A relationship between base saturation and BC/Al has also been defined (De Vries *et al.*, 1995).

The simulated values for base saturation for the scenarios A-C are shown in Figure 2. The results of SAFE, which is a multilayer model, are presented as weighted averages to produce results comparable with the two other models. In most cases, the calibrated models yielded consistent results for historical base saturation in 1900. At all sites, the model results indicate at least a slight decrease in base saturation until present day.

At each site, the predicted response depended on the future deposition scenario. The response to the different scenarios was consistent. As expected, scenario B, where 'Maximum Feasible Reductions' for S emissions and a 30 % reduction for NO_x emissions have been assumed (Table II), always resulted in the most significant improvement (Figure 2). For this scenario, either a stabilization of the current situation or an improvement generally was shown. The only clear exception was MAGIC and SMART simulations for Afon Hafren where continued soil acidification was indicated also for the B-scenario.

The A-scenario ('Best prediction') resulted in a slower response than the B-scenario. In many cases, this emission reduction for S was enough to stop continued soil acidification, although significant improvement was not always shown. For Afon Hafren, the simulations of MAGIC and SMART again indicate continued acidification (as well as SMART for Forellenbach) using this scenario.

The model results demonstrate that the selection of a certain emission reduction strategy influences also the time of ecosystem response (see Wright *et al.*, 1991). Dynamic models therefore are needed in addition to the steady-state techniques that presently are used for mapping of critical loads and their exceedance, for the assessment of emission reduction policies. However, the large amount of input data required by the dynamic models limits the possibility to use them on the regional scale.

The three models applied here gave, in most cases, consistent results (cf. Wright *et al.*, 1991). Although in many cases the models contain similar process descriptions, there are still important differences in the model structure, complexity, and calibration/validation techniques (Tiktak and Van Grinsven, 1995). The similarity of the predictions gives confidence in the use of such models for scenario assessments and other policy-oriented work.

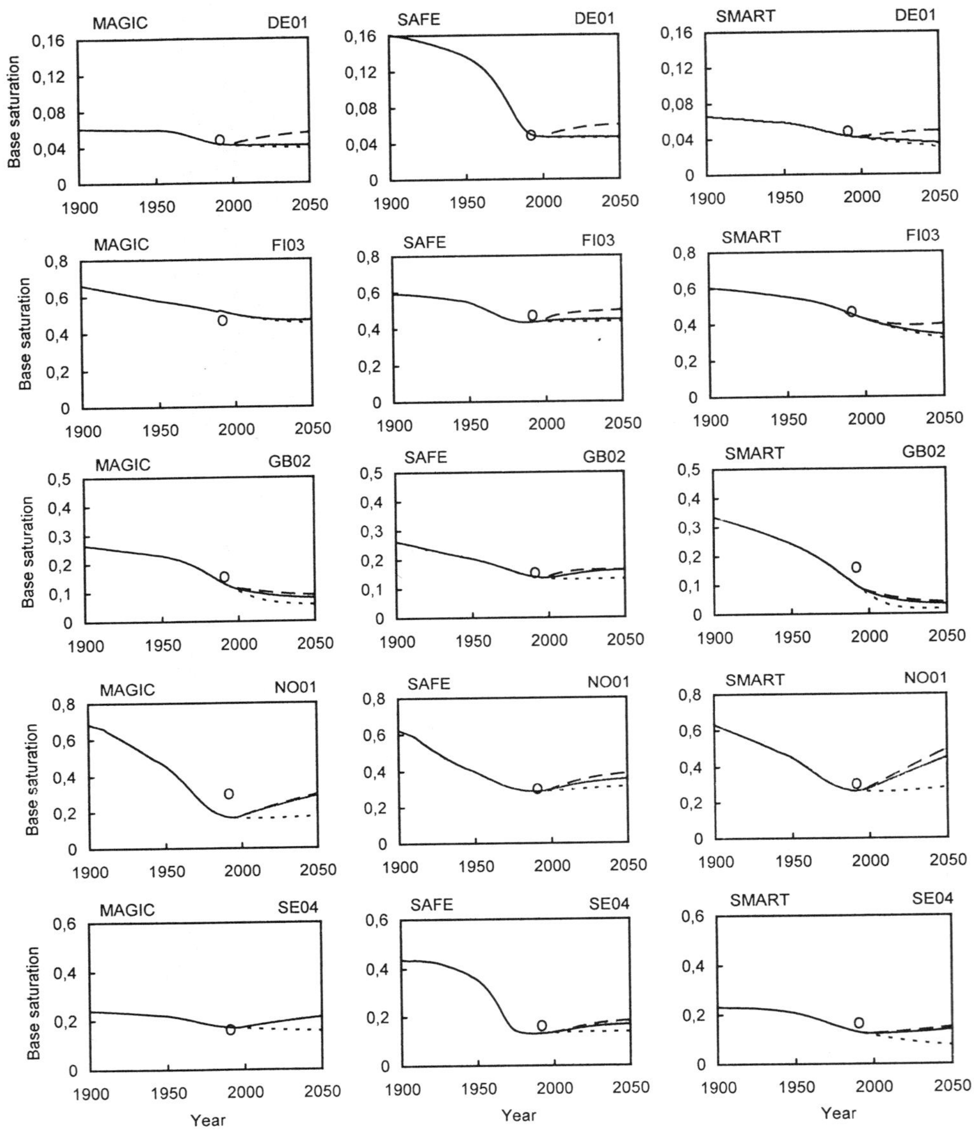

Fig. 2. Simulated and observed (circle) soil base saturation for Forellenbach (DE01), Hietajärvi (FI03), Afon Hafren (GB02), Birkenes (NO01) and Gårdsjön (SE04), using MAGIC, SAFE and SMART models and scenarios A-C (see Table II); scenario A = solid line, B = long dashes, C = short dashes.

Much work still is needed regarding the modeling of ecosystem processes, particularly regarding the effects of N deposition. All models applied in this study contain rather crude descriptions for N cycling and saturation. Different more detailed concepts are being developed and should be used also for scenario assessment, when properly tested and evaluated. It is also likely that ecosystem processes in the long term will be affected by future climate change. Synergistic effects between atmospheric deposition and climate change therefore also should be considered in future model development and scenario assessment (Wright and Schindler, 1995; Forsius *et al.,* 1997).

Acknowledgments

The Nordic Council of Ministers is gratefully acknowledged for financial support of this study. We would also like to thank Sophia Mylona and Erik Berge at EMEP/MSC-W for the use of transfer matrices for atmospheric pollutant transfer and data on historical S deposition. The ICP IM National Focal Points are acknowledged for providing additional unpublished data for the model applications.

References

Alveteg, M., Walse, C. and Warfvinge, P.: 1998, *Water, Air, Soil Pollut.* (accepted).
Asman, W. and Drukker, B.: 1988, *Atmos. Environ.* **22**, 725.
Bull, K.R.: 1995, *Water, Air, Soil Pollut.* **85**, 201.
Cofala, J. and Schöpp, W.: 1995, 'Assessing Future Acidification in Europe; Current State of the RAINS Model Development', Note prepared for the 15th meeting of the UN-ECE Task Force on Integrated Assessment Modelling, The Hague, The Netherlands.
Cosby, B.J., Hornberger, G.M. and Galloway, J.N.: 1985, *Water Resourc. Res.* **21**, 51.
De Vries, W., Posch, M. and Kämäri, J.: 1989, *Water, Air, Soil Pollut.* **48**, 349.
De Vries, W., Posch, M.,Reinds,G.J. and Kämäri, J.: 1994, *Water, Air, Soil Pollut.* **78**, 215.
De Vries, W., Posch, M., Oja, T., van Oene, H., Kros, H., Warfvinge, P. and Arp, P.: 1995, *Ecol. Model.* **83**, 283.
Forsius, M., Johansson, M., Posch, M., Holmberg, M., Kämäri, J., Lepistö, A., Roos, J., Syri, S. and Starr, M.: 1997, *Boreal Environ. Res.* **2**, 129.
Hedin, L.O., Granat, L., Likens, G.E., Buishand, T.A., Galloway, J.N., Butler, T.J. and Rodhe, H.: 1994, *Nature* **367**, 351.
Jenkins, A., Whitehead, P.G., Musgrove, T.J. and Cosby, B.J.: 1990, *J. Hydrol.* **116**, 403.
Johansson, M., Alveteg, M., Walse, C. and Warfvinge, P.: 1996, 'Derivation of Deposition and Uptake Scenarios', in M. Knoflacher, J. Schneider, and G. Soja (eds.), *International Workshop on Exceedance of Critical Loads and Levels, Conference papers BD.15/vol.15,* Federal Environment Agency, Vienna, Austria.
Kämäri, J., Posch, M., Kähkönen, A-M. and Johansson, M.: 1994, *Sci. Total Environ.* **160/161**, 687.
Kleemola, S. and Forsius, M. (eds.): 1996, *ICP Integrated Monitoring, 5th Annual Report, The Finnish Environment 27*, Finnish Environment Institute, Helsinki.

Løkke, H., Bak, J., Falkengren-Grerup, U., Finlay, R.D., Ilvesniemi, H., Nygaard, P.H. and Starr, M.: 1996, *Ambio* **25**, 510.

Mälkönen, E.: 1975, *Annual Primary Production and Nutrient Cycle in some Scots Pine Stands*, Communicationes Instituti Forestalis Fennia 84, Helsinki, Finland.

Marklund, L.G.: 1988, *Biomass Functions for Pine, Spruce and Birch in Sweden*, Swedish Agricultural University 45, Uppsala, Sweden.

Mylona, S.: 1993, *Trends of Sulphur Dioxide Emissions, Air Concentrations and Depositions of Sulphur in Europe since 1880*, EMEP/MSC-W Report 2, Oslo, Norway.

Sverdrup, H., Warfvinge, P., Blake L. and Goulding K.: 1995, *Agric. Ecosystems Environ.* **53**, 161.

Tiktak, A. and Van Grinsven, H.J.M.: 1995, *Ecol. Model.* **83**, 35.

Warfvinge, P. and Sverdrup, H.: 1992, *Water, Air, Soil Pollut.* **63**, 119.

Wright, R, Holmberg, M., Posch, M. and Warfvinge, P.: 1991, *Dynamic Models for Predicting Soil and Water Acidification: Application to Three Catchments in Fennoscandia*, Acid Rain Research Report 25, Norwegian Institute for Water Research, Oslo.

Wright, R.F. and Schindler, D.W.: 1995, *Water, Air, Soil Pollut.* **85**, 89.

PREDICTING FRESHWATER CRITICAL LOADS OF ACIDIFICATION AT THE CATCHMENT SCALE: AN EMPIRICAL MODEL

M.R. KERNAN, T.E.H. ALLOTT and R.W. BATTARBEE

Environmental Change Research Centre, University College London, London WC1H 0AP, United Kingdom

Abstract. Current applications of the critical loads concept are geared primarily toward targeting emission control strategies at a national and international level. Maps of critical loads for freshwaters have been produced in grid form based on water samples of representative sites within each grid square. However, the water chemistry data required to calculate freshwater critical loads are not always readily available at a national level and maps are therefore limited to catchments where such data exist. This paper describes the development of an approach that uses nationally available secondary data to predict freshwater critical loads for catchments lacking the appropriate water chemistry information. An empirical statistical model is calibrated using data from 78 catchments throughout Scotland. Water chemistry for each catchment has been determined. Each catchment is characterized according to a number of attributes. Redundancy analysis of these data shows clear relationships between catchment attributes and the critical load derived from the water chemistry. The key variables that explain most of the variation in critical load relate to soil, geology and land use within the catchment. Using these variables as predictors in a regression analysis, the critical load can be predicted across a broad gradient of sensitivity ($R^2_{adj} = 0.81$). The predictive power of the model was maintained when different combinations of explanatory variables were used. This accords the approach a degree of flexibility in that model parameterization can be geared toward availability of secondary data. There are limitations with the model as presently calibrated. However, the approach offers considerable scope for environmental managers to undertake national inventories of catchment sensitivity and specific assessments of individual catchments.

Keywords: acidification, catchments, freshwater critical loads, predictive model, Scotland, statistical modeling

1. Introduction

The development of the critical loads approach as applied to assessments of freshwater acidification has been reviewed widely (Henriksen *et al.*, 1990; CLAG Freshwaters, 1995; Bull, 1995). Critical loads maps for freshwater acidity have been produced that give some indication of the spatial distribution of freshwater sensitivity to acidification (Downing *et al.*, 1993; CLAG, 1995). However, the critical loads ascribed to each grid square in such maps are based on the chemistry of water samples (e.g., Kreiser *et al.*, 1993), and these data are not readily available at a national scale. Consequently, where the sensitivity of a large number of catchments across a wide geographical area needs to be assessed, it is necessary to undertake a costly water sampling program (e.g., Henriksen *et al.*, 1990; CLAG, 1995). Conversely, data relating to catchment characteristics more commonly are mapped on a national basis. This paper describes an empirical statistical model that quantifies the relationships between critical load and catchment characteristics in a predictive way, as a precursor to the development of a management tool for catchment scale critical load assessments.

Many authors have highlighted the link between the chemistry of surface waters and the characteristics of the catchments they drain (e.g., Reuss and Johnson, 1986; Hornung *et al.*, 1990a; Sverdrup *et al.*, 1992). Using these relationships, attempts have been made to identify where sensitive waters occur at regional and national levels (Edmunds and Kinniburgh, 1986; Hornung *et al.*, 1990a; Langan and Wilson, 1992; Hornung *et al.*, 1995). At a catchment scale, relationships have been quantified between surface water chemistry and geology (e.g., Duarte

Water, Air, and Soil Pollution **105:** 31–41, 1998.

and Kallf, 1989), soil (e.g., Rees *et al.,* 1989; Billett and Cresser, 1992) and land use (e.g., Hunsaker *et al.,* 1991) separately, or using an integrated approach (e.g., Lynch and Dise, 1985). The relationships quantified for individual catchments generally have been based on high resolution catchment specific model parameterization (e.g., Christopherson *et al.,* 1982; Billet and Cresser, 1992; Cosby *et al.,* 1995).

Previous attempts to predict freshwater critical loads nationally have not employed catchment-specific data (Hall *et al.,* 1995; Kernan, 1995). The current paper describes an intermediate approach between the low resolution, broadly generalized regional models and catchment specific models that can be applied only to catchments for which high resolution data are available. This approach allows estimates of sensitivity to be made for individual catchments using nationally available data. Although nitrogen (N) is recognized as an increasingly important component of acidification (Stoddard, 1994; Allott *et al.,* 1995), the critical load for sulfur (S) alone is used here as this model is robust and well established (Battarbee *et al.,* 1996). Critical loads models for total acidity (S+N) still are being developed and modified (e.g., Harriman *et al.,* 1995a).

2. Calibration Data

2.1. Water Chemistry

The calibration data set comprised calculated critical load (the response data) and catchment characteristics (the explanatory data) for a number of stream catchments. A total of 78 sites was selected, broadly distributed throughout Scotland across a range of soil and geology types. Values for a suite of chemical constituents were obtained using techniques described in Patrick *et al.* (1991) and these were used to derive critical loads according to the method of Battarbee *et al.* (1996). Although critical loads theory is underpinned by the assumption that critical loads are based on samples taken during a representative season (Forsius *et al.,* 1992), a single water sample was taken from each site to maximize the geographic extent of the data. This introduces an element of uncertainty to the model in that the temporal variation exhibited by streamwater chemistry reduces the likelihood that a single spot sample will be representative of mean conditions at a particular site. Higher streamflow following rainfall is characterized by a lower base-flow component. Run-off that by-passes the mineral soil horizons is less buffered than base-flow, which is in contact longer with cation exchange sites. This has implications for the calculation of critical loads. Peak-flow, for example, will be characterized by lower calcium concentrations (and therefore, lower critical loads), whereas at low-flow the critical load is likely to be over-estimated (CLAG, 1995). However, a number of studies indicate that although there may be substantial variations in chemistry associated with flow conditions, samples taken between mean and peak flow are more likely to be representative of mean conditions than samples taken at low flow (e.g., Langan and Harriman, 1993; CLAG, 1995). To avoid base flow conditions, the sampling strategy employed by the UK critical loads mapping program (CLAG, 1995) was adopted whereby spot samples were taken during October/November.

2.2. Catchment Characteristics

The catchment boundary of each site was digitized from 1:25,000 Ordnance Survey maps using a Geographical Information System (ARC/INFO). The digital catchments were then overlain onto a series of other digital data sets to characterize the catchments according to the percentage of different land cover, soil and geology types. The variables used to represent catchment processes and attributes are summarized in Table I.

Access to the national land cover database was made available by the Institute of Terrestrial Ecology. These data are derived using Landsat TM images (Fuller and Groom, 1983). Proportions of each land cover class were provided at 25 m^2 resolution and the original classification was aggregated to 6 classes (Table I, LC1-LC6).

The solid geology for each of the 10 km squares was digitized from 1:50,000 or 1:63,360 resolution maps published by the British Geological Survey. The geological formations were aggregated according to the system used by Edmunds and Kinniburgh (1986). Each catchment subsequently was characterised according to the percentage of each of the four buffering categories within its boundaries (Table I; G1-G4).

Digital soil maps were produced from maps at either 1:50,000 or 1:63,360 scale and a number of soil variables was derived for each catchment. A sensitivity classification applied to soil associations, based on the parent material used to define them (Langan and Wilson, 1991) and subsequently applied to soil series (Hornung *et al.,* 1995), was used to ascribe each polygon with a sensitivity class. A second classification was adopted where a soil critical load class was allocated to each soil map unit based on classes defined at a workshop held at Skokloster (Nilsson and Grennfelt, 1988). All soil map units in each catchment were allocated both a soil critical load (SCL) and a soil sensitivity class, the percentage of each comprising an individual variable. In addition to the classification variables, catchment-weighted values for soil water H^+ concentration and percent base saturation (% BS) were calculated for each catchment from soil profile data held at the Macauley Land Use Research Institute (MLURI). These data were used to assign a mean value for H^+ concentration and % BS to each soil type and, using the weighted averaging equation reproduced in Table I, it was possible to allocate a single figure for these variables to each catchment. A similar approach was applied to the lower limit of each critical load class giving a quantitative soil critical load value for each catchment. Such averaged catchment attributes commonly are used in dynamic models such as PROFILE (Sverdrup and Warvfinge, 1988) and MAGIC (Cosby *et al.,* 1995).

Catchments also were characterized by a number of other attributes gleaned from national datasets. Sulfur and N deposition, together with rainfall, were extracted from the CLAG critical loads database (CLAG, 1995). Distance from the sea was derived using ARC/INFO, as were catchment area and length of stream network. Altitude for the site and the highest point in the catchment were noted from the 1:25,000 Ordnance Survey maps.

3. Results

A redundancy analysis (RDA) (Van den Wollenburg, 1977; Israëls, 1992) was undertaken using CANOCO, Version 3.10 (ter Braak, 1987) to explore the relationships between catchment variables and critical load (CL). RDA can be viewed as a multivariate form of regression analysis whereby the response data are modeled as a function of one or more ordination axes,

TABLE I

Summary of catchment variables (Codes shown here are referred to in subsequent discussion)

Code	Variable	Source	Min.	Max.	Mean	Explanation
Area	Catchment area (km^2)	1	0.47	19.67	4.1	
S_len	Stream length (km)	1	1.27	84.80	9.1	Total length of stream network
Dist	Distance from sea (km)	1	0.02	76.05	24.5	Distance to nearest shore
Alt	Maximum altitude (m)	1	30	1130	514.7	Highest point in the catchment
Sdep	Total S deposition (keq H^+ ha^{-1} yr^{-1})	2	0.38	1.58	0.71	Modeled deposition for 20 km grid square
Ndep	Total N deposition (keq H^+ ha^{-1} yr^{-1})	2	0.53	1.58	0.96	Modeled deposition for 20 km grid square
Rain	Rainfall (mm yr^{-1})	2	709	3319	1750	Modeled rainfall for 20 km grid square
G1	% Geological sensitivity class 1	3	0	100	54.6	Granite and igneous rock, most meta-sediments, grits, quartz sandstones and decalcified sandstones, some Quaternary sands
G2	% Geological sensitivity class 2	3	0	100	24.3	Intermediate igneous rocks, metasediments free of carbonates, impure sandstones and shales, coal measures
G3	% Geological sensitivity class 3	3	0	100	15.7	Basic and ultrabasic igneous rocks, calcareous sandstones, most drift and beach deposits, mudstone and marlstones
G4	% Geological sensitivity class 4	3	0	85.8	5.4	Limestones, chalk, dolomitic limestones and sediments
SH	% High sensitivity soil	4	0	100	37.2	pH <4.5 ; base saturation <20%
SM	% Medium sensitivity soil	4	0	96.9	19.4	pH >4.5 < 5.5 ; base saturation >20% < 60%
SL	% Low sensitivity soil	4	0	96.8	10.8	pH <4.5 ; base saturation >60%
SCL1	% Soil critical load class 1	5	0	42.1	3.3	**Quartz, K-feldspar*: Granite, Quartzite : **0.1**
SCL2	% Soil critical load class 2	5	0	20.6	0.4	**Muscovite, Plagioclase, Biotite* (<5%): Granite, Gneiss; **0.5**
SCL3	% Soil critical load class 3	5	0	100	35	**Biotite, Amphibole* (<5%): Granodiorite, Greywacke, Schist, Gabbro : **1.0**
SCL4	% Soil critical load class 4	5	0	100	53.5	**Pyroxene, Epidote, Olivine*: Gabbro, Basalt: **2.0**
SCL5	% Soil critical load class 5	5	0	100	6	**Carbonates*: Limestone, Marl: **4.0**
H^+	H^+ concentration (H^+ :eq L^{-1})	6	2.93	29.3	13.9	▣Catchment weighted soil water H^+ concentration
%BS	% base saturation	6	4.76	76.2	24.2	▣Catchment weighted % base saturation
SCL	Soil critical load (keq H^+ ha^{-1} yr^{-1})	5	0	16	5.6	▣Catchment weighted soil critical load (based on lower class value)
LC1	% Land Cover Class 1	7	0	38	5.6	Water and built/bare ground
LC2	% Land Cover Class 2	7	0	89.7	17.1	Agricultural grass and arable
LC3	% Land Cover Class 3	7	0	25.4	2.9	Deciduous woodland
LC4	% Land Cover Class 4	7	0	74.2	5.8	Coniferous woodland
LC5	% Land Cover Class 5	7	0	100	41.6	Lowland semi - natural grass / moor
LC6	% Land Cover Class 6	7	0	100	27.1	Upland semi - natural grass moor

Key to sources: 1 = GIS/OS Map; 2 = CLAG Freshwaters; 3 = Kinniburgh and Edmunds (1986); 4 = Hornung *et al.*, 1995; 5 = Hornung *et al.*, 1994; 6 = MLURI national soil inventory; 7 = ITE Monks Wood.

* Italics indicate minerals controlling weathering. Otherwise explanation refers to parent material. Critical load shown in **bold** (keq H^+ ha^{-1} yr^{-1}).

▣Weighted averaging equation - $[x]_c = \Sigma A_s \cdot [x]_s / \Sigma A_s$ where $[x]_c$ is a value either for H^+, % BS or SCL for the catchment, A_s is the area of soil map unit s and $[x]_s$ is the value for H^+, % BS or SCL for soil map unit s. The denominator Σ is taken over soil map units s1......sn .

which are in turn a function of the explanatory data. The use of a single response variable, (CL), resulted in a single constrained axis (ter Braak, 1987), which is in effect the critical load. The eigenvalue for this axis was 0.914, indicating that 91.4 % of the variation in critical load is explained by the 28 catchment variables. An unrestricted Monte Carlo Permutation test (ter Braak, 1990) showed that this axis was significant at the 5 % level. Each variable is characterized by a biplot score (between -1 and 1 and equating to the correlation coefficient between the variable and the axis), which indicates the extent of the association (positive or negative) between that variable and critical load (see Figure 1).

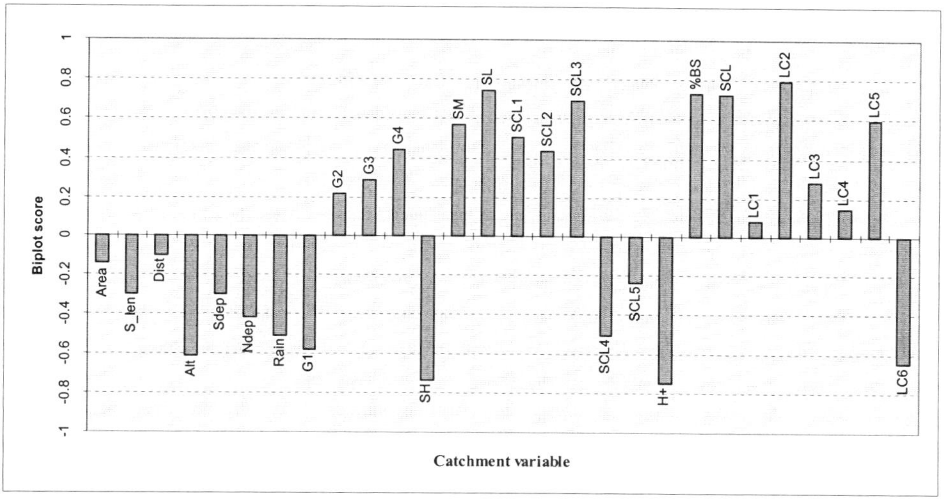

Fig. 1. Graph showing biplot scores for each explanatory variable (see Table I for catchment variable codes).

Those variables with high positive biplot scores have strong positive relationships with critical load (e.g., LC2, SL, % BS, SCL). Conversely, strong negative relationships are indicated by high negative scores (e.g., H^+, SL, LC6, Alt). However, the fact that the catchment data account for such a high proportion of the variation in critical load may partly be due to the number of explanatory variables. To develop a predictive model, it is necessary to identify which catchment variables independently explain significant fractions of the variation in critical load, allowing collinearity to be eliminated and reducing the possibility of spurious explanation associated with large numbers of predictors (Økland and Eilertson, 1994).

Forward selection using CANOCO (ter Braak, 1990) allowed the identification of a minimum set of catchment variables that explains the chemistry data almost as well as the full set. Variables were selected iteratively and the statistical significance of each tested using Monte Carlo permutation (ter Braak, 1990). The five variables indicated by the forward selection procedure (LC2, SCL, G4, G1, H^+) cumulatively explained 81 % of the variation in critical load, only 10 % less than when all 28 variables are included (Table II).

These five key catchment attributes subsequently were regressed onto critical load (Table III). Approximately 80 % of the variation in critical load is explained by the catchment parameters (significant at 95 % level). The regression diagnostics and examination of the residuals confirmed that the model fulfills the major assumptions required of multiple regression (Manly, 1992).

TABLE II
Results of forward selection on catchment variables.

Variable added	Cumulative variance exlained
LC2	.58
SCL	.69
G4	.76
G1	.79
H^+	.81
Variance explained by all 28 variables	0.91
No other significant variables	

TABLE III
Summary of multiple linear regression output with G1, G2, SCL, H^+ and LC2 as predictors using SAS/INSIGHT (SAS Institute Inc., 1993).

	Type III tests		Parameter estimates			95% confidence interval	
Predictor	*F*	*p*	Estimate	Tolerance	VIF	Lower	Upper
Intercept	n/a	n/a	0.8038	n/a	0	0.4474	1.1601
$Log_{10}(G1+1)$	9.5674	0	-0.1073	0.6206	1.611	-0.1765	-0.0382
$Log_{10}(G4+1)$	26.9283	0	0.2056	0.8461	1.182	0.1266	0.2846
$Log_{10}(SCL+1)$	7.5653	0.02	0.2625	0.5167	1.954	0.0414	0.4836
$\sqrt{H^+}$	5.6	0	-0.083	0.5018	1.993	-0.1427	-0.0228
$Log_{10}(LC2+1)$	33.6219	0	0.2133	0.5282	1.893	0.14	0.2867

Root mean squared error = 0.1698 $R^2 = 0.8090$ Adjusted $R^2 = 0.7958$

In an attempt to simplify the model, and consequently reduce the data requirements, an exercise was undertaken to assess whether the strength of the predictive relationship is substantially reduced by removing one or more of the predictors. The predictor that adds least to the fit of the model is SCL which has the lowest sum of squares (not reported). A second multiple regression, removing SCL as a predictor, gives an R^2_{adj} of 0.78 (RMSE= 0.1750). This represents s reduction of only about 1 % in explanatory power. SCL thus accounts for very little of the variation in critical load once the other variables are considered. In this second model H^+ had the lowest *F*- statistic of all the predictors and adds least to the fit of the model when the effects of all other parameters are considered. A third multiple regression was performed omitting H^+ and the resulting R^2_{adj} is 0.75 (RMSE =

0.1789). This model explains only 4 % less variation in critical load than the primary model. The omission of two predictor variables has not led to a substantial reduction in the fit of the model.

It also can be demonstrated that the choice of predictors need not be limited to those indicated by the forward selection procedure in RDA. Forward selection successively chooses variables that independently add most to the explained variation in critical load. However, it is possible that significant variables that are co-linear with those already selected will not be included. Regression analyses therefore were undertaken on a wider variety of explanatory variables selected on a more subjective basis. The explanatory power of the equations (R^2_{adj}) produced by regressing these variables onto critical load varies between 0.17 for simple linear regression with G4 as the predictor and 0.77 when the explanatory variables are % BS, LC2, G1 and G4 (Table IV). Taken with the R^2_{adj} values for the first three models, LC2 is the most significant single variable in terms of explaining critical load variation, confirming the results from the forward selection exercise. The implication here is that the percentage of arable land in a catchment most strongly influences the sensitivity of the surface waters in the calibration data set.

4. Discussion

Strong statistical relationships between catchment variables and critical load have been established, specifically with % arable land in the catchment (LC2), % granite and gneisses (G1), % limestones and chalk (G4), weighted soil critical load (SCL) and weighted soil water H^+ concentration (H^+). These relationships, which are stronger than those exhibited using lower resolution data (Hall *et al.*, 1995; Kernan, 1995), reflect some of the key processes operating within catchment systems and the attributes that determine their relative contribution to the buffering of surface waters. The importance of geology derives from the influence of bedrock mineralogy (particularly the presence of carbonate or weatherable silicate minerals) on primary mineral weathering processes and products. Rock types resistant to chemical weathering have smaller reserves of base cations available for weathering reactions, and consequently are less able to retain H^+ from acid deposition (Lynch and Dise, 1982). Therefore, areas overlying granitic or other highly siliceous geology with thin patchy soils (G1) tend to yield more sensitive drainage waters (Reuss *et al.*, 1987). Conversely, calcareous bedrock (G4) can buffer the most intense acid loading (Kramer, 1976) as a result of the rapid kinetics of $CaCO_3$ solution (Brakke *et al.*, 1990). With regard to soil, the importance of SCL and H^+ (and by implication % BS) in the model echoes work relating soil sensitivity to freshwater sensitivity (e.g., Langan and Wilson, 1992; Hornung *et al.*, 1995), which has demonstrated that soil sensitivity is a precursor to freshwater sensitivity. The base saturation of a soil provides a measure of the extent to which it has been acidified. Non-sensitive soils tend to have a high proportion of the cation exchange complex occupied by bases and are thus more capable of neutralizing incoming acid deposition than sensitive soils that may be dominated by exchangeable acidity. The soil water pH is also a guide to its acid neutralizing capacity as soil water H^+ is assumed to be in equilibrium with the exchange complex acidity (Langan and Wilson, 1992). In the models, high % BS and low H^+ concentration therefore are associated with well buffered drainage waters.

Table IV
Results of regression analyses on combinations of catchment attribute combinations.

Predictors	RMSE	$R^2_{adj.}$
LC2, G1, H^+	0.211	0.688
LC2, G4, H^+	0.193	0.738
LC2, G4	0.211	0.683
LC2,G1	0.233	0.615
H^+, G1, G4	0.214	0.677
% BS, LC2, G1, G4	0.182	0.766
% BS, LC2	0.226	0.639
% BS, LC5, G1	0.24	0.592
LC2	0.245	0.575
% BS	0.271	0.48
H^+	0.264	0.507
SCL	0.274	0.469
G1	0.313	0.304
G4	0.343	0.168

The most significant variable in all these analyses is the % arable/managed grassland (LC2) in the catchment. Streams and lakes in catchments dominated by arable land will have very high critical loads as they are more likely to be located in lowland areas on base rich soils. Even on more acid soils the practice of adding lime to increase pH and calcium concentrations will ameliorate the effects of acid deposition (Hornung *et al.,* 1990b). Thus, the importance of LC2 in the models may be due to in part to the broad gradient of critical load values in the calibration data set wherein the very high critical loads will be in agricultural, low lying catchments. However, the most sensitive sites, those in upland areas with poorly buffered soils, will contain no arable land and a predictive model based solely on LC2 will be of limited value for differentiating critical loads at these sites. Future model calibrations may usefully exclude lowland sites or increase the number of sites with very low critical loads as it is these more sensitive catchments which drive the need for emission reductions.

Although predicted critical load in the primary model (Table III) is based on five key catchment variables, several alternative models have been generated, using different combinations of predictors, yet having similar explanatory power. The predictors include a mixture of soil, geology and land use components. Whittaker (1984) reviewed a number of analyses, based on automatic selection procedures and concluded that the concept of a "best" regression equation is a fallacy. The present paper indicates that the choice of catchment variables used to predict critical load can be modified according to data availability or subjective *a priori* judgement of processes operating in a catchment, without significantly reducing predictive power.

The models described exemplify an approach to critical loads assessment on a national scale. The results of the statistical analyses suggest that, using this approach, it is feasible to use widely available catchment data to predict surface water critical loads. These nationally available data therefore have the potential to allow assessments of sensitivity to be made at the catchment scale for streams and lakes for which no water chemistry data exist, providing a management tool to facilitate, for example, impact assessment studies. Additionally, model performance (and thus the utility of the approach) may be improved, particularly with regard to sensitive sites, by the use of more detailed soil physical and chemistry data, now available for most soil map units/series, along with data relating to the drainage characteristics of soil types (HOST (Hydrology of Soil Types), Boorman *et al.,* 1995) that will allow the influence of hydrological pathways to be incorporated.

Acknowledgments

This paper forms part of the research undertaken as part of a NERC studentship (Ref GT4/92/17/R). Additional support was provided by DoE's CLAG Freshwater sub-group and by the NERC Environmental Diagnostics Freshwater Critical Loads Programme. We thank ITE for use of the Land Cover database, Simon Langan at MLURI for soils data, Ordnance Survey for permission to digitise OS maps, and BGS for permission to digitize geological maps. Thanks are also due to Steve Juggins for advice and Ron Harriman for undertaking the analytical chemistry.

References

Allott, T.E.H., Curtis, C.J., Hall. J., Harriman, R. and Battarbee, R.W.:1995, *Water, Air Soil Pollut.* **85**, 297.

Battarbee, R.W., Allott, T.E.H., Juggins, S. and Kreiser, A: 1996, *Ambio* **25**, 366.

Billet, M.F. and Cresser, M.S. : *Environ. Pollut.* **77**, 263.

Boorman, D,B., Hollis, J.M. and Lilly. A.: 1995, *Hydrology of Soil types: A Hydrologically-based Classification of the Soils of the United Kingdom Report No.26,* Institute of Hydrology, Wallingford.

Bull, K.R.: 1995, *Water, Air Soil Pollut.* **85**, 201.

Brakke, D.F., Henriksen, A., Norton, S.A.: 1990, *Verh. Internat. Verein. Limnol.* **24**, 146.

Christopherson, N., Seip, H.M. and Wright, R.F.: 1992, *Water Resour. Res.* **18**, 977.

CLAG: 1995, *Critical Loads of Acid Deposition for United Kingdom Freshwaters - Critical Loads Advisory Group, Sub-group Report on Freshwater,* Department of Environment, UK.

Cosby, B.J., Wright, R.F., Hornberger, G.M. and Galloway, J.N.: 1995 *Water Resour. Res.* **21**, 51.

Edmunds, W.M. and Kinniburgh, D.G.: 1986, *J. Geol. Soc.* **143**, 707.

Downing, R.J., Hettelingh, J.P. and de Smet, P.A.M.: 1993, *Calculation and Mapping of Critical Loads in Europe: Status Report 1993.* CCE, RIVM, Report 259101003. Bilthoven.

Duarte, C.M. and Kalff, J.: 1989, *Arch. Hydrobiol.* **115**, 27.

Forsius, M., Kämäri, J., and Posch, M.: 1992, *Environ. Pollut.* **77**, 185.

Hall, J.R., Wright, S.M., Sparks, T.H., Ullyet, J., Allott, T.E.H. and Hornung, M.: 1995, *Water, Air Soil Pollut.* **85**, 2521.

Harriman, R., Allott., Battarbee, R.W., Curtis, C., Jenkins, A. and Hall, J.R.: 1995, *Mapping and Modelling of Critical Loads. Report of a Workshop held at Grange-over-Sands, Cumbria.*

Henriksen, A., Kämäri, J., Posch, M., Lövblad., G., Forsius, M. and Wilander, A.: 1990, in *Intra- and Inter- Grid Variability of Critical Loads and their Exceedances. Miljörapport 1990:17.* Nordic Council of Ministers.

Hornung, M., Le-Grice, S., Brown, N. and Norris, D.: 1990a, in R.W. Edwards, Gee, A.S. and J.H. Stoner (eds.), *Acid Waters in Wales.* Kluwer Academic Publishers, Dordrecht.

Hornung, M., Brown, S.J., and Ranson, A.: 1990b, in R.W. Edwards, Gee, A.S. and J.H. Stoner (eds.), *Acid Waters in Wales.* Kluwer Academic Publishers, Dordrecht.

Hornung, M., Bull, K., Cresser, M., Hall, J.R., Loveland, P.J., Langan, S.J., Reynolds, B. and Robertson, W.H.: 1994, in R.W. Battarbee (ed.), *Acid Rain and its Impact: The Critical Loads Debate. Proceedings of a Conference held at the Environmental Change Research Centre, London.*

Hornung, M., Bull, K., Cresser, M., Ullyet, J., Hall, J.R., Langan, S., Loveland, P.J. and Wilson, M.J.: 1995, *Environ. Pollut.* **87**, 204.

Hunsaker, C.T., Levine, D.A., Timmins, S.P., Jackson, B.L. and O'Neill, R.V.: 1991 *Landscape Characterisation for Assessing Regional Water Quality. Report to the U.S Government. Contract No. DE-ACOS-840R21400*

Israëls, A.: 1992, *Stat. Applic.* **4**, 531.

Kernan, M.: 1995, *Water, Air Soil Pollut.* **85**, 2479.

Kramer, J.: 1976, *Proceedings of the 1st International Symposium on Acid Precipitation and the Forest Ecosystem,* US Department of Agriculture Forest Service.

Kreiser, A., Patrick, S.T., and Battarbee, R.W.: 1993, in Hornung, M. and Skeffington, R.A., (eds.), *Critical Loads: Concepts and Applications. ITE Symposium No.28,* London.

Langan, S.J. and Wilson, M.J.: 1992, *J. Hydrol.* **138**, 515.

Manly, B.J.F.: 1992, *The Design and Analysis of Research Studies.* Cambridge University Press, Cambridge.

Nilsson, J. and Grennfelt, P.: 1988, *Critical Loads for Sulphur and Nitrogen. UNECE/Nordic Council Workshop Report, Skokloster, Sweden*

Økland, R.H. and Eilertson, O. : 1994, *J. Veget. Sci.* **5**, 117.

Patrick, S.T., Juggins, S., Waters, D. and Jenkins, A.: 1991, *The United Kingdom Acid Waters Monitoring Network. Site Description and Methodology Report, Report to DoE and DoE (NI)*

Rees, R.M., Parker-Jervis, F. and Cresser, M.: 1989, *Water Res.* **23**, 511.

Reuss, J.O. and Johnson, D.W.: 1986, *Acid Deposition and the Acidification of Soils and Waters.* Springer Verlag, New York.

Reuss, J.O., Cosby, B.J. and Wright, R.F.: 1987, in S.A. Norton., S.E. Lindberg and A.L. Page (eds.), *Acid Precipitation Volume 4: Soils, Aquatic Processes and Lake Acidification.* Springer Verlag, New York.

SAS Institute Inc.: 1993, *SAS/INSIGHT® User's Guide, Version 6, Second Edition,* Cary, NC.

Stoddard, J.L.: 1994, in L.A. Baker, (ed.), *Environmental Chemistry of Lakes and Rivers*. American Chemical Society, No. 237.

Sverdrup, H. and Warfvinge, P.: 1988, *Water, Air Soil Pollut.* **38**, 387.

Sverdrup, H., Warfvinge, P., Rabenhorst, M., Janicki, A., Morgan, R. and Bowman M.: 1992, *Environ. Pollut.* **77**, 195.

ter Braak, C.J.F.: 1987, *CANOCO - a FORTRAN Program for CANOnical Community Ordination by [Partial] [Detrended] [Canonical] Correspondence Analysis, Principle Components Analysis (Version 3.1)*.

ter Braak, C.J.F.: 1990, *Update Notes: CANOCO Version 3.10.* Agricultural Mathematics Group, Wageningen.

Van den Wollenburg, A.L.: 1977, *Psychometrika* **42**, 207.

Whittaker, J.: 1984, *Appl. Stat.* **33**, 52.

EVIDENCE OF CURRENT SOIL ACIDIFICATION IN SPRUCE STANDS IN THE VOSGES MOUNTAINS, NORTH-EASTERN FRANCE

ETIENNE DAMBRINE[1], BENOIT POLLIER[1], ANNE POSZWA[1], JACQUES RANGER[1], ANNE PROBST[2], DANIEL VIVILLE[3], PHILIPPE BIRON[3], and ANDRE GRANIER[4]

[1]*INRA, Equipe Cycles Biogéochimiques, 54280 Seichamps, France;* [2]*CGS-CNRS, 67084 Strasbourg, France;* [3]*CEREG-ULP, 67083 Strasbourg, France;* [4]*INRA, Equipe Ecophysiologie, 54280 Seichamps, France*

Abstract. To demonstrate directly soil acidification under spruce stands in the Strengbach catchment, soils from two adjacent stands aged 40 and 90 years were sampled intensively in 1990 and resampled in 1996. Soils already were very acid in 1990. Between 1990 and 1996, soils had experienced a significant decrease in exchangeable Ca^{2+} and Mg^{2+} at all depths at the two sites except in the 0-10 cm layer, for which base saturation remained constant. Losses of Ca^{2+} calculated from a budget study at the same sites and from the comparison of exchangeable stores were similar. In contrast, the loss of Mg^{2+} from the exchange complex was higher than that computed from the field budget. Various reasons, including most probably uncertainties linked to the extremely low levels of Mg in the ecosystem, may explain this discrepancy. Since 1987, a general decrease of the $(Ca^{2+} + Mg^{2+})/Al^{3+}$ ratio in soil solution was observed at both sites. These results are consistent with present day acidification of soils poor in weatherable minerals under spruce stands in the Vosges Mountains.

Key words: Soil acidification, soil solution, spruce, Ca/Al ratio, Strengbach catchment, Vosges Mountains

1. Introduction

Soil acidification has been studied in Europe and North America for many years in the contexts of fish decline in low alkalinity lakes or coniferous forest decline on acidic substrates (Reuss and Johnson, 1986). In the Vosges Mountains (northeastern France), several studies have suggested that soil and stream acidification have occurred during this century as a result of acidic deposition and land use changes (cf. Landmann and Bonneau, 1995), but direct evidence has been lacking. In the Strengbach catchment, an integrated study of soil and stream acidification was initiated 10 years ago in the context of forest decline (Probst *et al.*, 1990). Mineral budgets calculated in 1990 after three years of investigation suggested that the soil under spruce forest was losing exchangeable base cations rapidly, and that given the low reserves of exchangeable base cations in the soils, losses could be detected by measuring the changes in soil base saturation after five years (Bonneau *et al.*, 1991). Therefore, soils were intensively sampled in 1990-91 and resampled in 1996. In addition, detailed investigations on soil mineralogy were initiated and weathering rates were calculated. Finally, soil solution monitoring was extended until 1996. In this paper, we compare results obtained by these different methods and discuss possible causes for discrepancies.

2. Materials and Methods

The sites studied are two spruce stands, 40 (S40) and 90 (S90) year old, located side by side at an elevation of 1,050 m. Mean annual air temperature is 5 °C and precipitation is 1,300

Water, Air, and Soil Pollution **105:** 43–52, 1998.

mm. Soils are coarse textured Typic Dystrochrepts underlain by a Ca-poor granite. Soil exchangeable base cation contents were very low and not statistically different between the two sites (Table I). At both sites, stands are second generation spruce plantations that have replaced the native mixed fir and beech forest. Needle yellowing and needle losses were first recorded at that site in 1983. Yellowing has been related to Mg deficiency, but Ca concentrations in the needles also are low (Le Goaster *et al.*, 1991). At S90, deficiency symptoms as well as needle loss (on average 30 %) have remained more or less stable since 1986. At S40, 20 % of the trees had lost needles and showed yellowing symptoms in 1988 (Le Goaster *et al.*, 1990). In spring 1988, a storm felled a large proportion of trees in the upper part of S40, including in the area where lysimeters were located. Broken trees were removed manually. Yellowing and needle loss increased significantly in autumn 1990 in response to a very severe drought. Since then, tree foliage has returned to the level observed in 1988.

TABLE I

Soil characteristics at S40 and S90; asterisks indicate significant differences between the two sites ($p = 0.05$); CEC = cation exchange capacity, BS = base saturation.

Layer or Depth (cm)	Gravel (% > 2 mm)	Density (g cm^{-3})	Clay (% < 2 mm)	C/N	pH (in water)	CEC cmol$_c$kg^{-1}	BS (%)
			Site S90				
L + F				23.9	3.8		
0 - 10 cm	45.7	1	20.2	15.5	3.5	13.3	7.5
10 - 30 cm	47.1	1.3	20.2	14.3	3.9	13.0	4.3
30 - 50 cm	49.9	1.3	19.9	13.9	4.2	10.9	4.2
50 - 70 cm	56.0	1.3			4.4	8.0	5.2
			Site S40				
L + F				25.4	4*		
0 - 10 cm	41.7	0.8	14.3*	16.7	3.5	11.1	8.4
10 - 30 cm	44.4	1.3	15.2*	16.5	3.8	11*	4.2
30 - 50 cm	54.1	1.1	17*	15.7 *	4.2	10	3.7
50 - 70 cm	61.4	1.4			4.4	7.7	4.8

Open field precipitation (P), temperature, light and wind speed, were monitored in a clearing adjacent to the studied stands. Transpiration (Lu *et al.*, 1995) and soil water content monitoring at each site were used to calibrate a water model (Biron, 1994). Beginning in December 1986 in S90 and December 1987 in S40, soil solution was collected monthly using eight (S90) or two (S40) zero tension lysimeters per sampling depth. Open field precipitation and throughfall also also collected. Solutions from replicate collectors were bulked on a volume basis, filtered (0.45 μm), analyzed for pH, and stored at 4 °C. Ca, Mg, K, Na, Si, Al, S were analyzed by ICP (Jobin Yvon 38+) and NH_4^+, NO_3^- and Cl^- by automated colorimetry (Technicon) generally within the week. Base cation dry deposition was computed using Na as a dry deposition index (Ulrich, 1983).

In March 1990, soils in S90 were sampled at five plots. At each plot (15 × 15 m), we sampled five replicates of the litter layer (L+F) using a 30 × 30 cm frame, and mineral soil at four depths using a hand auger (0-10, 10-30, 30-50, 50-70 cm). Soils in S40 were sampled in March 1991 using the same sampling scheme, but only three replicate samples were taken at each plot. Sampling plots in S40 were chosen in areas where trees had not been damaged by windthrow. The five (S90) or three (S40) replicates of mineral soil collected at each plot were bulked in the field. Litter and soil samples were air-dried, weighed, sieved, and the coarse fraction (>2 mm) was weighed. Soils were resampled at both sites in January 1996, and soils from S90 (0-10, 10-30 and 30-50 cm) were sampled again in March 1996. Exchangeable Ca^{2+}, Mg^{2+}, Mn^{2+}, Al^{3+} and H^+ were extracted from all soil samples in 1996 using 1 M KCl; extract concentrations of Ca^{2+}, Mg^{2+}, Mn^{2+} and Al^{3+} were analyzed by inductively coupled plasma emission spectrometry (ICP). Al^{3+} and H^+ were determined by titration (Espiau and Peyronnel, 1976). Exchangeable K^+ was displaced by 1 M NH_4Cl using the same procedure and analyzed by ICP. Cation exchange capacity was calculated by summing up exchangeable cations. For each soil layer, significant changes in exchangeable cations between 1990/91 and 1996 were checked by pairwise comparisons using Student's *t* test.

The annual increment of nutrients in biomass was determined in 1988 after a biomass and mineral content inventory (Le Goaster *et al.*, 1991). Mineralogical investigations were conducted on a representative soil profile to determine the quantitative mineralogy of the different particle size fractions (Fichter *et al.*, 1997a). Present weathering rate was calculated with the geochemical model PROFILE (Sverdrup and Warfvinge, 1988), using mineralogical and chemical data from the site (Fichter *et al.*, 1997b). The historical weathering rate also was calculated using isotitanium and isoquartz balances and dividing soil losses in base cations by 10,000 yr, i.e. the approximate age of the soil.

3. Results

3.1. Soil Exchangeable Cations

No significant change in the mass of L + F horizons was measured between 1990/91 and 1996 at either site. At S90, except in the 0-10 cm layer, exchangeable Ca^{2+} and Mg^{2+} concentrations were statistically lower in 1996 than in 1990/91 (Table II). Exchangeable Mn^{2+} concentrations decreased in all layers, but changes in exchangeable Al^{3+} and K^+ concentrations were not significant. At S40, exchangeable Mg^{2+} concentrations decreased significantly below 10 cm depth. Exchangeable Ca^{2+} concentrations decreased significantly only in the 10-30 cm layer, whereas exchangeable K^+ concentrations decreased significantly in the 10-30 cm and 50-70 cm layers. Exchangeable Al^{3+} concentrations decreased significantly in the 30-50 cm layer.

Concentrations of exchangeable Ca^{2+} and Mg^{2+} at S90 were lower in March than in January 1996, especially in the 0-10 cm layer, and differences in total soil exchangeable amounts were almost significant ($p = 0.07$ and 0.08 for Ca and Mg, respectively) (Poszwa 1996). Total amounts in exchangeable base cations in the top 70 cm of mineral soil were significantly lower ($p < 0.05$) in 1996 than in 1990/1991, except for exchangeable K^+ ($p = 0.8$) at S90 and exchangeable Ca^{2+} ($p = 0.1$) at S40 (Table III).

TABLE II

Mean concentrations of exchangeable cations and cation exchange capacity (CEC) ($cmol_c\ kg^{-1}$) at S90 and S40 and the difference in concentrations between 1996 and 1990 (or 1991); * $p < 0.05$, ** $p < 0.01$.

	Mn^{2+}		Mg^{2+}		Ca^{2+}		K^+		Al^{3+}		H^+		CEC	
Depth	90/91	diff.	90/91	diff.	90/91	diff.	90/91	diff.	90/91	diff.	90/91	diff.	90/91	diff.
S 90														
0-10 cm	0.27	-0.19**	0.19	-0.04	0.40	-0.08	0.40	0.13	9.89	-0.99*	2.12	0.13	13.27	-1.04*
10-30 cm	0.34	-0.21*	0.09	-0.03**	0.14	-0.06**	0.33	-0.04	11.12	-1.09*	0.92	0.01	12.94	-1.42*
30-50 cm	0.25	-0.18**	0.07	-0.04**	0.09	-0.05**	0.30	-0.02	9.67	-1.38	0.48	-0.28*	10.86	-1.95
50-70 cm	0.20	-0.14*	0.06	-0.04**	0.08	-0.04**	0.29	-0.04	7.14	-0.64	0.24	-0.08*	8.01	-0.98
S 40														
0-10 cm	0.03	0.01	0.20	-0.06	0.28	-0.11	0.46	-0.18	7.37	0.61	2.81	-0.63**	11.15	-0.36
10-30 cm	0.03	0.07	0.09	-0.03**	0.09	-0.05*	0.28	-0.09**	9.11	0.10	1.36	-0.47**	10.96	-0.47
30-50 cm	0.06	-0.01	0.06	-0.02**	0.07	-0.04	0.24	-0.04	9.02	-1.22*	0.56	-0.12	10.01	-1.45*
50-70 cm	0.06	0.00	0.05	-0.02	0.05	-0.01	0.27	-0.09*	7.03	-0.65	0.23	-0.01	7.69	-0.89

TABLE III

Comparison between the annual decrease (*: significant at $p < 0.05$) of exchangeable base cations between 1990-1991 and 1996, and the loss of base cations calculated from mass balance (deposition - drainage - biomass) during the same period, the weathering rate and the total soil store (0 - 70 cm) of exchangeable base cations in 1990/1991.

	S 40			*S 90*		
	Ca	Mg	K	Ca	Mg	K
			kg ha⁻¹ yr⁻¹			
Deposition - drainage	2.7	0.4	-3.1	-3.8	-0.4	-6.2
Biomass	14.3	1.9	11.3	5.1	0.5	2.4
Mass balance	-11.6	-1.5	-14.4	-8.9	-0.9	-8.6
Weathering rate (modeled/historical)	0.2/1.2	1.2/0.3	8.0/1.2	0.2/1.2	1.2/0.3	8.0/1.2
Loss of exchangeable cations	-7.6*	-2.8*	-27.6*	-7.6	-3.3*	-7.1
			kg ha⁻¹			
Exchangeable cation store	79.6	41.7	449.3	122.1	46.5	539.2

3.2. Time Variation of Soil Solution Composition

The temporal patterns (1988-1995) of concentrations in soil solutions at S40 reveal two distinct time periods (Figure 1). From 1988 to the autumn of 1990, NO_3^-, Ca^{2+} and Mg^{2+} concentrations were high and SO_4^{2-} concentration increased. Nitrate, Ca^{2+} and Mg^{2+} concentrations decreased abruptly in winter and spring 1990, then slightly and continuously toward low levels until the end of the study period, except for a brief increase in the autumn of 1994. Concentrations of Al^{3+} and H^+ remained stable or increased irregularly in comparison with the previous period. Sulfate concentration decreased slightly or increased (60 cm depth). Over the whole time period, the $(Ca^{2+} + Mg^{2+})/Al^{3+}$ ratio decreased at all depths, independently of the variations in anion concentration (Poszwa, 1996), as a result of both the decrease in base cations and the increase in Al^{3+} concentration. The temporal patterns (1987-1995) of concentrations in soil solutions at S90 reveal three time periods (Figure 2). From 1987 to winter 1988/89, peaks of Ca^{2+}, Mg^{2+} and Al^{3+} concentration corresponded to NO_3^- pulses. From 1989 to 1993, Ca^{2+}, Mg^{2+}, Al^{3+}, H^+, and NO_3^- concentrations decreased. From winter 1993 to the end of the study period, concentrations generally rose as peaks of Ca^{2+}, Mg^{2+}, Al^{3+}, H^+ and NO_3^- concentrations occurred. Over the whole time period, SO_4^{2-} cncentrations decreased. Over the study period, the $(Ca^{2+} + Mg^{2+})/Al^{3+}$ ratio of soil solutions decreased at the 10 cm depth, but no change occured at the 60 cm depth.

3.3. Budgets

Annual mean losses in total or exchangeable base cations from 1990/1991 to 1996 were calculated either from the flux budget, or from the decrease of soil exchangeable base cations (Table III). The decrease of exchangeable Ca^{2+} concentration was similar to losses calculated from the flux budget. This is in agreement with the very low weathering rate

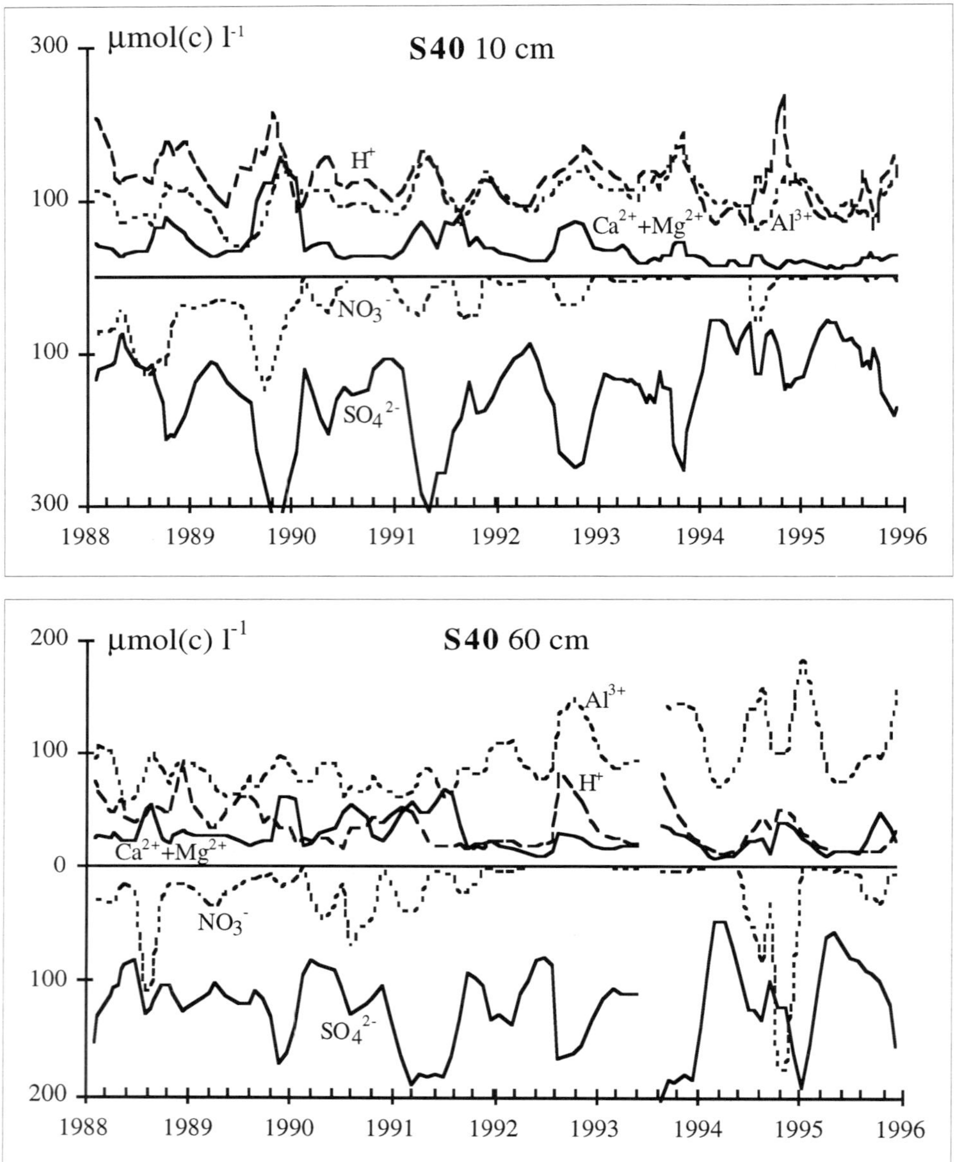

Fig. 1. Time variation (3rd order moving average) of soil solution composition at two soil depths (10, 60 cm) in S40 ($\mu mol_c L^{-1}$). Al^{3+} = 3 × (Al molar concentration). Notice that the *y*-axes for anions and cations are reversed.

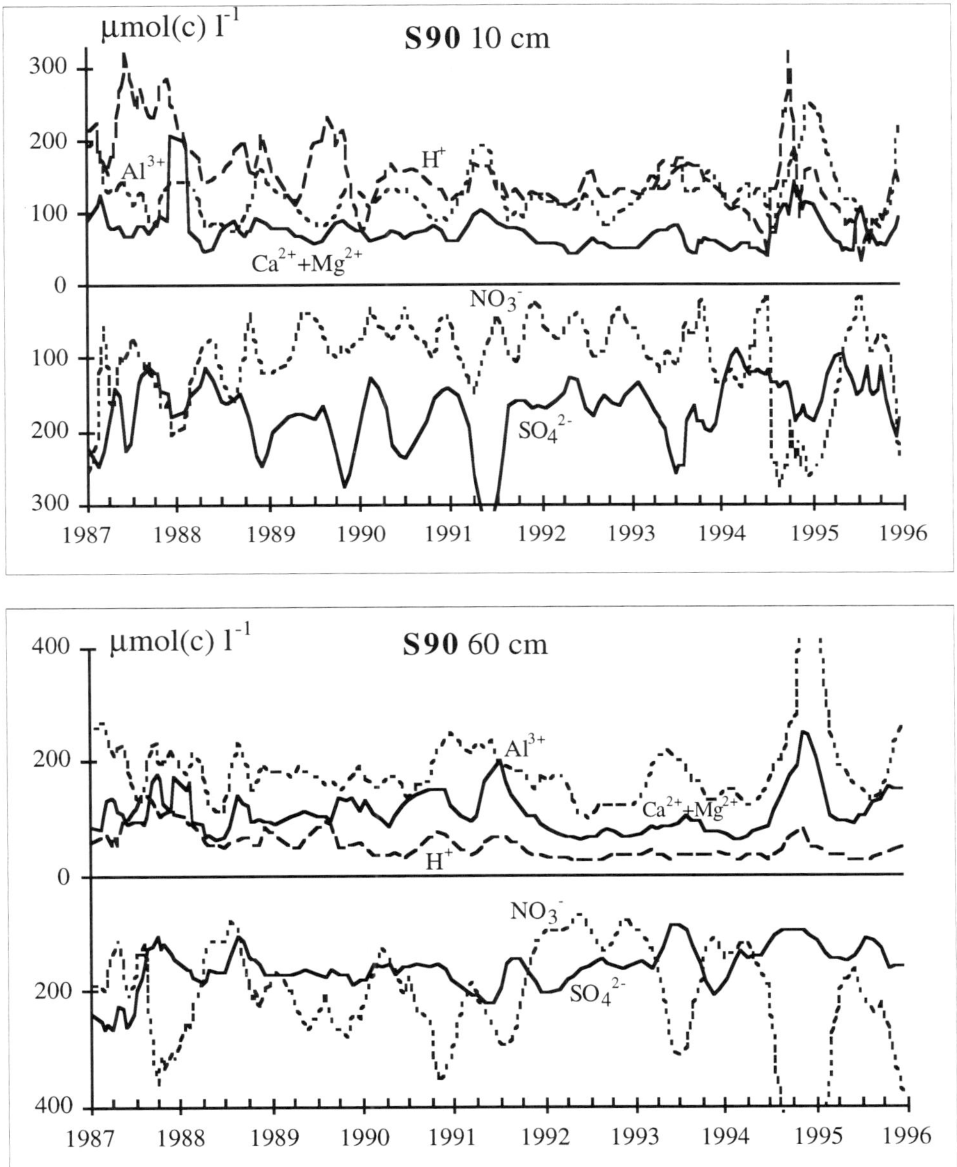

Fig. 2. Time variation (3rd order moving average) of soil solution composition at two soil depths (10, 60 cm) in S90 ($\mu mol_c L^{-1}$). Al^{3+} = 3 × (Al molar concentration). Notice that the *y*-axes for anions and cations are reversed.

computed for Ca^{2+}. The decrease in exchangeable Mg^{2+} concentration was 2 to 3 times higher than the Mg^{2+} loss computed from the flux budget. Moreover, a part of this loss should be compensated for by weathering. The decrease in exchangeable K^+ concentration at S40 was much higher than the flux of K^+ calculated from mass balance. If the modeled value for K^+ release by weathering was considered, the ion budget at S90 agreed with the fact that exchangeable K^+ was close to steady state (no significant change), but losses of exchangeable K^+ at S40 could not be explained.

4. Discussion

The different techniques used provided a strong evidence for an ongoing soil acidification. But discrepancies between the results remain, and might have several origins. First, uncertainties in the weathering rates are high. Release of Ca^{2+} depends on the accessibility of apatite inclusions in K- and Na-feldspars to roots and soil solutions, which is unknown. Incongruent transformation of micas into vermiculites, which releases K^+ and keeps Mg^{2+} in the structure of the vermiculite, is not taken into account by PROFILE, although it is presently occurring in the soil (Fichter *et al.*, 1997a). However, losses of exchangeable Mg^{2+} remain much higher than those computed from mass balance, even considering a Mg^{2+} release rate by weathering close to 0. On the other hand, because of the particular binding of K^+ in clay interlayers, the extraction by NH_4^+ is not the best way to measure changes in plant available K. Secondly, mineral immobilization has been taken in consideration only for wood. This simplification implicitly supposes that needle, branch and fine root biomass are in steady state. This assumption probably is acceptable for S90, where defoliation level has remained stable during the study period, but certainly is erroneous in S40, where yellowing and defoliation varied widely during the study period. An estimate of net annual uptake associated with leaf biomass and concentration can be obtained assuming that foliage biomass increased by the equivalent of one year litterfall, and that 20 % of the trees recovered a normal needle concentration between 1991 and 1996 (Le Goaster *et al.*,1991). These changes could contribute to explaining soil losses of 5.4, 0.4, 3.7 and 1.1 kg ha^{-1} yr^{-1} of Ca^{2+}, Mg^{2+}, K^+ and Mn^{2+} respectively over the study period. These amounts are insufficient to explain Mg losses from the soil exchange complex on the one hand, and would imply a high weathering rate for Ca^{2+} on the other hand. Ugolini (pers. comm., 1996) found large amounts of exchangeable Ca^{2+} in the coarse (>2 mm) particles at S90. Trees might have access to this source, which was not taken into account in the budget, but this access is probably limited as Ca^{2+} concentrations in the needles remain fairly low.

Nutrient accumulation or release by the litter layer also might contribute to the change in exchangeable cations. Net mineralization of the litter layer in S90 is suggested by high drainage of NO_3^-. Although we were not able to detect it, litter probably has accumulated in S40. Accumulation of K^+ in this layer might explain the decrease in exchangeable K^+ concentration. Another explanation for discrepancies in base cation losses estimated by different approaches is suggested by seasonal variation in exchangeable base cations. Although not statistically significant, exchangeable $Ca^{2+}$and Mg^{2+} concentrations were lower in March than in January, especially in the upper horizon, and large changes in exchangeable Mn^{2+} occurred also at several depths. Microbial immobilization is the most likely process to originate such changes.

At both sites, large and often abrupt temporal variations in NO_3^- concentrations in soil water occurred, reflecting shifts in NO_3^- deposition, nitrification and immobilization. High NO_3^- concentrations in 1987-90 may be attributed in the old stand to the mineralization of the large amount of litter that had fallen since the begining of forest decline, i.e. 1983, as well as to the poor health of the trees. Additionally, in the young stand, the storm that occurred in 1988 may have enhanced N mineralization. Soil rewetting after the severe drought in summer 1990 did not induce any further nitrification. The second NO_3^- peak, which occured in winter 1994/95, could not be related to any simple environmental parameter. The decrease in soil solution $(Ca^{2+} + Mg^{2+})/Al^{3+}$ over the study period probably illustrates the decrease in exchangeable base cations, as it did not result from a simple change in the anion concentration (Reuss and Johnson, 1986). However, this ratio did not decrease at 60 cm depth at S90 although exchangeable base cation decreased, while it decreased at 10 cm depth at S40 although exchangeable base cation remained unchanged. Additional work is needed to explain these discrepancies.

5. Conclusion

Globally, the ongoing soil acidification hypothesized from dynamic budget calculations and modeling was confirmed by the decrease of exchangeable base cation contents in the fine earth fraction as well as by the reduction in the ratio of $(Ca^{2+} + Mg^{2+})/Al^{3+}$ in soil solutions. However, rates of base cation losses estimated from these independant approaches were rather different. Using the same combination of techniques, similar results and discrepancies were obtained over a 10 year time period at the Solling site (FRG) (Matzner, 1988), while a better agreement was obtained by Richter *et al.* (1994) over 30 years at the Calhoun Experimental Forest (South Carolina). Risks of discrepancies increase as soils poor in exchangeable base cations and short time series are considered (Mandoli, 1997). Microbial cycles of nutrient immobilization or release directly or undirectly linked to root growth as well as annual variations in plant uptake and litterfall are the main likely causes for these discrepancies (Johnson *et al.*, 1991). Process-oriented investigations on the microbial immobilization of base cations should be developed to understand such changes.

References

Biron, P.: 1994, *Le Cycle de l'Eau en Forêt de Moyenne Montagne : Flux de Sève et Bilans Stationnnels,* Thèse, Université Strasbourg.

Bonneau, M., Dambrine, E., Nys, C. and Ranger, J.: 1991, *Sci. Sol* **29**, 125.

Espiau, P. and Peyronnel, A.: 1976, *Sci. Sol* **3**, 161.

Fichter, J., Turpault, M.P., Dambrine, E., Ranger, J.: 1997a, *Geoderma* (in press).

Fichter, J., Dambrine, E., Turpault, M. P. and Ranger, J.: 1997b, *Water, Air, Soil Pollut.* (in press).

Johnson, D.W., Cresser, M.S., Nilsson, S.I., Turner, J., Ulrich B., Binkley, D. and Cole, D.W.: 1991, *Proc. Royal Soc. Edinburg* **97B**, 81.

Landmann, G. and Bonneau, M.: 1995, *Forest Decline and Atmospheric Deposition Effects in the French Mountains*, Springer-Verlag, Berlin.
Le Goaster, S., Dambrine, E. and Ranger, J.: 1991, *Acta Œcologica* **12**, 771.
Le Goaster, S., Dambrine, E. and Ranger, J.: 1990, *Water, Air, Soil Pollut.* **54**, 269.
Lu, P., Biron, P., Breda, N. and Granier, A.: 1995, *Ann. Sci. For.* **53**, 117.
Matzner E.: 1988, *Der Stoffumsatz zweier Waldökosysteme im Solling*, Doctorarbeit, Universität Göttingen.
Mandoli, N.: 1997, *Boletin Investigacion Suelos España* **1**, 27.
Poszwa, A.: 1996, *Mise en Evidence de l'Acidification des Sols d'Aubure*, DEA, INRA Nancy.
Probst, A., Dambrine, E., Viville, D. and Fritz, B.: 1990, *J. Hydrol.* **116**, 101.
Reuss, J.O. and Johnson, D.D.: 1986, *Acid Deposition and the Acidification of Soil and Waters*, *Ecological Studies 59*, Springer-Verlag, Berlin.
Richter, D.D., Markevitz, D., Wells, C.G., Allen, H.L., April, R., Heine, P.R., Urrego, B.: 1994, *Ecology* **75**, 1463.
Sverdrup, H. and Warfvinge, P.: 1988, *Water, Air, Soil Pollut.* **38**, 387.
Ulrich, B.: 1983, 'Interaction of Forest Canopies with Atmospheric Consituents: SO_2, Alkali and Earth-Alkali Cations and Chloride', in B. Ulrich, and J. Pankrath (eds.), *Effects of Accumulation of Air Pollutants in Forest Ecosystems*, D. Reidel, Dordrecht, pp. 33-45.

A COMPARISON OF SOIL SENSITIVITY TO ACIDIFICATION BASED ON LABORATORY-DETERMINED SHORT-TERM ACID BUFFERING CAPACITY AND THE SKOKLOSTER CLASSIFICATION

MARK E. HODSON, SIMON J. LANGAN and DAVID G. LUMSDON

Macaulay Land Use Research Institute, Craigiebuckler, Aberdeen, AB15 8QH, Scotland

Abstract. The sensitivity of mineral soils to anthropogenically-induced acidification may be assessed using the Skokloster classification or by considering the short-term acid buffering capacity (STABC). The Skokloster classification is based on the assumption that the majority of acidification neutralization is due to base cation release from mineral weathering. It therefore considers the long-term neutralizing capacity of a soil on a scale of decades. The STABC of a soil is due largely to the adsorption and exchange of H^+ ions by soil components, and acts on the scale of single years. The United Kingdom maps generated to identify soils that are sensitive to acidification use the Skokloster classification. A comparison of the laboratory-determined STABC of the different horizons from soil profiles with the Skokloster classes of soils shows that there is generally no correlation between the two. Maps showing soil sensitivity to acidification based on measured STABC and Skokloster classes are markedly different. The STABC measure has the potential for quantitatively assessing a soil's sensitivity to acidification, while the Skokloster approach is empirical. It may be possible to combine the two approaches to produce a dynamic method of assessing soil recovery from acidification.

Keywords: Skokloster, acid buffering capacity, critical loads, acid deposition

1. Introduction

Under the second sulfur protocol to the UN-ECE convention on Long-Range Transboundary Air Pollution, the sensitivity of ecosystems to acidification was determined using a critical loads approach. At their simplest, critical loads can be assigned according to the Skokloster classes (Nilsson and Grennfelt, 1988). These classes are defined primarily on the basis of soil parent material (Table I) with the rationale that acid deposition is neutralized in soils through the release of base cations from mineral weathering. Because of this rationale, the Skokloster approach is not applicable to peats or organic horizons of mineral soils. In this study, only mineral soils (soils with < 43 % organic matter; Cope, 1973) are considered, and the organic horizons (F, H, L and O horizons) of the mineral soils are ignored. Using this scheme it is possible to assign empirically-based critical loads using existing databases (see Hornung *et al.*, 1995).

Although soil minerals represent a large store of potential acid neutralizing material, mineral weathering rates are relatively slow compared to rates of addition of acid through acid deposition. Because of this, mineral dissolution can consume a higher proportion of atmospherically deposited protons upon slow addition than upon quick addition of a fixed volume of strong acid (van Breemen, 1991). However, acid precipitation inputs occur as episodes and may influence soils and surface waters through storms or snow melt events (Morris and Thomas, 1987; Davis *et al.*, 1992). Therefore the Skokloster classes, while being potentially good predictors of long-term resistance to acidification, do not provide any information on the depletion of the exchange and adsorption capacity of the soil, i.e. short-term acidification.

Water, Air, and Soil Pollution **105:** 53–62, 1998.

The ability of a soil to consume protons present in acid deposition is due to a variety of processes, including mineral weathering reactions and the adsorption of protons by soil constituents such as organic matter, clay minerals and oxyhydroxides (van Breemen *et al.*, 1983; Bruggenwert *et al.*, 1991). In relation to the current work on the impact of acid deposition, it is the short-term acid buffering capacity (STABC), i.e. the abiotic acid buffering capacity of a soil other than by the consumption of protons through mineral dissolution, that is of importance.

TABLE I
The Skokloster soil classification.

Class	Normal parent rock	Mineral controlling weathering	
1	Granite, quartzite	Quartz, potassium feldspar	Most sensitive to
2	Granite, gneiss	Muscovite, plagioclase, biotite (<5%)	acidification
3	Granodiorite, greywacke, schist, gabbro	Biotite, amphibole (< 5%)	
4	Gabbro, basalt	Pyroxene, epidote, olivine (< 5%)	Least sensitive to
5	Limestone, marlstone	Carbonates	acidification

The Skokloster method is empirical, it determines soil sensitivity to acidification by assuming the weathering rate of soils with the same parent material to be the same, within a given range. To carry the debate and application of critical loads for soils further, it is desirable to have some method of actually measuring soil sensitivity to acidification. At the most detailed level involved in policy decisions, critical loads are assessed on a national basis, usually by determining the critical load of the dominant soil type within a grid square. Within Europe, these grid squares range from 1 to 37 km^2 in size (Posch *et al.*, 1995). It is impractical to try and measure the weathering rate of the soils within all such squares. For this reason, we have investigated the suitability of using the STABC of a soil, as determined using a batch equilibration technique in the laboratory, to indicate soil sensitivity to acidification. If such a measure were to be adopted within the critical loads community, it would be possible to obtain and measure the STABC of the dominant soil in each grid; therefore, the assessment of soil sensitivity to acidification could be based on a quantitative measure rather than the empirical, qualitative, Skokloster approach. For this reason, we determined the STABC of the A, B and C horizons of soils from 14 of the dominant soil associations in Scotland using a batch equilibration technique. These values provide a laboratory measurement against which empirically-based critical load estimates may be considered.

2. Materials and Methods

2.1. Laboratory Determination of Short-term Acid Buffering Capacity

For the present study, STABC is defined as that amount of protons required to reduce the field pH of one gram of soil measured against a background electrolyte of 0.01 M, pH 7 lithium perchlorate ($LiClO_4$) by 0.5 pH units after equilibration for seven days. Our definition of buffering capacity is similar to some published definitions (e.g., Federer and Hornbeck, 1985; "the number of moles of H^+ or OH^- that must be added to raise or lower the pH of 1 kg of soil by 1 pH unit"), but differs from others that define buffering capacity in terms of the number of moles of protons required to lower the pH of a mass of soil to a stated pH (e.g., Nätscher and Schwertmann, 1991, pH = 3; du Toit and Fey, 1994, pH = 3.8). The consequences of this difference are shown in Figure 1. To determine STABC, a batch equilibration technique was used. For each soil sample, 25 mL of 0.01 M $LiClO_4$ solution was added to 0.5 $\pm$ 0.1 g subsamples of soil. One mL of either distilled water or 0.001, 0.0025 or 0.004 M HNO_3 was then added. Samples were shaken continuously for 7 da. It was assumed that the STABC would have been exhausted after this time although some silicate minerals present in the soil would still have been reacting at a slow rate.

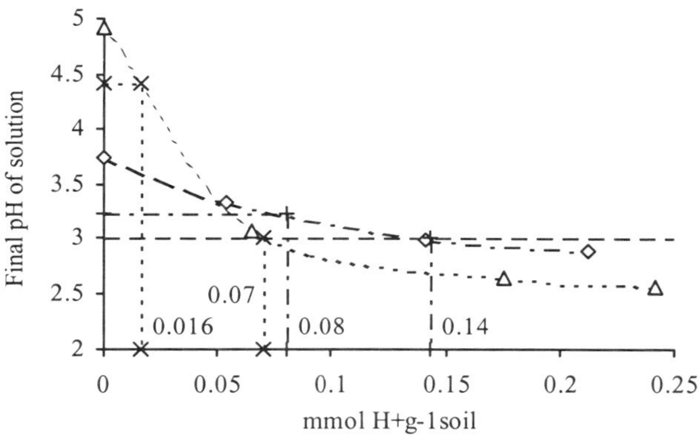

Fig. 1. Plot of final suspension pH against mmol H^+ g^{-1}_{soil} added to the soil for two samples from the 18 soil horizons initially used. If our definition of STABC is used, the soils have a STABC of 0.016 and 0.08 mmol H^+ g^{-1}_{soil}, respectively. If we had defined STABC as being the quantity of protons necessary to lower the pH to pH = 3, values of STABC would have been 0.07 and 0.014 mmol H^+ g^{-1}_{soil}. Thus, while absolute values of STABC are dependent on the definition of STABC, the relative STABC of different soils is, in this case, independent of the two different definitions. An advantage of our definition is illustrated if an equilibration curve is imagined that had flattened out at pH > 3. A value of the soil's STABC still could be determined using our definition. However, if we had defined STABC as the number of moles of proton required to lower the pH to pH = 3, then the soil would have an infinite STABC, which is unreasonable.

The pH of the solution subsequently was measured using a combination electrode and an Orion digital pH meter. The efficiency of the electrode varied between 97 and 100 %, based on commercial pH 4.00 $\pm$ 0.02 and pH 7.00 $\pm$ 0.02 buffers. Previous studies using this technique, but for different purposes, indicate that after 5 or 6 days the pH of the solution stabilizes (e.g., Lumsdon, 1996) and that the technique gives highly reproducible results. Randomly chosen samples from the present study were equilibrated three times and the resulting pH of the solution was the same to within 0.01 pH units. Plots of pH of the final solution against mmol of H^+ added per gram of soil were produced (e.g., Figure 1). Curve fitting functions available in the EXCEL spreadsheet package were used to generate equations that described smooth curves through the data points. The field pH of the soil was taken to be equal to that measured after the soil had equilibrated for seven days with distilled water. The best fit equations with an r^2 value of > 0.95 at a significance level of < 0.01 were used to calculate the STABC of the samples.

Measuring pH against a background electrolyte provides a reference condition for pH measurement. It also gives the solutions a constant ionic strength, which in turn fixes the ionic activity, so that adsorption of ions by soil constituents may be modeled at a future date. Thus, there are important advantages in using a background electrolyte. A potential drawback is that it may interact with the soil during the batch equilibration. In the present study, $LiClO_4$ was used as the background electrolyte to minimize any potential interactions (Rossotti, 1978; Bolton and Evans, 1996). Perchlorate ions (ClO_4^-) do not form strong complexes with other ions and at the low concentrations used, oxidation due to ClO_4^- will be negligible. Lithium ions are not specifically absorbed; although they may be adsorbed at exchange sites, Na^+, K^+ and H^+ are far more likely to be adsorbed (Toth, 1964; Wedepohl, 1969) and so, in our equilibrations, Li^+ adsorption effects are likely to be negligible.

The concentrations of acid were chosen in light of prior experience (e.g., Lumsdon, 1996) as being likely to generate suitable data points to construct the curves shown in Figure 1. Obviously, the conditions used do not mimic the natural world. The quantities of 0.5 g of soil and 25 mL of solution were chosen for operational convenience. Given that the results are normalized to mmol H^+ g^{-1}_{soil}, using different concentrations would not have altered the results, but the data points would have been less conveniently spaced for curve plotting.

2.2. SITE AND SAMPLE SELECTION

Initiall,y the STABC was determined on the A, B and C horizons of 18 uncultivated soil profiles from 14 of the dominant soil associations in Scotland, held in the Macaulay Land Use Research Institute soil archive (Table II). The soil associations chosen represent the spectrum of soil parent materials and geological units occurring in Scotland (Langan *et al.*, 1995). Three profiles from the Durnhill and Insch associations were used so that more than one profile from Skokloster classes 1 and 4 were used in the study. Details on the application of the Skokloster classification to soils have been published elsewhere (e.g., Nilsson and Grennfelt, 1988; Langan and Wilson, 1994) and will not be described in detail here.

To compare predictions about soil sensitivity to acidification on a national scale, soil profiles located at the intersects of a 20 km × 20 km grid superimposed onto a map of Scotland were used. Soils from these points were available for use from the Scottish National Soils Database archive held at MLURI. Of the 198 profiles identified, 114 were mineral soils, and as such deemed appropriate for the study. The soils come from a variety of ecosystems although none are from intensive agricultural land. Soils from a depth of 50 cm have been used as it was judged that at this depth the soil constituents would reflect the soil parent material and be free from wind-blown material. Also, critical loads are conventionally determined for soils at 50 cm depth. In all cases, the stored samples were subsampled using spinning rifflers to obtain representative samples.

TABLE II

Details of the 17 soils used in the first part of this study (soil types: PG Peaty gley, GBE Gleyed brown earth, HIP Humus iron podzol, BE Brown earth).

Association	Parent Material	% cover of Scotland	Skokloster Class	Acid Buffering Capacity (mmol H^+/g_{soil})			Soil Type
				A horizon	B horizon	C horizon	
Arkaig	Schists, gneisses, granulites, quartzites; principally from the Moine	16.22	2	0.018	0.006	0.000	PG
Balrownie	Mainly sandstones of Lower Old red sandstone age	1.83	3	0.026	0.010	0.006	GBE
Countesswells	Granites and granitic rocks	5.75	2	0.006	0.036	0.016	HIP
Durnhill	Quartzites and quartzose grits	1.6	1	0.010	0.010	0010	PG
			1	Expt. failure	0.010	0.003	PG
			1	0.007	0.005	No C	HIP
Ettrick	Lower Palaeozoic greywackes and shales	9.26	3	0.034	0.026	0.014	BE
Foudland	Slates, phyllites and other weakly metamorphosed argillaceous rocks	3.25	2	0.027	0.033	0.029	HIP
Hobkirk	Sandstones and marls of Upper Old Red Sandstone age	0.75	3	0.013	0.003	0.009	BE
Insch	Gabbros and allied igneous rocks	0.67	4	0.020	0.010	0.029	BE
			4	0.060	0.019	0.006	BE
			4	0.016	0.013	0.013	BE
Lochinver	Lewisian gneiss	4.67	2	0.017	0.020	0.020	PG
Rowanhill	Carboniferous sandstone, shales and limestone	3.05	3	0.034	0.023	0.037	BE
Sourhope	Old Red Sandstone intermediate lavas	1.71	3	0.030	0.023	0.020	BE
Strichen	Arenaceous schists and strongly metamorphosed argillaceous schists of Dalradian age	7.98	2	0.004	0.006	0.006	HIP
Tarves	Intermediate rocks or mixed acid basic rocks, both metamorphic and igneous	2.07	3	0.013	0.033	0.019	BE
Torridon	Torridonian sandstone and grits	2.25	2	0.006	0.007	0.004	PG

3. Results

There is very little correlation between the STABC and Skokloster class of either the 18 soil profiles used in the initial study or the soils located at the intersects of the 20 × 20 km grid, selected to provide a national coverage. For the first set of data, there is a correlation of $r^2 = 0.36$ ($p < 0.05$) between the STABC and Skokloster class of the A horizons. Variability in the values of the STABC and the Skokloster class of the soils taken from across Scotland are shown in Figure 2.

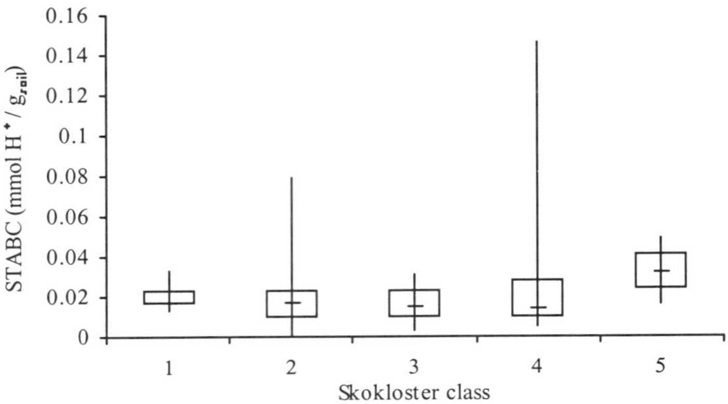

Fig. 2. Box and whisker plot showing variability of the STABC values of soils in the five different Skokloster classes for the soils taken from across Scotland.

Application of the ANOVA model to the initial 18 profiles shows that, statistically, there is no significant difference between the mean STABC values obtained for the A, B and C horizons and that there is no significant difference between the mean STABC values of the soils in the different Skokloster classes. The mean STABC of the A horizons of Brown earth soils is significantly greater ($p < 0.03$) than that of the A horizons of iron humus podzols and peaty gley soils.

Application of the two approaches on a spatial basis, using soils located at the intersects of a 20 × 20 km grid suggests that, for Scotland, important differences do exist between the results obtained. For example, the Skokloster maps indicate that the southern uplands (south Scotland) and Cairngorm plateau (northeast Scotland) are highly sensitive to acidification, whereas the STABC map does not. Given the lack of correlation between STABC and Skokloster class of the soils used to generate the soil sensitivity maps this is perhaps not surprising.

The relationships between STABC and a variety of other soil parameters for the 18 soil profiles used in the first part of this study were investigated. Taking the samples as a whole, the organic content of the soils (loss on ignition; LOI), correlated significantly with STABC ($r^2 = 0.47$, $p < 0.01$). The concentration of sand sized particles ($r^2 = 0.33$, $p < 0.05$), base saturation ($r^2 = -0.32$, $p < 0.05$), soil pH ($r^2 = 0.35$, $p < 0.05$), and concentration of oxalate extractable Fe ($r^2 = 0.31, p < 0.05$) also were significant. For A horizons, the concentration of sand ($r^2 = -0.70, p < 0.01$), silt ($r^2 = 0.61, p < 0.05$), and clay ($r^2 = 0.56, p < 0.05$) sized particles correlated well with STABC, together with organic content ($r^2 = 0.58, p < 0.05$) and the content of alkali feldspar ($r^2 = -0.61, p < 0.05$) and chlorite ($r^2 = 0.59, p < 0.05$). In B horizons, the concentration of silt sized particles ($r^2 = 0.58, p < 0.05$), plagioclase content ($r^2 = 0.54, p < 0.05$), oxalate extractable Fe ($r^2 = 0.49, p < 0.05$), and LOI ($r^2 = 0.51, p < 0.05$) correlate with STABC. There were no significant relationships between any of the measured soil properties and the STABC of the C horizons.

4. Discussion

The basic trends of the soil sensitivity data, produced using the Skokloster classification applied to the spatial dataset, are that soils developed from granites (concentrated in western Scotland and northeast Scotland) and from greywackes (concentrated in the Southern Uplands) are highly sensitive to acid deposition. This is similar to those trends seen in previous analyses using the Skokloster methodology, but applied to soil profiles selected using different sampling strategies (e.g., Langan and Wilson, 1994).

In using the STABC methodology, there are no discernible spatial patterns to the data. The distribution of the different STABC classes resembles neither the distribution of the Skokloster classes or the spatial variation in acid deposition concentrations or loadings. The lack of correlation between Skokloster classes and soil STABC implies that while the analysis using the Skokloster methodology may predict a soil's long term sensitivity to acidification, the analysis does not indicate which soils may potentially acidify in the short term due to the depletion of the soil's STABC.

Our definition of STABC results in the important conclusion that as soils become more acidic they possess a greater STABC and are therefore less sensitive to further acidification. Because of the logarithmic nature of the pH scale, a relatively acidic soil will require the addition of more protons to decrease the pH of the soil by a further 0.5 units than a less acidic soil. Thus the more acidic soil would have a larger STABC. A consequence of this definition is that the use of STABC as an indicator of soil sensitivity to acidification would direct attention to less acidic soils and suggests that it is these soils that are in more danger of acidification and require protection, rather than soils that are already acidic. In addition, if levels of acceptable soil pH are being set with respect to a species sensitivity to acidification, there is no reason to suppose that stating a pH below which organisms can not function is any better as a guide to policy than stating the fall in the soil pH which would be detrimental to that species.

Maximal UK acid deposition between 1986 and 1988 was about 100 mmol $H^+ m^{-2} yr^{-1}$, the maximal 5-day deposition being about 30 mmol $H^+ m^{-2} yr^{-1}$ (Warren Springs Laboratory, 1990). The lowest STABC that we measured in an A horizon was 0.004 mmol $H^+ g^{-1}_{soil}$. For such a horizon of thickness 10 cm and bulk density of 1.25 g cm^{-3}, there would be 1.25×10^5 g_{soil} m^{-2} with a STABC of 500 mmol $H^+ m^{-2}$. This is over 5 times the maximal UK acid deposition measured over a year and over 15 times the maximal acid deposition measured over five days, indicating that short-term acidification should not occur in soils. This accounts for the fact that short-term acidification, on the scale of several years, due to anthropogenic inputs has not been observed and is in agreement with the predictions of

McFee *et al.* (1977) and Federer and Hornbeck (1984). There are very few studies that have shown experimentally that short-term soil acidification rarely occurs. Furthermore, if it is assumed that the STABC is replenished solely by the supply of base cations by mineral weathering, it should be possible to combine the Skokloster and STABC approaches to develop a dynamic predictive model that would give a quantitative estimation of the rate of soil recovery from acidification. Where the STABC of a soil is known, the rate of soil acidification may be determined on the basis of the proton load in the deposition falling on that soil. If the weathering rates in the soil are also known (or are assumed to be equal to those proposed at the Skokloster meeting, Nilsson and Grennfelt, 1988), it should be possible to predict whether the STABC of a soil is being replenished or if a soil is still acidifying under a given deposition scenario:

$$STABC_y = \{([STABC_o \times r \times h] - [D \times y]) + (WR \times y)\} / (r \times h) \quad (1)$$

where:

$STABC_y$ = STABC after y years (mmol H^+ g^{-1}_{soil})
$STABC_o$ = initial measured STABC (mmol H^+ g^{-1}_{soil})
r = soil density (g m^{-3})
h = soil thickness (m)
D = mean annual proton load due to precipitation (mmol H^+ m^{-2})
y = number of years since the original STABC was measured
WR = protons neutralized by mineral weathering in mmol H^+ m^{-2} yr^{-1}.

The proton load used may be adjusted so that, for example, acidity falls at the rate agreed in the second sulfur protocol of the UN-ECE treaty on Long-Range Transboundary Air Pollution.

One important limitation to the approach described arises from the available soil data available nationally. The samples used in the current study were collected for the Scottish Soils National Database in the mid-1980s and they may thus already have been acidified. If this is so then our measurements of STABC do not indicate the soil's initial sensitivity to acidification. This does not, however, invalidate the theoretical basis of our technique. The measurements we have made indicate the soil's sensitivity to further acidification from its current status and it is this information that policy makers require when making decisions on acid emissions restrictions. If no measurements exist on a soil prior to acidification, it is not possible to legislate to reduce acidifying emissions so that a soil returns to its pristine state, as the conditions of that pristine state are not known. For the majority of soils this will be the case.

The results of the correlations indicate that the dominant controls on soil STABC are organic matter content, amorphous Fe oxide content and percentage of clay- and silt-sized soil particles. This is in broad agreement with other studies of the STABC of soils (e.g., Federer and Hornbeck, 1985; James and Riha, 1986; Sato and Ohkishi, 1993), which have concluded that STABC is due largely to adsorption of protons on exchange sites, primarily provided by organic matter. Clay- and silt-sized particles and amorphous Fe-oxides also provide exchange sites and may make a significant contribution to soil STABC if present in high enough concentrations.

5. Conclusion

The Skokloster classification and STABC both assess a soil's ability to buffer anthropogenically induced acidification. However the two techniques address different aspects of this buffering capacity, long-term supply of base cations and short-term adsorption of H^+, respectively. Our results indicate that, in the UK, soils should not acidify in the short term. When determining soil sensitivity to acidification, both the neutralizing capacity provided to the soil from mineral weathering and the buffering from the soils exchange and adsorption capacity should be considered. Measured STABC provides an objective measure of a soil's sensitivity to acidification, while the empirically based Skokloster methodology does not. Using the STABC methodology proposed here opens up the potential for predicting whether soils are recovering from acidification or continuing to acidify under different deposition scenarios.

Acknowledgments

MEH and SJL acknowledge funding from the National Power - PowerGen Joint Environmental Committee and SOEAFD. Additionally, we acknowledge helpful discussions with colleagues at MLURI. Funding for MEH to attend Biogeomon was supplied by The Distinguished Geologists Memorial Trust. The helpful input of two anonymous referees is also acknowledged.

References

Bolton, K.A. and Evans, L.J.: 1996, *Can. J. Soil Sci.* **76**, 183.
Bruggenwert, M.G.M., Hiemstra, T. and Bolt, G.H.: 1991, 'Proton Sinks in Soil Controlling Soil Acidification', in B. Ulrich and M.E. Sumner (eds.), *Soil Acidity*, Springer-Verlag, New York, pp. 8-27.
Cope, D.W.: 1973, *The Soils of Gloucestershire 1,* Harpenden, Soils Survey Record.
Davis, T.D., Tranter, M., Wigington, Jr., P.J. and Eshleman, K.N.: 1992, *J. Hydrol.* **132**, 25.
duToit, B. and Fey, M.V.: 1994, *15th World Congress of Soil Science Proceedings*, Volume 5b.
Federer, C.A. and Hornbeck, J.W.: 1985, *Water, Air, Soil Pollut.* **26**, 163.
James, B.R. and Riha, S.R.: 1986, *J. Environ. Qual.* **B15**, 229.
Hornung, M., Bull, K.R., Cresser, M., Hall, J., Langan, S.J., Loveland, P. and Smith, C.: 1995, *Environ. Pollut.* **90**, 301.
Langan, S.J. and Wilson, M.J.: 1994, *Water, Air, Soil Pollut.* **75**, 177.
Langan, S.J., Sverdrup, H.U. and Coull, M.: 1995, *Water, Air, Soil Pollut.* **85**, 2497.
Lumsdon, D.G.: 1996, *Appl. Geochem.* **11**, 285.
McFee, W.W., Kelly, J.M. and Beck, R.H.: 1977, *Water, Air, Soil Pollut.* **7**, 401.
Morris, E.M. and Thomas, A.G.: 1985, *J. Glaciol.* **31**, 190.
Nätscher, L. and Schwertmann, U.: 1991, *Geoderma* **48**, 93.

Nilsson, J. and Grennfelt, P.: 1988, *Critical Loads for Sulphur and Nitrogen,* UN-ECE/Nordic Council Workshop Report, Skokloster, Sweden, Nordic Council of Ministers NORD 1988:97.
Posch, M., de Smet, P.A.M., Hettelingh, J.-P. and Downing, R.J.: 1995, *Calculation and Mapping of Critical Thresholds in Europe, Status report 1995*, Co-ordination Centre for Effects, RIVM, The Netherlands.
Rossotti, H.: 1978, *The Study of Ionic Equilibria: An Introduction,* Longman, London and New York.
Sato, K. and Ohkishi, H.: 1993, *Ambio* **22**, 232.
Toth, S.J.: 1964, 'The Physical Chemistry of Soils', in F.E. Bear (ed.), *Chemistry of the Soils*, American Chemical Society Monographs, Chapman and Hall Ltd., London.
van Breemen, N., Mulder, J. and Driscoll, C.T.: 1983, *Plant and Soil* **75**, 283.
van Breemen, N.: 1991, 'Soil Acidification and Alkalinization', in B. Ulrich and M.E. Sumner (eds.), *Soil Acidity*, Springer-Verlag, New York, pp. 1-7.
Warren Springs Laboratory: 1990, *Acid Deposition in the UK, 1986 - 1988, Third Report of the UK Review Group on Acid Rain*, U.K. Department of the Environment.
Wedepohl, K.H.: 1969, *Handbook of Geochemistry*, vol. 2, Springer-Verlag, Berlin.

PROTON BUDGETS OF FOREST ECOSYSTEMS ON VOLCANOGENOUS REGOSOLS IN HOKKAIDO, NORTHERN JAPAN

HIDEAKI SHIBATA[1], MACHIKO KIRIKAE[2], YUMIKO TANAKA[3], TOSHIO SAKUMA[2] and RYUSUKE HATANO[2]

[1]*University Forest Uryu, Faculty of Agriculture, Hokkaido University, 250 Aza-Tokuda, Nayoro 096, Japan*
[2]*Laboratory of Soils, Faculty of Agriculture, Hokkaido University, N 9 W 9, Kita-ku, Sapporo 060, Japan*
[3]*University Forest Tomakomai, Faculty of Agriculture, Hokkaido University, Aza-Takaoka, Tomakomai, 053, Japan*

Abstract. The proton budgets of deciduous and coniferous forest ecosystems on volcanogenous regosols in Hokkaido, northern Japan, were studied by measuring the biogeochemical fluxes (atmospheric deposition, canopy leaching, vegetation uptake and leaching from soil) at each site during a three year period. The proton budgets were developed for individual compartments of the ecosystem: vegetation canopy, organic and mineral soil layers. At both sites, atmospheric S deposition was the dominant proton source in the vegetation canopy. In organic horizons, dissociation of weak acids (bicarbonate and/or organic acids) and vegetation uptake of base cations were the dominant proton sources, and the net mineralization of base cations was the dominant proton sink. Atmospheric acid deposition was almost neutralized in the forest canopy and organic horizon. At both sites, weathering and/or ion exchange of base cations and protonation of weak acids (mainly bicarbonate) were the dominant proton sinks in the mineral soil. In both organic and mineral soil, internal proton sources (mainly vegetation uptake of base cations and dissociation of weak acids) exceeded external proton sources, indicating that acid deposition was not the main driving force of soil acidification in the studied forest ecosystems.

Keywords: acidification, atmospheric deposition, forest canopy, Japan, leaching, proton budgets

1. Introduction

The biogeochemical dynamics of a forest ecosystem are composed of external budgets (atmospheric deposition and stream discharge) and internal cycling (litterfall, mineralization and nutrient uptake by vegetation, etc.) of certain elements. Although atmospheric deposition is an important vector for nutrient input to a forest ecosystem, acidification of atmospheric deposition may cause degradation of terrestrial ecosystems (Johnson and Lindberg, 1992). Proton budget studies have been used to investigate the relationship between soil acidification and elemental cycling (Ulrich *et al.*, 1980; Driscoll and Likens 1982; van Breemen *et al.*, 1983; 1984; 1986; Pačes, 1985; Binkley and Richter, 1987; Matzner 1989; Verstraten *et al.*, 1990; Bredemeier *et al.*, 1990; Binkley, 1992). Although many studies have focused on the proton budget of whole ecosystems (watershed scale), it is important to study the proton budget of individual compartments (canopy, forest floor and mineral soil) of a forest ecosystem to understand the interrelationship between soil and vegetation (van Breemen *et al.*, 1986). Recently, airborne anthropogenic acidic pollutants in the Asian region have increased to become comparable with those in northern Europe and America (Arndt and Carmichael, 1995). However, data on elemental fluxes based on field observation are rare in this region (Sato and Takahashi, 1996).

Quantification of proton budgets in a forest ecosystem is difficult because observation of elemental fluxes in each compartment (atmosphere, vegetation and soil) is technically problematic, especially for dry deposition, hydrological flux in the soil and relative nutrient

Water, Air, and Soil Pollution **105:** 63–72, 1998.

uptake by vegetation from the soil. We have developed proton budgets for the canopy, and for organic and mineral soil, adopting some assumptions based on field observation. The objective of this study was to clarify the proton budget in each compartment of two different forest ecosystems on the same soil type in northern Japan based on field observation of biogeochemical fluxes.

2. Materials and Methods

2.1. Study Site

This study was conducted in natural deciduous broad-leaved (site DF) and planted evergreen coniferous (site CF) stands located at the Horonai River Watershed in the University Forest, Tomakomai, Faculty of Agriculture, Hokkaido University (40°40' N, 141°36' E). Mean annual precipitation is 1189 mm (Shibata and Sakuma, 1996). Mean annual temperature is about 6.4 °C. The dominant species at site DF were mongolian oak (*Quercus mongolica* var. *grosseserrata*), mono maple (*Acer mono*), heart-leaved hornbeam (*Carpinus cordata*), Japanese linden (*Tilia japonica*) and other species (Shibata and Sakuma, 1996). Eastern white pine (*Pinus strobus*) and Korean nut pine (*Pinus koraiensis*) dominated at site CF. The dominant soil in the study area was volcanogenous regosols (Andic Udipsamments, Soil Survey Staff, 1994), consisting mainly of clastic pumice and sand (Sakuma, 1987). The volcanic parent materials of this soil were deposited in A.D. 1739 (C horizon) and A.D. 1667 (2C horizon). Table I shows the physical and chemical properties of the studied soil. Soil pH was moderate (5.3-6.2) at both sites. Although the contents of exchangeable base cations were relatively low at both sites, the total contents of basic cations were relatively high. The saturated hydraulic conductivity of the mineral soil was extremely high ($10^{-1} \sim 10^{-2}$ cm s^{-1}), suggesting that surface runoff during precipitation was almost negligible (Table I).

2.2. Field Observation of the Biogeochemical Process

The methods used for collection of precipitation, throughfall and stemflow followed those of Shibata and Sakuma (1996) and Shibata *et al.* (1995). Soil leachate was collected using a circular plate tension lysimeter (about 700 cm^2, 5-7 kPa) at each site (Shibata *et al.*, 1995; Shibata, 1996). Two lysimeters were installed below the O and A horizons and installed within the 2C horizon (about 60 cm), respectively. Field observations were conducted from May 1990 to April 1993. The methods of chemical analysis followed those of Shibata *et al.* (1995) and Shibata (1996).

2.3. Methods for Calculation of Elemental Fluxes

Elemental fluxes for wet deposition, throughfall and stemflow were calculated from the ionic concentrations and the observed amount of water in each sample, as described by Shibata and Sakuma (1996) and Shibata (1996). Dry deposition fluxes were estimated from the elemental budget of the forest canopy using Na as a conservative inert tracer (Matzner, 1989). This estimation method for dry deposition may cause some underestimation of

TABLE I
Physical and chemical properties of volcanogenous regosol at the study sites

Horizon	Depth (cm)	B.D.[a] (Mg cm^{-3})	pH (H_2O)	Exch. Cations (cmol$_c$ kg^{-1})[b] Ca	Mg	K	Na	CEC[c] (cmol$_c$ kg^{-1})	*Ks*[d] (cm s^{-1})	Total cations (mol$_c$ kg^{-1})[e] Ca	Mg	K	Na
Site DF (deciduous forest)													
A	0-6	0.56	5.3	4.43	0.75	0.27	0.14	18.5	1.1×10^{-1}	2.5	2.4	0.1	0.4
C	6-21	0.95	5.7	0.19	0.04	0.04	0.03	2.5	1.9×10^{-1}	2.7	2.7	0.1	0.4
2AC	21-26	0.97	6.0	0.39	0.07	0.04	0.04	2.1	7.6×10^{-2}	3.0	2.4	0.1	0.4
2C	26-50+	1.03	6.2	0.12	0.03	0.04	0.04	1.2	5.7×10^{-1}	2.9	2.7	0.1	0.4
Site CF (coniferous forest)													
A	0-7	0.74	5.5	2.60	0.47	0.18	0.11	17.4	1.0×10^{-1}	2.3	2.3	0.1	0.4
C	7-18	1.08	5.8	0.51	0.10	0.06	0.05	2.2	2.2×10^{-1}	2.8	2.8	0.1	0.4
2AC	18-27	1.09	6.0	0.55	0.07	0.04	0.05	2.1	3.6×10^{-2}	2.1	2.1	0.1	0.4
2C	27-68+	1.11	6.0	0.15	0.02	0.03	0.04	1.9	5.3×10^{-1}	2.6	2.6	0.1	0.4

[a] bulk density
[b] exchangeable cations extracted by 1M ammonium acetate (pH 7)
[c] cation exchangeable capacity (extraction by 1 M ammonium acetate at pH 7 and subsequently 10 % NaCl solution)
[d] saturated hydraulic conductivity
[e] total cations detected by fluorescence X-ray analyzer

gaseous dry deposits (NO_x, NH_4^+, SO_x), which are partly absorbed by tree leaves directly. Biomass and annual net primary production of vegetation had been investigated earlier at each site by Sakuma *et al.* (1994). Standing parameters of vegetation at each site are shown in Table II. Elemental flux associated with vegetation uptake was calculated using the method described by Sakuma *et al.* (1994) and Shibata (1996). For estimation of the relative fluxes of vegetation uptake from the organic and mineral soil horizons, we investigated the distribution pattern of fine roots (< 2 mm) using cylindrical soil cores (200 cm^3), taking 10 replicates in each horizon (Table III). The relative vegetation uptake of each element was calculated from the relative distribution of the fine roots measured as dry weight at each site under the assumption that the relative uptake of elements by vegetation was proportional to the fine root distribution. Furthermore, we assumed that the ratio of NH_4^+ to NO_3^- uptake by vegetation was equal to that of the ionic concentration of NH_4^+ and NO_3^- in soil leachate from each horizon. Because these assumptions may cause some errors for vegetation uptake fluxes of some elements, further detailed studies of tree root function in relation to water and nutrient uptake were needed.

TABLE II

Standing parameters of vegetation at each site (Sakuma *et al.*, 1994).

Site	Number (# ha^{-1})	Height (m)	DBH (cm)	Basal area (m^2 ha^{-1})	Biomass (Mg ha^{-1})
DF	4,300	5.5	6.5	26	243
CF	1,200	7.1	13.5	25	89

Elemental fluxes in leachate from soils containing roots were calculated from the ionic concentration of the soil leachate from the 2C horizon (60 cm depth, Figure 1) and the amount of water discharge from the soil (Shibata 1996). The amount of discharge water (*DW*) was calculated by assuming that the leaching flux of Cl^- was equal to that of the atmospheric input flux (wet and dry deposition, Table IV) of Cl^- (IN_{Cl}, Equation 1):

$$DW = IN_{Cl} / C_{Cl} \tag{1}$$

where C_{Cl} is the Cl^- concentration in the soil leachate at 60 cm depth. The water flux of the organic horizon was estimated from *DW* (Equation 1) and ratio of fine root distribution (Table III), as mentioned below. Water uptake flux by vegetation (*Wup*) was assumed to be equal to the difference between water inputs to soil (*Win*, throughfall + stemflow) and discharge water (*DW*, Equation 1) from rooting zone (Equation 2):

$$Wup = Win - DW \tag{2}$$

The calculated *Wup* at site DF and CF was 384 and 670 mm, respectively. Since the distribution ratio of fine roots in the organic horizon at sites DF and CF was 26 % and 43 %, respectively (Table III), we calculated the water uptake flux by vegetation from the organic horizon at respective sites as 100 mm and 288 mm for DF and CF site. Water flux

TABLE III
Fine root distribution at each soil horizon.

Horizon	Dry mass ($Mg\ ha^{-1}$) Site DF	Site Cf
Organic soil	2.8 (0.26)	2.6 (0.43)
A	4.7	2.3
C	1.1	0.5
2AC	2.0	0.7
2C	0.03	0.01
Mineral Soil	7.8 (0.74)	3.5(0.57)
Sum	10.6 (100)	6.1 (100)

Note: values in parenthesis represent ratios of dry weight to total weight

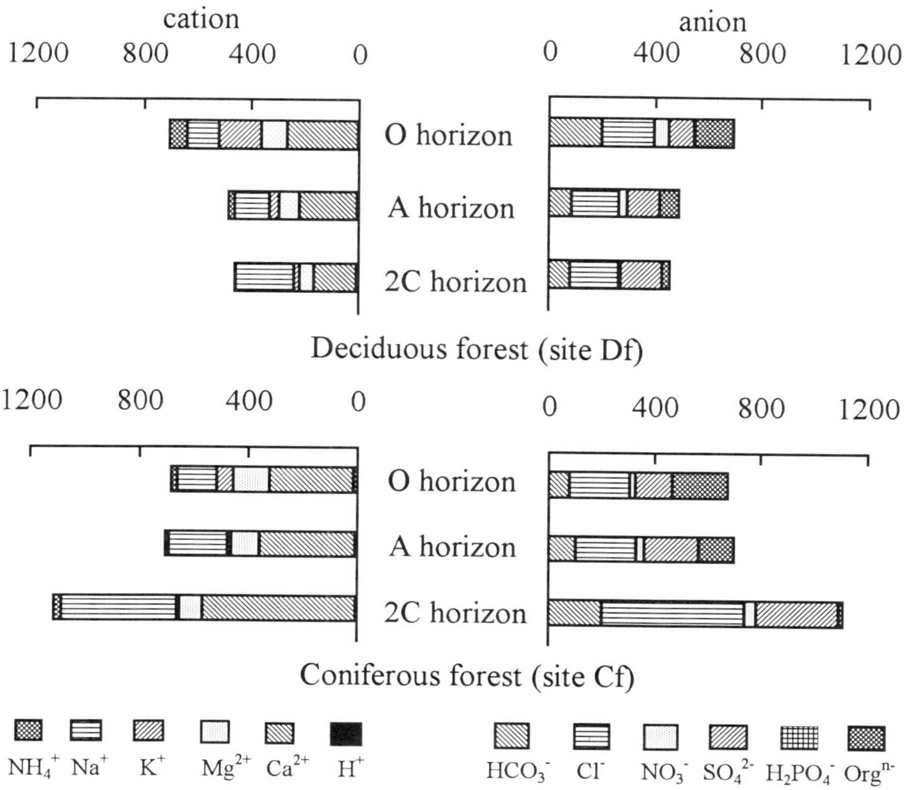

Fig. 1. Annual volume-weighted mean ionic concentrations ($\mu mol_c\ L^{-1}$) in soil leachate from O, A and 2C horizons at deciduous (site DF) and coniferous (site CF) forests from May 1990 to April 1993.

from the organic horizon was calculated from the difference between the water uptake flux by vegetation from the organic horizon and water input flux to the soil due to throughfall and stemflow (*Win*). Elemental fluxes (kmol$_c$ ha^{-1} yr^{-1}) of the leachate from the organic and mineral soil were calculated from annual volume-weighted mean ionic concentration (mol$_c$ L^{-1}) in the soil leachate (Figure 1) and calculated water fluxes (mm yr^{-1}), as mentioned above at each horizon.

3. Results and Discussion

3.1. Annual Elemental Fluxes

Table IV shows the annual elemental fluxes of atmospheric (wet and dry) deposition, throughfall, stemflow and soil leachate from the rooted zone (60 cm depth) at each site. Although each elemental flux of wet deposition at site CF was almost equal to that at site DF , the fluxes of dry deposition at site CF were more than ten times higher than those at site DF, reflecting the difference in capture efficiency of each canopy for atmospheric dry deposition (Shibata and Sakuma, 1996). This difference in dry deposition fluxes was responsible for the difference in ionic fluxes due to throughfall and stemflow between the two sites. The output flux of most elements from the rooting zone by soil leachate at site CF was larger than at site DF. Total leaching fluxes of base cations from the rooting zone at sites DF and CF were 3.46 and 6.41 kmol$_c$ ha^{-1} yr^{-1}, respectively, exceeding total inputs (throughfall and stemflow) at both sites (DF, 2.92; CF, 4.85 kmol$_c$ ha^{-1}yr^{-1}, Table IV). The difference in elemental fluxes between inputs (throughfall and stemflow) and the output

TABLE IV

Annual elemental fluxes (kmol$_c$ ha^{-1} yr^{-1}) of atmospheric (wet and dry) deposition, throughfall, stemflow and soil leachate from rooted soil at each site.

	H^+	Ca^{2+}	Mg^{2+}	K^+	Na^+	NH_4^+	Cl^-	NO_3^-	SO_4^{2-}	$H_2PO_4^-$	Alk.
					Deciduous forest (Site DF)						
Wd	0.66	0.58	0.27	0.11	1.03	0.16	1.24	0.18	0.66	0.00	0.73
Dd	0.08	0.07	0.03	0.01	0.13	0.02	0.16	0.02	0.08	0.00	0.09
Tf+Sf	0.13	0.78	0.41	0.56	1.17	0.19	1.29	0.15	0.80	0.00	1.00
Lc	0.05	1.22	0.40	0.18	1.66	0.03	1.40	0.05	1.19	0.01	0.86
	De-				*Coniferous forest (Site CF)*						
Wd	0.64	0.57	0.25	0.13	0.95	0.15	1.24	0.14	0.59	0.00	0.72
Dd	0.99	0.88	0.39	0.20	1.47	0.23	1.92	0.22	0.91	0.00	1.11
Tf+Sf	0.39	1.01	0.76	0.66	2.42	0.24	2.60	0.31	1.53	0.00	1.04
Lc	0.02	3.34	0.49	0.07	2.51	0.17	3.16	0.26	1.83	0.01	1.30

Wd; wet deposition, Dd; dry deposition, Tf; throughfall, Sf; stemflow, Lc; leachate from rooting zone, Alk.; Alkalinity includes bicarbonate and/or organic acids (inorganic cations inorganic anions).

(leachate from rooted soil) of the soil system at sites DF and CF was 0.54 and 1.56 $kmol_c$ ha^{-1} yr^{-1}, respectively, and it appeared that these fluxes were related to the internal cycling of base cations (vegetation uptake, ion exchange and weathering etc.) within the soil system, as discussed below. Table V shows the annual elemental fluxes associated with vegetation uptake and leaching from each soil horizon. The total uptake of base cations by vegetation at sites DF and CF was 8.4 and 3.0 $kmol_c$ ha^{-1} yr^{-1}, respectively (Table V). At site DF, vegetation uptake from the mineral soil was larger and leaching flux was smaller compared to that from organic soil. It is suggested that the vertical decrease in the leaching flux of base cations at site DF was strongly affected by the larger uptake by vegetation from mineral soil. On the other hand at site CF, vegetation uptake of base cations from both the organic and mineral soil were smaller than those at site DF and little affected the leaching flux of base cations from each soil horizon (Table V).

TABLE V

Annual elemental fluxes ($kmol_c$ ha^{-1} yr^{-1}) of leaching and vegetation uptake in organic and mineral soil horizons.

	Vegetation uptake				Leaching fromsoil			
	Base cations	NO_3^-	NH_4^+	SO_4^{2-}	Base cations	NO_3^-	NH_4^+	SO_4^{2-}
			Deciduous forest (Site DF)					
Organic horizon	2.2	0.5	0.6	0.1	6.7	0.6	0.7	1.0
Mineral horizon	6.2	2.0	1.4	0.3	3.5	0.05	0.03	1.2
Sum	8.4	2.5	2.0	0.4	10.2	0.65	0.73	2.2
			Coniferous forest (Site CF)					
Organic horizon	1.3	0.8	0.8	0.2	6.3	0.2	0.2	1.4
Mineral horizon	1.7	1.4	0.8	0.2	6.4	0.3	0.2	1.8
Sum	3.0	2.2	1.6	0.4	12.7	0.5	0.4	3.2

Base cations: Ca^{2+} + Mg^{2+} + K^+ + Na^+

3.2. PROTON BUDGET

Based on the assumption that elemental reactions were at a steady state within the soil solution phase, the net elemental flux (F_{soil}) from the soil solid phase to the solution phase can be calculated from the elemental budget at the soil solution phase, as in Equation (3):

$$F_{soil} = F_{leach.} + F_{up} - F_{in} \tag{3}$$

F_{leach}: elemental flux due to soil leachate from each soil horizon

F_{up}: elemental flux due to vegetation uptake from each soil horizon

F_{in}: input flux of elements to each soil horizon, e.g. F_{in} for organic horizon was the flux due to throughfall and stemflow; F_{in} for mineral soil was the flux due to leachate from organic horizon.

A proton budget was calculated for the organic and mineral soil horizon at each site from the elemental budget in each horizon, i.e., the net elemental flux into the soil solution (F_{soil}) and vegetation uptake flux from the soil solution (F_{up}). The proton budget for the vegetation canopy at each site was calculated from the elemental budget within the canopy, i.e., atmospheric deposition and throughfall plus stemflow. Table VI refers the proton budgets for the individual compartments (canopy, organic and mineral soil) at each site.

TABLE VI

Proton budget (kmol$_c$ ha^{-1} yr^{-1}) of the canopy (Cp) organic (Os) and mineral soil (Ms) at the deciduous (site DF) and coniferous (site CF) forest on volcanogenous regosol.

	H^+ source							H^+ source						
	H^{+a}	Base[b]	NH_4^{+c}	NO_3^{-d}	Ani[e]	Alk[f]	Sum	H^{+a}	Base[b]	NH_4^{+c}	NO_3^{-d}	Ani[e]	Alk[f]	Sum
	Deciduous forest (Site DF)													
Cp	0.7	0.0	0.0	0.0	0.1	0.2	1.0	0.1	0.7	0.0	0.1	0.1	0.0	1.0
Os	0.1	2.2	0.6	1.0	1.1	2.7	7.7	0.0	6.0	1.1	0.5	0.1	0.0	7.7
Ms	0.0	6.2	1.4	1.4	0.5	0.0	9.5	0.1	3.0	0.7	2.0	0.8	2.9	9.5
	Coniferous forest (Site CF)													
CP	1.7	0.0	0.1	0.0	0.0	0.0	1.8	0.4	0.0	0.0	0.1	0.6	0.7	1.8
OS	0.4	1.3	0.8	0.7	0.0	1.8	5.0	0.1	2.7	0.8	0.8	0.6	0.0	5.0
Ms	0.1	1.7	0.8	1.5	1.7	0.0	5.8	0.0	1.9	0.7	1.4	0.2	1.6	5.8

a) input of free H^+ (includes dry deposition), b) uptake of base cations (Ca+Mg+K+Na) by vegetation, c) ammonium uptake by vegetation and absorption to soil, d) nitrification of N, e) mineralization of Cl^-, SO_4^{2-} and $H_2PO_4^-$, f) dissociation of bicarbonate and organic anions, g)output of free H^+, h)weathering and ion exchange reaction of base cations, i) desorption of absorbed ammonium and mineralization of ammonium, j) nitrate uptake by vegetation, k) uptake of Cl^-, SO_4^{2-} and $H_2PO_4^-$ by vegetation and absorption of these anions to soil, l) protonation of bicarbonate and organic anions.

In both canopies, the main source of free H^+ was from atmospheric deposition and was mainly by airborne sulfuric acid (Table IV). The difference in the capture efficiency for acidic dry deposition of the two vegetation canopies (Table IV) strongly affected the difference in free H^+ flux as a proton source in the canopy. Leaching of base cations from the canopy was the major proton sink at site DF, whereas at site CF leaching of base cations from the canopy was not evident proton sink. This difference in the mechanism of proton neutralization at the canopy may be related to the smaller uptake of base cations by coniferous, compared with deciduous vegetation (Table V). In organic horizons, dissociation of weak acids (bicarbonate and/or organic anions) and vegetation uptake of base cations were the major proton sources at both sites. These proton sources at the organic horizon were closely related to the biological activity (respiration, production of organic anions, etc.) of roots and associated microorganisms in the organic horizon. Mineralization of base cations in organic soil was the major proton sink, and the flux at site DF was greater than that at site CF. In both mineral soils, vegetation uptake of base cations was a major part of the proton source (DF, 65 %; CF, 29 %), especially at site DF. The largest proton sink in the mineral soils was weathering and/or ion exchange reaction of base cations, followed by protonation of weak acid (mainly HCO_3^-, Figure 1) and NO_3^- uptake by vegetation. These results indicate that the internal cycling of base cations within a soil-vegetation system strongly influences the proton budget of organic and mineral soil. Volcanogenous regosol, young relatively unweathered soil has a relatively large pool of base cations in the parents material (Table I), relating that the importance of internal cycling of base cations in the total proton budget.

3.3. Effect of External and Internal Proton Sources on Soil Acidification

Van Breemen *et al.* (1984) defined external proton sources as the sum of net NH_4^+ input, net NO_3^- output and net free H^+ and SO_2 input. Internal Proton sources are defined as the sum of net uptake of base cations by vegetation, net non-nitrogen mineralization of anions and dissociation of weak acids. In acidic soil in the northeastern United States and western Europe, external Proton sources often exceed internal sources, resulting in soil acidification due to Al dissolution and SO_4^{2-} retention (van Breemen *et al.*, 1984; Bredemeier *et al.*, 1990).

Although external protons supplied as free H^+ in atmospheric deposition were the major proton source in both canopies (Table VI), free H^+ flux was quite small in comparison with total proton sources in organic and mineral soil at both sites, indicating that acid deposition was mostly neutralized in both the canopy and the organic horizon. Although the net NH_4^+ input to the whole soil-vegetation system was positive at both sites (DF, 0.15; CF, 0.21 $kmol_c\ ha^{-1}\ yr^{-1}$, Table IV), net NO_3^- output was negative (DF, -0.15; CF, -0.10 $kmol_c\ ha^{-1}\ yr^{-1}$, Table IV), indicating that the amount of external protons supplied by N transformation was small in comparison with the internal proton source in the soil-vegetation system (Tables V, VI). In organic and mineral soil, the internal proton source composed mainly of vegetation uptake of base cations was larger than the external proton source at both sites. The role of base cation uptake by vegetation as a proton source in the Asian region is relatively unstudied. Although Sato and Takahashi (1996) have reported that vegetation uptake is a negligible proton source in mixed forest on non-calcareous residual regosols in central Japan, they studied proton budgets at the watershed scale. In the rooting zone, biological activity (especially for vegetation uptake of base cations) played an important role in the proton budgets of the soil-vegetation system.

4. Conclusion

In the studied forests, atmospheric acid deposition was mainly neutralized in the canopy and organic soil. In organic and mineral soil, the external proton source was rather small in comparison with the total proton source, so that acidification of mineral soil was caused mainly by internal proton sources due to vegetation uptake of base cations and dissociation of weak acids in the soil-vegetation system.

Acknowledgments

This study was conducted with the special support of the staff of the University Forest-Tomakomai, Faculty of Agriculture, Hokkaido University, and we are grateful to Ms. Atsuko Tomita and Professor K. Ishigaki and all the staff of this institution.

References

Arndt, R.L. and Carmichael, G.R.: 1995, *Water, Air, Soil Pollut.*, **85**, 2283.
Binkley, D. and Richter, D.: 1987, *Adv. Ecol. Res.*, **16**, 1.
Binkley, D.: 1992, 'H^+ Budgets', in D.W. Johnson and S.E. Lindberg (eds.), *Atmospheric Deposition and Forest Nutrient Cycling*, Springer-Verlag, New York, pp. 450-466.
Driscoll,C.T. and Likens,G.E.: 1982, *Tellus*, **34**, 283.
Johnson,D.W. and Lindberg, S.E.: 1992, *Atmospheric Deposition and Forest Nutrient Cycling*, Springer-Verlag, New York.
Matzner, E.:1989, 'Acidic Precipitation: Case Study Solling', in D.C. Adriano and M. Havas (eds.), *Acidic Precipitation Volume 1*,., Springer-Verlag, New York, pp. 39-83.
Pačes T.: 1985, *Nature*, **315**, 31.
Sato, K. and Takahashi, A.: 1996, *Water, Air, Soil Pollut.*, **88**, 313.
Sakuma T.: 1987, *Res. Bull. Exp. For. Hokkaido Univ.*, **44**, 749. (in Japanese with English summary)
Sakuma T., Tomita, A., Shibata, H. and Tanaka, Y.: 1994, *Jpn. J. Soil Sci. Plant Nutr.*, **65**, 677, 684.
Shibata, H., Satoh, F., Tanaka, Y. and Sakuma, T.: 1995, *Water, Air, Soil Pollut.*, **85**, 1119.
Shibata, H., and Sakuma, T.: 1996, *Soil Sci. Plant Nutr.*, **42**, 1.
Shibata, H.: 1996, Ph.D. Thesis, Hokkaido University, 135p.
Soil Survey Staff: 1994, *Keys to Soil Taxonomy*, U.S.Department of Agriculture, Washington, D.C.
Urlich B., Mayer, R. and Khanna, P. K.: 1980, *Soil Sci.*, **130**, 193.
van Breemen, N., Mulder, J. and Driscoll, C.T.: 1983, *Plant and Soil*, **75**, 28.
van Breemen, N., Driscoll, C.T. and Mulder, J.: 1984, *Nature*, **307**, 599.
van Breemen, N., De Visser, H.B. and van Grinsven, J.J.M.: 1986, *J. Geol. Soc.*, **143**, 659.
Verstraten, J.M., Dopheide, J.C.R., Duysings, J.J.H.M., Tietema, A. and Bouten, W.: 1990, *Plant and Soil*, **127**, 61.

GIS ANALYSIS OF SURFACE WATER CHEMISTRY SUSCEPTIBILITY AND RESPONSE TO INDUSTRIAL AIR POLLUTION IN THE KOLA PENINSULA, NORTHERN RUSSIA

OLGA RIGINA

Institute of Northern Ecology Problems (INEP), 184200 Apatity, Murmansk reg., RUSSIA
Swedish University of Agricultural Sciences (SLU), S-901 83 Umeå, Sweden

Abstract. The Kola Peninsula (Figure 1) is the most industrially developed and urbanized region in the Russian North. The main pollution sources are the large smelters *Severonickel* and *Pechenganickel*, which are responsible for > 80 % of SO_2 emission and nearly 100 % of the Ni and Cu emission in the region. The heterogeneous structure of the hydrologic network and geochemistry across the region cause natural variability in water chemistry. Due to moist and cool climate, large territories on the Kola are sensitive to acid precipitation, but around the smelters, geological peculiarities, erosion and alkaline dust emissions increase buffering capacity of waters to acidification. Lake and river monitoring implemented from 1989-1993 by the laboratory of water ecosystems, Institute of Northern Ecology Problems was a basis for GIS-analysis (ARC/VIEW-2.1) of the major constituents of water chemistry attributed to anthropogenic load in the region. After correction for sea salt, acid-neutralizing capacity (ANC) of waters was calculated. The highest concentrations of anthropogenic non-sea sulfates were confined to the districts with high cation concentrations ($Na^+ + K^+ + Ca^{2+} + Mg^{2+}$) that reduced negative effects. The most vulnerable regions were found in the northern tundra and high mountain tundra.

Keywords: geographic information system, Kola Peninsula, metal pollution, surface water chemistry

1. Introduction

The Kola Peninsula is the north-west extremity of Russia (Lat. 66°20'-69°57'; Long. 28°25'-41°26'). A large part of the territory lies inside the Polar Circle, and the average air temperature is 0 °C. The Kola sub-Arctic has a weak capacity for self-remediation and low resistance to unfavorable influences due to its low level of mass- and energy-exchange. At the same time, the Kola is subjected to high anthropogenic influences attributed to long-term metallurgical activity (Figure 1). The main sources of pollution are the two large nickel smelters: *Severonickel* in Monchegorsk and *Pechenganickel* in Zapolyarny, that together release more than 80 % of the SO_2 (>513 kt yr^{-1}) and nearly 100 % of Ni (3.7 kt yr^{-1}) and Cu emissions (2.6 kt yr^{-1}) in the region. The total territory of the Kola Peninsula exceeding the critical deposition rate of S (0.3 g S m^{-2} yr^{-1}) occupies over 150,000 km^2 (Touvinen *et al.*, 1993). Deposition of heavy metals is more local because of gravitational settling. Sulfates on the Kola North are the major pollutant factor, causing acidification of waters (Moiseenko, 1994). The surface waters are subjected to waterborne pollutant flows as well as airborne pollution of catchments.

Geographic Information Systems (GIS) have become an important and widespread tool in ecological applications, owing to their ability to integrate different sources of information and to reveal new, latent relationships for evaluation of environmental impact. The objective of this study was to use GIS-analysis to evaluate spatial patterns in the major constituents of water chemistry attributed to anthropogenic load in the Kola Peninsula.

Water, Air, and Soil Pollution **105:** 73–82, 1998.

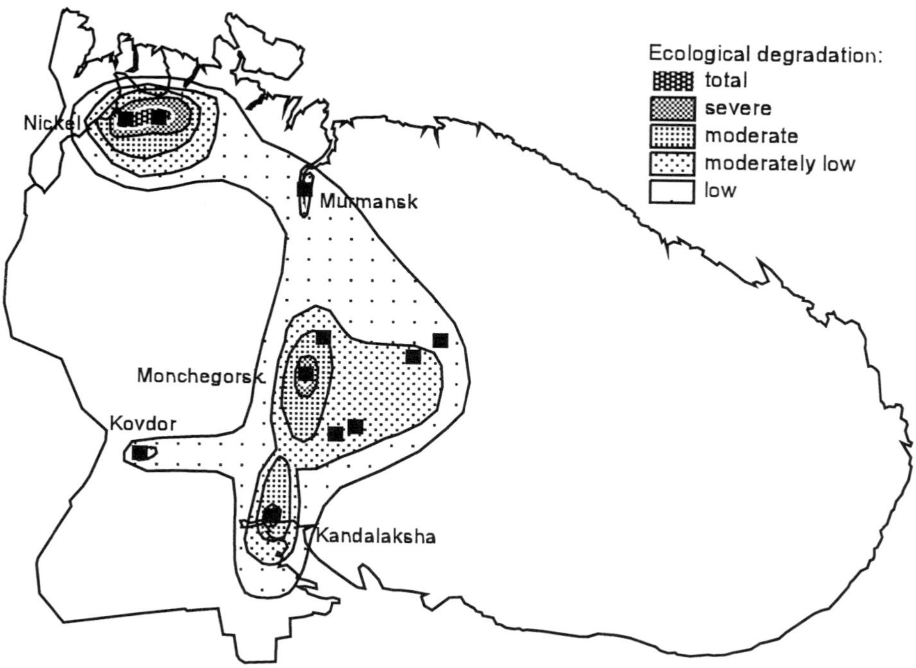

Fig. 1. The sources of pollution and degradation of ecosystems on the Kola Peninsula (after Kryuchkov and Makarova, 1989).

2. Material and Methods

There are 111,609 lakes on the Kola peninsula. According to the ratio between the base cations, the surface waters on the Kola belong to bicarbonate-sodium or bicarbonate-calcium type (Moiseenko *et al.*, 1996). The heterogeneous structure of hydrology and geochemistry across the region contribute to natural variability of the surface waters. A considerable proportion of the lakes are referred to as 'sensitive lakes' due to the underlying acidic rocks.

Beside the geology, the climatic conditions in the Arctic regions enhance the sensitivity of the territory to acidification (Moiseenko, 1994) because: 1) the supply of waters is mainly controlled by precipitation, 2) atmospheric deposition accumulates as snow-pack during the long winter (6-7 months), and acid snowmelt is released into catchment area by surface runoff during the short spring flood period, and 3) runoff does not contact the soil, because the soils are deeply frozen almost year around, including during snowmelt.

To assess surface-water acidification in the Kola Peninsula, 370 lakes and lake-river systems were monitored extensively by the Institute of Northern Ecology Problems during 1989-1993. Samples were taken during the August-October period for catchments with areas less than 10 km^2 (small lakes with atmospheric input) and during May-June for catchments with areas greater than 10 km^2 (lake-river systems). The databases were analyzed separately. Sampling design and laboratory performance were reported by Moiseenko (1994). After correction for sea-salt, acid-neutralizing capacity ($\mu eq\ L^{-1}$) of the surface waters was calculated as difference between non-marine base cations and strong acid anions (Henriksen *et al.*, 1992; Moiseenko *et al.*, 1996):

$$ANC_1 = Ca^{2+} + Mg^{2+} + Na^+ + K^+ - SO_4^{2-} - NO_3^-$$

$$\text{or } ANC_2 = HCO_3^- + A^{n-} - H^+ - Al^{n+},$$

where HCO_3^- is the bicarbonate concentration, derived by Gran-citration, $H^+ = 10^6 \times 10^{-pH}$, Al^{n+} is the sum of all positively-charged Al species, and A^{n-} is the concentration of organic anions, calculated empirically from the concentration of organic matter: $A^{n-} = POB \times (4.7 - e^{(-0.32*POB)})$.

A map of the Kola Peninsula (1:1,000,000) (DCW, 1993) including layers of coast line, populated places, drainage net and elevation, was used as a basis for GIS presentation. A geological map of the Kola Peninsula (1:8,000,000), as well as spatial modeling results for S deposition from local sources (Rigina and Baklanov, 1998), were digitized in corresponding projection. Sulfur deposition was predicted by a numerical trajectory model for meso-scale, which took into account the locations and parameters of emission sources, the vertical structure of wind and the influence of topography.

Concentrations of water chemistry constituents were displayed as point information across the region by means of ARC/View for a single field (ANC) or one field versus another [HCO_3^- vs. SO_4^{2-}; ($Ca^{2+} + Mg^{2+}$) vs. ($Na^+ + K^+$)]. During spatial analysis, the databases on water chemistry were linked to spatial information on bedrock and modeled S deposition by means of a spatial join (point-in-polygon analysis; Anon., 1996) within ARC/VIEW, which intersected the water chemistry point data with polygons representing bedrock and S deposition classes. Statistical analysis of the information obtained was implemented using MINITAB.

To make the analysis meaningful, the data were sorted in order of increasing rock acidity. The latter was determined by taking into account contents of SiO_2 as the simplest index of acidity against the buffering ability (content of $Na^+ + K^+$ and soluble $CaCO_3$) for each class. Nine bedrock classes ordered from least to most acid were distinguished, and these ordinal values were applied to a geological map for the Kola peninsula to generate a map of rock sensitivity by means of ARC/VIEW. The percentage of each class from the total territory of the Kola Peninsula was calculated using the statistical tool of ARC/VIEW.

3. Results

Over most of the territory $Ca^{2+} + Mg^{2+}$ prevails over $Na^+ + K^+$ (Figure 2), the exception being the central part with the mountain massifs (the Khibiny, Lovozero) formed by apatite-nepheline syenites. However, K^+ prevails in coastal areas. The lowest content of base cations occurs in the tundra zone, formed by hard rocks of granitic gneiss formations. Here the waters are expected to be most vulnerable to acid load.

Changing the ratio of molar concentration of the dominant anions (HCO_3^-/SO_4^{2-}) toward prevailing sulfates reflects a tendency toward acidification. Sulfates dominate in

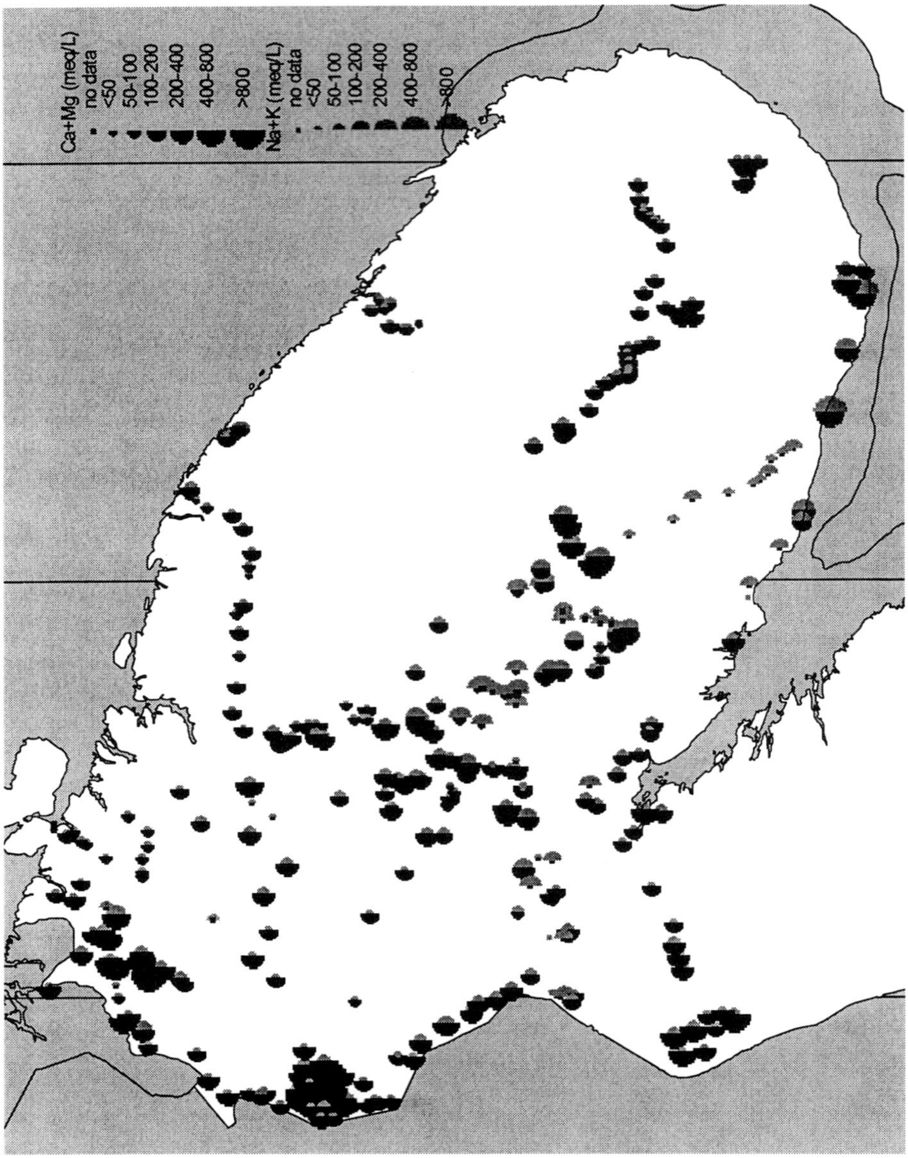

Fig. 2. Territorial distribution of base cations ($Ca^{2+} + Mg^{2+}$) versus ($Na^{+} + K^{+}$) ($\mu eq\ L^{-1}$) in surface waters of the Kola peninsula.

water, and values of alkalinity reach a critical level (HCO_3 < 50 μeq L^{-1}) over a considerable part of the territory (Figure 3), even in areas remote from the smelter emission sources. The highest concentrations of SO_4^{2-}, nonetheless, are attributed to the industrial centers. Equal concentrations of HCO_3^- and SO_4^{2-} in lakes indicates a transitional state regarding acidification.

Acid-neutralizing capacity of waters is the most widely used criterion for water acidification. Low ANC values (the background ANC for Fennoscandia is 20 μeq L^{-1}; Henriksen *et al.,* 1990) are widespread across the region (Figure 4) and indicate the most vulnerable areas, which are elevated territories with acid bedrock (Figure 5). Statistical analysis indicates that > 60 % of rocks on the Kola are acidic (Table I).

TABLE I
Bedrock classes by increasing acidity.

#	% of area	Definition
1	2	Upper Proterozoic sedimentary and felsic volcanogenic rock
2	4	Post Svecokarelian granitoid
3	7	Mainly metasediment and mafic volcanic rock and felsic volcanic rock
4	9	Granulite in Lapland
5	11	Svecokarelian metasediment and volcano-genic rock, mafic and ultramafic rock
6	13	Archaean supracrustal rock (greenstone
7	16	Lower Svecokarelian metasediment and volcanogenic rock
8	18	Archaean granitoid gneiss, granitoid, mafic-ultramafic intrusion and anorthosite
9	20	Carbonatites and nepheline syenites

Although around the smelters, SO_4^{2-} prevails as the dominant ion in waters, the acid-neutralizing capacity remains high (ANC > 100 μeq L^{-1}). This might be due to high base cation concentrations in the lakes around the smelters, possibly coupled with alkaline aerosol inputs. The maximal values of ANC (>200 μeq L^{-1}) occur at Pechenga, Monchegorsk and Kovdor, and are attributed to alkaline dust emissions into the atmosphere at these industrial centers. The lowest ANC values occur in the northern and western parts of the Kola Peninsula (northern tundra and high mountain tundra with acid bedrock), which are vulnerable to acid loads.

The sum of base cations ($Na^+ + K^+ + Ca^{2+} + Mg^{2+}$) was related to ANC_2 (for August-October samples, [$Na^+ + K^+ + Ca^{2+} + Mg^{2+} = 42.93 + 0.94 \times ANC_2$], r = 0.93, n = 93, p = 0.000; for Man-June samples, [$Na^+ + K^+ + Ca^{2+} + Mg^{2+} = 43.8 + 1.05 \times ANC_2$], r = 0.73, n = 86, p = 0.000). No correlation was found between ANC_2 and SO_4^{2-} concentrations for the either sample period (r = -0.104, n = 91 for August-October; r = 0.251, n = 85 for May-June). However, SO_4^{2-} concentrations in the August-October samples fell into nine discrete groups distinguished by rock acidity (Figure 6), inferring that geology may be an important factor in the snow-free period. Besides, geological influence is more important in small than in large catchments. In the May-June period, while catchments are exposed to released contaminants that have accumulated during long winter, pollution is a major factor.

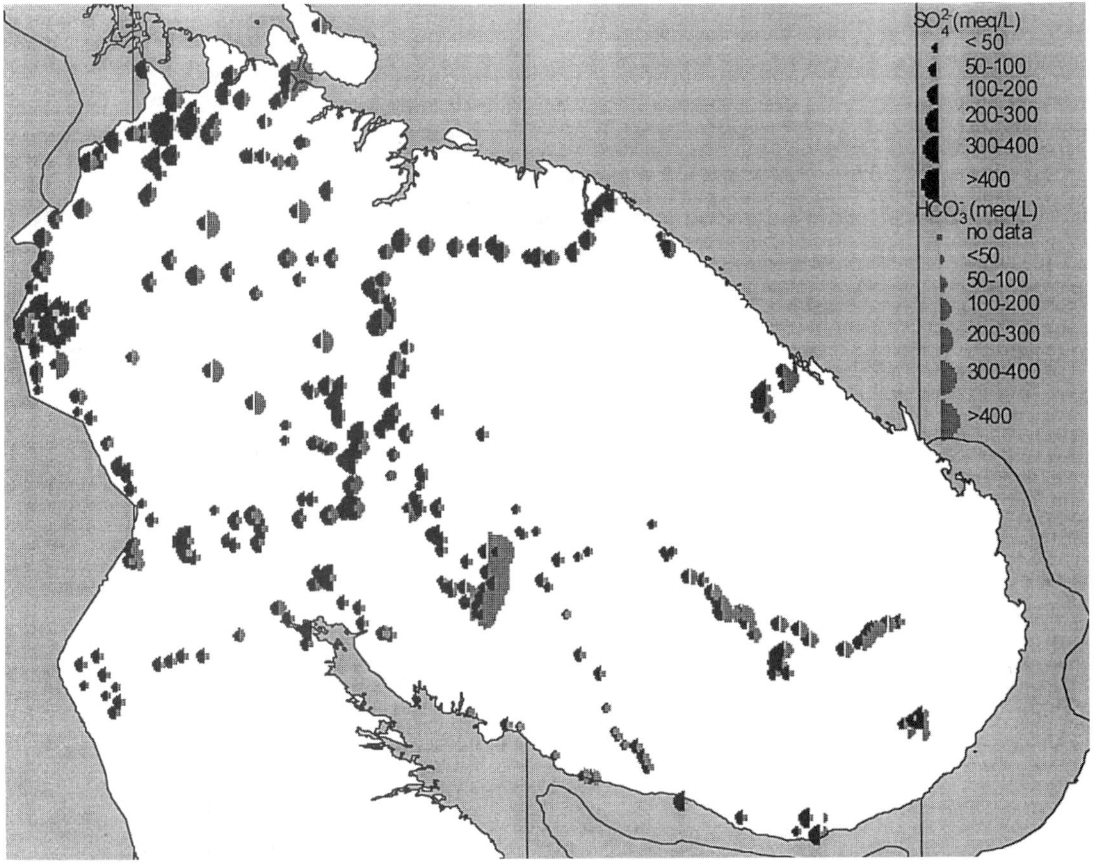

Fig. 3. Territorial distribution of sulfates versus bicarbonates (μeq L^{-1}) in surface waters of the Kola peninsula.

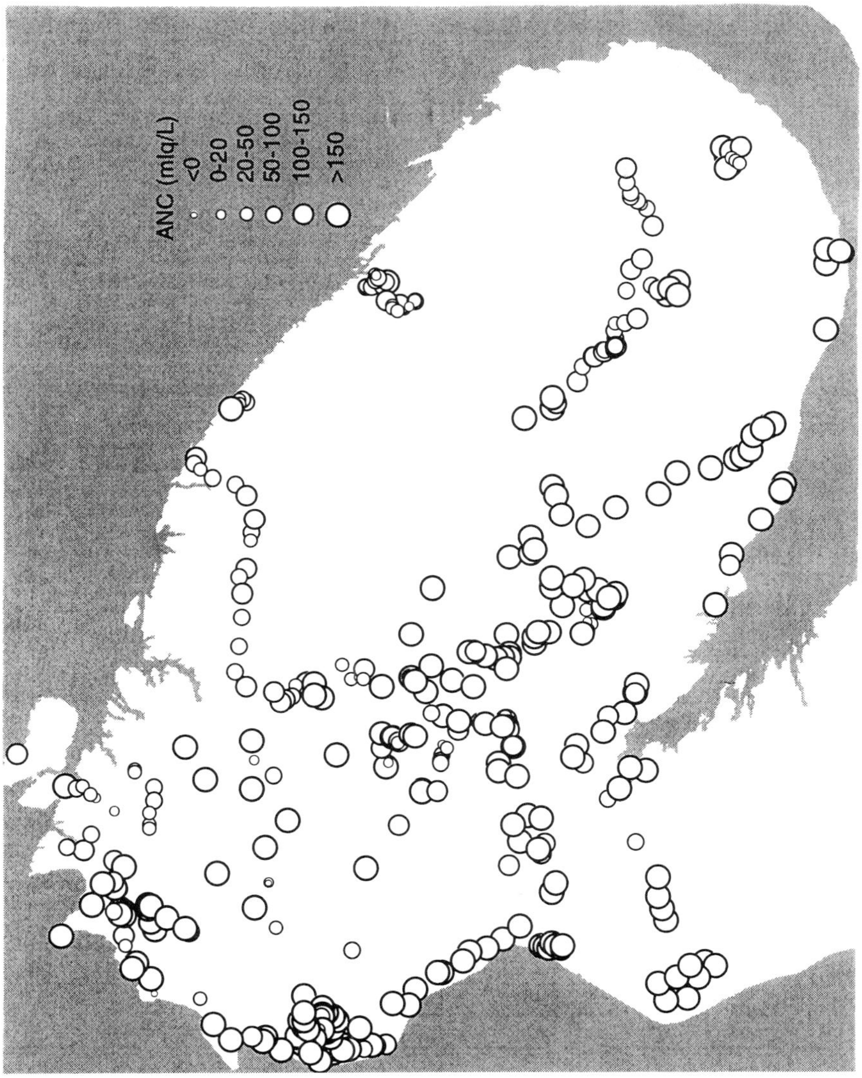

Fig. 4. Acid neutralizing capacity ($ANC_1 = Ca^{2+} + Mg^{2+} + Na^{+} + K^{+} - SO_4^{2-} - NO_3^{-}$), $\mu eq\ L^{-1}$ in surface waters of the Kola peninsula.

Fig. 5. Classes of underlying rocks on the Kola Peninsula by increasing acidity (1-less acid, 9-most acid).

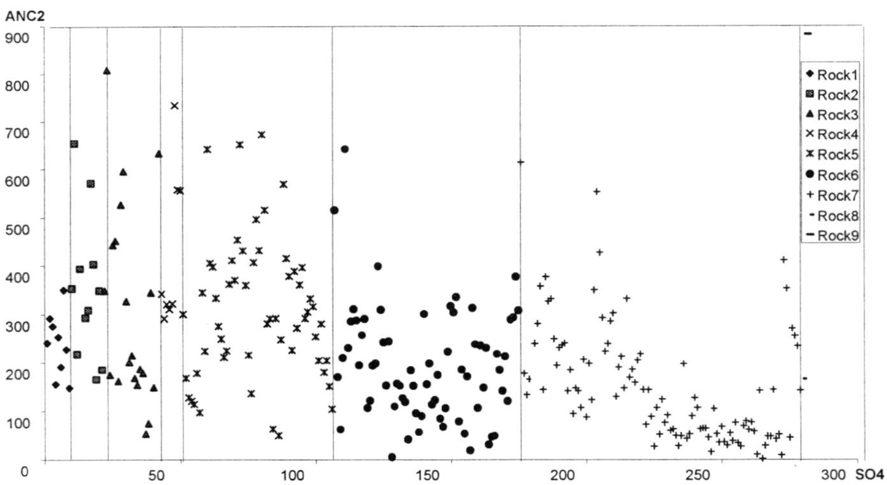

Fig. 6. SO_4^{2-} versus acid-neutralizing capacity ($ANC_2 = HCO_3^- + A^{n-} - H^+ - Al^{n+}$), μeq L^{-1} in surface waters, August-October samples by underlying rock classes (rock 1-less acid, rock 9 -most acid, Table I).

The SO_4^{2-} deposition pattern (Figure 7 in Rigina and Baklanov, 1998) showed a good correlation with monitoring data of SO_4^{2-} in surface water during the August-October period across the Kola peninsula (r = 0.813, n = 283, p = 0.000). A high correlation between S deposition and surface water SO_4^{2-} concentrations implies that pollution on the Kola peninsula originates from local sources and that airborne pollution dominates over waterborne.

4. Conclusions

The GIS approach revealed relationships between various constituents of surface water chemistry. The highest concentrations of anthropogenic non-sea sulfates were confined to the districts with high cation concentrations ($Na^+ + K^+ + Ca^{2+} + Mg^{2+}$), which minimizes negative effects. Sulfates dominate over HCO_3^- across a considerable part of the territory, even in areas remote from the emission sources. The GIS approach displayed variability of surface water chemical constituents across the region and revealed regional sensitivity to acidification. The maximal values of acid-neutralizing capacity of water (ANC > 200 µeq L^{-1}) were attributed to alkaline emissions into the atmosphere at the industrial centers: Pechenga, Monchegorsk, and Kovdor. Low values of acid-neutralizing capacity of water were widespread, and represent areas that are most vulnerable to anthropogenic loads (northern tundra and high mountain tundra with acid bedrock). The GIS approach ascertained new relationships between constituents of water chemistry and other information about the region, available in geographical presentation. Spatial variability of major constituents of water chemistry is influenced by anthropogenic load as well as by geological and climatic factors. The GIS allowed a mathematical model of spatial distribution of S deposition to be verified using monitoring data for surface water SO_4^{2-} concentrations. A high correlation between S deposition and surface water SO_4^{2-} concentrations implies that pollution on the Kola peninsula originates from local sources and that airborne pollution dominates over waterborne. The GIS served as statistical tool: over 60 % of the rocks on the Kola are acid, increasing sensitivity of the territory to acidification.

Acknowledgments

The author is very grateful to Prof. T. Moiseenko, the laboratory of water ecosystems of INEP for providing with data and supervision. The author greatly appreciates help with GIS by F. Stamm, from the Kiruna GIS-laboratory and P. Södeberg and B. Näslund from the Swedish University of Agricultural Sciences. A final thanks is to Drs A. Vinogradov and A. Baklanov from the Kola Scientific Center for fruitful discussions. The author is very grateful to the anonymous referees and Prof. K. Wieder for their valuable reviews.

References

Anonymous: 1996, *ArcView GIS - The Geographical Information System for Everyone, Manuals*, Environment Systems Research Institute, New York, USA.

DCW: 1993, *The Digital Chart of the World for use with ARC/INFO - Data Dictionary*, Environment Systems Research Institute, New York, USA.
Dowling, R., Hettelingh, J.-P. and de Smet, P. (eds.): 1993, *Convention on Long-range Transboundary Air Pollution - Calculation and Mapping of Critical Loads in Europe: Status Report 1993*, ISBN No. 90-6960-047-1, Bilthoven, the Netherlands.
Henriksen, A., Kämäri, J., Posch, M., Lovblad, G., Forsius, M., Wilander, A.: 1990, *Critical Loads to Surface Waters in Fennoscandia*, Helsinki, Nordp. 124..
Henriksen, A., Kämäri, J., Posch, M., Lovblad, M., Wilander, A.: 1992, *Ambio* **21**, 356.
Kryuchkov, V. and Makarova, T.: 1989, *Aerotechnogenic Influence on Ecosystems in the Kola Peninsula*, Kola Science Center, Apatity (in Russian).
Moiseenko, T.: 1994, *Ambio* **23**, 418.
Moiseenko, T., Rodyushkin, I., Dauvalter, I., Kudryavtseva, L.: 1996, *Forming of Quality of Surface Waters and Bottom Sediments under Athropogenic Loads on Catchments of Arctic Basin*, Kola Science Center, Apatity. (in Russian)
Rigina, O. and Baklanov, A.: 1998, *Water, Air, Soil Pollut.* **102**, this volume.
Tuovinen, J-P., Laurila, T., Lättilä, H., Ryaboshapko, A., Brukhanov, P. and Korolyev, S.: 1993, *Atmos. Environ.* **27**, 1379.

EFFECTS OF ATMOSPHERIC SEA-SALT DEPOSITION ON SOILS AND FRESHWATERS IN NORTHEAST SCOTLAND

CATHERINE C. WHITE, RICHARD P. SMART and MALCOLM S. CRESSER

Department of Plant and Soil Science, University of Aberdeen, Aberdeen, Scotland AB24 3UU

Abstract. The majority of Scottish upland soils are particularly sensitive to acid deposition because of their low weathering rates. The compositions of the exchangeable base cations of such soils in the United Kingdom are dominated by sea salt inputs rather than by mineral weathering inputs of base cations. Catchments with low mineral weathering rates are also those particularly susceptible to freshwater acidification. Therefore, catchments exhibiting a high sea salt effect should also exhibit the most acid waters under base flow and storm flow conditions. A field evaluation study based on 61 catchments in NE Scotland has shown that this is indeed the case. River water pH under both base flow and high flow conditions is correlated stronhly with the relative contribution of Na^+ to the sum of Ca^{2+}, Mg^{2+} and Na^+. From these results, an attempt is being made to produce a quantitative signature of weathering for the soils within the catchment upstream of the sampling point. Representative soil samples from the LFH, AE, B and C horizons and on 4 different parent materials have been obtained from the surrounding catchments to validate the above results for associated soil solutions. Sampling took place on upland moorland podzols under *Calluna vulgaris*. Tension lysimeters were used to sample the soil solutions so that their chemistry could be compared with that of the relevant river water.

Keywords: acidification, river water chemistry, sea salt deposition, Scotland, soil solution chemistry

1. Introduction

Soil chemistry and catchment hydrology play important roles in controlling riverwater chemistry. The relationship between susceptibility of ground waters to acidification and weathering rates of bedrock minerals was recognized in the UK in the early 1980s, and was used to provide qualitative susceptibility maps (Kinniburgh and Edmunds, 1984). More recently, a similar approach has been applied to UK surface waters, although this also took into account modifying effects of soil mineralogy and land use (Hornung *et al.*, 1995). The semi-quantitative results were mapped using a five-point sensitivity scale. The critical loads approach is being used in Europe for soils and surface waters to quantify the thresholds of acid deposition above which damage is predicted to occur according to present knowledge (Hettelingh *et al.*, 1991; de Vries *et al.*, 1993; Posch *et al.*, 1995; Hornung and Skeffington, 1993). It employs empirical models to provide quantitative estimates of weathering rates as a basis for evaluating neutralization capacity. This concept presupposes that the same amount of alkalinity generated naturally, *i.e.* in the absence of anthropogenic acidifying pollutant deposition, has no significant natural beneficial effect upon soil, a questionable hypothesis. Uncertainty is associated with weathering rate values obtained because of soil spatial variability and the simplifying assumptions which must be made; this may be more of a problem when using default values built into the model to calculate weathering rates, because site-specific data are inadequate. Different approaches are used for calculation of critical loads of surface waters, so the critical loads for rivers and adjacent soils often differ considerably, as might be expected.

This paper suggests an alternative, unifying system for classifying susceptibility to acidification. Many regions of northern Europe, including the UK, are subject to a strong

Water, Air, and Soil Pollution **105:** 83–94, 1998.

maritime influence. In such regions, and where soil mineral weathering rates are low and soils unfertilized, maritime-derived base cations have more effect than base cations from biogeochemical weathering on the composition of exchangeable soil base cations (White *et al.*, 1996). This is reflected directly in the base cation ratios of associated drainage waters, which become increasingly dominated by Na^+ rather than by the ions released by weathering. Data for 61 catchments and sub-catchments in northeastern Scotland have been used to test the hypothesis that the extent of Na^+ dominance at any point in a river should provide a quantitative indication of the upstream weathering rate, encompassing the effects of the diverse bedrock types present. Thus Na^+ dominance should be indicative of the susceptibility of the river concerned to acidification under diverse discharge conditions.

Evidence for sea salt deposition effects upon upland catchments may be found both in regional soil survey data and in studies of causes of acid episodes in upland rivers and streams. Seeking direct evidence for acid deposition effects upon mineral soils, Barton *et al.* (1994) measured B horizon soil pH for Scottish *Calluna vulgaris* moorland podzol profiles. Within defined altitude and slope constraints, they examined correlations between soil pH and acid deposition fluxes for soils derived from quartzites, non-calcareous sandstones, and granites, which were thought to be likely to be acidification sensitive. Only when the compensatory effects of deposition fluxes of base cations, including maritime-derived Na^+, were taken into account were significant correlations obtained. For soils from all horizons of similar Scottish moorland podzols derived from granites, it was later demonstrated that soil pH correlated with critical load exceedance only when atmospheric sea salt deposition was included in critical load and exceedance calculations (White *et al.*, 1996). Studies of chemical changes, especially changes in Ca^{2+}:Na^+ ratio, during incubation of moist soils have provided additional evidence for sea-salt deposition effects upon Scottish upland soils with low mineral weathering rates (Elias and Cresser, 1997).

It is well documented that precipitation episodes with high sea salt concentrations in the rain lead to acid flushes in UK upland rivers (Langan, 1989; Soulsby, 1995). Because the additional H^+ ions in such flushes come from soil cation exchange sites, Na^+ and Mg^{2+} ions must be being retained on cation exchange sites. This Na^+ and Mg^{2+} retention is an ameliorative effect of sea salt inputs upon soil base saturation and pH is to be expected in the longer term, countering the short-term mobile anion effect on pH of drainage water.

The soils most likely to show direct evidence of strong sea salt influence are ombrotrophic peat soils, in which biogeochemical mineral weathering is an insignificant process. Cresser *et al.* (1997) examined the direct effects of precipitation composition on the chemistry of streams draining from peats in western and northeastern Scotland. The ratios of Na^+:Ca^{2+}:Mg^{2+} were virtually identical in river water and in precipitation, with strong Na^+ dominance reflecting the maritime influence. Without any base cation inputs from weathering, the peat cation exchange sites equilibrate with both precipitation and drainage water, resulting in identical base cation ratios. On the other hand, when mineral weathering started to make a significant contribution to the exchangeable base cation composition, the ratios of base cations in precipitation and in river water started to differ. The above discussion suggests that the extent of dominance of the contribution of Na^+ to $\Sigma(Na^+ + Ca^{2+} + Mg^{2+})$ in water in a river compared with its dominance in precipitation should reflect quantitatively weathering rates upstream of the sampling point; high Na^+ dominance indicating a low upstream weathering rate unless Na^+ release from Na^+-rich minerals such as albite is substantial. Weathering rates, moreover, provide a direct measure of resistance to acidification. Thus, strong Na^+ dominance should be indicative of a river susceptible to natural and anthropogenic acidification. The primary objective of the present study was to test the hypothesis that the ratio $Na^+/(Na^+ + Ca^{2+} + Mg^{2+})$ in river water at a given site in an area with a strong maritime influence upon precipitation chemistry provides an index of weathering rate, as reflected in H^+ concentrations in river water under diverse flow regimes.

2. Materials and Methods

Samples from 61 sites in the 2,100 km^2 of the River Dee catchment (Figure 1) were used to test the hypothesis. The headwater streams of the Dee are at an altitude of about 1,250 m, about 105 km west of Aberdeen in northeastern Scotland; the river flows eastward into the sea at Aberdeen. Its catchment has diverse parent materials, including significant amounts of dolomite, limestone, calc-schists and serpentine, as well as granites, granodiorites and associated glacial tills. Its general characteristics have been summarized recently in another publication (Langan *et al.*, 1997). Sufficient sites were selected to provide a good representation of the parent materials present.

Samples were collected every 14 days over a period of six months and analyzed immediately upon return to the laboratory. Due to the flashy nature of discharge from Scottish upland catchments, a substantial number of samples must be taken over a long time period to characterize river water both under base flow and high flow conditions (Cresser *et al.* 1984). The mean values of six Na^+, Ca^{2+}, Mg^{2+} and H^+ concentrations were used to characterize base flow waters, using data for the samples with the highest Ca^{2+} concentrations (Reid *et al.*, 1981). All such samples were collected under very low flow conditions after periods of dry weather. The single set of storm flow samples corresponded to acid episodes, and were all collected at high discharge on the same sampling day during heavy rainfall; they gave the lowest observed pH value for each site.

Representative soil samples from the LFH, AE, B and C horizons and on four different parent materials (dolomite, diorite, granite and quartzite) were collected in early summer in triplicate from surrounding catchments to investigate how closely the river water ion ratio results were related to those of the associated soil solutions. Sampling took place on upland moorland podzols under *Calluna vulgaris*. The soils were analyzed chemically and small tension lysimeters were used to extract soil solution so that its chemistry could be compared with that of the relevant river water chemistry.

3. Results

3.1. A Novel Index of Susceptibility of Rivers and Their Catchments to Acidification

The sea salt effect on river waters should be greatest where the water drains soils with low weathering rates. Thus, the sampling sites may be ranked in order of increasing relative contribution of Na^+ to $\Sigma(Na^+ + Ca^{2+} + Mg^{2+})$ in river waters. This should be equivalent to ranking them in order of catchment weathering rates. The results thus ranked for both the

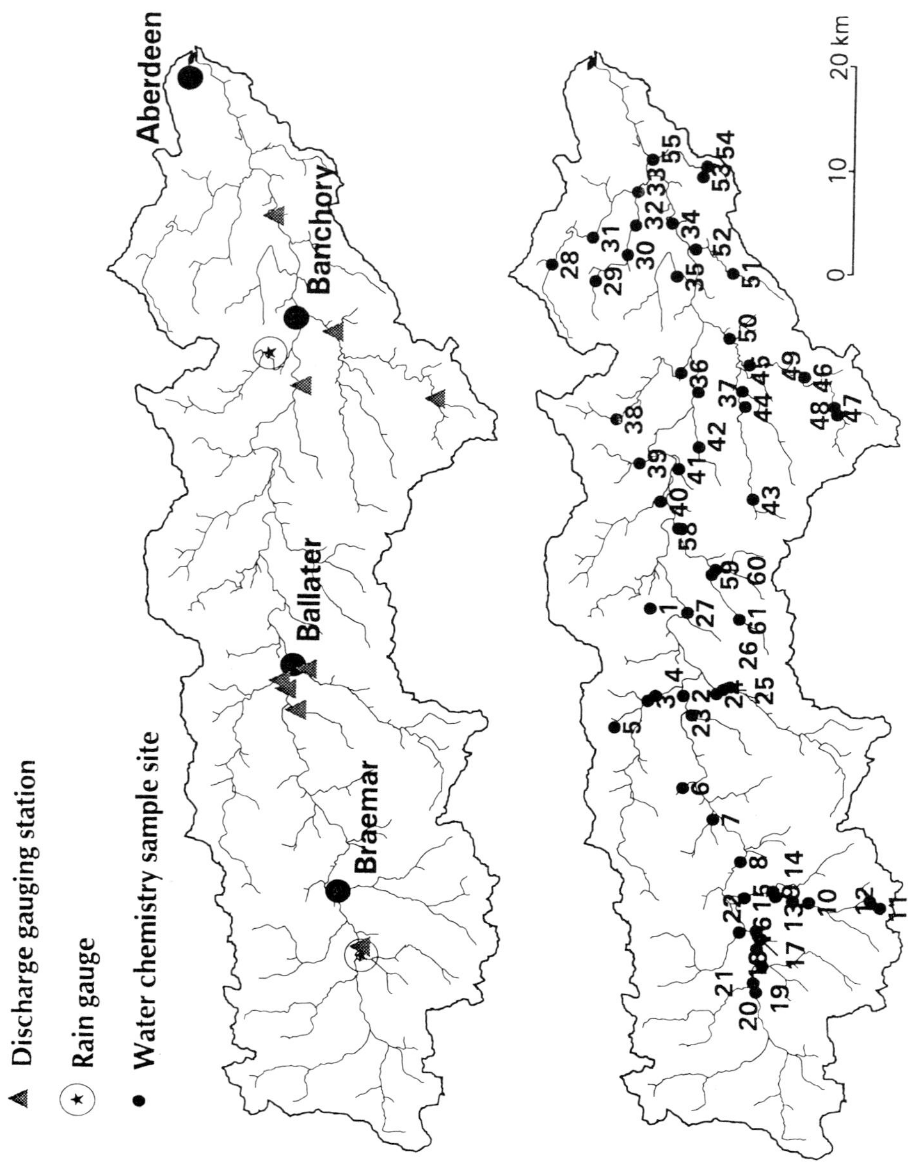

Fig. 1. Map of the River Dee catchment showing the 61 sampling points.

base flow (means of six values) and storm flow data (single values) are shown in Figure 2. It is informative to compare the ranking order with the parent material data (Figure 2). The low ranked sites are from the main river channel (e.g., 17, 15) or are from catchments containing limestone (10, 7, 5, 9, 41) or dolomite (16, 12, 19, 13). Highly ranked catchments are associated with granite (60, 45, 48, 1, 43), granite and quartzite (22, 58, 20) or granite and undifferentiated schist (49, 46, 47). Thus, the Na^+ ranking system is indeed indicative of upstream weathering rates, as hypothesized.

The catchments with the highest relative Na^+ concentrations should also have the lowest pH values. Linear regression analysis confirmed that the relationships between river water pH and the relative contribution of Na^+ to $\Sigma(Na^+ + Ca^{2+} + Mg^{2+})$ were significant at $p < 0.001$ (Figure 3). Fitting simple quadratic functions to the data improved the r^2 values even further, to 0.638 for base flow and 0.719 for storm flow samples (Figure 3). The relationships were examined also using a single set of base flow data only. Even then they were still highly significant, and base flow data could still be used to predict quantitatively maximum storm flow H^+ concentrations during acid flushes.

3.2. Effects of Sea Salt Deposition on Soil Solution

The above results suggest that there is a relationship between the magnitude of the sea salt effect in river waters and upstream geology. This clearly suggests that many of the acidification-sensitive (low weathering rate) soils from which the waters have drained are being influenced strongly by sea salt inputs. The percent relative contributions of Na^+, Ca^{2+} and Mg^{2+} on the soil cation exchange sites, as assessed by extraction with 1 M ammonium acetate, to the sum of all three is shown for each parent material and for each horizon (Figure 4). Soil weathering rates become less dominant the more acidic the soil becomes, and Figure 4 shows the soil exchangeable Na^+ increases in importance the more acidic the parent material. Therefore, the more acidic the soil the greater the sea salt effect. However, the corresponding results for the soil solution of these soils showed much less variation between the different parent materials (Figure 5). The percent relative contribution of Na^+, Ca^{2+} and Mg^{2+} to the sum of all three in the soil solution is very similar to the corresponding relative contributions for precipitation (i.e., Na^+ ranges from 60 -70 %, Ca^{2+} from 5 -15 % and Mg^{2+} from 20 - 25 %).

Initially it had been expected that a direct relationship would exist between the river water chemistry and the soil solution data. It had been assumed that base flow chemistry would be best reflected by the B and C horizon soil solution chemistry and that the LFH and AE horizon soil solutions would reflect high flow chemistry. The results from a number of river water sites (both base flow and high flow) and their means were compared with the relevant results from the soil solutions of the four parent materials. The most comparable results were found for the granite (Figure 6). However, overall the results were not comparable to the extent anticipated, with sea salts having more of an effect on soil solution chemistry than expected in the base-rich soils. This suggests that the fluxes of elements from weathering are small compared with those from precipitation, at least as far as the upper 50 cm or so of the soil are concerned. Thus, the base cation ratios are similar for precipitation and soil solution. The fact that the river water samples show more diverse effects of bedrock type on base cation ratios than expected from the ratios for soil solutions suggests that groundwater inputs from depths greater than 50 cm, where weathering rates

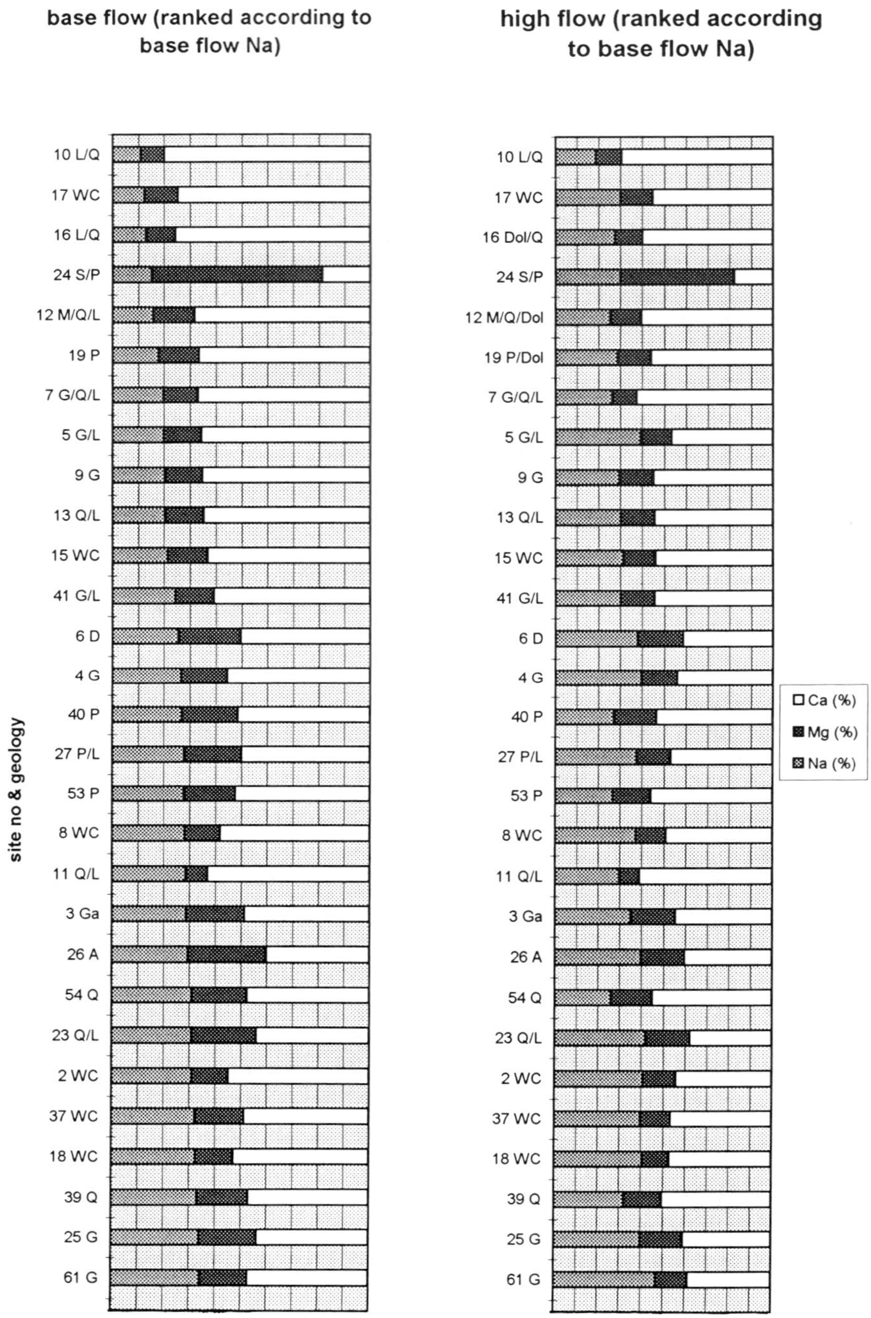

Fig. 2. The percent relative contribution of Na^+, Ca^{2+} and Mg^{2+} to the sum of all three for base flow (left) and storm flow (right) conditions. The data are ranked according to relative contribution of base flow Na^+. For the relevant site location and site geology see Fig. 1.

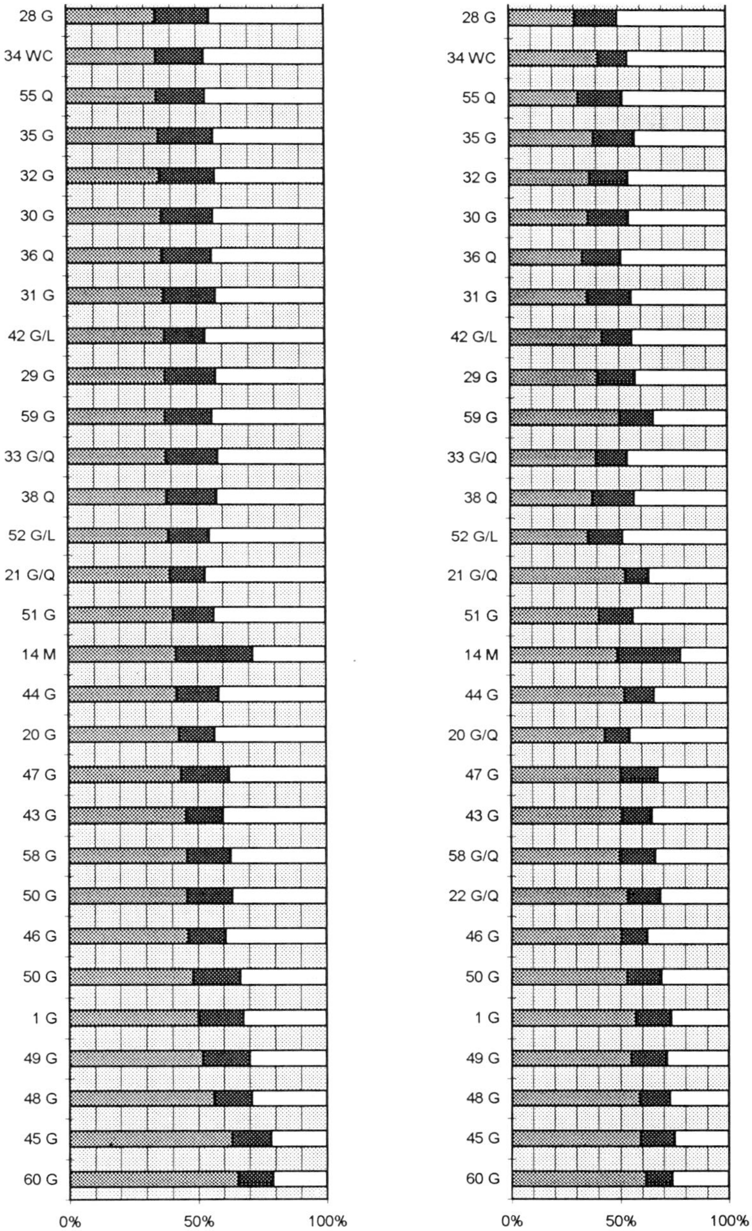

G = granite; L = limestone; Q = quartzite; P = pelite; M = mica schist; Ga = gabbro
D = diorite; A = amphibolite; S = serpentinite; WC = whole catchment; Dol = dolomite.

Fig. 2. (continued)

are higher, play a substantial role in regulation of the River Dee water quality. With hindsight, this is perhaps not surprising. Soil solution compositions require more thorough investigation, especially their variation with depth.

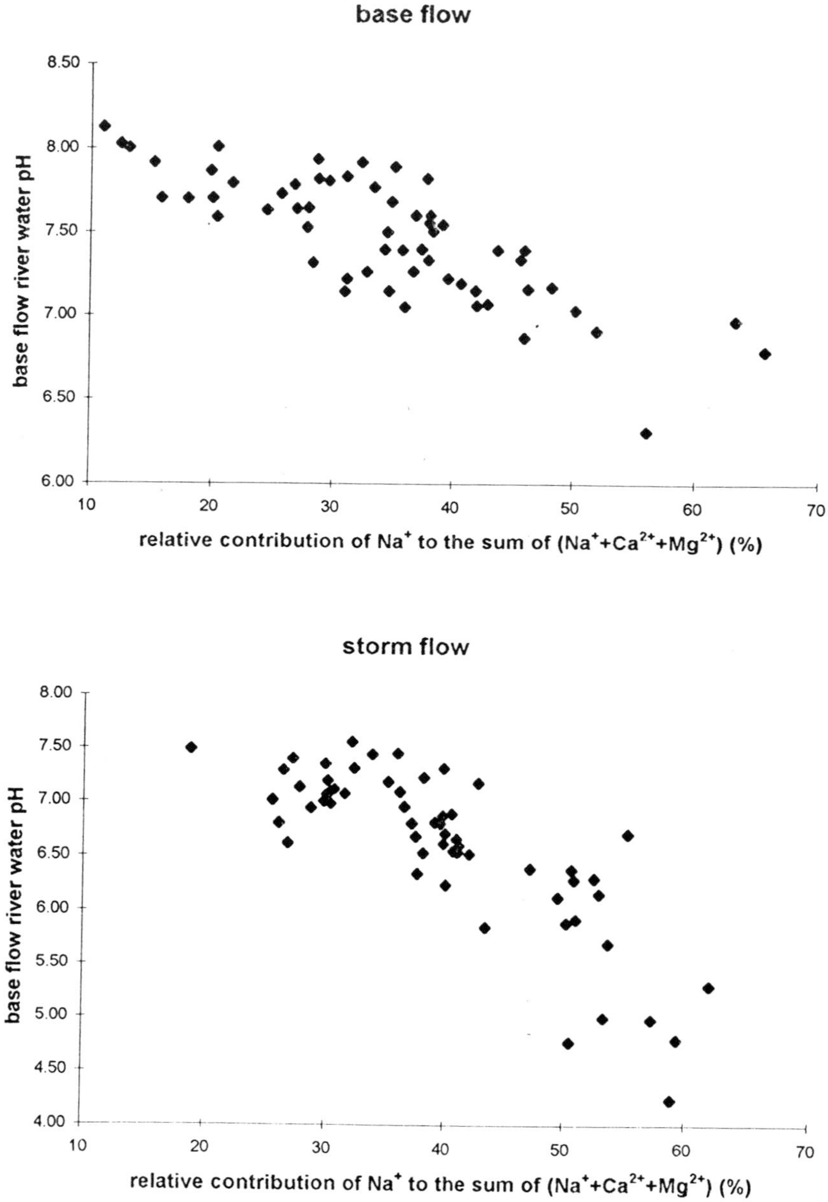

Fig. 3. Relationships between river water pH and relative contribution of Na^+ to $\Sigma(Na^+ + Ca^{2+} + Mg^{2+})$ for samples from sub-catchments of the River Dee catchment, for mean values of six base flow samples (upper plot) and for a single set of storm flow samples (lower plot).

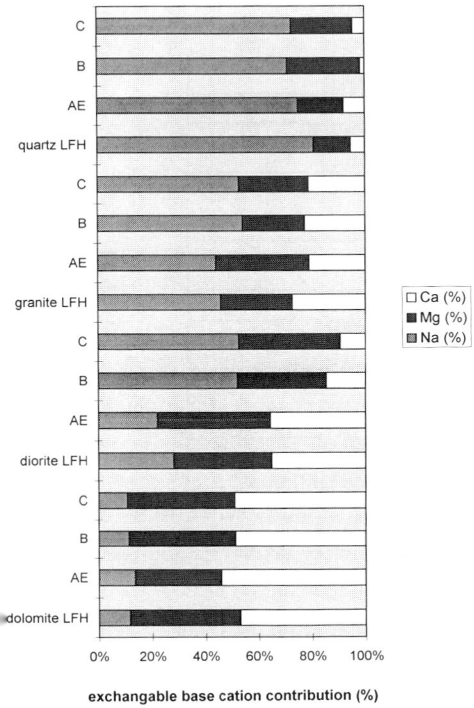

Fig. 4. The percent relative contribution of exchangeable Na^+, Ca^{2+} and Mg^{2+} in the soil to the sum of all three for each parent material and for each horizon.

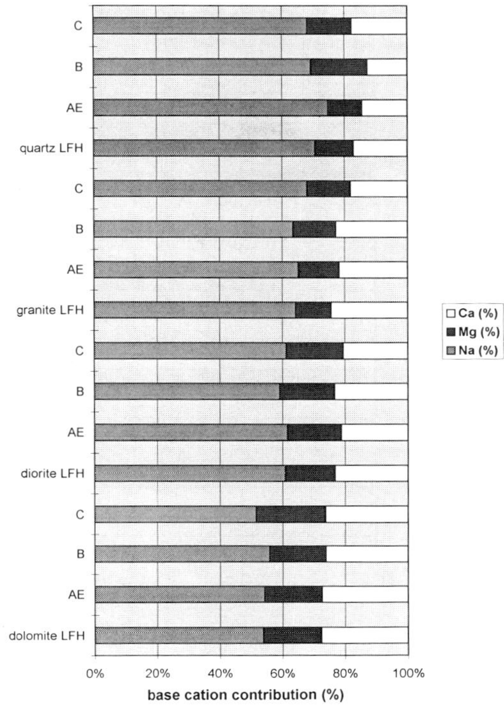

Fig. 5. The percent relative contribution of Na^+, Ca^{2+} and Mg^{2+} to the sum of all three for the soil solutions for each parent material and for each horizon.

base flow granite

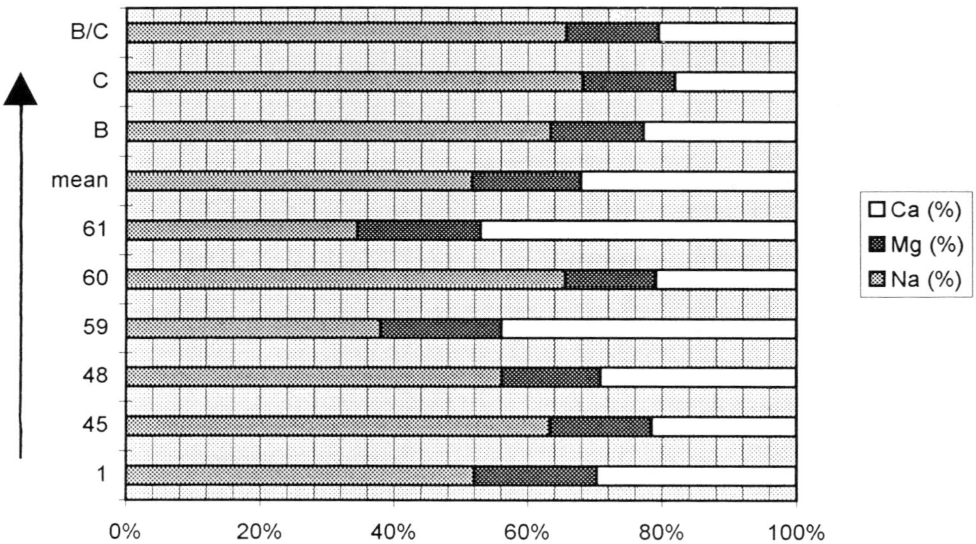

high flow granite

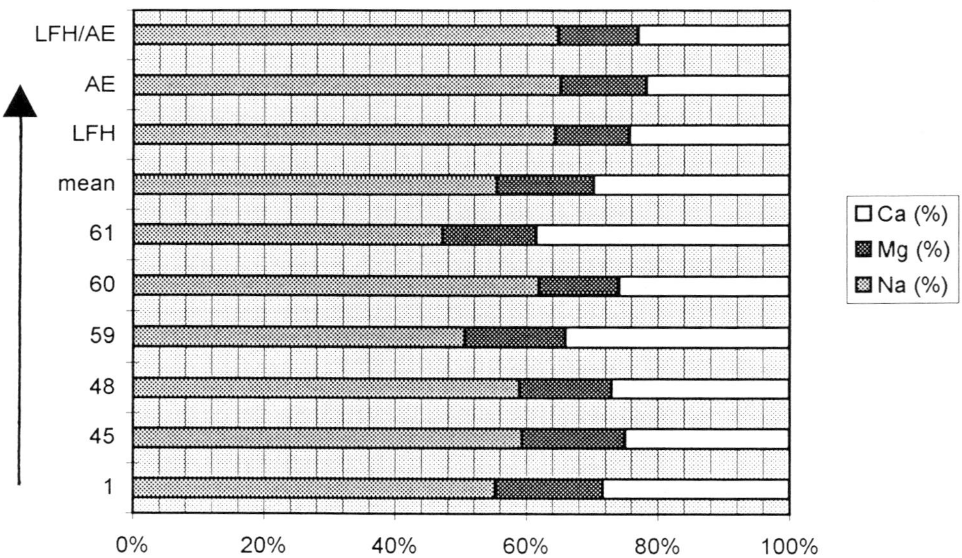

Fig. 6. The % relative contributions of Na^+, Ca^{2+} and Mg^{2+} to the sum of all three for base flow (top) and high flow (bottom) river water samples (site nos. 61, 60, 59, 48, 45, 1). Also included, for comparisons, are results for the corresponding granite soil individual B and C horizon soil solutions and their mean (to compare with base flow river water data) and for the individual LFH and AE horizon soil solutions and their mean (to compare with high flow river water data).

4. Conclusions

The use of a Na^+ dominance index has several advantages over existing approaches to quantifying critical loads. The latter can be applied simply to individual soil types within a catchment. However, for catchments with several diverse soil types it is difficult, and certainly both data- and labor-intensive, to adequately represent an overview of a catchment's soils. Using the proposed approach, however, providing an integrated overview for a catchment is very simple. We currently are investigating methods for converting relative contributions of Na^+ to weathering rates, and hence to critical load values. For a river, relative position on the Na^+ dominance index provides a quantitative measure of river water pH under extreme flow conditions, a parameter of direct ecological relevance.

The proposed catchment weathering index applies to both soils and surface waters. Potentially it is very cost effective, because analysis of a single base flow river water sample is all that is required to classify a catchment, once an appropriate calibration has been constructed for a given region. However, further work will be needed to test the regional extent over which calibrations may be applied legitimately. While the relationship has been tested provisionally over an area of only 2,100 km^2, it has been evaluated using a diverse range of single and mixed parent materials. It therefore seems likely that the approach could be applied successfully on a larger scale, and this needs to be tested experimentally. Indeed, because the proposed index is a quantitative measure of integrated catchment soil/rock weathering rates that affect the river water chemistry, it should be directly related to alkalinity, but this too needs to be tested experimentally.

Acknowledgments

We are grateful to the N.E.R.C. and the UK Department of the Environment, Transport and the Regions for financial support for this work.

References

Barton, D., Hope, D., Billett, M.F. and Cresser, M.S.: 1994, *Appl. Geochem.*, **9**, 127.

Cresser, M.S., Creasey, J. and Edwards, A.C.: 1984, *Proceedings of the International Symposium on Hydrochemical Balances of Freshwater Systems*, Uppsala, Sweden, IAHS Pub. No. 150, pp. 187-202.

Cresser, M.S., Dawod, A.M. and Rees, R.M.: 1997, *Water Res.*, **31**, 2253.

de Vries, W. Posch, M., Reinds, G.J. and Kamari, J.: 1993, *Critical Loads and Their Exceedance on Forest Soils in Europe*, Rep. No. 58, DLO Winand Staring Centre, Wageningen.

Elias, E.A. and Cresser, M.S.: 1997, *Geoderma*, **75**, 171.

Hettelingh, J.-P., Downing, R.J. and de Smet, P.A.M.: 1991, *Mapping Critical Loads for Europe*, CCE Tech. Rep. No. 1, RIVM, Bilthoven.

Hornung, M. and Skeffington, R.A. (eds.): 1993, *Critical Loads: Concepts and Applications*, ITE Symposium No. 28, HMSO, London.

Hornung, M., Bull, K.R., Cresser, M.S., Ullyett, J., Hall, J.R., Langan, S., Loveland, P.J. and Wilson, M.J.: 1995, *Environ. Pollut.*, **87**, 207.
Kinniburgh, D.G. and Edmunds, W.M.: 1984, *The Susceptibility of UK Groundwaters to Acid Deposition*, Report to the Department of the Environment, BGS, Wallingford.
Langan, S.J.: 1989, *Hydrol. Process.*, **3**, 25.
Langan, S.J., Wade, A.J., Smart, R., Edwards, A.C., Soulsby, C., Billett, M.F., Jarvie, H.P., Cresser, M.S., Owen, R. and Ferrier, R.C.: 1997, *Sci. Total Environ.*, **194/195**, 419.
Posch, M., de Smet, P.A.M., Hettelingh, J.-P. and Downing, R.J. (eds.): 1995, *Calculation and Mapping of Critical Thresholds in Europe, Status Report 1995*, Co-ordination Centre for Effects, RIVM Rep. No. 259101004, RIVM, Bilthoven.
Reid, J.M., MacLeod, D.A. and Cresser, M.S.: 1981, . *J. Hydrol.*, **50**, 129.
Soulsby, C.: 1995, *Hydrol. Process.*, **9**, 183.
White, C.C., Cresser, M.S. and Langan, S.J.: 1996, *Sci. Total Environ.*, **177**, 225.

ACIDIFICATION OF FRESHWATER WETLANDS: COMBINED EFFECTS OF NON-AIRBORNE SULFUR POLLUTION AND DESICCATION

LEON P.M. LAMERS, SILVIA M.E. VAN ROOZENDAAL and JAN G.M. ROELOFS

Research Group Environmental Biology, Department of Ecology, University of Nijmegen, Toernooiveld 1, 6525 ED Nijmegen, The Netherlands

Abstract. In recent decades, SO_4^{2-} concentrations have increased in groundwater and surface water of freshwater wetlands. For many minerotrophic peatlands, S originating from SO_4^{2-}-polluted groundwater and surface water is a more significant source of SO_4^{2-} than the actual atmospheric deposition of S compounds. Lowered groundwater tables in wetlands, as a result of either natural or anthropogenic desiccation, may cause acidification because of concomitant geochemical oxidation processes. The impact of the enhanced availability of reduced S compounds, due to preceding SO_4^{2-} pollution, on these processes was tested in a mesocosm experiment, using soil cores including vegetation from a mesotrophic wet meadow. The soils had been maintained in waterlogged condition for seven months, using two environmentally relevant SO_4^{2-} concentrations (2 and 4 mmol L^{-1}). The groundwater table was reduced in two successive steps: 10 cm below soil surface, and complete desiccation. Control pretreated soils did not show a decrease in soil pH during desiccation, due to adequate buffering by bicarbonate. However, both SO_4^{2-}-pretreated groups showed a significant drop in pH (from 6.5 to 4.5) caused by additional sulfide oxidation, leading to high SO_4^{2-} concentrations (10 and 16 mmol L^{-1}, respectively). Cation exchange and acidification-related solubilization processes induced the mobiliztation of base cations and potentially phytotoxic metals like Al. Nutrient concentrations in soil moisture were influenced strongly by SO_4^{2-} pretreatment, showing distinct patterns for P, N and K. Therefore, S polluted groundwater and surface water may severely increase the sensitivity of wetlands to desiccation. The results are discussed in relation to wetland management.

Key words: sulfur, sulfate, desiccation, acidification, nutrients, water quality, pollution, wetlands

1. Introduction

The effects of increased sulfur input from atmospheric deposition on the biogeochemistry and ecological functioning of wetlands have received much attention (*e.g.* Gorham, 1958, 1976; Johnson, 1979; Schindler *et al.*, 1980; Cook *et al.*, 1986; Reuss *et al.*, 1987). However, in many minerotrophic freshwater wetlands, SO_4^{2-} concentrations in surface and sediment water have increased recently from less than 0.2 mmol L^{-1} to concentrations of 0.5 mmol L^{-1} and higher (over 3 mmol L^{-1}; Lamers *et al.*, in press). It is unlikely that these values can be attributed solely to enhanced atmospheric S input. The affected wetlands receive large amounts of S from groundwater and surface water, sources contaminated by anthropogenic SO_4^{2-} dumping and weathering of marine geological deposits. Allochthonous river water, containing much higher SO_4^{2-} concentrations than the autochthonous water, is used to compensate for desiccation (Roelofs, 1991; Smolders and Roelofs, 1993; Koerselman *et al.*, 1993). Furthermore, the influx of groundwater from surrounding agricultural land supplies high amounts of SO_4^{2-} (Kölle and Schreeck, 1982; Stuyfzand, 1993; Jansen and Roelofs, 1996).

In earlier studies, the eutrophication observed in peatlands receiving only low amounts of nutrients could be attributed to increased SO_4^{2-} loading from river water (Roelofs, 1991; Smolders and Roelofs, 1993). In anoxic soils and sediments, increased SO_4^{2-} reduction

Water, Air, and Soil Pollution **105:** 95–106, 1998.

rates are known to cause eutrophication via sulfide-induced phosphate mobilization and/or enhancement of the mineralization rate (Ohle, 1954; Boström *et al.* 1982; Caraco *et al.*, 1989; Roelofs, 1991; Smolders and Roelofs, 1993; Lamers *et al.*, in press). A considerable amount of the resultant sulfide precipitates with metals (in particular with iron), thereby increasing the pool of reduced S in soils and sediments.

Desiccation of wet soils and sediments may be accompanied by acidification because of proton production by chemical processes, such as chemical oxidation of reduced Fe and sulfide, and by oxygen-dependent microbiological processes, like enhanced nitrification rates and oxidation of Fe and sulfide. Acidification due to oxygen availability during drought is well known from research on acid SO_4^{2-} soils (van Breemen, 1975) and acid mine drainage (Barton, 1978; Harries and Ritchie, 1983; Wieder, 1989), in areas containing high amounts of iron monosulfide (FeS) and/or pyrite (FeS_2). Acidification due to desiccation also has been shown for poorly buffered pools (Vangenechten *et al.*, 1981; Schuurkes *et al.*, 1988), in which the acid neutralizing capacity (ANC) of the sediment was insufficient to prevent the drop in pH caused by oxidation of accumulated airborne S.

In recent decades, groundwater levels in many wetlands worldwide have dropped by a few decimeters up to several meters. This is a result of extensive hydraulic operations for agricultural purposes and of increased water extraction. Consequently, the extent, duration and frequency of desiccation have increased in many wetlands and in pastures with vegetation types dependent upon a high ground water table. Furthermore, global climate change is thought to have a serious negative impact on the amount of precipitation during summer (IPCC, 1995). Desiccation therefore has become one of the principal problems for wetland conservation. The increased availability of iron sulfides, as indicated above, may severely enhance the sensitivity to dehydration, concomitantly affecting the ecological functioning of these wetlands during drought.

To evaluate the combined effects of preceding high S loads and desiccation, both on soil acidification and on general biogeochemistry, a mesocosm experiment was set up, using soil cores with vegetation from a minerotrophic wet meadow. The cores had been maintained in waterlogged condition for seven months with three different environmentally relevant SO_4^{2-} concentrations (Lamers *et al.*, in press) before they were subjected to desiccation. It was hypothesized that the increased S loads severely influence soil biogeochemistry during desiccation.

2. Materials and Methods

Soil cores were collected from a mesotrophic wetland meadow in the nature reserve 'De Bruuk' near Nijmegen, The Netherlands (51° 45'N; 5° 58'E). The upper 10 cm, used for the experiment, included moderately decomposed peat containing loam and living roots, classified as Rhizic Hydromoder (Green *et al.*, 1993). The vegetation was dominated by *Carex nigra* and could be considered Caricion nigrae (Ellenberg, 1988). In total, 18 sods (diameter 18 cm, depth 12 cm), including vegetation, were cut to fit tightly into plastic containers. Perforations in the bottom of the containers were covered from inside with plastic gauze to prevent loss of soil, and each container was suspended in another larger plastic container (12 L; Figure 1). The containers were placed in a climate control chamber, with a light level of 110 $\mu E\ m^{-2}\ s^{-1}$, a daily photoperiod of 12 hr, an ambient air temperature of 20 °C, and an air humidity of 50-60 % saturation. Each experimental flow-through unit received water from its own stock container through black silicon tubes, at a flow speed of 10 L wk^{-1} maintained by peristaltic pumps. The water level in the outer container could be manipulated by changing the overflow level (Figure 1). During both the preceding waterlogging period of seven months and the subsequent phases presented here, three different SO_4^{2-} concentrations, added as Na_2SO_4 (0, 2 and 4 mmol L^{-1}, indicated as [0], [S2], [S4]), were used in addition to the basic composition (Table I). Each experimental group consisted of six replicates, with random assignment to the three SO_4^{2-} concentration treatments. During pretreatment, the inflow tubes were placed in the centre of the top of the inner container, enabling a vertical water flow through the soil (Figure 1).

TABLE I
Chemical composition of the basic medium used. For the SO_4^{2-} treatments, either 2,000 or 4,000 µmol L^{-1} Na_2SO_4 was added.

Constituent	Concentration (µmol L^{-1})	Constituent	Concentration (µmol L^{-1})
Ca	2,000	K	5.0
Mg	400	Mn	0.9
Na	2,000	Zn	0.7
HCO_3^-	2,000	Cu	0.2
Cl	4,885	B	0.8

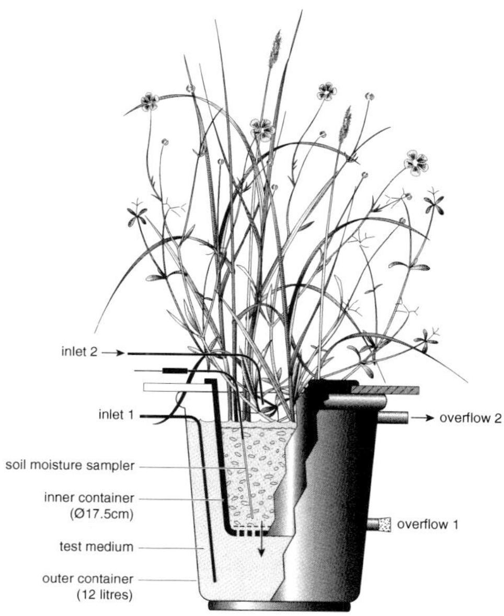

Fig 1. Experimental set-up for one unit. The water level in the soil could be manipulated by using different overflow levels.

Three soil moisture samplers (Rhizon SMS-10 cm; Eijkelkamp Agrisearch Equipment, The Netherlands) were placed in each container, and the collected samples were pooled in order to collect a representative pore water sample. Variation among the subsamples was moderate (standard deviation, at most, 10-15 % of the mean for all parameters measured). The removal of the sods did not result in any experimental artifacts; further details provided by Lamers *et al.* (in press). After the groundwater table had been reduced to 10 cm below soil surface (-10 cm) for nine weeks, to simulate moderate desiccation, the soils were further dehydrated for six weeks by removing the medium from the outer container. After this period, the water table was restored to -10 cm for three weeks and subsequently to the original waterlogged (pretreatment) situation.

Soil redox potential measurements were carried out in triplicate in each pot at 5 cm depth, using a mV-meter with a platinum wire electrode and a Ag/AgCl (3 M KCl) reference electrode. Values were converted to the potential relative to the normal hydrogen reference electrode (E_h). The pH was determined with a standard Ag/AgCl electrode and alkalinity was estimated by titrating part of the sample to pH 4.2 using 0.01 M HCl. After adding a few grains of citric acid to prevent precipitation of metal ions, color at 450 nm was measured for colorimetric background correction. The samples were stored in iodated polyethylene bottles (100 mL) at -28 °C until further analysis. The concentration of free sulfide was determined in a 10 mL subsample fixed immediately after collection with sulfide antioxidant buffer containing NaOH, Na-EDTA, and ascorbic acid (Van Gemerden, 1984). A sulfide ion-specific Ag electrode and a double junction calomel reference electrode were used (Roelofs, 1991). The concentrations of ortho-phosphate (SRP, soluble reactive P), NO_3^- (and NO_2^-), NH_4^+, and Cl^- (used as a tracer for solute concentration due to dehydration) in the pore water samples were measured colorimetrically with Technicon AA II systems, using ammonium-molybdate (Henriksen, 1965), hydrazine sulfate (Technicon, 1969), salicylate (Kempers and Zweers, 1986) and ferriammoniumsulfate (Technicon, 1969), respectively. The data were corrected for color caused by humic acids. Potassium was determined by flame photometry (Technicon Flame Photometer IV). Total concentrations of Ca, Mg, Fe, Al, Mn, Zn, and S in the moisture samples were determined by inductively coupled plasma emission spectrometry (Jarrell Ash IL Plasma-200).

A repeated measures analysis was used to examine the response to treatments (Potvin *et al.*, 1990), using the SAS procedure GLM, model one-way ANOVA, for repeated measures (SAS, 1989). Data were log-transformed to make the variances less dependent on the sample means and to fit a normal distribution. Differences at a given time were analyzed by *a posteriori* Tukey tests at the 0.05 confidence limit. For clarity of presentation, the means and standard errors presented in the figures represent the non-transformed data.

3. Results

During desiccation, soil redox potential (E_h) increased from values around -150 mV to 400 mV, without an effect of SO_4^{2-} treatment (Figure 2; Table II). In the middle of the dry period, between weeks 15 and 20, no pore water samples could be taken because of dehydration. During the initial waterlogging, soil pH in both SO_4^{2-}-treated soils was higher than the control value ($p < 0.05$), and after water the table reduction to -10 cm it decreased slightly. There were, however, no significant differences from the control treatment. At the beginning of the dry period, the pH of both SO_4^{2-} treatments dropped to values of 4.5-5. This persisted even two weeks after the waterlogged situation had been restored (significantly lower than [0]). The pH of the controls, however, barely decreased, fluctuating between 6.4 and 5.9. During the initial waterlogging, there was a significant difference between the alkalinities of the soil moisture for the three groups, with values being higher for the SO_4^{2-}-treated soils. After reducing the water table to -10 cm, alkalinity rapidly declined in all groups. During the initial phase of complete desiccation, alkalinity was below 100 μeq L^{-1} (Figure 2). After renewed waterlogging, alkalinity recovered so that by week 29, the value for [S4] was significantly higher than for the control.

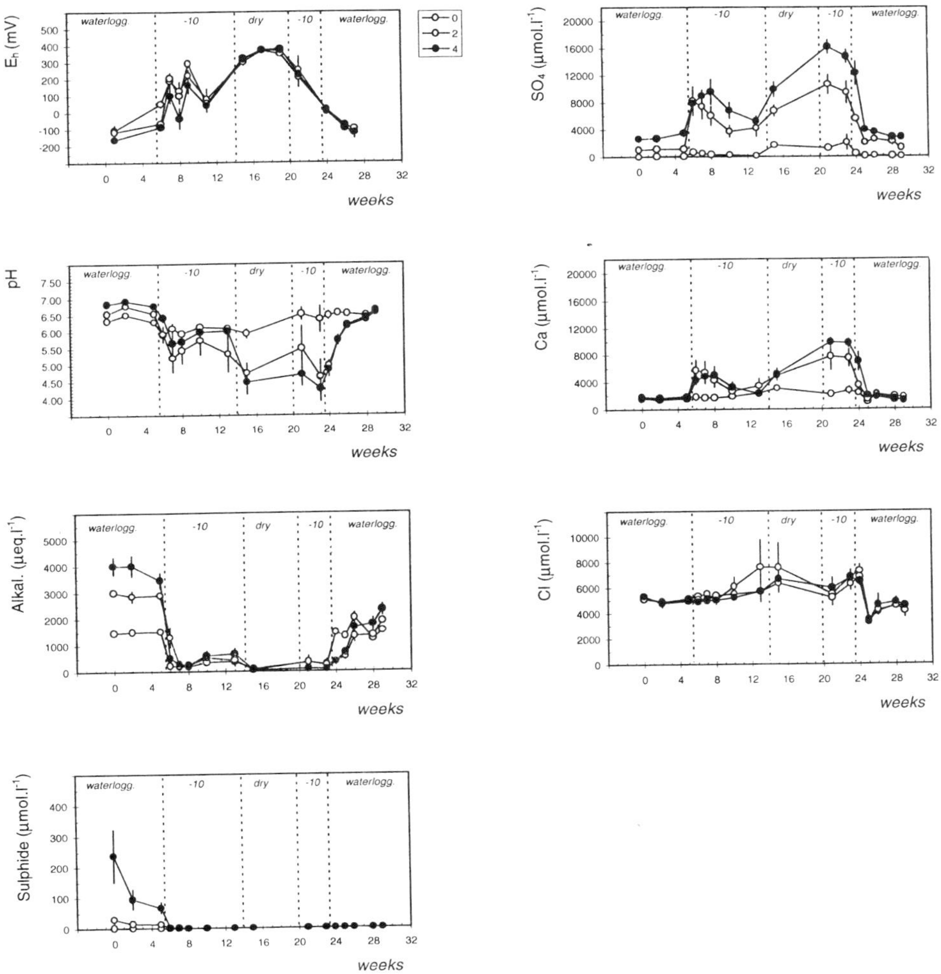

Fig. 2a. Soil pore water characteristics (means ± standard errors; n = 6), during consecutivewaterlogging (waterlogg.), water level reduction to 10 cm below soil surface (-10), complete desiccation (dry), restored water table at -10 cm (-10), and renewed waterlogging (waterlogg.) with either 0, 2 or 4 mmol L^{-1} SO_4^{2-} (indicated 0, 2, 4; white, grey, and black markers, respectively).

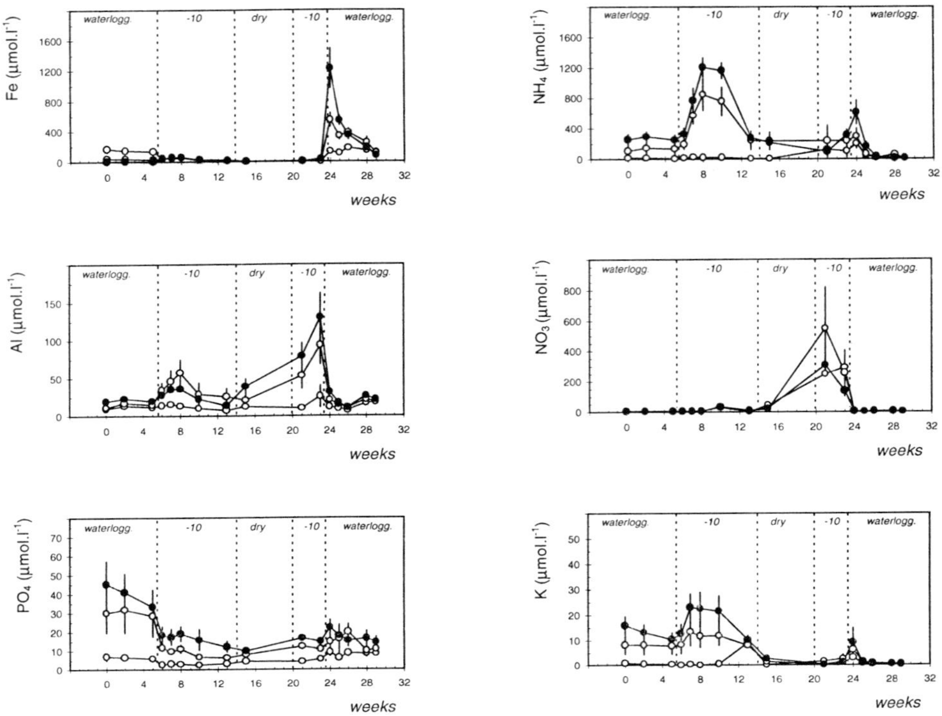

Fig. 2b. Soil pore water characteristics (means ± standard errors; n = 6), during consecutivewaterlogging (waterlogg.), water level reduction to 10 cm below soil surface (-10), complete desiccation (dry), restored water table at -10 cm (-10), and renewed waterlogging (waterlogg.) with either 0, 2 or 4 mmol L^{-1} SO_4^{2-} (indicated 0, 2, 4; white, grey, and black markers, respectively).

Chloride, being relatively uninfluenced by soil biogeochemistry, was used as a tracer for solute concentration by dehydration. Figure 2 and Table II show that there were no significant differences between the three groups. Chloride concentrations rose from 5,000 to 7,000 μmol L^{-1} during the period in which sampling was possible. After rehydration of the soils, Cl^- concentrations returned to pretreatment levels after a brief decline (Figure 2).

TABLE II

Repeated measures analysis of variance (general linear models) of the effects of SO_4^{2-} addition on several pore water characteristics, as shown in figure 1 and 2 (except for Mg, Mn, Zn). For treatment df = 2, and for time df = 15, except for E_h for which df = 11. *F*-values are given, with their level of significance: [a]$p \leq 0.01$, [b]$p \leq 0.001$, [c]$p \leq 0.0001$; ns, not significant.

	Independent variable		
Dependent variable	SO_4^{2-}-treatment	Time	Interaction
Alkalinity	6.96 [a]	23.51 [b]	2.73 [a]
pH	8.43 [a]	17.14 [c]	4.75 [c]
SO_4^{2-}	898.45 [c]	56.47 [c]	5.6 [c]
E_h	2.45 ns	59.25 [c]	1.24 ns
Cl^-	0.50 ns	15.51 [c]	0.62 ns
Sulfide	68.46 [c]	230.61 [c]	50.75 [c]
ortho-P	18.53 [b]	9.45 [c]	2.67 [a]
NH_4^+	21.46 [c]	13.5 [c]	4.27 [c]
NO_3^-	0.12 ns	90.5 [c]	0.84 ns
K	10.79 [a]	17.28 [c]	7.30 [c]
Fe	0.45 ns	89.74 [c]	10.18 [c]
Ca	12.16 [a]	16.52 [c]	3.47 [c]
Mg	8.28 [a]	28.90 [c]	6.54 [c]
Al	26.42 [c]	10.72 [c]	1.31 ns
Mn	15.36 [b]	6.52 [c]	5.24 [b]
Zn	54.01 [c]	42.61 [c]	5.60 [c]

During waterlogging, SO_4^{2-} concentrations in the pore water of [S2] and [S4] were 1,000 and 2,500-3,000 µmol L^{-1}, respectively (Figure 2; Table II). After water table reduction, the concentrations rapidly increased to about 8,000 µmol L^{-1} for both [S2] and [S4], after which they decreased again. During complete desiccation and subsequent restoration of the water table to -10 cm, SO_4^{2-} concentrations rose to about 10,000 and 16,000 µmol L^{-1}, respectively (both significantly higher than for [0]). In the controls, lowering of the water table to -10 cm and the subsequent complete desiccation also increased SO_4^{2-} concentrations. Renewed waterlogging restored the values to their pre-drought concentrations for all groups. The concentration of sulfide was significantly enhanced by the SO_4^{2-} treatment during the first waterlogging period, with significant differences between the three groups (Figure 2, Table II). After lowering the water table and subsequent desiccation, no detectable amounts of sulfide (i.e., concentration < 1 µmol L^{-1}) were noted. Even after renewed waterlogging, free sulfide could not be detected .

Calcium concentrations for [S2] and [S4] showed a significant peak of 6,000 µmol L^{-1} after the first reduction of the water table, and increased to 7,500 and 10,000 µmol L^{-1} after complete desiccation and the restoration of the water table to -10 cm, respectively

(Figure 2). The same trend was observed for magnesium, with concentrations rising significantly to 1,500 and 2,000 μmol L^{-1}, respectively (data not shown).

Iron concentrations in pore water were significantly depressed by both SO_4^{2-} treatments during the initial waterlogged situation (Figure 2). During desiccation, the values decreased (to less than 5 μmol L^{-1}) and then differences disappeared. After renewed waterlogging, iron concentrations showed a peak, with values significantly higher for both [S2] and [S4]. The concentrations of Al did not follow the same pattern. Compared to controls, significant peaks of Al were noted after reducing the water table and complete desiccation for [S2] and [S4]. The control soils did not show this strong response to desiccation, although a small peak of the Al concentration was detected after complete desiccation (Figure 2). The same trend holds for Mn and Zn concentrations (data not shown). Manganese in [S2] and [S4] soils increased from 6 to 40-50 μmol L^{-1} after reducing the water table to -10 cm, and further to 30-70 μmol L^{-1} following complete desiccation. Zinc concentrations in these groups rose from 0.7 μmol L^{-1} to 5-10 μmol L^{-1} during -10 cm, and to 20 μmol L^{-1} after complete desiccation.

The dynamics of the major nutrients showed distinct patterns (Figure 2). Ortho-P concentrations were significantly higher for [S2] and [S4] during initial SO_4^{2-} treatment. Desiccation decreased the concentrations and there were few significant differences between the groups. Ammonium concentrations were also higher ($p < 0.05$) in the SO_4^{2-}-treated soils during waterlogging. Lowering the water table to -10 cm significantly raised NH_4^+ concentrations in [S2] and [S4] soils, as compared to controls. The control soils did not show such a response. During complete desiccation and rehydration the differences disappeared, although [S4] showed a peak. Nitrate concentrations markedly increased after restoring the water table to -10 cm after desiccation (no significant differences between groups). Waterlogging rapidly returned concentrations to levels as low as 5 μmol L^{-1}. Potassium concentrations were significantly higher for [S2] and [S4] during waterlogging and water table reduction. During complete desiccation and the subsequent period, K concentrations decreased to low levels in all treatments.

4. Discussion

The pretreatment of increased SO_4^{2-} supply during waterlogging stimulated sulfate reduction, as indicated by SO_4^{2-} disappearance from the medium, alkalinity generation and production of free sulfide (Figure 2). Sulfate treatment had a dose-dependent effect on the mobilization of ortho-P, NH_4^+, and K. This type of nutrient mobilization, without input of nutrients, is termed internal eutrophication so as to distinguish it from eutrophication by allochthonous nutrients (Roelofs, 1991). Free Fe concentrations were depressed strongly as a result of precipitation with sulfide (see also discussion in Lamers *et al.*, in press).

The initial step of reducing the water table to -10 cm was taken to evaluate the effect of slight desiccation, as often occurs in wet meadows (including the one from which the sods were collected) during summer, on soil biogeochemistry. Soil redox potential rose by 200-250 mV, comparable to field data during water table drawdown to -10 cm (data not shown), and SO_4^{2-} was rapidly generated in [S2] and [S4], to levels twice the maximum SO_4^{2-} treatment concentration of 4,000 μmol L^{-1}. Chloride concentrations showed that this could not be attributed to concentration effects (Figure 2), but must have been caused by oxidation of FeS and/or FeS_2:

$$4\ FeS + 9\ O_2 + 10\ H_2O \rightarrow 4\ Fe(OH)_3 + 4\ SO_4^{2-} + 8\ H^+.$$

Iron ion concentrations remained low as a result of iron oxidation to iron hydroxides. After rehydration, these Fe compounds were reduced again, showing peaks after one week for both SO_4^{2-}-treated groups. The highest peak was found for [S4], for which the source of oxidized Fe originating from FeS_x was highest because of a higher sulfate reduction rate during pretreatment (Lamers *et al.*, in press). The responses of the three groups were contrary to those during the beginning of the pretreatment, showing lower peaks for [S2] and [S4] due to rapid FeS_x precipitation (Lamers *et al.*, in press).

For the control, the effects on pH by acid production due to FeS_x oxidation were limited, but the bulk of soil moisture alkalinity (almost entirely caused by HCO_3^-) was rapidly consumed. Nitrate concentrations did not increase sufficiently to account for the observed consumption of alkalinity by proton generation through nitrification. However, the increase in SO_4^{2-} concentration from 50 to 650 μmol L^{-1} for the control does explain the observed HCO_3^- consumption (see equation above). In [S2] and [S4] soils, not only was HCO_3^- consumed, but also there was mobilization of Ca and Mg, indicating cation exchange against protons at the soil adsorption complex and dissolution of $CaCO_3$.

During desiccation, HCO_3^- was consumed completely in all three groups by the oxidation of FeS_x and the absence of alkalinity generating processes like denitrification. Whereas SO_4^{2-} concentrations increased to 2,000 μmol L^{-1} for controls, high values up to 16,000 μmol L^{-1} were measured in [S2] and [S4]. These values are similar to those measured in drainage water originating from desiccated peatlands (Heathwaite, 1990). The observed drought-related SO_4^{2-} production and alkalinity consumption also has been described for poorly buffered moorland pools (Vangenechten *et al.*, 1981; Schuurkes *et al.*, 1988; Van Dam, 1988) and mineral rich fens (Kemmers and Jansen, 1988; Boeye and Verheyen, 1994).

In the minerotrophic peaty soils of the mesocosms, acid neutralization by cation exchange at the soil adsorption complex (as indicated by further mobilization of Ca and Mg) was inadequate to prevent soil acidification down to pH 4.5 in the SO_4^{2-}-pretreated soils. This induced the mobilization of metals like Al, Zn, and Mn to potentially toxic concentrations. In the controls, however, no effects of desiccation on soil pH could be detected and Al, Mn, and Zn concentrations remained low. Base cations were mobilized, but to a much lesser extent than in the SO_4^{2-}-pretreated soils. Apparently, the ANC of minerotrophic soils, not subject to SO_4^{2-} pollution, was adequate to prevent pH changes and the subsequent severe mobilization of metals. By definition, however, there was acidification because ANC decreased (van Breemen *et al.*, 1983). Van Haesebroeck *et al.* (1997) also found no significant effect of drought on the pH of pore water in rich fen soils, although the pH of water-extracted fresh soil at the end of the drought period had slightly decreased.

The fact that NH_4^+ and K concentrations showed a peak for [S2] and [S4] during water table reduction to -10 cm (Figure 2) presumably indicated exchange of both monovalent cations against protons at soil binding sites and/or a temporal stimulation of the mineralization rate. For freshwater marshes, an increase of the mineralization rate during moderate desiccation has been shown (De Groot and Van Wijck, 1993). The decrease in K concentrations for all treatments during complete desiccation may suggest jarosite ($KFe_3(SO_4)_2(OH)_6$) formation (van Breemen, 1976).

Nitrification was stimulated only after complete desiccation, as shown by the strong increase in NO_3^- concentrations of all soils after rehydration. Subsequent waterlogging probably suppressed nitrification and stimulated denitrification, leading to low NO_3^- concentrations (Berendse *et al.*, 1994). The decrease in ortho-P concentrations during desiccation most probably can be explained by increased binding of phosphate to oxidized Fe(III) compounds, such as iron phosphate, iron(hydr)oxide-phosphate and humic-iron-phosphate complexes (Patrick and Khalid, 1974; Khalid *et al.*, 1977; Richardson, 1985; Golterman, 1984, 1988).

5. Implications for Wetland Management

From the outcome of this experiment it may be concluded that the present S pollution in many freshwater wetlands is a threat not only by inducing internal eutrophication, but also by increasing the sensitivity of wetlands to desiccation, another general threat to wetland functioning and biodiversity. The increased pool of FeS_x will stimulate drought-related soil acidification, influencing both ANC and pH, and will cause an increase of potentially toxic metals like Al in pore water. The combination of increased S loading with a growing risk of the occurrence of desiccation is therefore an important cause for concern. Furthermore, the high SO_4^{2-} concentrations generated by desiccation and the subsequent runoff following rehydration will inevitably cause internal eutrophication in reduced peaty soils and sediments where the water is discharged.

The extent to which drought sensitivity increases depends on soil ANC, the pre-drought levels of both SO_4^{2-} and reducible organic compounds (regulating sulfate reduction rates), and on the amount of Fe available for sulfide precipitation. Particularly in areas that possess a large Fe pool in the soil (e.g., because of actual or former groundwater discharge), SO_4^{2-} pollution will seriously enhance FeS_x storage and the related response to desiccation. Therefore, for management and restoration of wetlands, much importance should be devoted to the preservation or restoration of the original hydrology and hydrochemistry of deteriorating wetlands.

Acknowledgments

The authors would like to thank Mr. J. Eygensteyn for his help with chemical analyses, Mrs. J. Van Westreenen for preparation of the first figure, and Prof. J.M. Van Groenendael, Dr. C. Van der Drift, and Mr. B. Kelleher for critically reading the manuscript. We also thank Mr. M. Bolten for his kind permission to collect the soil cores from the nature reserve 'De Bruuk'.

References

Barton, P.: 1978, in J.O. Nriagu (ed.), *Sulfur in the Environment, Part II: Ecological Effects*, John Wiley and Sons, New York, pp. 324-358.
Berendse, F., Oomes, M.J.M., Altena, H.J. and De Visser, W.: 1994, *J. Appl. Ecol.* **31**, 40.
Boeye, D. and Verheyen, R.F.: 1994, *J. Vegetation Sci.* **5**, 553.
Boström, B. Jansson, M. and Forsberget, C.: 1982, *Arch. Hydrobiol.* **18**, 5.
Caraco, N.F., Cole, J.J. and Likens, G.E.: 1989, *Nature* **341**, 316.
Cook R.B., Kelly, C.A., Schindler, D.W. and Turner, M.A.: 1986, *Limnol. Oceanogr.* **31**, 134.
De Groot, C.J. and Van Wijck, C.: 1993, *Hydrobiologia* **252**, 83.
Ellenberg, H.: 1988, *Vegetation Ecology of Central Europe*, Cambridge University Press, Cambridge.
Golterman, H.L.: 1984, *Hydrobiologia* **91**, 169.
Golterman, H.L.: 1988, *Hydrobiologia* **159**, 149.
Gorham, E.: 1958, *Phil. Trans. Royal Soc. London Ser. B* **241**, 147.
Gorham, E.: 1976, *Water, Air, Soil Pollut.* **6**, 457.
Green, R.N., Trowbridge, R.L. and Klinka, K.: 1993, *Towards a Taxonomic Classification of Humus Forms,* Forest Science Monograph 29, Bethesda, MD.
Harries, J.R. and Ritchie, A.I.M.: 1983, *Water, Air, Soil Pollut.* **19**, 155.
Heathwaite, A.L.: 1990, *Water, Air, Soil Pollut.* **49**, 159.
Henriksen, A.: 1965, *Analyst* **90**, 29.
IPCC: 1995, *Second Assessment Report on Climate Change*, Intergovernmental Panel on Climate Change.
Jansen, A.J.M. and Roelofs, J.G.M.: 1996, *Ecol. Engineering* **7**, 279.
Johnson, A.H.: 1979, *Science* **206**, 834.
Kemmers, R.H. and Jansen, P.C.: 1988, *Agr. Water Manag.* **14**, 399.
Kempers, A.J. and Zweers, A.: 1986, *Commun. Soil Sci. Plant Anal.* **17**, 715.
Khalid, R.A., Patrick, W.H.J. Jr. and DeLaune, R.D.: 1977, *Soil Sci. Soc. Am. J.* **41**, 305.
Koerselman, W., Van Kerkhoven, M.B. and Verhoeven, J.T.A.: 1993, *Biogeochemistry* **20**, 63.
Kölle, W. and Schreeck, D.: 1982, in *Impact of Agricultural Activities on Groundwater*, Internat. Symp. IAH Vol. XVI, Prague, pp. 191-202.
Lamers, L.P.M., Tomassen, H.B.M. and Roelofs, J.G.M.: In press, *Environ. Sci. Technol.*
Ohle, W.: 1954, *Vom Wasser* **21**, 13.
Patrick, W.H. and Khalid, R.A.: 1974, *Science* **186**, 53.
Potvin, C., Lechowicz, M.J. and Tardif, S.: 1990, *Ecology* **71**, 1389.
Reuss, J.O., Cosby, B.J. and Wright, R.F.: 1987, *Nature* **329**, 27.
Richardson, C.J.: 1985, *Science* **228**, 1424.
Roelofs, J.G.M.: 1991, *Aquat. Bot.* **39**, 267.
SAS: 1989, *SAS/STAT Users Guide*, SAS Institute Inc. Cary, N.C, Version 6.
Schindler, D.W., Wageman, R., Cook, R.B., Ruszcynski, T. and Prokopowich, J.: 1980, *Can. J. Fish. Aquat. Sci.* **37**, 342.
Schuurkes, J.A.A.R., Kempers, A.J. and Kok, C.J.: 1988, *J. Freshwater Ecol.* **4**, 369.
Smolders, A. and Roelofs, J.G.M.: 1993, *Aquat. Bot.* **46**, 247.

Stuyfzand, P.J.: 1993, *Hydrochemistry and Hydrology of the Coastal Dune Area of the Western Netherlands*, Ph.D. Thesis, Free University of Amsterdam, The Netherlands.
Technicon: 1969, *Technicon Auto Analyzer Methodology, Industrial Method 11 and 33-69w*, Technicon Corporation, New York.
van Breemen, N.: 1975, *Soil Sci. Soc. Amer. Proc.* **39**, 1153.
van Breemen, N.: 1976, *Genesis and Solution Chemistry of Acid Sulfate Soils in Thailand.* Agric. Res. Rep. 848, Pudoc, Wageningen, The Netherlands.
van Breemen, N., Mulder, J. and Driscoll, C.T.: 1983, *Plant Soil* **75**, 283.
van Dam, H.: 1988, *Freshwater Biol.* **20**, 157.
Van Gemerden, H.: 1984, *Arch. Mikrobiol.* **139**, 289.
Vangenechten, J.H.D., Bosmans, F. and Deckers, H.: 1981, *Hydrobiologia* **76**, 149.
Van Haesebroeck, V., Boeye, D., Verhagen, B., and Verheyen, R.F.: 1997, *Biogeochemistry* **37**, 15.
Wieder, R.K.: 1989, *Wetlands* **9**, 299.

TRACE GAS EMISSIONS FROM A NORTH WALES FEN - ROLE OF HYDROCHEMISTRY AND SOIL ENZYME ACTIVITY

H. KANG, C. FREEMAN and M.A. LOCK

School of Biological Sciences, University of Wales, Bangor, LL57 2UW, United Kingdom

Abstract. Fluxes of methane and nitrous oxide from a wetland were compared to hydrochemistry (NO_3^- and SO_4^{2-}) and soil enzyme activities (β-glucosidase) in an attempt to predict gas fluxes using chemical and enzymatic activities in north Wales, UK. In a one-year survey, the enzyme activities ranged from 0.028 to 0.065 μmol g^{-1} min^{-1}, while NO_3^- and SO_4^{2-} concentrations ranged between 0-0.78 and 5-200 mg L^{-1}, respectively. Methane and nitrous oxide emissions varied between 0.32-240 and 0.28-5.48 mg m^{-1} da^{-1}, respectively. The field survey was followed by laboratory-based manipulation experiments, from which pilot mathematical models were constructed, and related to the field data. The variation of methane emission was attributed to changes in SO_4^{2-} concentration and temperature at the site. In contrast, NO_3^- concentration, water table height, and β-glucosidase activity were major controlling factors for nitrous oxide emission. The models explained more than 80 % of the variation observed in the field study.

Keywords: β-glucosidase, fen, gas emission, methane, nitrous oxide, soil enzymes, Wales, wetland

1. Introduction

Wetlands are significant sources of greenhouse gases such as CH_4 and N_2O (Clymo, 1983; Hemond, 1983), which are 25 and 200 times, respectively, more radiatively active than CO_2 on a molar basis (Rodhe, 1990). The atmospheric concentrations of the gases are increasing at about 0.8 % for CH_4 and about 0.25 % for N_2O per year (Ramanathan, 1988; Pearman, 1988). Various anthropogenic and natural impacts could accelerate or decelerate these emissions by influencing biological processes in wetlands (Gorham, 1991).

Gas emissions from wetlands have been recognized to be controlled by various physico-chemical factors. Hydrology, soil temperature, type and amount of substrates, and water chemistry (e.g., SO_4^{2-} or NO_3^-) have been reported as the main variables controlling methane emission from wetlands (Crill *et al.*, 1991). For N_2O emission, hydrology, NO_3^- supply and carbon (C) supply have been delineated as important aspects in a field scale study (Groffman, 1991). The reasons are that: 1) denitrification is associated with anaerobic conditions, 2) NO_3^- is an electron acceptor during denitrification, and 3) most denitrifying organisms are heterotrophic and thus require an organic C source. Temperature has been reported to have minimal effects compared to the other three factors in field studies. For example, Skiba *et al.* (1994) observed that soil temperature did not significantly affect N_2O emissions from upland soils, and yet 30 % of the variation was explained by NO_3^- and soil moisture contents. In the present study, we adopted a β-glucosidase assay, as an indicator of the rate at which labile organic C was introduced into the wetland soil. It has generally been accepted that the hydrolysis of polymers is the rate-limiting step in mineralization and consequently C supply to microorganisms (Chróst, 1991). As β-glucosidase mediates the final and critical step in cellulose decomposition, measurement of its activity could be useful in assessing the rate at which C becomes available in wetlands. Beta-glucosidase activity has been widely studied in relation to decomposition and C cycling in forests (Sinsabaugh

Water, Air, and Soil Pollution **105:** 107–116, 1998.

et al., 1992) and wetlands (Freeman *et al.*, 1996).

Our experiments were divided into two parts. First, a one-year field survey of gas emission, hydrochemistry, and soil enzyme activity was conducted in a calcareous fen in north Wales. Second, we manipulated various physico-chemical factors to determine the relationship between those factors and the gas emission from the peat samples. Based on these laboratory manipulations, we set up pilot mathematical models to which the field data waere applied. Our objective was to relate trace gas emissions to the above variables.

2. Methods and Materials

2.1. Study Site

The study site is located on Anglesey island, north Wales, UK (UK Grid SH 497 826). The dominant vegetation of the area is *Festuca rubra* and *Juncus acutiflorus*. The soil organic matter content (loss-on-ignition) was 87.9 % (n = 4, standard error = 0.2). The loss-on-ignition method may overestimate organic matter in calcareous mineral soils due to high carbonate content, but has been employed widely for organic soils (Karam, 1993). The water pH ranged from 5.7 to 6.2 during the sampling period. The area is classified as a calcarious fen, with high concentrations of K^+ (0-4.1 mg L^{-1}), Mg^{2+} (2.9-4.9 mg L^{-1}), Ca^{2+} (61.4-104.7 mg L^{-1}), and Na^+ (3.1-19.5 mg L^{-1}) in the pore water.

2.2. Field Study

Samples were collected on a monthly basis from January to December, 1996. The gas samples were collected using a closed-chamber technique (Freeman *et al.*, 1994a). After a 1-hr collection, gas samples were withdrawn into gas-tight syringes (SGE). Five replicates were taken on each sampling occasion. Approximately 10 mL water samples were taken using syringes and water samplers placed at 10 cm depth (Freeman *et al.*, 1994b). On the day of sampling, the water samples were filtered with 0.2 μm filters and maintained at 2 °C until chemical analysis. Five samples were taken from beside each of the gas samplers. For enzyme analysis, four replicates of about 6,000 cm^3 of surface peats (0-10 cm depth) were taken. All the samples were kept at 4 °C until the analysis, which was carried out within a week of the sampling day.

The gas samples were analyzed using an Ai Cambridge model 92 Gas Chromatograph equipped with Porapak QS column (Freeman *et al.*, 1994b), while the water samples were load into a Dionex 2000i ion chromatograph with AS4A anion column to determine SO_4^{2-} and NO_3^- concentrations (Freeman *et al.*, 1993). In the enzyme assay, 1 mL of peat core was added to 7 mL of 400 μM methylumbelliferyl-β-D-glucopyranoside solution and homogenized in a Seward Colworth model 400 stomacher. The samples were incubated for 1 hr at field temperature, using either an incubator or water bath with dip cooler. Field pH levels were maintained in the reaction. After the incubation, 1.5 mL of reactant were pipetted into centrifuge vial and centrifuged at 10,000 rpm (7,200 g) for 5 min. Then 0.5 mL of the supernatant were transferred to 2.5 mL of deionized water in a vial. Fluorescence was determined with a Perkin-Elmer LS50 fluorometer at 450 nm emission and 330 nm excitation wavelengths with slit setting 2.5. Calibration curves were made using 0-200 μM of MUF free acid solution in soil slurry on each sampling occasion to correct interference of phenolics and adsorption to soil matrix (Freeman *et al.*, 1995).

2.3. Laboratory-Based Manipulation Experiment

Peat samples were collected from the surface (0-10 cm), as the upper layer of wetland soil contains most of the microbial activity (Gammelgaard *et al.*, 1992; Freeman *et al.*, 1995). One mL of field-moist peat sample (0.15 g of dry mass) was placed in a 30 mL glass bottle, followed by addition of 10 mL of treatment solution. For SO_4^{2-} (applied as Na_2SO_4), concentrations of 0, 10, 100, 250, 500 mg L^{-1} were used. For NO_3^- (applied as $NaNO_3$) solutions with 0.1, 1, 10, 20 mg L^{-1} concentrations were used. Ionic strength was not controlled in this experiment. For the effect of β-glucosidase on the gas emissions, 3 mL of β-glucosidase (Sigma, EC 3.2.1.21) solution were added to the soil. To ensure the emissions were not NO_3^--limited, 7 mL of NO_3^- solution (5 mg L^{-1}) were applied before the enzyme addition. The amounts of β-glucosidase applied were 50 units (3.57 mg L^{-1}), 25 units, 12.5 units, and 0 unit (one unit of the enzyme hydrolyzes 1 μmole of *p*-nitrophenol glucoside at pH 5 and 36 °C). After a 30-min stabilization, the bottles were closed with air-tight lids with rubber septa. Then, the bottles were incubated at 10 °C for 2 hr to accumulate emitted gases in the head space, followed by gas collection with gas-tight syringes after passing a needle through the cap. To study the effect of temperature on CH_4 emission, four incubation temperatures (2, 5, 8, 16 °C) were adopted using 1:10 (v/v) soil slurries prepared with deionized water. For each treatment, three replicates were employed.

3. Results

3.1. Field Study

Soil temperature and water table height changes are shown in Figure 1. The temperature ranged from 2 to 14.5 °C, showing a maximum in July. Water table height decreased below the surface (-8 ~ -6 cm) during summer (July - September). During the other months, the level was close to or higher than the soil surface.

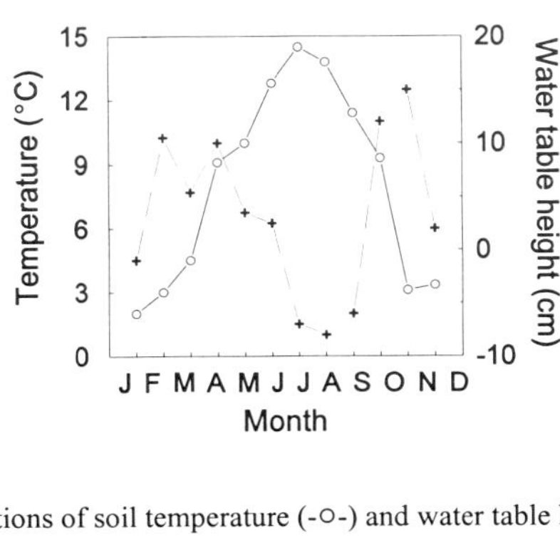

Fig. 1. Annual variations of soil temperature (-○-) and water table height (-✚-) in the field.

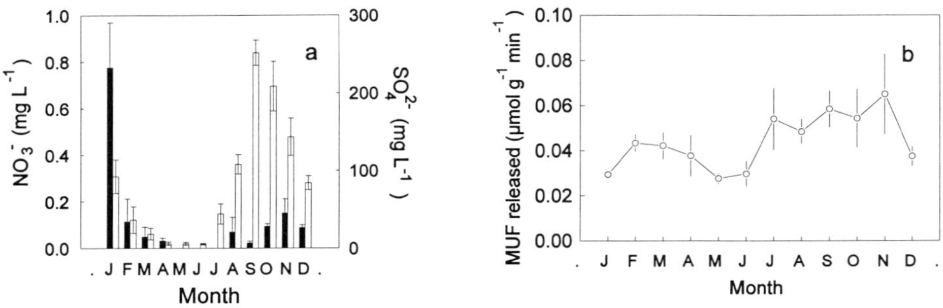

Fig. 2. Annual variations of nitrate concentration (shaded bar) and sulfate concentration (unshaded bar) (a), and β-glucosidase activity (b) in the field (n=5 for nitrate and sulfate, n=4 for β-glucosidase; mean ± se)

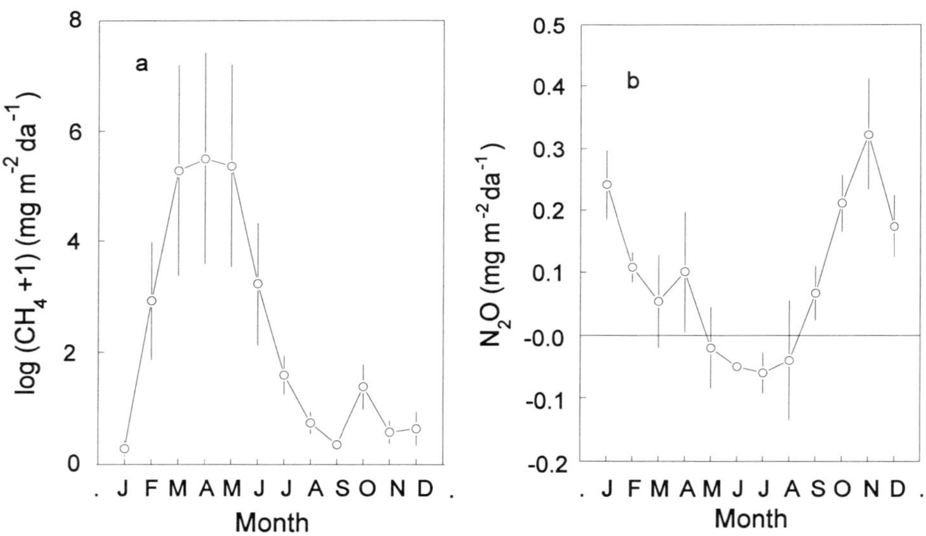

Fig. 3. Annual variations of methane (a), and nitrous oxide (b) emissions in the field (n=5; mean ± standard error).

Sulfate concentration in the water decreased continuously until June (Figure 2a) and increased sharply to about 250 mg L^{-1} in September, after which the concentration decreased to about 80 mg L^{-1} by December. Nitrate concentration in the water fluctuated between 0 and 0.15 mg L^{-1}, except for January when a high concentration (0.77 mg L^{-1}) was detected (Figure 2a). During May to July, NO_3^- concentrations were below the minimum detection threshold.

For glucosidase activity, two broad peaks of the activity were discernable (Figure 2b); one appeared over February to April and the other occurred over July to November. The activity ranged from 0.028 μmol g^{-1} min^{-1} to 0.065 μmol g^{-1} min^{-1}.

Annual variation of the CH_4 emission observed in the field had one large peak over March to May, and another small one in October (Figure 3a). Nitrous oxide emission decreased from January to March (Figure 2b). After a slight increase in April, there was no N_2O emission from May to August. The emission increased from September, peaking in November.

3.2. LABORATORY-BASED MANIPULATION EXPERIMENTS

Sulfate addition lowered CH_4 emission by 91 % ($p < 0.05$) (Figure 4a). Simple linear regression between logarithmic SO_4^{2-} concentration and logarithmic CH_4 emission showed a negative correlation ($r^2 = 0.703$). The response of CH_4 emission to temperature showed an approximate hyperbolic curve, with the highest emission at 8 °C and lower at 2 °C and 16 °C (Figure 4b). Although the response could be expressed with two equations (e.g., one linear and one second order equation), we assumed one second order equation for simplicity.

The N_2O emission increased linearly with increasing β-glucosidase addition (Figure 5a). Nitrous oxide emission showed linear relation with logarithmic NO_3^- concentration (simple linear regression, $r^2 = 0.996$) (Figure 5b). At 10 mg L^{-1}, saturation of the nitrate addition (e.g., no further increase of N_2O emission by increase of NO_3^- addition) was observed.

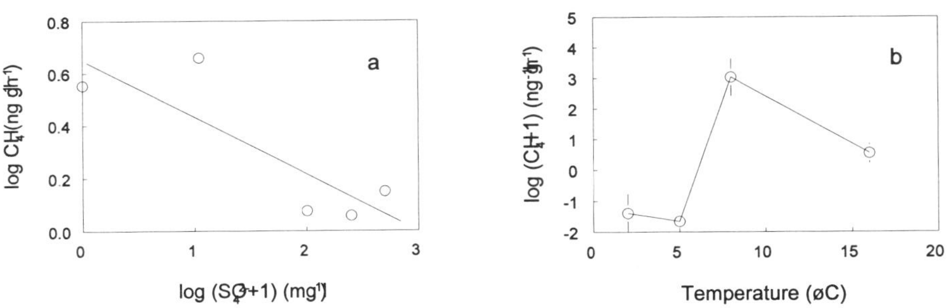

Fig. 4. Effects of sulfate (a), and temperature (b) on CH_4 emission (n=3).

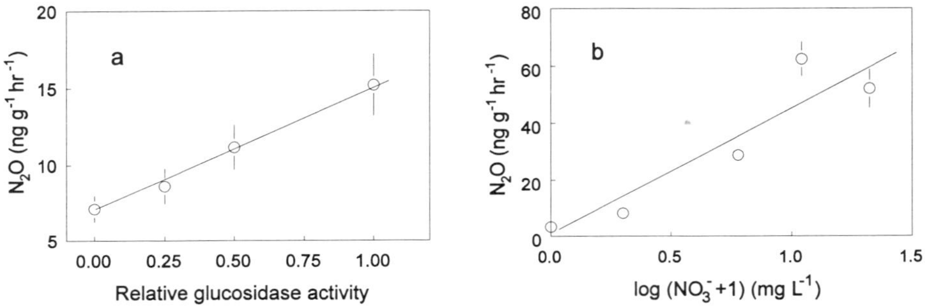

Fig. 5. Effect of β-glucosidase activity (a), and nitrate (b) on N_2O emission (n=3; mean ± se).

3.3. Mathematical Model

3.3.1. *Methane*

Based on the results from the laboratory-based experiments, a pilot model was constructed for CH_4 emission. The premise was that logarithmic CH_4 emission has a negative relationship with logarithmic SO_4^{2-} concentration, a hyperbolic function in relation to temperature, and a linear correlation with water table height. While the effects of SO_4^{2-} and temperature were based on the manipulation experiment, the effect of increased water table height was assumed as a positive effect on the emission based on earlier field studies (Freeman *et al.*, 1993). The relationship is expressed as:

$$\log(CH_4+1) = a\times(16-T)\times T + b\times\log(SO_4^{2-}) + c\times(\text{water table height}) + d$$

where, a, b, c, and d represent optimization coefficients for the model and T is temperature.

To determine the optimization coefficients of the model, mean values for each month were applied and multiple linear regression carried out. The water table height did not improve the fit of the model: the r^2 was 0.876 when water table effect was included. As a result, we propose the following equation for the model:

$$\log(CH_4+1) = 0.052\times(16-T)\times T - 1.15\times\log(SO_4^{2-}) + 4.31$$
$$(r^2 = 0.875, p = 0.000, n = 12)$$

3.3.2. *Nitrous oxide*

In case of the N_2O, NO_3^- concentration, water table height, and β-glucosidase activity were used for the model. The assumption was that N_2O emission has a positive linear correlation with logarithmic NO_3^- concentration, water table height, and glucosidase activity. Again, NO_3^- and glucosidase effects were based on the manipulation experiments, while the effect of water table height was assumed from earlier studies (Freeman *et al.*, 1993).

$$N_2O = a' \times \log(NO_3^- + 1) + b' \times (\text{water table height}) + c' \times (\text{glucosidase}) + d'$$

where, a', b', c', and d' represent optimization coefficients for the model.

The same approach as the CH_4 model was applied; 12-month field data were fitted into the equation to determine the coefficients. The result was as follows:

$$N_2O = 0.53 \times \log(NO_3^- + 1) + 0.0094 \times (\text{water table height}) + 3.99 \times (\text{glucosidase}) - 0.17$$
$$(r^2 = 0.852, p = 0.001, n = 12)$$

4. Discussion

4.1. Methane

Methane emissions from wetlands often have been found to positively correlate with soil temperature (Dise *et al.*, 1993). The main reasons have been reported to be the activation of methanogens by temperature increases, and the weak temperature dependence of CH_4 consumption (Dunfield *et al.*, 1993). However, poor correlations between temperature and CH_4 emission also have been reported, suggesting the mechanisms controlling CH_4 flux to be more complex than often is assumed. For example, Wilson *et al.* (1989) proposed a step function rather than simple positive correlation for temperature effects on CH_4 emission from a temperate swamp. It also has been found (Dunfield *et al.*, 1993) that the temperature dependence of CH_4 consumption varies inconsistently between sites within the range of temperatures observed in the present study (i.e., 2-16 °C). The approximately hyperbolic curve observed in our manipulation experiment (Figure 1b) was confirmed in a temperature gradient bar experiment (a laboratory-based experiment), where the fen soil (as used in this study) showed a similar hyperbolic response (Dowrick, Ph.D. thesis, in prep.). Therefore, we assumed a hyperbolic response of CH_4 emission in the model over the given temperature range. The falling CH_4 emission at 16 °C may indicate that the temperature optimum for CH_4 oxidation is higher than that for methanogenesis in the given temperature range. In the manipulation experiment, a 2-hr incubation was adopted to determine the effect of temperature on gas emission, as a 2-hr incubation represents the maximum period over which production rates approach linearity.

Inhibitory effects of SO_4^{2-} on CH_4 emission have been widely reported (Yavitt *et al.*, 1987; Freeman *et al.*, 1994a). Under conditions of abundant SO_4^{2-}, sulfate-reducing bacteria may outcompete methanogens for organic substrates leading to the suppression of methanogenesis (Ormeland, 1988). We observed a similar response in the manipulation experiment (Figure 1a), as well as in the field. The high concentration (250 mg L^{-1}) of SO_4^{2-} in the winter season is found rarely in freshwater wetlands. We speculate that stormy rain from the sea in the winter would affect the SO_4^{2-} concentration at the site, which is near the coast.

Waterlogging can lower oxygen availability and redox potential, both of which favor methanogenesis (Gottschalk, 1986). In contrast, water table drawdown would suppress CH_4 emission, mainly due to a lowering of the anaerobic zone and an increasing depth over which CH_4 oxidation could occur (Whalen and Reeburgh, 1990; Freeman *et al.*,1993). However, in the present study, the model based on two factors (SO_4^{2-} and temperature) explained 85 % of the variation in the field, and water table height did not improve the model. This contrasts with a mathematical model proposed for CH_4 emission from peatlands in Minnesota (Dise *et al.*,1993), in which temperature and water table heights were the main controlling factors.

The remaining unexplained variation (15 %) potentially could be attributed to the effects of the vegetation in the field. Vascular plants influence CH_4 emissions in various ways. First, translocation of photosynthate to the rhizosphere has been reported, which may be used as a substrate for methanogenesis (Millela *et al.*, 1995). Second, vegetation also may facilitate CH_4 transport from the anaerobic zone to the atmosphere (Whiting and Chanton, 1992), or transport oxygen into the rhizosphere (Conrad, 1989). Incorporation of the vegetation effect could increase the predictability of the model. For example, Yavitt (1997) assessed the role of plants in CH_4 emission by experimentally removing the shoots in *Typha latifolia* wetlands.

In this study, CH_4 emissions ranged from 0.32 to 240 mg m^{-2} da^{-1}. Comparable values of CH_4 emission from northern fens are 0-75 mg m^{-2} da^{-1} in Wales (Freeman *et al.*, 1994a), 96-188 mg m^{-2} da^{-1} (Crill *et al.*, 1988) and 11-767 mg m^{-2} da^{-1} (Dise, 1993) in the U.S., and 0-343 mg m^{-2} da^{-1} and 0-1,976 mg m^{-2} da^{-1} in Canada (Moore and Knowles, 1990; Barlett and Harriss, 1993).

4.2. Nitrous Oxide

The environmental conditions in an undisturbed peatland are considered to be highly favorable for N_2O production through denitrification (Hemond, 1983). However, the contribution from nitrification also could be significant at certain sites (Firestone and Davison, 1989). Soil moisture content, NO_3^- supply, and C supply have been reported as important factors in field scale investigations of N_2O emission (Groffman, 1991). High soil water content induces anaerobic conditions, which prompts vigorous denitrification (Groffman, 1994). However, many wet sites have little denitrification due to NO_3^- limitation. Nitrate supply has been widely reported as a limiting factor until it reaches a concentration of 10 mg L^{-1} (Groffman, 1994), although some other studies have suggested that first-order kinetics would still exist at concentrations higher than 25 mg L^{-1} (Reddy and Patrick, 1984). In our manipulation study, NO_3^- concentration above 10 mg L^{-1} seems to be non-limiting for N_2O emission.

Dissolved organic carbon (DOC) has been used widely to assess C availability regarding N_2O emission (Bijay-Singh *et al.*, 1988; Swerts *et al.*, 1996). This approach might be of limited value in relation to wetlands for several reasons. First, the DOC pool easily could be changed by decomposition and uptake by microorganisms. It has been reported that the generation of readily utilizable products (e.g., monomeric C compounds) is slower than assimilation by microorganisms in aquatic ecosystems (Sinsabaugh *et al.*, 1997). Second, it has been observed that recalcitrant organic matter (e.g., phenolics) represents a large proportion of the DOC pool in wetlands (Hemond, 1990; Freeman *et al.*, 1994b), and these have been reported to be inhibitory to microorganisms (Freeman and Lock, 1992; Wetzel, 1992). Thus, even under a high concentration of DOC, denitrification could be limited by C availability.

In the present study, we measured β-glucosidase activity to assess the effect of C supply on N_2O emission. The addition of β-glucosidase data into the model increased the model r^2 from 0.708 to 0.852. If other enzymes (e.g., xylosidase or cellobiohydrolase) involving C cycles are included in the model, inclusion could increase the sensitivity of the model still further. In our study, enzyme activity ranged from 0.028 μmol g^{-1} min^{-1} to 0.065 μmol g^{-1} min^{-1}, which is lower than reports from other wetland soils (Freeman *et al.*, 1995; Freeman *et al.*, 1996). The range of N_2O emission (-0.06-0.32 mg m^{-2} da^{-1}) in this study is similar to that determined from a Finnish fen (0.02-0.12 mg m^{-2} da^{-1}) and from Dutch and Polish fens (0.1-0.8 mg m^{-2} da^{-1}) (Nykänen *et al.*, 1995; Verhoeven *et al.*, 1996).

In this study, we have attempted to elucidate possible mechanisms that contribute to temporal variation in trace gas emission in the field. Our approach used laboratory-based experiments to construct mathematical models to relate external variables to those emissions. With further refinement, the models proposed in this study may be of additional value by providing a helpful tool for predicting the possible impact of climatic change and acid deposition on wetland trace gas emission.

References

Barlett, K.B. and Harriss, R.C.; 1993, *Chemosphere*, **26**, 261.

Bijay-Singh, Ryden, J.C. and Whitehead, D.C.: 1988, *Soil Biol. Biochem.* **20**, 737.

Chróst, R.J.: 1991, 'Environmental Control of the Synthesis and Activity of Aquatic Microbial Ectoenzymes', in R.J. Chróst (ed.), *Microbial Enzymes in Aquatic Environments*, Springer Verlag, New york, pp. 29-59.

Clymo, R.S.: 1983, 'Peat', in D.W. Goodall (ed.) *Ecosystems of the World 4A; Mires: Swamp, Bog, Fen, and Moor,* American Society for Microbiology, Washington, D.C., pp. 219-235.

Conrad, R.: 1989, 'Controls of Methane Production in Terrestrial Ecosystems', in M.O. Andreae, and D.S. Schimel (eds.), *Exchange of Trace Gases between Terrestrial Ecosystems and the Atmosphere,* John Wiley & Sons Ltd., NY, pp. 39-58.

Crill, P.M., Harriss, R.C. and Bartlett, K.B.: 1991, 'Methane Fluxes from Terrestrial Wetland Environments', in J.E. Roger and W.B. Whitman (eds.), *Microbial Production and Consumption of Greenhouse Gases,* American Society for Microbiology, Washington, D.C., pp. 91-110.

Crill, P.M., Bartlett, K.B., Harriss, R C., Gorham, E., Verry, E.S., Sebacher, D.I., Mazdar, L. and Sanner, W.: 1988, *Global Biogeochem. Cycles* **2**, 317.

Dise, N.B.: 1993, *Global Biogeochem. Cycles* **7**, 123.

Dise, N.B., Gorham, E. and Verry, E.S.: 1993, *J. Geophys. Res.* **98**, 10583.

Dunfield, P., Knowles, R., Dumont, R. and Moore, T.R.: 1993, *Soil Biol. Biochem.* **25**, 321.

Firestone, M.K. and Davison, E.A.: 1989, 'Microbiological Basis of NO and N_2O Production and Consumption in Soil', in M.O. Andrease , and D. S. Schimel (eds.), *Exchange of Trace Gases between Terrestrial Ecosystems and the Atmosphere,* John Wiley & Sons Ltd., NY, pp. 7-21.

Freeman, C. and Lock, M.A.: 1992, *Appl. Environ. Microbiol.* **58**, 2030.

Freeman, C., Lock, M.A. and Reynolds, B.: 1993, *Biogeochemistry* **19**, 51.

Freeman, C., Hudson, J., Lock, M.A., Reynolds, B. and Swanson, C.: 1994a, *Soil Biol. Biochem.* **26**, 1439.

Freeman, C., Lock, M.A. and Reynolds, B.: 1994b, *Verh. Internat. Verein. Limnol.* **25**, 1318.

Freeman, C., Liska, G., Ostle, N. J., Lock, M. A., Reynolds, B., and Hudson, J.: 1996, *Plant Soil* **180**, 121-127.
Freeman, C., Lock, M.A., Hughes, S., Reynolds, B. and Hudson, J.A.: 1997, *Environ. Sci. Technol.* **31**, 2438.
Gammelgaard, A., Freeman, C. and Lock, M.A.: 1992, *Soil Biol. Biochem.* **25**, 505.
Gorham, E.: 1991, *Ecol. Applic.* **1**, 182.
Gottschalk, G.: 1986, *Bacterial Metabolism,* Springer-Verlag, NY.
Groffman, P.M.: 1991, 'Ecology of Nitrification and Denitrification in Soil Evaluated at Scales Relevant to Atmospheric Chemistry', in J.E. Roger and W.B. Whitman (eds.), *Microbial Production and Consumption of Greenhouse Gases*, American Society for Microbiology, Washington DC, pp. 91-110.
Groffman, P.M.: 1994, *Biogeochemistry* **1**, 15.
Hemond, H.: 1983, *Ecology* **64**, 99.
Hemond, H.: 1990, 'Wetlands as the Source of Dissolved Organic Carbon to Surface Waters', in E.M. Perdue, and E.T. Gjessing (ed.), *Organic Acids in Aquatic Ecosystems*, John Wiley & Sons, Chichester, pp. 301-313.
Karam, A.: 1993, 'Chemical Properties of Organic Soil', in M.R. Carter (ed.), *Soil Sampling and Methods of Analysis*, Lewis Publishers, Boca Raton, FL, pp. 459-471.
Millela, C., Sundh, I., Svensson, B.H. and Nilsson, M.: 1995, *Biogeochemistry* **28**, 93.
Moore, T.R. and Knowles, R.: 1990, *Biogeochemistry* **11**, 45.
Nykänen, H., Alm, J., Lång, K., Silvola, J. and Mrtikainen, P.J.: 1995, *J. Biogeogr.* **22**, 351.
Ormeland, R.S.: 1988, 'Biogeochemistry of Methanogenic Bacteria', in A.J.B. Zehnder (ed.), *Biology of Anaerobic Micro-organisms*, Wiley, NY, pp. 641-705.
Pearman, G.I.: 1988, 'Greenhouse Gases: Evidence for Atmospheric Changes and Anthropogenic Causes', in G.I. Pearman (ed.), *Greenhouse Planning for Climate Change,* CSIRO, Australia, pp. 3-21.
Ramanathan, V.: 1988, *Science* **240**, 293.
Reddy, K.R.. and Patrick, W.H.: 1984, *CRC Crit. Rev. Environ. Control* **13**, 273.
Rodhe, H.: 1990, *Science* **248**, 1217.
Sinsabaugh, R.L., Antibus, R.K., Linkins, A.E., McClaugherty, C.A., Rayburn, L., Repert, D. and Weiland, T.: 1992, *Soil Biol. Biochem.* **24**, 743.
Sinsabaugh, R.L., Findlay, S., Franchini, P. and Fischer, D.: 1997, *Limnol. Oceanogr.* **42**, 29.
Skiba, U., Fowler, D., Smith, K.:1994, *Environ. Monit. Assess.* **31**, 153.
Swerts, M., Merckx, R. and Vlassak, K.: 1996, *Plant Soil* **181**, 145.
Verhoeven, J.T.A., Keuter, A., van Logtestijn, R., van Kerkhoven, M.B. and Wassen, M.: 1996, *J. Ecol.* **84**, 647.
Wetzel, R.G.: 1992, *Hydrobiologia* **229**, 181.
Whalen, S.C. and Reeburgh, W. S.: 1990, *Nature* **346**,160.
Whiting, G.J. and Chanton, J.P.: 1992, *Global Biogeochem. Cycles* **6**, 225.
Wilson, J.O., Crill, P.M., Bartlett, K.B., Sebacher, D.I., Harriss, R.C. and Sass, R.L.: 1989, *Biogeochemistry* **8**, 55.
Yavitt, J.B., Lang, G.E. and Wieder, R.K.: 1987, *Biogeochemistry* **4**, 141.
Yavitt, J.B.: 1997, *Wetlands* **17**, 394.

ESTIMATING RAINY SEASON NITROUS OXIDE AND METHANE FLUXES ACROSS FOREST AND PASTURE LANDSCAPES IN COSTA RICA

W.A. REINERS[1], M. KELLER[2] and KENNETH G. GEROW[3]

[1]*Department of Botany, The University of Wyoming, Laramie, Wyoming 82071-3165, USA*
[2]*Institute of Tropical Forestry, USDA Forest Service, Rio Piedras, Puerto Rico 00928-2500*
[3]*Departments of Statistics, and Zoology and Physiology, The University of Wyoming Laramie, Wyoming 82071-3165, USA*

Abstract. The objectives of this research were to estimate exchanges of CH_4 and N_2O, both radiatively active gases, between soil and atmosphere on hilltop, slope and swale hillslope positions of northeastern Costa Rica; and to assess the importance of accounting for topography in making areal estimates across hilly terrain. Emission rates from soils were measured during the rainy season on three hillslope positions of both actively grazed pastures and primary forests. Emission rates from pasture and forest sites were significantly different for both gases. Differences between slope positions, though notable, were not significantly different for CH_4, but were significantly different for N_2O. The forest landscape was partitioned with GIS methods into hilltop, slope and swale topographic positions. The calculated areas for each of these were multiplied by their respective emission rates to calculate overall flux from the entire forested area of 618 ha. Nitrous oxide flux ranged from 16,181 to 17,100 g N d^{-1}. Similarly, CH_4 flux ranged from -6,201 to -6,658 g CH_4 d^{-1}. Errors associated with both estimating mean emission rates for each hillslope position and judgmental errors in partitioning the landscape into hillslope positional classes are important to making landscape-scale estimates of flux.

Keywords: Costa Rica, geographic information system, landscape-scale, methane flux, nitrous oxide flux

1. Introduction

Atmospheric concentrations of nitrous oxide (N_2O) are rising at an average rate of 0.8 ± 0.2 ppbv yr^{-1} (Khalil and Rasmussen, 1992). This is of global concern because of N_2O's role in the atmospheric radiative balance and as an interactant in stratospheric ozone dynamics. Methane (CH_4), another radiatively active gas, has been increasing in the atmosphere since 1993 at an annual rate of about 8 ppbv (Houghton et al., 1996). Soils of the moist tropics are both a source and a sink for CH_4 (Keller et al.,1990), but represent the single largest global source of N_2O (Keller et al., 1986; Matson and Vitousek, 1990; Bouwman and Taylor, 1996). Further, emission rates from tropical soils appear to be increasing, particularly where land use has changed from native vegetation to agriculture, and most especially where N-fertilizer is applied (Bouwman et al., 1993; Erickson and Keller, 1997). For these reasons, estimation of regional N_2O fluxes from tropical regions where fluxes are high has become a central goal for improving global budget estimates.

Estimating regional gas fluxes between land and atmosphere by scaling up from small sampling units over heterogeneous areas is fraught with potential error (King, 1991; Rastetter et al., 1992; Kicklighter et al., 1994; White and Running, 1994). Land use, soil types, topography and site history are obvious sources of variability. For example, land use change in northeastern Costa Rica has led to a complex mosaic of primary forest, pasture, abandoned pasture, secondary forest, and various kinds of agricultural fields (Reiners et al., 1994; Keller and Reiners, 1994). We have reported on fluxes of N_2O, NO and CH_4 for some of the land use types comprising this mosaic over an annual cycle (Keller et al., 1993; Keller and Reiners, 1994). In general, N_2O fluxes are higher in primary forest than in active or abandoned pastures, or than in secondary forests, while CH_4 consumption is greater in forest than other land use types. These fluxes show significant seasonal and successional variations as well, so that generalizations require consideration of the time of year and history of individual sites.

Our previous work in northeastern Costa Rica was confined to convex sites--broad ridges or hilltops--specifically to reduce variability due to hillslope position. Without gas exchange data for other hillslope positions, however, these results cannot be extrapolated over landscapes with full confidence. The effects of topography on biogenic trace gas fluxes through associated variables such as soil texture and drainage have been documented by other investigators. Spatial effects have been shown for cultivated fields (Folorunso and Rolston, 1984), temperate old fields (Robertson et al., 1988), temperate forests (Groffman and Tiedje, 1989a,b; Groffman et al., 1992, 1993b; Bowden et al., 1993), prairie (Groffman et al., 1993a; Groffman and Turner, 1995), shortgrass steppe (Parton et al., 1988; Mosier et al., 1996), tropical forests (Livingston et al., 1988; Bowden et al., 1992), sagebrush steppe (Matson et al., 1991), desert (Peterjohn and Schlesinger, 1991) and subalpine meadow-forest (Mosier et al., 1993).

The primary objective of this research was to estimate fluxes of N_2O and CH_4 over a Costa Rican forested landscape based on gas emission measurements along hillslope positions and to apply results to a representation of the hillslope distributions for the research area. The secondary objective was to assess the importance of accounting for terrain variability in estimating regional fluxes--a crucial issue for scaling up local measurements to regional and larger areas.

This research was performed on, and adjacent to, La Selva Biological Station, 10°26' N, 84°0' W, in Sarapiqui Canton of Heredia Province, northeastern Costa Rica. This region ranges in elevation from 35 to 137 m on Pleistocene lava flows partially covered with Pleistocene and Holocene lahars and alluvium (Sanford et al., 1993). The climate is classified as Tropical Wet Forest (Sanford et al., 1993). The soils to which this research was restricted are all oxic Humitropepts (Reiners et al., 1994), and the native vegetation of this region is wet tropical forest (Hartshorn, 1983).

The upland landscapes on which this research was conducted are decomposable into hilltop, slope and swale positions. Hilltops are mostly flat-topped or gently convex surfaces, slopes range between 20 and 45 %, and the swales, while concave in cross-section, are gently sloping in longitudinal section. Field observations of water movement across hillslopes during storms indicated that surface drainage moves off hilltop shoulders, flows down slopes and accumulates temporarily in swales. Observations of surface organic matter indicated some fluvial and colluvial movement from slopes into the swales. Based on these observations, together with the fact that inorganic nitrogen is derived through mineralization of organic nitrogen, we inferred that on a typical hillslope profile, soil water, organic carbon and inorganic nitrogen were least abundant on slopes, greatest in swales, and intermediate on hilltops.

From our understanding of variables driving N_2O and CH_4 fluxes (Davidson, 1991; Conrad, 1996; Galchenko et al., 1989; Born, 1990; Davidson et al., 1992; Li et al., 1992; Bender and Conrad, 1994) and on our own observations (Keller and Reiners ,1994; Parsons et al., 1993) we know that CH_4 consumption is greatest in well-drained soils and production is greatest in organic-C-rich, poorly drained soils. Therefore, we hypothesized that CH_4

consumption would be greatest on slopes, production greatest in swales, and that hilltops would be intermediate between these sites of consumption and production. Similarly, because net N_2O emissions are driven by labile C and inorganic-N supply under anaerobic conditions, we hypothesized that N_2O emissions would be least on slopes where drainage is greatest and from which organic litter and organic-N are partially removed by fluvial and colluvial processes, and would be greatest in swales. Again, we hypothesized that N_2O flux would be intermediate on the hilltops where organic matter, organic-N and soil moisture levels would be intermediate.

2. Methods

2.1. Determining Topographic Effects on Emission Rates

Three sites were located within or adjacent to the La Selva Biological Station for each of the actively grazed pasture and forest land use types, all on the Matabuey soil association (Sollins et al., 1993). To avoid pseudoreplication, all sites had independent histories or locations. These are the same forest and active pasture sites detailed by Reiners et al., (1994) and from which an annual cycle of environmental measurements and trace gas fluxes were reported by Keller and Reiners (1994).

At each of the six sites, sample areas were designated on hilltop, mid-slope and swale positions. The hilltop sites were identical to those measured in 1990-1991 (Keller and Reiners, 1994). Mid-slope and swale sites were located downslope of the original measurement areas. At the pasture sites, the areas actually sampled were fenced to avoid direct cattle effects. To compensate for lack of grazing, the grass was periodically cut and the cuttings removed.

Sampling locations were fixed along four, parallel 15 m (in the pastures) and four, 30 m (in forest sites) transect lines on the hilltop positions as described by Keller and Reiners (1994). At the mid-slope and swale positions, single 30 m transect lines were established along contour lines and measurements were taken at random points along these lines during each measurement period.

All sites were sampled twice in each of July and August of both 1992 and 1993 for a total of four sample periods in the rainy season for this area (see Keller and Reiners [1994] for seasonal effects). The order in which the sites were sampled in any of the four months was random. Eight sampling points for N_2O and CH_4 were permanently fixed in random positions along the hilltop transects of each site (two per transect). On mid-slope and swale positions, however, eight sampling points were located randomly along transect lines for each sampling period. The static, vented field chamber method (Hutchinson and Livingston, 1993) was used to measure N_2O and CH_4 fluxes, as described in Keller and Reiners (1994). Chamber bases were enclosed with a chamber top for 28 min for each measurement during which gas samples were removed with nylon syringes at five time intervals.

Eight soil cores (0-10 cm) were removed from random points along the transect lines and stored in polyethylene bags for transport to the laboratory. These samples were divided into two aliquots for measurement of soil moisture (oven-dry weight), soil NO_3^-, and soil NH_4^+ ion concentrations as described by Keller and Reiners (1994).

Nitrous oxide was measured in the syringe samples with a Shimadzu GC-8 gas chromatograph with ECD detector. Methane was measured with a Shimadzu mini-2 FID gas chromatograph. Both were analyzed within 36 hr of collection. Storage tests on all syringes were repeated several times through the year and were found to maintain gas concentrations for at least 36 hr to within 2 %. Syringes failing to meet this criterion during period testing were replaced. Details on processing and gas standards are reported in Keller and Reiners (1994). Gas fluxes were calculated from linear regressions of the changes in concentration with time within the enclosed chambers. Positive changes indicated net emission of gas from the soil; negative changes indicated net consumption of gas by the soil. Limits of precision for flux measurements based on the 95 % confidence interval of the slopes of the linear regressions were approximately 0.5 ng N cm^{-2} h^{-1} for N_2O and 0.5 mg CH_4 m^{-2} d^{-1} for CH_4.

Soil NO_3^- and NH_4^+ ion concentrations were determined from filtered (Whatman 42) 2M KCl extracts after 1 hr agitation (Keeney and Nelson ,1982). Extracts were refrigerated prior to analysis. The extracts were analyzed using standard colorimetric methods on a Technicon II Autoanalyzer at the La Selva Biological Station. Nitrite (NO_2^-) and NO_3^- were summed for the reported concentrations of NO_3^-. Soil samples were oven-dried at 105 ± 5 °C for 48 hr for soil moisture measurements.

Because the chamber locations were fixed on the hilltop sites, they were not randomly distributed with respect to the four sampling events. Therefore, we were not able to use time as a variable. Alternatively, we used a split plot design in which all eight replicates for all four times were averaged to give means for individual sites. Thus, the replicates (3 per position for each of pasture and forest) for analysis were grand means for each position at each site. The split plot ANOVA tested for significant differences between land use types, between slope positions, and interactions between land use types and slope positions. When interactions were not significant, their sums of squares were joined with error sums of squares for a recalculation of the significance of position effects. Our hypotheses concerning slope position effects on CH_4 and N_2O emission rates were quite specific; an overall ANOVA is less powerful where alternative hypothesis about effects are concerned. Therefore, we performed one-tailed paired *t*-tests on gas emission rates following the ANOVA. Software used was produced by Minitab Inc. (1992). For those cases in which position on slope was significant according to ANOVA, we performed Bonferonni range tests.

2.2. Determining Hillslope Positions Across the Landscape

The landscape for which areal gas fluxes were estimated was limited to La Selva Biological Station property where high resolution, digital, topographic data were available for landscape modeling, and to forested areas on Jaguar and Matabuey soils representing the vegetation and soils on which forest field measurements were made. This area totaled 618 ha. Unfortunately, there are no digital elevation data for the pasture sites that lay outside the station boundary so that estimation of areal fluxes was restricted to forested landscapes of La Selva. Gas emission measurements were made only on the Matabuey soil type, an oxic Humitropept developed on the Taconazo lava flow (andesitic/basaltic) (Reiners et al., 1994). However, the Jaguar soil type is pedologically indistinguishable from the Matabuey (Sollins et al., 1993) although derived from a different source, the Esquina lava flow (andesitic). Some areas under Matabuey and Jaguar soil types are vegetated with high-graded secondary rather than primary forest. Given the small differences in gas emissions measured in an earlier study between fully cut-over secondary forest and primary forest (Keller and Reiners, 1994), and given the limited amount of disturbance in the mapped area, the primary forest rates were used for the entire area.

Delineation of hillslope positions is laborious and highly subjective when done by hand from contour lines on hard copy maps. Furthermore, the best topographic maps available in this case were at 1:50,000 scale with 20 m contour intervals of dubious quality. Therefore, we followed a terrain analysis procedure using high quality GIS data archived for La Selva Biological Station. GIS data included: 1) station boundaries, 2) streams, 3) soil types, and 4) a 100 m × 50 m lattice elevation data set augmented with other data for stream intersections, etc. for a total of 8,515 points used to model topography. No error estimations accompanied these coverages, but data were based on recent, professional ground surveys.

The land surface was modeled with two techniques, by kriging and by creating a triangulated irregular network (TIN). For the kriging approach, the nearest 12 sample points within a 200 m radius were used to interpolate the z value at each mesh point. After making a series of trial calculations, the exponential model was chosen to best represent the surface. For the TIN approach, a triangulated network was created with a proximal tolerance of 5 m to exclude points falling within the specified distance of other points. With both approaches, a 5 m mesh point spacing was specified. Both approaches were done with standard Arc/Info commands (ESRI, Redlands, California, USA).

The Fels and Matson (1996) algorithm to compute landscape position was applied to both surface models using a 19 × 19 neighborhood. With a 5 m cell size, the neighborhood was 95 m on a side and equal to the average ridge to valley distance in the east-west direction. The landscape position values for each surface model were then classified into one of three categories: hilltop, slope or swale. In both cases, the thresholds between classes were determined by carefully examining profiles of elevation and landscape position and adjusting each of the hillslope position thresholds so that the desired features were selected. In practice, for a given threshold, some desired features were excluded, while some undesired features were included. We attempted to optimize the thresholds to minimize undesired features in each hillslope position. For the surface modeled using kriging, the landscape positions were classified as: $-1.000 \leq$ ridge ≤ -0.035; $-0.035 <$ slope ≤ 0.030; $0.030 <$ swale ≤ 1.000. For the surface modeled using TIN, the landscape position classification was: $-1.000 \leq$ ridge ≤ -0.070; $-0.070 <$ slope ≤ 0.045; $0.045 <$ swale ≤ 1.000. Because the thresholds for the landscape position values calculated for the two surface models were significantly different, the landscape position values for the surface modeled using TIN were also classified using the thresholds for the surface modeled using kriging.

One resulting terrain model derived with the TIN function is shown in Figure 1. The areas of hilltops, slopes and swales, so defined, were calculated and multiplied by mean gas fluxes from the respective topographic positions to estimate spatially integrated rainy season gas fluxes through the forested parts of La Selva on Matabuey and Jaguar soils.

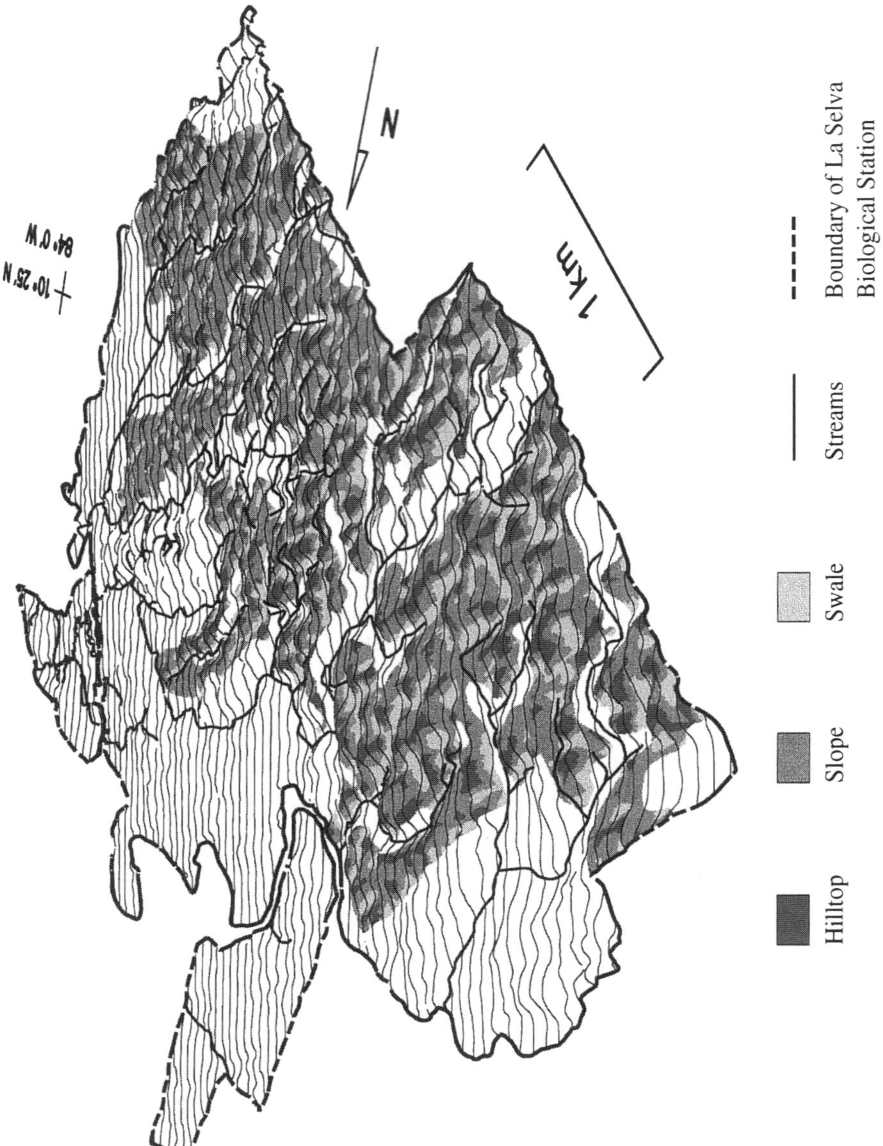

Fig. 1. Oblique view of the La Selva property (hatched pattern) within which the defined landscape is shown in shaded patterns representing hilltop slope and swale hillslope positions. This representation is derived from the TIN1 model (see text).

3. Results

3.1. Gas Emission Rates

The means and standard deviations for soil moisture, NH_4^+, NO_3^-, and emission rates of N_2O and CH_4 are given in Table I. All variables except NH_4^+ were significantly different between pasture and forest sites. Forest sites were wetter, had higher NO_3^- concentrations, higher N_2O fluxes and, on average, consumed CH_4, whereas pasture sites, on average, produced CH_4.

TABLE I

Means (n = 3; ± standard deviations) of soil moisture (% dry mass), NH_4^+-N (mg N kg^{-1} dry soil), NO_3^--N (mg N kg^{-1} dry soil), N_2O emission rate (ng N cm^{-2} h^{-1}) and CH_4 emission rate (mg m^{-2} d^{-1}) for hilltop, slope and swale locations of primary forest and actively grazed pasture sites. Within each column, values that are significantly different from one another are noted by different superscripts.

Position	Soil moisture	Soil NH_4^+-N	Soil NO_3^--N	N_2O flux	CH_4 flux
		Pasture sites			
Hilltop	0.70 ± 0.01^a	8.74 ± 0.52^a	1.87 ± 1.29	4.34 ± 5.28	-0.14 ± 0.52
Slope	0.68 ± 0.03^a	7.50 ± 2.71^a	0.87 ± 0.40	1.12 ± 0.31^a	1.13 ± 0.52
Swale	0.73 ± 0.05^b	15.47 ± 5.25^b	1.08 ± 0.57	2.54 ± 0.78^b	1.09 ± 0.49
Grand Mean	0.70 ± 0.02^a	10.57 ± 2.37	1.27 ± 0.47^a	2.67 ± 2.75^a	0.69 ± 0.02^a
		Forest sites			
Hilltop	0.79 ± 0.03^a	8.58 ± 2.09	5.79 ± 1.72	4.85 ± 0.80	-0.76 ± 0.10
Slope	0.72 ± 0.03^b	8.75 ± 0.18	5.66 ± 1.02	10.11 ± 3.56^a	-1.18 ± 0.68
Swale	0.83 ± 0.10^c	10.15 ± 3.43	7.50 ± 2.77	20.84 ± 17.08^b	-1.11 ± 0.49
Grand Mean	0.78 ± 0.04^b	9.16 ± 0.88	6.32 ± 0.88^b	11.94 ± 8.71^b	-1.02 ± 0.30^b

$p < 0.025$ for soil moisture inter-land use comparison, $p < 0.05$ for soil moisture inter-position comparison, $p < 0.05$ for NH_4^+-N inter-position comparison, $p < 0.005$ for inter-land use comparison of NO_3^--N, $p < 0.1$ for inter-land use comparison of N_2O, $p < 0.01$ for inter-land use comparison of CH_4, $p < 0.013$ for slope vs swale position comparison.

Among the three environmental variables, only soil moisture and NH_4^+ concentrations were significantly different among slope positions by a general ANOVA F-test. In the pasture, soil moisture was less on hilltops and slopes than on swale positions; in the forest, soil moisture was least on slopes, intermediate on hilltops, and greatest in swales. Ammonium was significantly higher in swales compared with hilltop and slope positions in the pasture sites only.

Nitrous oxide emission rates varied by a factor of 3.9 between slope positions in the pasture sites (hilltop/midslope), and by a factor of 4.3 in the forest sites (swale/hilltop), but none of these differences was significant by the ANOVA F-test (Table I). Methane emission rates varied by a factor of 9.1 in the pasture sites (midslope/hilltop), and by a factor of 1.6 in the forest sites (midslope/hilltop), but none of these slope position differences was significant by the ANOVA. The interaction between land use type and slope position was significant for CH_4 emission. In pastures, emissions tended to increase in a positive direction in a down-slope direction, whereas in forests, emissions tended to increase in a negative direction in a down-slope direction.

Our research hypotheses for CH_4 and N_2O emissions were, in both cases, expressed by the following ranking of means: lowest--slopes, middle--hilltops, highest--swales. Methane can be both consumed or produced from soil so that "low" emission rate would mean a production rate close to zero, or even negative. For CH_4 emission, the relevant column of Table I shows little evidence in favor of our hypothesis. Only the slope/swale relationship in pasture sites is in the right direction (more CH_4 production on slopes than swales), but not significantly so. For N_2O, our hypothesis fared better. The relevant column of Table I reveals that hilltops do not follow our suggested pattern, but that the relationships between slopes and swales are supportive in both pastures and forests. A one-tailed paired t-test on log-transformed means (swale minus slope within each site) yielded a p-value of 0.013 in support of our hypothesis (mean difference 0.687, standard error 0.213).

3.2. Areal Emission Estimates

Estimates of areas in the three slope positions on Matabuey and Jaguar soils, and the gas fluxes associated with those positions are given in Table II. Both the Krige and TIN1 models classified 54-56 % of the area into hillslope positions, and 20-25 % of the area into each of hilltops and swales. The TIN2 model divided the area nearly evenly into hilltops, slopes and swales.

Even for CH_4, which did not have significantly different rates among hillslope positions (Table I), the means were still the best predictors for describing hillslope positional effects. Therefore, emission means for both of the two gases were multiplied by areas of each of the hillslope positions as derived from each of the three terrain models to calculate areal flux estimates over the defined landscape. Total fluxes for the designated areas, and the percentages of map area fluxes contributed by each topographic position with the intermediate estimates pertaining to the individual gases, are given in Table II.

4. Discussion

4.1. Comparisons with Previous Estimates

It is impossible to make comparisons of our slope and swale results with those of other studies as such measurements have never been published for Costa Rica or anywhere else of which we are aware. However, we can compare hilltop values of the present study with rainy season (April through December) soil conditions and gas emissions published earlier for the same sites (Keller and Reiners, 1994). Twelve months of rainy season soil concentration data from the forest hilltop sites from the earlier work averaged 4.05 and 8.85 mg N kg^{-1} dry soil for NH_4^+ and NO_3^-, respectively, compared with 8.58 and 5.79 mg N kg^{-1} dry soil in the rainy season of this current study. The same comparisons for active pasture hilltop sites are 4.46 and 1.71 mg N kg^{-1} dry soil in the earlier study for NH_4^+ and NO_3^-, respectively, versus 8.74 and 1.87 mg N kg^{-1} dry soil in the current work. Soil NH_4^+ concentrations tended to be higher in the current than the previous study, but NO_3^- was quite comparable between the two study periods.

TABLE II

Estimated areas and gas fluxes for hillslope positions based on three different terrain classification models: one based on kriging, a second based on the TIN process with its own positional definitions (TIN1), and a third based on the TIN process, but with the krige positional definitions (TIN2)(see text for GIS methods). Areas modeled are restricted to forested Matabuey and Jaguar soil types of La Selva Biological Station. Nitrous oxide emission rates are in ng N cm^{-2} hr^{-1}; CH_4 emission rates are in mg CH_4 m^{-2} d^{-1}. Areal fluxes for N_2O are g N da^{-1}; for CH_4 are g CH_4 da^{-1}. Percentages of total flux are calculated only for the case having the median estimate of total areal flux. Standard errors of area flux are calculated as the standard error of the flux rate (Table I) multiplied by the product of flux rate and area.

Landscape position	Landscape model	Area (ha)	% total area	Average flux rate	Total area flux	Percent total flux	Std. error area flux
			Nitrous oxide				
Hilltop	Krige	159	26	4.85	1,851		176
	TIN1	127	21	4.85	1,478		141
	TIN2	229	37	4.85	2,666	16	254
Slope	Krige	335	54	10.11	8,128		1,652
	TIN1	347	56	10.11	8,420		1,712
	TIN2	205	33	10.11	4,974	30	1,011
Swale	Krige	124	20	20.84	6,202		2,934
	TIN1	144	23	20.84	7,202		3,408
	TIN2	184	30	20.84	9,203	55	4,354
			Methane				
Hilltop	Krige	159	26	-0.76	-1,208		92
	TIN1	127	21	-0.76	-965		73
	TIN2	229	37	-0.76	-1,740	28	132
Slope	Krige	335	54	-1.18	-3,953		1,315
	TIN1	347	56	-1.18	-4,095		1,362
	TIN2	205	33	-1.18	-2,419	60	805
Swale	Krige	124	20	-1.11	-1,376		351
	TIN1	144	23	-1.11	-1,598		407
	TIN2	184	30	-1.11	-2,042	21	52

Comparisons of gas fluxes for hilltop sites can be made the same way. Nitrous oxide emission rates in the same hilltop sites averaged over nine months in the rainy season in the earlier study were 8.58 and 2.07 ng N cm^{-2} hr^{-1} in the forest and pasture, respectively, compared with 4.85 and 4.34 ng N cm^{-2} hr^{-1} in the present study. Similarly, CH_4 emission rates in the same hilltop sites averaged over nine months in the rainy season in the earlier study were -1.05 and 0.52 mg m^{-2} da^{-1} in the forest and pasture, respectively, compared with -0.76 and -0.14 mg m^{-2} da^{-1} in the present study. While some of these differences are as much as two-fold, the absolute differences are remarkably small, given the possible weather differences on sampling dates between years, and given the range of sampling variation encountered in this kind of work (see standard errors in Table I).

4.2. Slope Position Effects

The data supported our hypothesis that swales were the wettest positions in both pasture and forest sites. As hypothesized, the slope positions were the driest positions in forest sites, but they were not drier than hilltops in the pasture sites. As hypothesized, NH_4^+ concentrations were highest in swales in the pastures, but the differences were not significant in the forest sites. The results for NO_3^- concentrations did not support our hypotheses in either forest or pasture sites.

Six specific slope position hypotheses were posited for each of the two gases (2 sites × 3 comparisons). In the case of CH_4, expectations were met in three of the six cases but none of the differences was significant. In the case of N_2O, expectations were met in four of six cases, two of these being significant.

4.3. Areal Estimates

Area fluxes by hillslope position varied considerably depending on the terrain classification model used, but the sum of total area fluxes over the entire defined landscape varied only by 5 % for N_2O, and by 7 % for CH_4 (range/median total) (Table II). The Krige-based landscape model can be considered an intermediate classification of the terrain in having intermediate areal percentages of hilltop and slope estimates. The following discussion is based on results for that model. With the Krige model, hilltops comprised 26 % of the land area, but contributed only 11 % of the total N_2O flux and 18 % of the CH_4 flux. Conversely, swales which comprised only 20 % of the land area by the Krige model, contributed 38 % and 21 % of total N_2O and CH_4 fluxes, respectively. Clearly, if estimations of trace gas fluxes over the entire landscape were based on only one hillslope position, they would improperly estimate landscape fluxes. If, for example, we were to multiply our hilltop average N_2O flux of 4.85 ng N cm^{-2} hr^{-1} by the total area of 618 ha, the estimated flux from the total area would be only 7,193 g da^{-1}, rather than the total of 16,181 g da^{-1} estimated by taking landscape variation into account according to the Krige model. This 2.24-fold difference underlies the importance of assessing topographic variations for making areal flux estimates in this environment.

4.4. ERROR SOURCES FOR ESTIMATING AREAL GAS FLUXES

Sources of error inherent in this and similar studies include: random errors in field sampling, transport, storage and transfer of gas samples; error in estimating mean emission rates for specific hillslope positions, errors in the GIS data, and judgmental error in decomposing the landscape into slope positions. Errors inherent in actual chromatographic analyses are themselves very small. On the other hand, errors inherent in flux calculations for individual plot measurements (see methods) are large; for example, 10 % and 66 %, respectively, of mean flux values for N_2O and CH_4, for forest hilltop sites. These are significant sources of error but are likely to be randomly distributed among the samples. Variances associated with estimating hillslope position emission averages are given as standard deviations in Table I. For forest data, standard deviations range from 16 to 82 % of the means for N_2O, and from 13 to 44 % of the means for CH_4. Because these means and standard deviations are based on only three replicates, it is impossible to evaluate normality and to know whether the error is random or systematic.

We have no way to evaluate errors in the original GIS data, but we can evaluate the effects of our judgement in defining slope position the three terrain classification models. The ranges between the three estimates for each slope position areal flux can be compared with their associated standard errors derived from sampling variance (Table II). For both N_2O and CH_4, in all cases but one, the range of areal flux within slope position was greater than any individual standard error. The one exception was the swale estimate for N_2O. Therefore, in almost all cases, the choice of a partitioning model is more important than sampling variance in estimating areal fluxes for individual slope positions.

The situation is reversed, however, when estimating areal fluxes from the entire landscape (Table III). The ranges in total area flux estimations were determined by subtracting the flux rate derived from the landscape partitioning model having the minimum flux from the partitioning model having the maximum flux. For N_2O, the difference is 919 g N da^{-1}; for CH_4, the difference is 457 g CH_4 da^{-1}. Both of these differences are smaller than their respective pooled standard errors of the means calculated as the square root of the sum of the variances divided by the square root of the replicates (three). Thus, at the level of the entire landscape, between-site sampling variation is a more important source of potential error than is the choice of landscape partitioning model.

TABLE III
Combinations of viable slope position models producing maximum ranges of N_2O-N and CH_4 gas flux for the total study area.

Relative flux rate	Landscape model	Total flux (g da^{-1} ± std. error)
N_2O maximum	TIN1	17,100 ± 3,816
N_2O intermediate	TIN2	16,843 ± 4,477
N_2O minimum	Krige	16,181 ± 3,372
CH_4 maximum	TIN1	-6,658 ± 1,423
CH_4 intermediate	Krige	-6,537 ± 1,364
CH_4 minimum	TIN2	-6,201 ± 967

From these analyses, we find that the relative importance of error contributed by either average hillslope position emission rate estimation or choice of landscape partitioning, depends on the level of generalization examined. At the level of estimating mean emission rates for hillslope positions, the terrain partitioning factor was usually more important, but at the level of estimating fluxes over the entire landscape level, positional mean emission estimation was always larger. In general, both are important. For N_2O, the standard error is 14.3 % of the intermediate flux mean, while the landscape partitioning range (919) is

14 % of the intermediate flux estimate (Table III). For CH_4, the standard error is 10.6 % of the intermediate flux mean, while the landscape partitioning range (457) is 7 % of the intermediate flux estimate. Clearly, future studies of this type must make every effort to produce both the best hillslope position estimate possible, as well as the best possible relationship between actual emission rates and classes of hillslope conditions, however defined.

Acknowledgments

This research was supported by NSF grants BSR88-07911, BSR91-08035, and DEB 94-9408802; and by EPA grant DW12935643-01 from the Environmental Research Program, Athens, GA. We thank Christine Pfeifer for her performance and assistance in the initial organization of the study, and M. Mitre, M. Nunez, R. Vargas, D. Villegas and G. Wright for their assistance in the field and laboratory. We also thank Jennifer Powers at the La Selva Biological Station for assisting in acquisition of GIS data and Robert Thurston for performing the GIS analyses.

References

Bender, M. and Conrad, R.: 1994, *Biogeochemistry* **27**, 97.

Born, M., Dorr, H. and Levin, I.: 1990, *Tellus, Series* B **42B**, 2.

Bouwman, A.F., Fung, I., Matthews, E. and John, J.: 1993, *Global Biogeochem. Cycles* **7**, 557.

Bouwman, A.F. and Taylor, J.A.: 1996, *Global Biogeochem. Cycles* **10**, 307.

Bowden, W.B., McDowell, W.H., Asbury, C.E. and Finley, A.M.: 1992, *Biogeochemistry* **18,** 77.

Bowden, R.D., Castro, M.S., Melillo, J.M., Steudler, P.A. and Aber, J.D.: 1993, *Biogeochemistry* **21,** 61.

Conrad, R.: 1996, *Microbiol. Rev.* **60,** 609.

Davidson, E.A.: 1991, 'Fluxes of Nitrous Oxide and Nitric Oxide from Terrestrial Ecosystems', in J.E. Rogers and W.B. Whitman, (eds.), *Microbial Production and Consumption of Greenhouse Gases: Methane, Nitrogen Oxides, and Halomethanes,* American Society of Microbiology, Washington, D.C., pp. 219-235.

Davidson, E.A., Matson, P.A., Vitousek, P.M., Riley, R., Dunkin, K., Garcia-Mendez, G. and Maass, J.M.: 1992, *Ecology* **74,** 130.

Erickson, H.E. and Keller, M.: 1997, *Soil Use and Mgmt.* (in press).

Fels, J.E. and Matson, K.C.: 1996, 'A Cognitively-based Approach for Hydrogeomorphic Land Classification Using Digital Terrain Models,' in *Proceedings of the Third International Conference on Integrating GIS and Environmental Modeling*, National Center for Geographic Information and Analysis, Santa Barbara, CA. [CD]

Folonoruso, O.A. and Rolston, D.E.: 1984, *Soil Sci. Soc. Amer. J.* **48**, 1214.

Galchenko, V.F., Lein, A. and Ivanov, M.: 1989, 'Biological Sinks of Methane', in M.O. Andreae and D.S. Schimel (eds.), *Exchange of Trace Gases Between Terrestrial Ecosystems and the Atmosphere*, John Wiley & Sons, Ltd., Chichester, pp. 59-71.

Groffman, P.M. and Tiedje, J.M.: 1989a, *Soil Biol. and Biochem.* **21,** 613.
Groffman, P.M. and Tiedje, J.M.: 1989b, *Soil Biol. and Biochem.* **21**, 621.
Groffman, P.M., Tiedje, J.M., Mokma, D.L. and Simkins, S.: 1992, *Landscape Ecology* **7,** 45.
Groffman, P.M., Rice, C.W. and Tiedje, J.M.: 1993a, *Ecology* **74,** 855.
Groffman, P.M., Zak, D.R., Christensen, S., Mosier, A. and Tiedje, J.M.: 1993b, *Ecology* **74,** 1579.
Groffman, P.M. and Turner, C.L.: 1995, *Landscape Ecology* **10**, 255.
Hartshorne, G.S. 1983, 'Plants', in D.H. Janzen (ed.), *Costa Rican Natural History*, University of Chicago Press, Chicago, pp. 118-157.
Houghton, J.T., Neira Filho, L.G., Callander, B.A., Harris, N., Kattenberg, A. and Maskell, K. (eds.), *Climate Change 1995. The Science of Climate Change*, Cambridge University Press, Cambridge.
Hutchinson, G.L. and Livingston, G.P.: 1993, 'Use of Chamber Systems to Measure Trace Gas Fluxes', in L.A. Harper, A.R. Mosier, J.M. Duxbury and D.E.Rolston (eds.), *Agricultural Ecosystem Effects on Trace Gases and Global Climate Change*. American Society of Agronomy, Madison, pp. 63-78.
Khalil, M.A.K. and Rasmussen, R.A.: 1992, **97**, 14,651.
Keeney, D.R. and Nelson, D.W.: 1982, 'Nitrogen - Inorganic forms', in A.L. Page, R.H. Miller and D.R. Keeney (eds.), *Methods of Soil Analysis. Part 2. Chemical and Microbiological Properties*, American Society of Agronomy, Madison, pp. 643-709.
Keller, M., Kaplan, W.A. and Wofsy, S.C.: 1986, *J. Geophys. Res.* **91**, 11,791.
Keller, M., Mitre, M.E. and Stallard, R.F.: 1990, *Global Biogeochem. Cycles* **4**, 21.
Keller, M., Veldkamp, E., Weitz, A.M. and Reiners, W.A.: 1993, *Nature* **365**, 244.
Keller, M., and Reiners, W.A.: 1994, *Global Biogeochem. Cycles* **8,** 399.
Kicklighter, D.W., Melillo, J.M., Peterjohn, W.T., Rastetter, E.B., McGuire, A.D., Steudler, P.A. and Aber, J.D.: 1994, *J. Geophys. Res.* **99**, 1303.
King, A. W.: 1991, 'Translating Models Across Scales in the Landscape', in M.G. Turner and R.H. Gardner (eds.), *Quantitative Methods in Landscape Ecology*, Springer-Verlag, New York, pp. 479-517.
Li, C., Frolking, S. and Frolking, T.A.: 1992, *J. Geophys. Res.* **97,** 9777.
Livingston, G.P., Vitousek, P.M. and Matson, P.A.: 1988, *J. Geophys. Res.* **93D2,** 1593.
Matson, P.A. and Vitousek, P.M.: 1990, *BioScience* **40,** 667.
Matson, P.A., Volkmann, C., Coppinger, K. and Reiners, W.A.: 1991, *Biogeochemistry* **14**, 1.
Minitab, Inc.: 1996, *Release 11 for Windows*, Minitab, Inc., State College, PA.
Mosier, A.R., Klemedtsson, L.K., Sommerfeld, R.A and Musselman, R.C.: 1993, *Global Biogeochem. Cycles* **7**, 771.
Mosier, A.R., Parton, W.J., Valentine, D.W., Ojima, D.S., Schimel, D.S. and Delgado, J.A.: 1996, *Global Biogeochem. Cycles* **10,** 387.
Parsons, W.F.J., Mitre, M.E., M. Keller, M.E. and Reiners, W.A.: 1993, *Biogeochemistry* **22,** 179.
Parton, W.J., Mosier, A.R. and Schimel, D.S.: 1988, *Biogeochemistry* **6,** 45.
Peterjohn, W.T. and Schlesinger, W.H.: 1991, *Soil Sci. Soc. Amer. J.* **55**, 1694.
Rastetter, E.B., King, A.W., Cosby, B.J., Hornberger, G.M., O'Neill, R.V. and Hobbie, J.E.: 1992, *Ecol. Applic.* **2**, 55.

Reiners, W.A., Bouwman, A.F., Parsons, W.F.J. and Keller, M.: 1994, *Ecol. Applic.* **4,** 363.
Robertson, G.P.,Huston, M.H., Evans, F.C. and Tiedje, J.M.: 1988, *Ecology* **69**, 1517.
Sanford, R.L., Jr., Paaby, P., Luvall, J.C. and Philips, E.: 1993, 'The La Selva Ecosystem: Climate, Geomorphology, and Aquatic Systems', in L.A. McDade, K.S. Bawa, H.A. Hespenheide, and G.S. Hartshorne (eds.), *La Selva: Ecology and Natural History of a Tropical Rain Forest*, University of Chicago Press, Chicago, pp. 19-33.
Sollins, P., Sancho, F., Mata, R., and Sanford, R.L.: 1993, 'Soils of La Selva', in L.A. McDade, K.S. Bawa, H.A. Hespenheide, and G.S. Hartshorne (eds.), *La Selva: Ecology and Natural History of a Tropical Rain Forest*, University of Chicago Press, Chicago, pp. 34-53.
White, J.D. and Running, S.W.: 1994, *J. Vegetation Sci.* **5,** 687.

GIS-BASED EXTRAPOLATION OF LAND USE-RELATED NITROUS OXIDE FLUX IN THE ATLANTIC ZONE OF COSTA RICA

ROEL A.J. PLANT

Laboratory of Soil Science and Geology, Wageningen Agricultural University, P.O. Box 37, 6700 AA Wageningen, The Netherlands

Abstract. I estimated the areal nitrous oxide (N_2O) flux from 281,347 ha of Costa Rican lowland covered with primary and secondary forest, pastures and banana plantations by linking the DeNitrification-DeComposition (DNDC) model with a Geographic Information System (GIS). Generalized soil, texture and land use maps were overlaid to yield unique combinations of N_2O flux control factors. Overlay patches were associated with the nearest of seven available meteorological stations. Monte Carlo-based sensitivity analysis was used to identify DNDC's key driving variables and required map attributes. Clay content, initial soil organic carbon (SOC), bulk density, and pH were selected as key driving variables. For 217 patch classes, DNDC simulations were carried out with climate data for seven different years. The estimated average areal flux was 6.8 kg N_2O-N ha^{-1} yr^{-1}. Possible applications of the GIS-DNDC interface presented include estimation of long-term areal flux dynamics from a changing land use mosaic, and prediction of areal fluxes resulting from alternative land use scenarios.

Keywords: nitrous oxide, land use change, GIS, simulation modeling, regional modeling, Costa Rica

1. Introduction

The known sources and sinks of nitrous oxide (N_2O), a greenhouse gas and stratospheric ozone destructor, are imbalanced (Prather *et al.*, 1995) and atmospheric N_2O concentrations are increasing at a rate of 0.25 - 0.31 % yr^{-1} (Prinn *et al.*, 1990). Tropical soils have been recognized as the largest known N_2O source. Soils of tropical forests alone may produce 20 - 41 % of the total annual production of 14.7 Tg N_2O-N (1 Tg = 10^{12} g) (Matson and Vitousek, 1990).

Land use controls soil N_2O emissions (Keller and Reiners, 1994), and therefore land use changes resulting from human activity may significantly alter regional and global N_2O budgets (Luizao *et al.*, 1989; Keller *et al.*, 1993). In the future, land use planning may be used effectively, in principle, to mitigate N_2O emissions. Much of the land use change over recent decades has involved conversion of tropical forests to high-input agriculture and low-input pasture. Conversion to low-input pasture is especially common in South and Central America (Hecht, 1992) and causes N_2O fluxes to increase by a factor of 5 - 8 initially (Keller *et al.*, 1993). About a decade after deforestation, however, fluxes decline below primary forest levels.

In high-input tropical agriculture, many crops receive N fertilizer in excess of plant demand (Keller and Matson, 1994). Current global estimates of N_2O flux from fertilized agricultural soils are based on studies in temperate zones (Eichner, 1991). However, the few studies on N_2O flux from tropical agriculture conducted so far (Garcia-Mendez *et al.*, 1991; Matson *et al.*, 1996; Veldkamp and Keller, 1997) suggest that this source type may be severely underestimated.

To assess effects of land use changes on N_2O emissions, short-term flux measurements at plot or field scales must be aggregated over longer temporal cycles and larger spatial

Water, Air, and Soil Pollution **105:** 131–141, 1998.

units. However, extrapolation is thwarted by the great spatial and temporal variability in small-scale N_2O fluxes (Folorunso and Rolston, 1984). Moreover, areal flux estimates often are based on few measurements because sampling logistics are complicated. Current N_2O budgets, therefore, may be highly inaccurate (Matson *et al.*, 1989).

Although several extrapolation methods exist, a lack of rules to guide extrapolation presently is a key conceptual problem (Schimel and Potter, 1995). The simplest approach to obtain an areal flux estimate is to sample an area randomly at a given time. Because of the extreme space-time variability of N_2O flux, a large number of samples may be required to achieve an acceptable level of error. Therefore, this method is of limited practical use.

To increase sampling efficiency, ecological variations may be used to stratify a study area. This "flux × area" approach employs ecological variations as the single dominant factor controlling N_2O flux (e.g., Matson and Vitousek, 1990). For all ecosystem groups, representative fluxes are calculated from a limited number of measurements. The sum of "flux × area" products yields the areal N_2O flux estimate. This approach is straightforward, yet potentially inaccurate; ecosystem groups still may be heterogeneous with respect to soil and climate. The method is useful, however, if no geographically explicit information on ecosystem distribution is available. For example, Keller and Matson (1994) used tabular agricultural census data on land use distribution to explore effects of 40 years of land use change on areal N_2O flux in the Atlantic Zone of Costa Rica.

A more comprehensive extrapolation method employs a Geographic Information System (GIS) to overlay multiple factors controlling N_2O flux, and a process-based model to simulate fluxes for all different, unsampled factor combinations (e.g., Burke *et al.*, 1990; Schimel *et al.*, 1990; Potter *et al.*, 1996).

This paper presents a GIS-based assessment of areal N_2O flux from tropical forest, low-input pastures and high-input banana plantations in the Northern Limón Province, Costa Rica. The assessment employed the DeNitrification-DeComposition (DNDC) simulation model (Li *et al.*, 1992a,b, 1994a,b), interfaced with an extant GIS on soils, climate, and land use (Wielemaker and Vogel, 1993; Stoorvogel, 1995). Primary goals were: 1) to optimize the GIS-DNDC link by identifying key driving variables, and 2) to compare the GIS-based areal flux estimate with a more straightforward "flux × area" estimate.

2. Materials and Methods

2.1. DNDC Model

Simulation models have been developed to simulate soil processes regulating N_2O production and emission at the microsite scale (Focht, 1974; Leffelaar and Wessel, 1988), the field scale (Li *et al.*, 1992a,b), and the ecosystem scale (Parton *et al.*, 1996; Potter *et al.*, 1996). A key issue in selecting a model for linkage with a GIS is the model's ability to simulate short-term flux dynamics (Schimel and Potter, 1995). From microsite studies, it is well known that flux responses to short-term variations in soil wetting and drying are critical. Therefore, the model should feature short-term representations of microbial growth and electron transfer reactions. On the other hand, the model also should use inputs that are readily available from a standard GIS on soils.

DNDC (Li *et al.*, 1992a; Li *et al.*, 1994a,b) is among the models that satisfy both requirements and thus can, in principle, be interfaced with a GIS. DNDC is an integrated one-dimensional model of field-level C and N dynamics in arable crop-soil systems, with a strong focus on denitrification and N oxide emissions. A soil climate submodel calculates hourly soil water content, soil temperature profiles, and transpiration. Daily rates of soil respiration in three active pools, mineralization, nitrification, ammonification, ammonia volatilization and ammonium adsorption are simulated by a decomposition submodel. A denitrification submodel calculates hourly production and emission of NO, N_2O, and N_2. A crop/land use submodel calculates daily plant N uptake, litter and root recycling, and accounts for C and N additions through fertilization and manure amendment. DNDC features numerous internal and external parameters. The readily available external parameters, however, are known to adequately cover the major factors that influence regional variations in N_2O emissions (Li *et al.,* 1992b).

2.2. GIS - Map Generalizations

The spatial distribution of DNDC's key driving variables was represented by three coverages; a land use map providing information on crop/vegetation type and management characteristics, a soil map providing data on soil physical and chemical characteristics, and a texture map providing data on soil hydraulic parameters. The maps were created using data from a 1:150,000 soil survey for the Northern Atlantic Zone of Costa Rica stored in a regional GIS (ARC/INFO, ESRI, Redlands, CA) (Wielemaker and Vogel, 1993). The GIS recently was extended with a 1992 land use layer and climate data from 14 meteorological stations (Stoorvogel, 1995). For this study, an area covering 281,347 ha was selected from the GIS (Figure 1A).

In the GIS attribute database, mapping units on the original soil coverage are described by one to five non-georeferenced terrain units. Terrain units are unique combinations of terrain properties, including a soil series. Initially, 75 different soil series were identified and described by 300 representative profiles (Wielemaker and Vogel, 1993); 123 representative profiles were analyzed for basic soil physical and chemical properties. Stoorvogel (1995) developed a new structure for this GIS attribute database that includes a rule base for generalizations at different hierarchical levels (i.e., mapping unit, pedon, and soil horizon).

Using this rule base, soil series were generalized to eight functional soil groups at the pedon level, each having a similar agricultural potential (Stoorvogel, 1995). At the mapping unit level, a generalization by dominant soil group (i.e., the functional soil group covering the largest fraction of the mapping unit) was carried out. A similar generalization was made to derive a soil texture map. A soil texture class (Wielemaker and Vogel, 1993) was linked to the soil series at the pedon level. After grouping the soil series by texture class, mapping units were generalized by dominant texture class. The original classification of the 1992 land use coverage features "primary forest," "secondary forest," and "banana plantation" classes (Stoorvogel, 1995). The five different pasture classes were generalized to one single "pasture" class. All other land use types were reclassified as "other use."

A

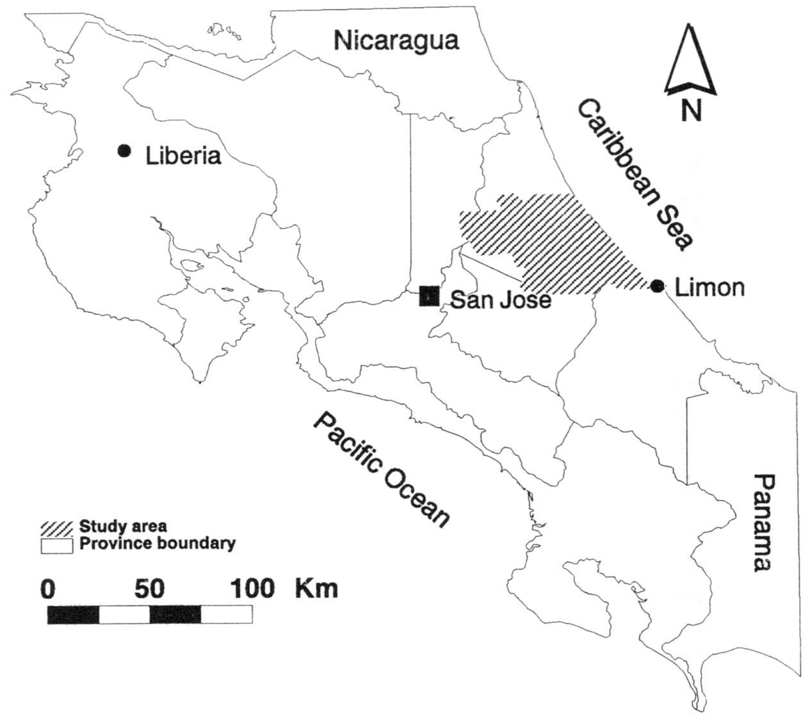

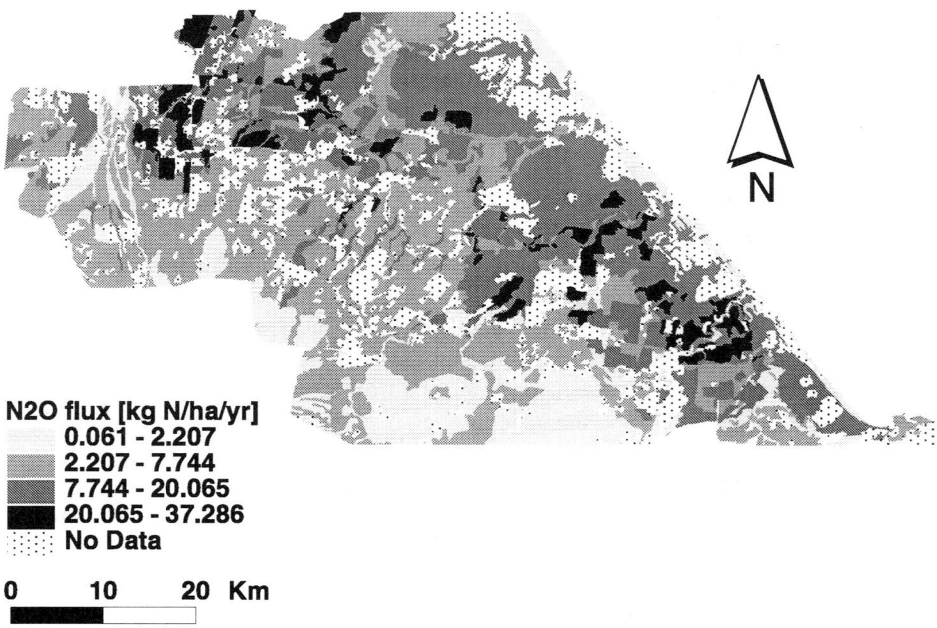

Fig. 1. Study area (A) and average regional distribution of N_2O flux (1982 - 1988) (B).

2.3. Sensitivity Analysis

To set up attribute tables for the generalized maps, knowledge about the sensitivity of DNDC's driving variables is required. Ideally, all key driving variables, i.e. model inputs modulating major variations in simulated N_2O flux, should match entries in the GIS attribute database. In practice, however, only a few key driving variables may be readily available. If additional measurements and knowledge about functional relationships are available, missing key driving variables can, in principle, be calculated using pedo-transfer functions (Bouma, 1989).

Key driving variables were identified by Monte Carlo-based sensitivity analysis. Sensitivity analyses were carried out for all selected land use types. Version 1.1 of the UNCSAM software package (Janssen *et al.*, 1992) was used for samplings and analyses. UNCSAM draws samples from the statistical distributions of input variables using the Latin Hypercube sampling technique. Resulting input-output sets are summarized by ordinary and rank regression analysis. I used uniform distribution functions, parameterized with a minimum and maximum value, for all 11 parameters (cf. Table II) and ignored correlations between inputs. For all inputs, 100 samplings and corresponding one-year model runs were carried out. Based on the decision tree given by Janssen *et al.* (1992), the Standardized Regression Coefficient (SRC) was selected as an appropriate sensitivity measure.

2.4. GIS - Map Attributes

The GIS attribute database does not provide information on crop and management characteristics. Therefore, typical land use scenarios were created for selected land use types. To simulate forest emissions, DNDC was extended with a routine that accounts for water and N uptake by a forest cover, and monthly turnover of fine roots, litter and wood (i.e.,

TABLE I
Parameter settings for primary and secondary forest.

Parameter name	unit	value
ratio C:N leaves	[-]	66
ratio C:N wood	[-]	200
ratio C:N fine roots	[-]	45
turnover rate litter[†]	[1/month]	0.047
turnover rate wood[†]	[1/month]	0.004
turnover rate fine roots[†]	[1/month]	0.139
fraction leaves[†]	[-]	0.03
fraction wood[†]	[-]	0.96
fraction fine roots[†]	[-]	0.01
water requirement	[mm H_2O Mg C^{-1}]	52.6
C in full-grown forest[†]	[Mg ha^{-1}]	220

[†] Veldkamp, 1993

TABLE II

Sensitivity analysis results for primary forest (FP), secondary forest (FS), pasture (PA), and banana (BB banana -under plants; BC banana - between plants). R-square coefficients reflect the goodness of fit of the linear regression equation predicting N_2O flux from the independent variables (n = 100).

				Rank of Standardized Regression Coefficient (SRC)				
Parameter	unit	min	max	FP	FS	PA	BB	BC
Clay fraction	[-]	0.05	0.95	1	1	2	1	1
Bulk density	[Mg m^{-3}]	0.2	1.2	2	2	1	2	2
Fraction passive SOC‡	[-]	0.2	0.9	3	3	3	5	3
Initial SOC‡	[-]	0	0.1	4	4	4	10	4
Initial WFPS†	[-]	0.3	0.9	5	5	5	3	7
Initial NH_4^+	[ppm]	0	15	6	6	6	8	5
pH	[-]	3	8	7	7	7	4	6
Initial NO_3^-	[ppm]	0	15	8	8	9	7	9
WFPS† at field capacity	[-]	0.5	0.9	9	9	8	6	8
Saturated conductivity	[cm min^{-1}]	0.008	1.056	10	10	10	11	10
WFPS† at wilting point	[-]	0.1	0.4	11	11	11	9	11
R^2				0.52	0.49	0.40	0.49	0.57

† water-filled pore space; ‡ soil organic carbon

stems, branches and coarse roots). All turnover calculations are based on daily N uptake rate, which, in DNDC, is the key process linking vegetation growth with soil status (Li *et al.*, 1994b). The routine calculates monthly amounts of C returned to the soil from the daily N uptake rate using partitioning coefficients, C:N ratios and turnover rates (Table I). The initial amount of N in the vegetation is calculated from the forest age and the average amount of C in the full-grown forest (Table I) (Armson, 1979). The forest routine was tested for a primary and a secondary Costa Rican forest site (Keller and Reiners, 1994). For primary forest, a typical age of 150 yr was assumed. The typical age of secondary forest was set to 25 yr. Results (not shown) indicate that average simulated fluxes are within the range of measured fluxes. However, the observed seasonal N_2O flux patterns (Keller and Reiners, 1994) were poorly captured.

Fluxes of N_2O from pasture were simulated using parameters for traditional low-productive pasture. More than 50 % of the study area's tropical forest acreage cleared during the last 40 years has been put under this type of pasture (Veldkamp, 1993). No distinction was made between grazed and ungrazed pastures. Because pastures are rarely fertilized in the study area (Veldkamp *et al.*, in press), simulations were performed without N fertilizer input.

To simulate N_2O fluxes from banana plantations, I applied the modeling scheme I proposed earlier (Plant *et al.*, submitted). In short, the scheme consists of separate DNDC runs for under plant conditions where 360 kg N ha^{-1} yr^{-1} are added in 13 applications and decomposing litter on the soil surface is absent, and between plant conditions where no fertilizer is applied and plant residue is left to decompose on the soil surface. Simulated fluxes were weighted by the fraction of area influenced by the fertilizer application.

Functional soil groups are proxies for key driving soil physical and chemical variables. Likewise, texture classes are proxies for soil hydraulic parameters. Thus, lookup tables featuring parameter records for entries on the maps are required. Based on sensitivity analysis results, I selected key driving variables for the soil map attribute table. Means were calculated using the Soil Representative Analyses Table (SRAT) available from the GIS attribute database (Wielemaker and Vogel, 1993). No data were available for peat soils.

Variables not selected as key driving variables were assigned a representative value, derived from published sources, per land use type. For forest and pasture soils, parameters given by Reiners *et al.* (1994) were used. Soil characteristics given by Veldkamp and Keller (1997) were used for banana. If soil parameters were unavailable from published sources, I used DNDC defaults.

The generalized soil, texture, and land use maps were overlaid in ARC/INFO. Using polygon centroids and coordinates of seven selected meteorological stations, seven climate zones were created by assigning a nearest station identifier to all 3,180 patches on the overlay. A rainfall N concentration of 0.1 mg L^{-1}, which corresponds to 4 kg N ha^{-1} yr^{-1} with an annual rainfall of 4,000 mm, was assumed throughout the study area (Keller, pers. comm.).

In theory, $8 \times 15 \times 5 \times 7 = 4{,}200$ patch classes exist, but actually 304 occurred. After masking patch classes with land use class "other use" and/or soil code "peat", 217 classes remained for which one-year DNDC simulations were performed. To account for effects of inter-annual climate variability (Plant *et al.*, submitted), all DNDC simulations were carried out with 7 different annual climate records (1982 - 1988). Areal fluxes were calculated using the following equation

$$\hat{F} = \sum_{i=1}^{N} g(s_i, l_i, c_i) a_{s_i, l_i, c_i} \qquad (1)$$

where g is the function describing the relationship between soil parameters (s), land use parameters (l), climate parameters (c), and N_2O flux (i.e., DNDC), a_{s_i, l_i, c_i} is the fraction of area occupied by the ith combination of soil, land use and climate parameters (i.e., patch class), and N is the number of patch classes.

3. Results

3.1. Sensitivity Analysis

Based on the rankings given in Table II, clay content, bulk density and initial SOC were selected as key driving variables. In addition, pH was selected because this variable ranked as the fourth sensitivity source for banana plant base conditions. Although initial WFPS ranked higher than pH for this land use type, initial WFPS was not selected as a key variable because it is not a static soil attribute. Information on passive SOC was available neither from the SRAT, nor from published sources. Therefore, 70 % passive SOC was assumed for all land use types (Veldkamp, 1993). DNDC defaults were used for texture class attributes because they appeared to be not very sensitive.

3.2. Regional Distribution of N_2O Flux

The regional distribution of simulated fluxes is shown in Figure 1B. Fluxes ranged from 0.1 to 37.3 kg N_2O-N ha^{-1} yr^{-1}. The patterns in Figure 1B are, as expected, most obviously related to the land use mosaic. Towards the east side of the study area, simulated fluxes are somewhat higher. This possibly can be attributed to the abundance of moderately well and poorly drained soils in this area. No relationship between regional flux distribution and climate zones was observed.

Using equation (1), an average areal flux of 6.8 kg N_2O-N ha^{-1} yr^{-1} was calculated. Simulated areal fluxes ranged from 6.2 kg N_2O-N ha^{-1} yr^{-1} (1982 climate data) to 7.3 kg N_2O-N ha^{-1} yr^{-1} (1984 climate data) (standard deviation 0.4). Using the extent of the study area (281,347 ha), the total annual loss was estimated to be 1.8 - 2.1 Gg N_2O-N yr^{-1} (1 Gg = 10^9 g).

4. Discussion

4.1. Estimating Areal Fluxes

To illustrate the value of GIS-based areal flux estimation, I compared results with a "flux × area" estimate. Average measured fluxes for selected land use types were calculated using published data on 15 Costa Rican sites (Table III). On all sites, monthly surveys of flux over at least one year were conducted. Forest and pasture sites (Keller *et al.*, 1993; Keller and Reiners, 1994) were on similar Inceptisols. Banana sites (Veldkamp and Keller, in press) were both on Inceptisols and Andisols. To calculate a pasture average, only flux measurements on old active pastures were used. Fractions of area covered by each land use type were derived from the 1992 land use coverage. Summation of "flux × area" products yielded an areal flux of 4.5 kg N_2O-N ha^{-1} yr^{-1}, whereas the GIS-based estimate was 6.8 kg N_2O-N ha^{-1} yr^{-1}. Thus, the more explicit description of soil and climate variability within land use types modulated conditions favoring extremely high fluxes. Although patches featuring these extreme conditions may be small, they may strongly affect the areal flux estimate.

The difference between averaged representative flux measurements and simulated average fluxes is most evident for banana. The simulated average flux for banana (Table III) exceeds the two-sample average by a factor 2.4. This large difference may be caused by the contribution of fluxes from banana plantations on soils with both high bulk density and high SOC.

The great advantage of simulation analysis is that it allows flux estimates for a very large number of control factor combinations. Moreover, effects of inter-annual climate variations can be incorporated. The latter would hardly be possible in the field.

TABLE III

"Flux × area" and GIS-based areal flux estimate (value in parenthesis is standard deviation).

Land use type	Area (%)	# of field sites	# of delineated areas in GIS	Field Average (kg N_2O-N ha^{-1} yr^{-1})	Simulated Average (kg N_2O-N ha^{-1} yr^{-1})
Primary forest	20	4	264	6.4 (2.4) †‡	4.7 (4.1)
Secondary forest	14	3	719	3.7 (1.1) †	4.9 (3.6)
Pasture	36	6	1021	3.7 (3.6) ‡	4.4 (3.6)
Banana plantation	14	2	283	9.5 (4.8) §	22.6 (8.7)
Other use	16	0	894	-	-
Areal flux				4.5	6.8

† Keller and Reiners (1994); ‡ Keller et al. (1993); § Veldkamp et al. (in press)

The GIS-DNDC interface presented can be used to estimate past areal fluxes, and long-term areal flux dynamics, if digital land use maps are available for multiple decades. For Costa Rica, areal photo coverages on a national scale exist for 1952, 1960 and 1984. These photographs provide a sound basis for mapping past land use mosaics.

Another useful application may be the analysis of future land use scenarios. Trace gas emissions may become increasingly important in regional land use planning in the future, especially when high-input agriculture is becoming more common. Future land use scenarios may feature crops like manioc, maize, ornamentals, palmito, and papaya. At present, most common tropical crops are implemented in DNDC. However, model validation for each individual land use type may not always be possible because field measurements are lacking.

4.2. Improving the Extrapolation

The extrapolation presented can be improved in several ways. First, fluxes from recently cleared pastures were omitted. Young pastures are known to contribute significantly to the areal N_2O flux. Keller and Matson (1994) did include young pasture as a distinct land use class, and estimated an areal flux for 1993 of 10.0 kg N_2O-N ha^{-1} yr^{-1}. However, no spatial information on past land use is available as yet.

A second improvement would concern the forest routine. Although average simulated and measured forest fluxes (4.8 and 5.1 kg N_2O-N ha^{-1} yr^{-1}, respectively, Table III) are within the same range, a more explicit representation of forest biogeochemistry may better account for functional differences between primary and secondary forest. In the DNDC simulations presented (Figure 1B), primary forest fluxes systematically were underestimated, while fluxes from secondary forest systematically were overestimated.

Third, the base map generalizations introduced aggregation errors that may propagate to higher scale levels. Because soil properties were lumped by soil and texture class, within-class variations were averaged out. A more sophisticated approach would employ statistical descriptions of the distributions of key driving variables and correlations within each soil and texture class. Monte Carlo simulation could be used to calculate statistical expectations (i.e., means) of N_2O flux per class.

Acknowledgments

I thank Nico van Breemen, Johan Bouma, Steve Frolking and one anonymous reviewer for critical comments on earlier versions of this manuscript. I am grateful to the people of the WAU Research Programme on Sustainability in Agriculture, who provided the GIS coverages. Michael Keller and Ed Veldkamp are gratefully acknowledged for providing flux data. I also thank Changsheng Li, who helped in developing the forest routines. This work was supported by the Netherlands Foundation for the Advancement of Tropical Research through grant W77-150. The participation of Changsheng Li was supported by the National Science Foundation through grant DEB94-20612 (to the University of Wyoming).

References

Armson, K.A.: 1979, *Forest Soils: Properties and Processes*, University of Toronto Press, Toronto.
Bouma, J.: 1989, in B.A. Stewart (ed.), *Advances in Soil Science* 9, Springer-Verlag, New York, pp. 77-213.
Burke, I.C., Schimel, D.S., Yonker, C.M., Parton, W.J., Joyce, L.A. and Lauenroth, W.K.: 1990, *Landscape Ecol.* **4**, 45.
Eichner, M.J.: 1990, *J. Environ. Qual.* **19**, 272.
Focht, D.D.: 1974, *Soil Sci.* **118**, 173.
Folorunso, O.A. and Rolston, D.E.: 1984, *Soil Sci. Soc. Am. J.* **48**, 1214.
Garcia-Mendez, G., Maass, J.M., Matson, P.A. and Vitousek, P.M.: 1991, *Oecologia* **88**, 362.
Hecht, S.B.: 1992, in T.E. Downing, S.B. Hecht, H.A. Pearson and C. Garcia-Downing, (eds.), *Development or Destruction: the Conversion of Tropical Forest to Pasture in Latin America*, Westview, Boulder, Colorado, pp. 7-25.
Janssen, P.H.M., Heuberger, P.S.C. and Sanders, R.: 1992, *Report No. 959101004*, National Institute of Public Health and Environmental Protection, Bilthoven, The Netherlands.
Keller, M. and Matson, P.A.: 1994, in R.G. Prinn (ed.), *Global Atmospheric-Biospheric Chemistry*, pp. 103-118.
Keller, M. and Reiners, W.A.: 1994, *Global Biogeochem. Cycles* **8**, 399.
Keller, M., Veldkamp, E., Weitz, A.M. and Reiners, W.A.: 1993, *Nature* **365**, 244.
Leffelaar, P.A. and Wessel, W.W.: 1988, *Soil Sci.* **146**, 335.
Li, C., Frolking, S.E., Harriss, R.C. and Terry, R.E.: 1994a, *Chemosphere* **28**, 1401.
Li, C., Frolking, S. and Harriss, R.: 1994b, *Global Biogeochem. Cycles* **8**, 237.
Li, C., Frolking, S. and Frolking, T.A.: 1992a, *J. Geophys. Res.* **97**, 9759.
Li, C., Frolking, S. and Frolking, T.A.: 1992b, *J. Geophys. Res.* **97**, 9777.
Luizao, F., Matson, P., Livingston, G., Luizao, R. and Vitousek, P.: 1989, *Global Biogeochem. Cycles* **3**, 281.
Matson, P.A. and Vitousek, P.M.: 1990, *Bioscience* **40**, 667.
Matson, P.A., Billow, C., Hall, S. and Zachariassen, J.: 1996, *J. Geophys. Res.* **101**, 18533.

Matson, P.A., Vitousek, P.M. and Schimel, D.S.: 1989, in M.O. Andreae and D.S. Schimel (eds.), *Exchange of Trace Gases Between Terrestrial Ecosystems and the Atmosphere*, Dahlem Workshop Report, Wiley & Sons, Chichester, pp.97-108.

Parton, W.J., Mosier, A.R., Ojima, D.S., Valentine, D.W., Schimel, D.S., Weier, K. and Kulmala, A.E.: 1996, *Global Biogeochem. Cycles* **10**, 401.

Plant, R.A.J., Veldkamp, E. and Li, C.: 1997, *Global Biogeochem. Cycles* (submitted)

Potter, C.S., Matson, P.A., Vitousek, P.M. and Davidson, E.A.: 1996, *J. Geophys. Res.* **101**, 1361.

Prather, M., Derwent, R., Ehhalt, D., Fraser, P., Sanhueza, E. and Zhou, X.: 1995, in J.T. Houghton, L.G. Meira Filho, J. Bruce, H. Lee, B.A. Callander, E. Haites, N. Harris and K. Maskell (eds.), *Climate Change 1995*, Cambridge University Press, pp. 73-126.

Prinn, R., Cunnold, D., Rasmussen, R., Simmonds, P., Alyea, F., Crawford, A., Fraser, P. and Rosen, R.: 1990, *J. Geophys. Res.* **95**, 18369.

Reiners, W.A., Bouwman, A.F., Parsons, W.F.J. and Keller, M.: 1994, *Ecol. Applic.* **4**, 363.

Schimel, D.S. and Potter, C.S.: 1995, in P.A. Matson and R.C. Harriss (eds.), *Biogenic Trace Gases: Measuring Emissions from Soil and Water (Methods in Ecology Series)*, Blackwell Science, pp. 358-383.

Schimel, D.S., Parton, W.J., Kittel, T.G.F., Ojima, D.S. and Cole, C.V.: 1990, *Climatic Change* **17**, 13.

Stoorvogel, J.J.: 1995, Ph.D. Thesis, Wageningen Agricultural University, Wageningen, The Netherlands .

Veldkamp, E.: 1993, Ph.D. Thesis, Wageningen Agricultural University, Wageningen, The Netherlands.

Veldkamp, E. and Keller, M.: 1997, *J. Geophys. Res.* **102**, 15889.

Veldkamp, E., Keller, M. and Núñez, M.: 1998, *Global Biogeochem. Cycles* (in press)

Wielemaker, W.G. and Vogel, A. (Eds.): 1993, Atlantic Zone Programme CATIE-UAW-MAG, Guapiles, Costa Rica, Report No. 22.

SYNTHESIS OF NITROGEN POOLS AND FLUXES FROM EUROPEAN FOREST ECOSYSTEMS

N.B. DISE[1], E. MATZNER[2] and P. GUNDERSEN[3]

[1]*Dept. of Earth Sciences, The Open University, Milton Keynes, MK76AA, United Kingdom.*
[2] *Institute of Terrestrial Ecosystem Research, University of Bayreuth (BITÖK), Bayreuth, Germany*
[3] *Danish Forest and Landscape Research Institute, Hoerholm Kongevej 11, 2970 Hoerholm Denmark*

Abstract. To investigate which ecosystem parameters determine the risk and magnitude of nitrate leaching we compiled data from published and unpublished sources on dissolved inorganic nitrogen (DIN: $NO_3^- + NH_4^+$) in throughfall, DIN leaching loss in runoff or seepage water, and other ecosystem characteristics from 139 European forests. Not all data were available for all sites: 126 sites had at least one year's data on DIN inputs and DIN leaching loss; 40-50 sites had some data on soil chemistry and/or vegetation pools of N. DIN inputs in throughfall range between <1 and about 70 kg N ha^{-1} yr^{-1} and the losses with seepage or runoff range between < 1 and 50 kg N ha^{-1} yr^{-1}. Retention of N within the ecosystem increases with increasing DIN deposition and increasing proportion of NH_4^+ in deposition. The amount of N in needles and litterfall shows a significant linear relationship with throughfall deposition of DIN, whereas the C:N ratio of the organic (OH) horizon is uncorrelated to the level of throughfall-DIN flux. About 50 % of the variability in DIN leaching loss can be explained by the flux of DIN in throughfall. Alternatively, about 60 % of the variability in DIN leaching loss can be explained in a two-variable multiple regression combining the C:N ratio of the organic soil and the pH of the mineral soil. The survey data suggest that leaching of DIN from forest ecosystems in Europe is related in part to current DIN deposition and in part to the longer-term internal ecosystem N status as reflected in the chemistry of the humus and acidification status of the soil.

Keywords: dissolved inorganic nitrogen, Europe, forest ecosystems, nitrogen leaching, nitrogen saturation

1. Introduction

The response of forest ecosystems in Europe to chronic N deposition is highly variable. An understanding of what underlies this variability is necessary in order to set targets for the emission of anthropogenic N compounds. Nitrogen input to many forest ecosystems in Europe exceeds assimilation capacity, resulting in export of dissolved inorganic N (DIN: $NO_3^- + NH_4^+$) in runoff or seepage water (nearly always dominated by NO_3^-). However, the relationship between DIN input and DIN loss is not straightforward; different forests exposed to similar DIN deposition loads can retain vastly different amounts of N (e.g., Dise and Wright, 1995; Gundersen, 1995). Understanding how forest ecosystems process nitrog, and under what circumstances they release significant amounts of DIN in leachate, is necessary to set reasonable targets for the emission of anthropogenic N compounds.

Manipulative experiments on forested plots or catchments, such as the NITREX network (Wright and van Breemen 1995), are valuable to precisely identify those components of terrestrial ecosystems that respond most rapidly to changes in N deposition, and how they respond. Advantages of such networks include using similar methodologies across sites, and sampling coordination to avoid gaps in the data. However, network studies may be biased toward processes of short-term importance, and because one can never be certain that the few sites chosen are "typical," extrapolation of the results across the region is problematic.

Water, Air, and Soil Pollution **105:** 143–154, 1998.

Regional surveys can consider many more sites, but less intensively than experimental studies. General patterns revealed in surveys can confirm that important processes identified through experimental manipulations are widely applicable. Surveys also may be useful to identify simple parameters for regional assessment of risk. A disadvantage of surveys is that data may not be directly comparable among sites, especially if assembled from other studies. Furthermore, the survey approach is empirical, precluding inferences about causality.

In many ways, the strengths of each approach compensate for the weaknesses of the other. Therefore, both together are vital for developing a workable understanding of regional N dynamics. In this study, we combine several European databases to examine broad patterns of forest ecosystem response to N deposition and to address the following questions: 1) do regional survey results from forest ecosystems across Europe confirm the general pattern of important processes identified from manipulation experiments (particularly the NITREX network); 2) on a regional basis, is there a relationship between N deposition and the internal ecosystem storage and cycling of N; 3) on a regional basis, is there a relationship between N deposition and other forest ecosystem components; and 4) can the output of inorganic N from forest ecosystems across Europe be predicted from forest ecosystem characteristics that are easily measured?

2. Methods

Using a minimum criterion of one year's data on DIN input (throughfall or bulk deposition), a database of 139 European forest ecosystems was compiled (Table I). Most of the measurements are averages of several years within the period 1984-94, although some long-term data also are used (for example, Solling, Germany, where there is no significant long-term trend in throughfall DIN from the early 1970's to the present). Half of the sites are catchments where runoff is measured in an outflow stream, and half are plots where seepage water is measured in lysimeters. About 20 % of the sites are dominated by deciduous trees, the rest are coniferous forests. DIN input was estimated from throughfall data (recognizing that throughfall DIN flux encompasses N derived from wet plus dry deposition from the atmosphere along with N contributed from crown leaching, less N removed by direct canopy assimilation). For 27 sites where throughfall fluxes were not measured, they were estimated by multiplying bulk precipitation DIN fluxes by the ratio of throughfall flux to bulk precipitation DIN flux averaged cross similar sites in the region. For NO_3^--N this scaling factor ranged from 1.2 to 2 and for NH_4^+-N it ranged from 1 to 2.5. Organic-N deposition data generally were unavailable.

Other information assembled from the data included: the throughfall input of major inorganic ions (70 sites); the N concentration of current-year needles (37 sites), of litterfall (17 sites), and of the organic (OH) horizon (19 sites); the C:N ratio of the organic horizon (42 sites), the pH (in water) of the A horizon (38 sites) and the B horizon (46 sites), and the stand age (100 sites). Because the data came from many different sources, it is likely that the methodology, time of year of sampling and even exact definitions of some of the ecosystem components vary among the different studies. This variability adds an unknown amount of error in detecting trends in the data.

TABLE I
Overview of database compilation

Database	# of sites	Dates	# of deciduous sites	Reference
ENSF	53	69-87	7	Hauhs *et al.*, 1989
UNECE-IM	29	91-94	5	Forsius *et al.*, 1996
ECOFEE	34	85-94	15	Gundersen, 1995
NITREX/EXMAN	11	84-93	0	Tietema and Beier, 1995
M&G	14	73-93	0	Matzner and Grosholz, 1997

We first examine the major patterns of DIN input and loss across Europe. We then consider the relationship between internal ecosystem parameters and the input or output of DIN to empirically evaluate the processing of N in forest catchments from needle to runoff, on a regional basis, and identify the major statistical correlates to DIN input and output. Simple linear regressions and multiple linear regressions are used to assess the strength of linear relationship among variables; untransformed variables are used except in cases where a transformation of the dependent variable greatly improves the model fit (R^2 value and variance homogeneity). Pearson correlation coefficients also are calculated for some variables. Finally, we discuss similarities and differences between the empirical analysis and the NITREX manipulations, and suggest how these results may contribute to understanding how forest ecosystems process N deposition.

3. Results

3.1. Overview

DIN inputs in throughfall range from <1 to about 70 kg N ha^{-1} yr^{-1}. Throughfall input of both NO_3^--N (Figure 1) and NH_4^+-N(Figure 2) is highest along a wide band across central Europe encompassing sites in the Netherlands, Germany, the Czech Republic, Slovakia, Poland and Hungary. Lower input of DIN occurs in Wales, Denmark and southern Scandinavia, and the lowest of all in the sites in Ireland, central and northern Scandinavia, southern Europe and Russia. Although the overall pattern is similar between deposition of NO_3^--N and deposition of NH_4^+-N (i.e., sites with high NO_3^--N deposition generally receive high NH_4^+-N deposition), the range in NH_4^+-N deposition is much greater than that of NO_3^--N deposition. Nitrate deposition reaches a maximum of only about 15 kg N ha^{-1} yr^{-1}, whereas NH_4^+-N deposition exceeds 50 kg N ha^{-1} yr^{-1} in places.

Loss of DIN in seepage or runoff (dominated by NO_3^--N) ranges from < 1 to 50 kg N ha^{-1} yr^{-1}. In all but a few sites, NH_4^+-N export is a very small proportion of total DIN export. DIN loss follows the same general pattern as DIN input, with the highest leaching losses in central Europe, lower leaching in southern Europe and negligible leaching across most of Scandinavia (Figure 3).

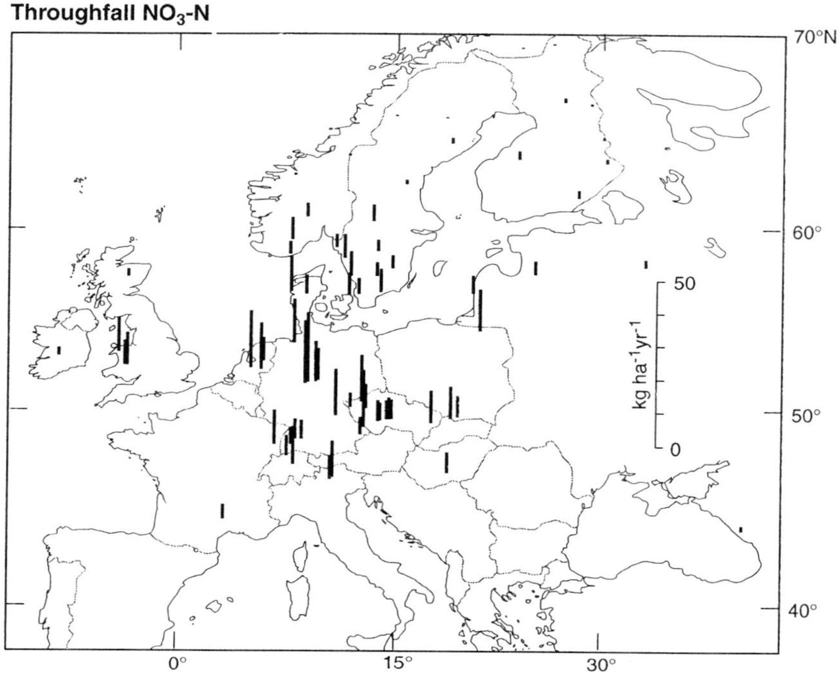

Fig. 1. Throughfall NO_3^--N, survey sites; data are the average of at least one year of regular measurement.

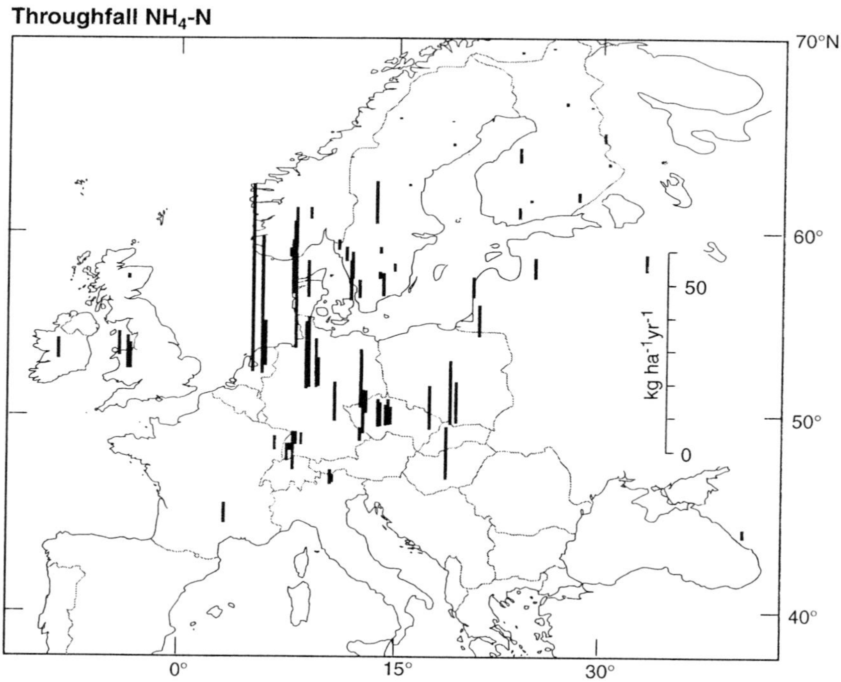

Fig. 2. Throughfall NH_4^+-N, survey sites.

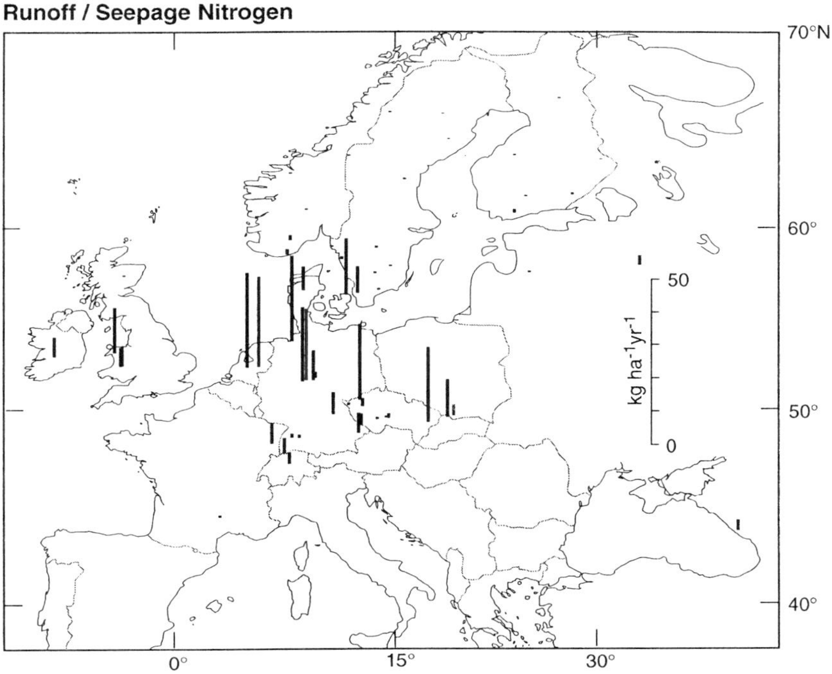

Fig. 3. Runoff/seepage DIN, survey sites.

Overall, about 50 % of the variability in DIN loss (DIN_{out}) can be explained by DIN input (DIN_{in}):

$$DIN_{out} = 0.50 \times DIN_{in} - 1.46 \qquad r^2 = 0.49, N = 126, p = 0.000$$

The model is not greatly changed if the data are restricted to sites for which throughfall input of N is known ($r^2 = 0.44$, N = 97, $p = 0.0000$), only conifer sites ($r^2 = 0.44$, N = 85, $p = 0.0000$), or only hardwood sites ($r^2 = 0.77$, N = 17, $p = 0.0000$). These smaller models also have the same slope (0.5) as the regression using all of the data.

Based on the regression model, one would estimate that about half of the DIN in throughfall is lost to seepage water or runoff and the rest is stored in the catchment. However, the actual data are highly variable, with the proportion of DIN retained ranging from nearly 100 % of input to a net annual DIN release (DIN in output > DIN in input) (Figure 4). Below inputs of about 10 kg N ha^{-1} yr^{-1}, DIN leaching is very low, with nearly all sites exporting less than 2 kg N ha^{-1} yr^{-1} in seepage water or runoff. Above this, and up to inputs of as high as 30 kg N ha^{-1} yr^{-1}, some sites retain nearly all the input DIN whereas others leach 100 % or more of the input. These results confirm as a general European pattern the results of other studies on subsets of the data (e.g., Dise and Wright, 1995; Gundersen, 1995).

Throughfall DIN at high deposition sites (> 30 kg N ha^{-1} yr^{-1}) is dominated by NH_4^+-N (Figure 4, sites in triangles). Although all of these high-input sites leach DIN, N retention can be extremely high - up to 50 kg ha^{-1} yr^{-1} (84 % retention). Reflecting this, regression equations estimate nearly three times as much inorganic N loss for a given input of NO_3^--N as for the same input of NH_4^+-N (regression equations are: $DIN_{out} = 1.54 \times NO_3^-\text{-}N_{in} - 4.272$, $r^2 = 0.46$, N = 120, $p = 0.000$, and $DIN_{out} = 0.56 \times NH_4^+\text{-}N_{in} + 1.34$, $r^2 = 0.38$, N = 120, $p = 0.000$).

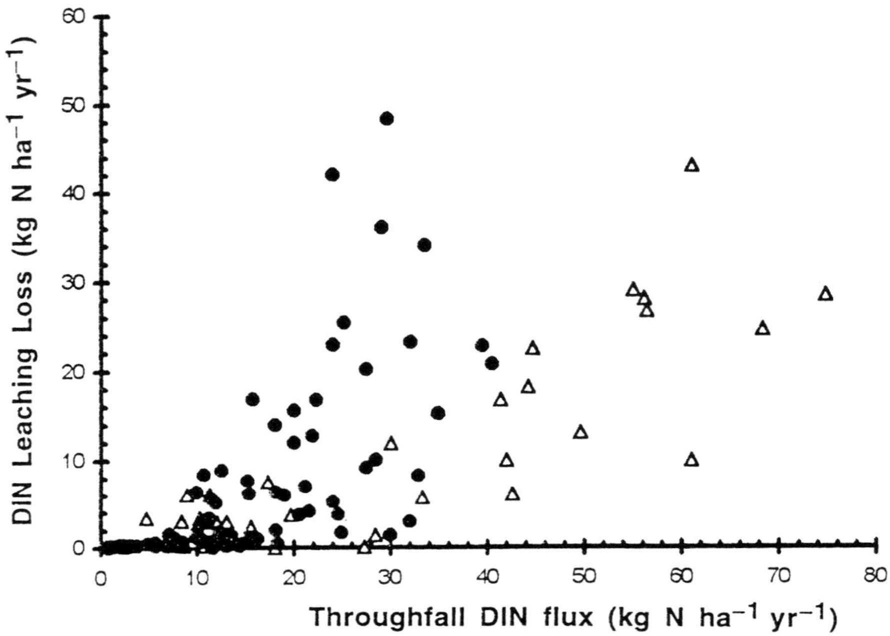

Fig. 4. Runoff/seepage DIN versus throughfall-DIN. Sites with throughfall DIN dominated by NH_4^+-N (>65 % of total inorganic N as NH_4^+-N) are shown in open triangles.

3.2. Ecosystem Processing of Input Nitrogen

3.2.1. Bulk Precipitation to Throughfall

Bulk precipitation of NH_4^+-N ranges from below 1 kg ha^{-1} yr^{-1} (in northern parts of Scandinavia) to around 17 kg ha^{-1} yr^{-1} (in the Netherlands). These inputs are enhanced by up to five times in throughfall through addition of dry deposited NH_4^+ particles or NH_3 aerosols. Bulk precipitation of NO_3^--N ranges from below 1 kg ha^{-1} yr^{-1} (in parts of Scandinavia) to about 11 kg ha^{-1} yr^{-1} (in northern Germany). Nitrate-N in throughfall is enhanced by a maximum of three times over wet deposition, through either addition of dry deposition or leaching from foliage. In about 20 % of the sites, throughfall DIN is lower than bulk precipitation of DIN, possibly due to direct canopy assimilation of NO_3^--N or NH_4^+-N (Brumme *et al.,* 1992; Eilers *et al.,* 1992). In most of these (17 % of all sites), throughfall NH_4^+-N is less than bulk precipitation NH_4^+-N, in 10 % of all sites both NH_4^+-N and NO_3^--N are lower, and in 3 % only NO_3^--N is lower. Sites receiving very low levels of DIN in bulk precipitation (<4 kg N ha^{-1} yr^{-1} total) show negligible amounts of DIN in throughfall.

3.2.2. Forest Ecosystem Parameters Related to Throughfall DIN

The N concentration of needles, litterfall and the organic horizon all show significant linear relationships with both the NO_3^--N and the NH_4^+-N flux in throughfall. The concentration of N in current-year needles from 37 sites (conifer only) in the database averages 14.4 mg N g^{-1} dry wt, with a range between 9 and 23 mg g^{-1}. Needle N concentration shows the strongest relationship to throughfall NH_4^+-N flux ($r^2 = 0.40$, N=37, $p = 0.000$, Figure 5A).

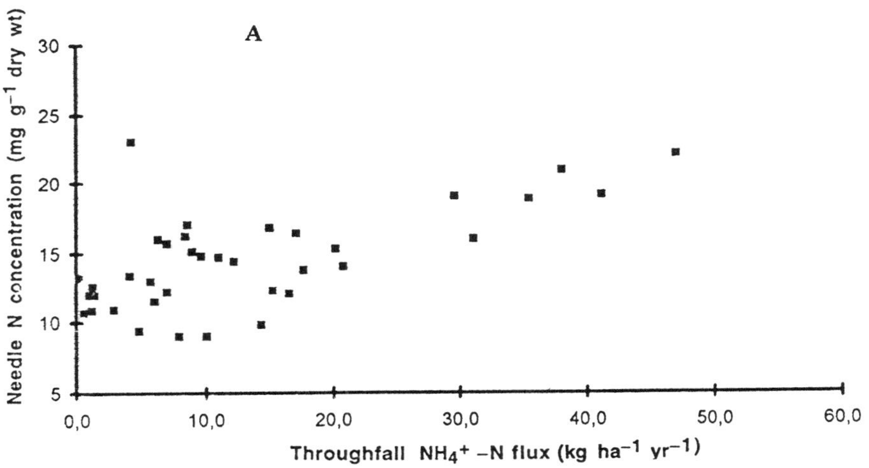

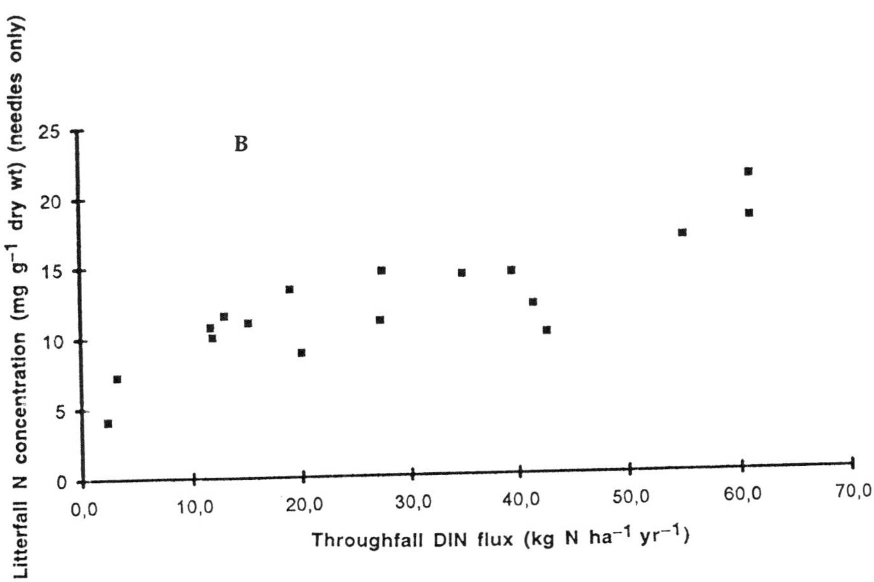

Fig. 5. Relationships between N concentrations of needles and litterfall (mg N g^{-1} dry wt) and throughfall flux of inorganic N (kg N ha^{-1} yr^{-1}). A) Needle N concentration versus throughfall NH_4^+-N, r^2=0.40. B) Litterfall N concentration versus throughfall DIN, r^2 = 0.71.

The N concentration of needles in litterfall from 17 sites averages 12.3 mg N g^{-1} dry wt, with a range between 4 and 21 mg g^{-1} dry wt. Considering only those sites for which both sets of data are available for the same year (N = 16), litterfall-N concentration is reduced by an average of 23 % from the N concentration of needles on the tree (mean of 15.4 mg g^{-1} dry wt needles versus mean of 11.9 mg g^{-1} dry wt litterfall). Litterfall N concentration shows the strongest relationship to throughfall DIN flux ($r^2 = 0.71$, N = 17, $p = 0.0001$, Figure 5B). Organic horizon N also shows a significant, but much weaker relationship to throughfall DIN flux ($r^2 = 0.29$, N = 19, $p = 0.02$). Needle N, litterfall N and organic horizon N are all significantly linearly related to DIN loss in seepage ($r^2 = 0.23$, $p = 0.005$ for needle N; $r^2 = 0.55$, $p = 0.001$ for litterfall N; $r^2 = 0.29$, $p = 0.02$ for organic horizon N).

3.2.3. Organic Horizon C:N and Mineral Soil pH

The C:N ratio of the organic horizon is uncorrelated to the amount of DIN in throughfall ($r = 0.01$, $p = 0.97$), but shows a weak linear relationship with DIN leaching loss ($r^2 = 0.10$, N = 42, $p = 0.04$, Figure 6). Overall, DIN loss is very low, less than 5 kg N ha^{-1} yr^{-1} for sites with C:N ratios above 30, up to 15 kg N ha^{-1} yr^{-1} for sites with C/N ratios above 26, and up to 45 kg N ha^{-1} yr^{-1} for sites with C:N ratios above 17 (Figure 6). However, the variability in the data is very high and there are only five sites with C:N ratio above 30. A stronger linear relationship between humus C:N ratio and leachate N flux was shown by Matzner and Grosholz (1997) using a subset of 17 sites, mostly in Germany.

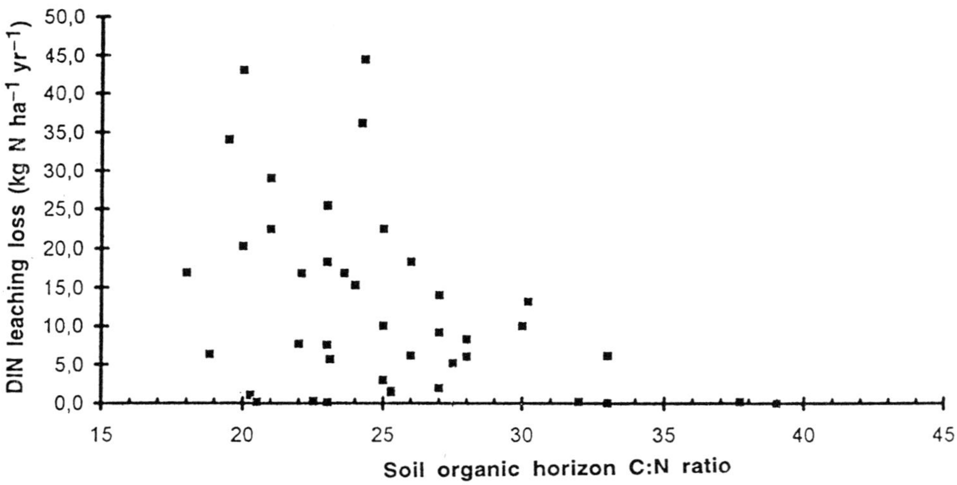

Fig. 6. Runoff/seepage DIN versus C:N ratio of organic layer, survey sites, $r^2 = 0.10$.

There is a significant linear relationship between soil pH (B horizon) and DIN input ($r^2 = 0.38$, N = 56, $p = 0.000$) and between soil pH and DIN output ($r^2 = 0.21$, N = 47, $p = 0.001$, Figure 7). Log-transforming values for DIN output improves the fit of the latter model ($r^2 = 0.30$, N = 46, $p = 0.001$). Above a level of about pH 4.2, DIN loss is very low

(Figure 7): of the 23 sites represented, only one leaches more than 8 kg N ha^{-1} yr^{-1}. By contrast. DIN loss in the 24 sites with soil pH below pH 4.2 ranges from 0 to 45 kg N ha^{-1} yr^{-1}. Soil pH in the A horizon shows the same broad trends as B-horizon pH.

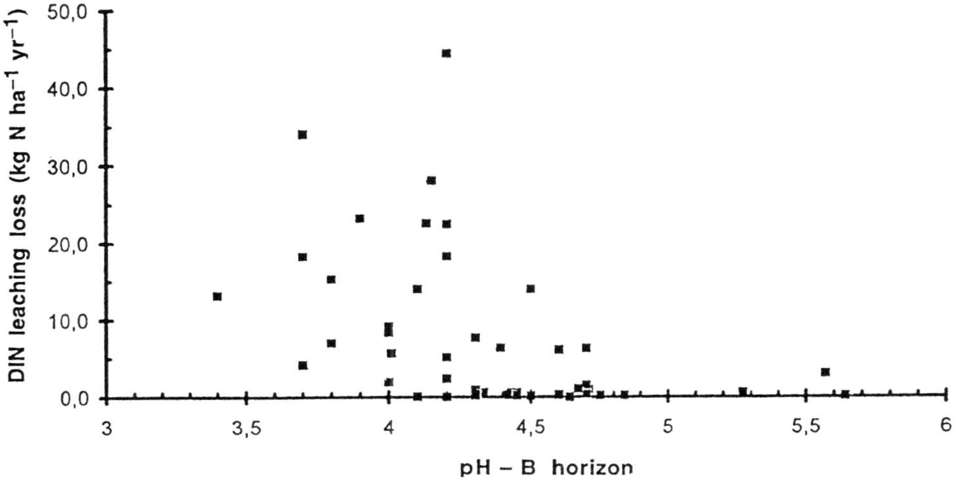

Fig. 7. Runoff/seepage DIN versus pH B horizon (H_2O), survey sites, $r^2 = 0.21$.

The regression model that explains the most amount of variability in DIN loss from the data in the survey combines B horizon soil pH (pH_B) and organic horizon C:N ($C:N_{org}$). A log transformation on DIN_{out} gives the best-fit model:

$$\log DIN_{out} = 7.795 - 1.151\times(pH_B) - 0.09\times(C:N_{org}); \quad R^2 = 0.61, N = 22, p < 0.001$$

Substituting DIN input for pH also produces a highly significant regression model, but with lower R^2:

$$\log DIN_{out} = 0.030\times(DIN_{in}) - 0.061\times(C:N_{org}) + 1.493; \quad R^2 = 0.50, N = 41, p < 0.002$$

4. Discussion

4.1. Inputs and Outputs of DIN

This study indicates that about half of the variability in the seepage flux of DIN from forested catchments across Europe can be explained by the flux of DIN in throughfall. In addition to DIN input, a combination of organic C:N ratio and mineral soil pH are significant in a multiple regression model of DIN loss. The significance of soil pH is unclear: low soil pH may be an effect of enhanced long-term N deposition (via base cation removal by NO_3^- leaching), a cause of NO_3^--N loss from ecosystems (due to effects of soil acidity and reactive Al on rooting depth or plant nutrition imbalances), or both (or neither). The most likely explanation is that low pH is an effect of NO_3^--N leaching rather than a cause, and that mineral soil pH differences among sites in part reflect differences in cumulative N deposition. Because of the high level of unexplained variability in the data, and thus wide confidence intervals, however, the regression models derived from these data are not useful as predictive tools to estimate the flux of DIN from forest ecosystems.

High-DIN input sites dominated by NH_4^+-N in throughfall retain proportionally more N in the catchment than those receiving lower DIN input, but in roughly equal amounts of NO_3^--N and NH_4^+-N (Figure 4). Elevated needle N concentration at these sites suggests some luxury uptake of NH_4^+-N (Figure 5A) and enhanced N storage in vegetation. A preferential storage of NH_4^+-N also was found in a NITREX site that was fertilized with both NH_4^+-N and NO_3^--N (Emmett *et al.,* 1995). Several factors might account for an enhanced retention of NH_4^+-N over NO_3^--N. Most of the forests in the survey are conifers, especially Norway spruce (*Picea abies* Karst), a species that preferentially takes up NH_4^+-N over NO_3^--N (Berger, 1995). Ammonium may be nitrified, immobilized by microbes or taken up by plant roots before transport to deeper layers occurs. In addition, the mobility of NH_4^+ in forest soils is less than that of NO_3^- because NH_4^+ is subjected to cation exchange (Matschonat and Matzner 1995).

It is likely that much of the additional variability in the data is due to site history. Factors such as the outbreak of insect pests, storm damage or, especially, site management (e.g., liming, thinning, changes in species composition, soil preparation before planting, etc.) will affect the N dynamics of a site and may cause DIN leaching at low DIN inputs or N retention at high DIN inputs. We lack the detailed information from each site to evaluate such factors. Another source of error is the difficulty in estimating actual DIN inputs, together with year-to-year variability in DIN inputs and losses -- for many sites only one or two years of data were available.

4.2. Forest Ecosystem Response to DIN Deposition

Consistent with the hypothesis put forth by Aber *et al.* (1989), the NITREX manipulation experiments (Wright and van Breemen, 1995) showed that different forest ecosystem components respond at different rates to long-term DIN addition or DIN removal (Tietema *et al.,* 1995, Gundersen *et al.,* 1998). Over the course of the NITREX manipulations (up to 10 years) at different sites, the DIN content of runoff and soil solution changed rapidly, in some sites within a year. Needle N concentration, litterfall N concentration and root biomass changed in some sites after 3-4 years. No response was detected within the duration of the experiment, however, in soil chemical status (C:N, N concentration, pH). Presumably, these "slow-response" variables (Gundersen *et al.,* 1998) change over much longer time periods.

Some of the changes induced in the vegetation by increasing or decreasing DIN input in NITREX have spatial parallels across the gradient of DIN input in this study. These include a significant increase in foliar N concentration and litterfall N concentration with increasing DIN in throughfall. However, we also found significant but weak statistical relationships between "slow-response" soil parameters, such as the N concentration of organic soil and soil pH, that did not change over the duration of the NITREX treatments. In addition, parallels can be drawn between the temporal NITREX response and the strength of association between throughfall DIN and forest ecosystem components on a regional basis. In this study, the r^2 value is relatively high for the relationship between input DIN and vegetation/litterfall storage of N ($r^2 = 0.4$-0.7), and low between input DIN and soil storage of N ($r^2 = 0.1$-0.3).

Although not directly linked to current N deposition in this study, organic horizon C:N probably also reflects in part a long-term ecosystem response to elevated N deposition. Indeed, a decrease in the C:N ratio of the forest floor over the last several decades has been observed in German forest ecosystems receiving elevated DIN deposition over that period (von Zezschwitz, 1985; Hildebrand, 1994). A decrease of the C:N in humus might be caused by decreased C:N in litter (as in the above studies) or by increased rates of immobilization of mineral N by microorganisms (Tietema, 1993).

5. Conclusions

The temporal progression of N saturation as suggested by Aber *et al.* (1989), and in part confirmed by the NITREX experiments, is also shown spatially across a gradient of DIN in the survey sites through significant linear relationships between throughfall DIN flux and either DIN loss in leachate, needle N concentration, litterfall N concentration, or (weakly) the N concentration of organic soil. The survey results and the NITREX results are consistent with the concept that enhanced DIN deposition has increased the internal ecosystem storage and cycling of N in European forests. An additional response to chronically enhanced DIN input appears to be acidification of the mineral soil.

The empirical models link leaching ofDIN from forest ecosystems in Europe to current N deposition, soil acidification status, and the long-term internal ecosystem N status as reflected in the chemistry of the humus. The data suggest that three broad criteria are necessary, but not in themselves sufficient, to leach NO_3^- from coniferous forests. These are: 1) high DIN flux in throughfall (above about 10 kg ha^{-1} yr^{-1}), 2) low organic layer C:N (below about 30), and 3) low mineral soil pH (below about 4.3). Although such trends can be identified, the data inevitably have large errors. Further improvements in data quality are necessary for a more complete description of regional N dynamics across Europe.

Acknowledgments

Funding was provided by the German Bundesministerium für Bildung, Wissenschaft, Forschung und Technologie (BMBF), BEO 51-0339476B.

References

Aber, J.D., Nadelhoffer, K.J., Streudler, P. and Melillo, J.: 1989, *BioScience* **39**, 378.
Berger, A.: 1995, *Bayreuther Forum Ökologie* **23**, 270.
Brumme, R., Leimcke, U. and Matzner, E: 1992, *Plant and Soil* **142**, 273.
Dise, N.B. and Wright, R.F.: 1995, *Forest Ecol. Manage.* **71**, 153.
Eilers, G., Brumme, R. and Matzner, E.: 1992, *Forest Ecol. Manage.* **51**, 239.

Emmett, B.A., Brittain, S.A., Hughes, S., Görres, J., Kennedy, V., Norris, D., Rafarel, R., Reynolds, B. and Stevens, P.A.: 1995, *Forest Ecol. Manage.* **71**, 45.
Forsius, M., Vuorenmaa, J. and Kleemola, S.: 1996, 'Assesment of Nitrogen Processes at ICP IM sites', in S. Kleemola and M. Forsius, (eds.), *ICPM-IM Programme Annual Synoptic Report*, Finnish Environment Agency, Helsinki, pp. 25-38.
Gundersen, P.: 1995, *Water, Ai,r Soil Pollut.* **85**, 1179.
Gundersen, P., Emmett, B.A., Kjonaas, O.J., Koopmans, C.J. and Tietema, A.:1998, *Forest Ecol. Manage.*, in press.
Hauhs, M., Rost-Siebert, K., Pačes, T., Vigerust, B. and Raben, G.:1989, 'Summary of European Data', in J.L. Malanchuk and J. Nilsson (eds), *The role of Nitrogen in the Acidification of Soils and Surface Waters*, Council of Nordic Ministers, Miljörapport **89**, 10: 5.1-5.37.
Hildebrand, E. E.: 1994, *Allg. Forst. Zeitscrift* **49**, 99.
Matschonat, G. and Matzner, E.: 1996, *Z. Pflanzener. Bodenk.* **159**, 505.
Matzner, E. and Grosholz, C.: 1997, Forst. Cbl. **116**, 39.
Tietema, A.: 1993, *Biogeochemistry* **20**, 45.
Tietema, A., et al.: 1995, *Water, Air, Soil Pollut.* **85**, 1623.
Tietema, A. and Beier, C.: 1995, *Forest Ecol. Manage.* **71**, 143.
von Zezschwitz, E.: 1985, *Forstw. Cbl.* **104**, 205.
Wright, R.F. and van Breemen, N.: 1995, *Forest Ecol. Manage.* **71**, 163.

THE BEHAVIOR OF NITROGEN ISOTOPES IN ACIDIFIED FOREST SOILS IN THE CZECH REPUBLIC

F. BUZEK, J. ČERNÝ and T. PAČES

Czech Geological Survey, Klárov 3, 118 21 Prague 1, Czech Republic

Abstract. The effects of atmospheric deposition on N cycling in acidified soils were studied at three spruce and one beech forested sites in the Czech Republic. Nitrogen content and $\delta^{15}N$ were monitored in bulk and throughfall precipitation, needles, leaves, soils and soil solutions. Changes in soil NO_3^- production, effect of admixing of atmospheric N in spruce forest and N consumption in deciduous forest are described using changes in ^{15}N fractionation of mineralized N in soil. Admixing of atmospheric NH_4^+ can be identified at low concentrations of exchangeable NH_4^+. The $\delta^{15}N$ ratio of atmospheric NO_3^- input is on average by 2 ‰ less negative than the δ^5N ratio in soil water; admixing changes the $\delta^{15}N$ of soil NO_3^- detected in lysimeters.

Keywords: ammonium, Czech Republic, natural abundance ^{15}N, nitrate, nitrogen cycling, forest soil acidification

1. Introduction

Nitrate is rarely leached from temperate and boreal coniferous forests at sites with low atmospheric N input. In these forests, rich mycorrhizal flora uses N directly in litter and returns it to plants bypassing the microbial population. In central Europe, however, NO_3^- ions in soils are not fully consumed because of high atmospheric N input, decreasing denitrification capacity in acidified soil, and land-use changes replacing deciduous trees by conifers. Conifers have at least 10 times lower capacity to consume NO_3^- than do deciduous trees or herbaceous plants. Consequently, the forests cannot cope with high NO_3^- atmospheric deposition, which results in a higher NO_3^- production in central European soils (Schulze, 1996).

Natural abundance ^{15}N is used in this study to examine qualitatively or semi-quantitatively processes of N cycling in forest ecosystems. The advantage of natural abundance studies is their easy application at any site without additional work and costs connected with labeling; the disadvantage is a relatively low natural variation in N isotope ratios, which sometimes results in an overlap between the source and receiver pool signals making it difficult to estimate the rates of N transformations (Nadelhoffer and Fry, 1994).

The objectives of the ^{15}N monitoring were: 1) identification of atmospheric N in N compounds of soil solutes, and 2) separation of natural ^{15}N enrichment caused by biological transformation from the effects caused by variations in atmospheric input.

2. Materials and Methods

2.1. Sampling Site

Three sampling sites, Načetín (770 m a.s.l.), Červená Jáma (800 m a.s.l.) and Jezeří (550 m a.s.l.) are located in the Erzgebirge mountains at the Czech-German border near the city of Most. This region belongs to the "Black Triangle" - the most polluted and most acidified

Water, Air, and Soil Pollution **105:** 155–164, 1998.

region of Europe. The highest annual input of acid emissions occurred during the period of 1978 to 1982 (108 kg S ha^{-1} yr^{-1}, 13 kg of N ha^{-1} yr^{-1}; Pačes, 1985). The acidification gradually has declined. The atmospheric input during our investigation period of 1995 - 1996 was 35 kg S ha^{-1} yr^{-1} and 12 kg of N ha^{-1} yr^{-1} (Černý *et al.*, 1998). The fourth site, Salačova Lhota (557 to 744 m a.s.l.), is located in the Bohemian-Moravian Upland (160 km southeast of the Erzgebirge) where the input of acidic emissions has been comparatively low. The input was 21 kg S ha^{-1} yr^{-1} and 8.6 kg of N ha^{-1} yr^{-1} during the 1975 - 1983 period (Pačes, 1985). The input slightly increased to 26 kg S ha^{-1} yr^{-1} and 11 kg of N ha^{-1} yr^{-1} during our investigation period (Fottová, 1996). The bedrock and soils are very similar at all four sites. Prevailing bedrock is gneiss and quartzite. Soils range from Dystric Cambisols to Cambic Podzols.

Načetín, Červená Jáma and Salačova Lhota are covered with 40 to 60 year old spruce (*Picea abies*) forest. Jezeří is covered with 100 year old beech forest (*Fagus sylvatica*). In the Erzgebirge, the spruce forest is damaged by acid emissions to the extent of a complete dieback. Most of the area has been deforested. Our spruce sites represent just remnants of the original spruce forests. The spruce forest at Salačova Lhota appears healthy; however, slight symptoms of decline have been observed since the beginning of our observation in 1976.

The soils in the Erzgebirge sites have pH_{KCl} from 2.6 to 4.2 and the cation exchange capacity (CEC) is very low (0.13 to 1.6 meq kg^{-1}). These properties are not significantly different from the less acidified site of Salacova Lhota where pH_{KCl} varies from 2.7 to 3.4 and CEC varies from 0.27 to 1.6 meq kg^{-1}.

The Salačova Lhota catchment has an area 1.68 km^2, fully covered by spruce. The concentration of NO_3^--N in runoff is less than 0.2 mg L^{-1}. The beech site of Jezeří is situated in a 2.61 ha catchment on the eastern slope of the Erzgebirge mountains facing one of the most industrialized regions in the Czech Republic. The average NO_3^--N concentration in runoff is less than 0.1 mg L^{-1}. A small subcatchment (0.29km^2), Červená Jáma, in the upper part of the Jezeří catchment is spruce-forested. The average NO_3^--N concentration in runoff is about 1 mg L^{-1} (Fottová, 1996).

2.2. Sample Collection

Rainfall and throughfall were sampled using polyethylene funnels. Cumulative monthly samples were collected. Thymol was added to collectors to minimize microbial activity in the water samples during the sampling period (Gillett and Ayres, 1991). Before the determination of ^{15}N, rain and throughfall samples were examined for contamination with bird droppings by testing the samples for phosphate; contamined samples were rejected. Two sets of suction lysimeters were installed in 1995 at a depth of 90 cm at the Načetín stand. The lysimeters were placed at an increasing distance from a tree (numbered from Cal 1 to Cal 4). The area where the lysimeters were installed is covered with *Callamagrostis villosa*.

Soil samples were taken from five profiles in the vicinity of the lysimeters and from three profiles at each of the other sites (Salačova Lhota, Červená Jáma and Jezeří). Three levels (0-5 cm, 5-15 cm, and 15-40 cm or 15-80 cm), corresponding to O, A and Bv or C soil horizons, were sampled.

Soil samples were placed into polyethylene bags and frozen on the site at -80 °C using dry ice in cooling boxes. After transportation to the laboratory, the samples were stored in a refrigerator at -16 °C. The samples were divided into subsamples used for water content determination and N extraction. The rest was dried at about 100 °C. The soil fraction below 2 mm was ground in a ball mill and used for the determination of cation exchange capacity, pH, elemental composition and $\delta^{15}N_{tot}$. An extraction of N from 500g of fresh soil was performed in 1L of 1M KCl solution at room temperature by stirring for 30 min. The extract was filtered through a GF/C Whatman glass prefilter anda 0.45μm WCN Whatman fine filter. Filtrate volume was measured. The solution was checked for NH_4^+ and NO_3^- content, preserved by an addition of thymol and stored in a refrigerator.

About 500 mL samples of atmospheric precipitation were concentrated on Dowex 50W and Dowex 1x cation and anion exchange resins, separated into NH_4^+ and NO_3^-, checked for NH_4^+ and NO_3^- content, preserved by addition of thymol, and stored.

Solution aliquots were alkalized with MgO and steam distilled (Devarda alloy was added to solutions for NO_3^- analysis) into a small excess of diluted H_2SO_4. Ammonium sulfate was dried and stored for $\delta^{15}N$ analyses.

Fresh litter samples and spruce needles were dried at 100 °C, ground in a ball mill and analyzed for elemental and N isotope composition. The samples of soil water, bulk rainfall and throughfall were preserved by thymol, chemically analyzed and stored for N isotope measurement.

Isotope analyses were performed by flash combustion in Fisons 1108 elemental analyzer coupled to Mat 251 isotope ratio mass spectrometer via open-split interface ConFlo I in continuous flow regime. Sample size was adjusted to contain from 0.5 to 1 mg of N. External reproducibility of $\delta^{15}N$ measurements was 0.15 ‰, overall reproducibility for the steam distilled samples was about 0.4 ‰.

3. Results and Discussion

3.1. Atmospheric Input

Measured $\delta^{15}N$ ratios in bulk deposition varied from -6 to +5 ‰ for NO_3^- and from -13 to +3 ‰ for NH_4^+. Througfall ^{15}N ratios varied from -6 to +23.5 ‰ for NO_3^- and from -6 to +4.5 ‰ for NH_4^+ (Figure 1). The NO_3^- concentration and its $\delta^{15}N$ ratio are negatively correlated. A similar trend is observed for NH_4^+. Such trends may indicate mixing between two end members representing different sources, e.g., a local source and a distant source of the N compounds. The average $\delta^{15}N$ ratio of NO_3^- in precipitation is less negative than the $\delta^{15}N$ ratio in soil NO_3^- (-2 to +2 in comparison with -5 to -3; see below). Two throughfall samples collected in December 1995 and May 1996 have extremely positive $\delta^{15}N$ ratios for NO_3^- (12.3 and 23.5 ‰). We do not know the origin of these nitrates. The sample from December may be related to a fractionation effect on NOx during foggy and inversion atmospheric conditions that prevail during the month.

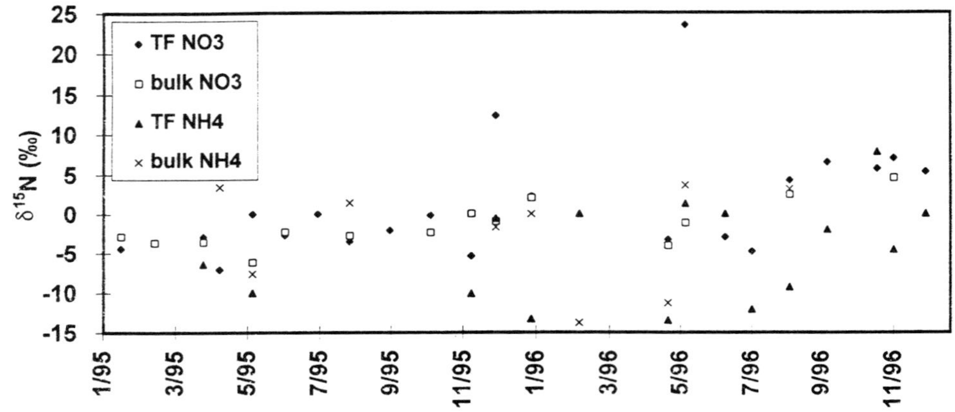

Fig.1. $\delta^{15}N$ of NH_4^+ and NO_3^- in rainfall (bulk) and throughfall (TF) from January 1995 to December 1996 at Načetín.

3.2. Plant Material

Elemental and isotopic composition of the organic N input was examined in 1- and 2-year old needles and leaves. Their C/N mass ratio varied from 37 to 41; $\delta^{15}N$ ratios of needles ranged from -3.9 to -4.2 ‰ and $\delta^{15}N$ of leaves ranged from -3.7 to -4.2 ‰. The variations are related to the N content and age of the needles. A similar trend was observed by Gebauer and Dietrich (1993). The isotope fractionation due to N assimilation is very low (Hübner, 1986; Gebauer and Schulze, 1991). Therefore, we assumed that the $\delta^{15}N$ ratios of assimilated N (NH_4^+, NO_3^-, or DON) are similar to the average $\delta^{15}N$ ratios for needles and leaves (-4 ‰).

3.3. Soil

Soil at Načetín has been exposed to high acid deposition for at least 40 years. The C/N mass ratio in the soil of Načetín differs from a comparable forest soil in the relatively clean area of Salačova Lhota (Figure 2). The C/N mass ratio in the less polluted area is controlled by organic input. The mass ratio of the input is about 39. The mass ratio decreases from 29 in the O horizon to 15 in the B horizon. The C/N ratio at Načetín is very low in the O horizon (about 15) and increases to about 25 in the A and Bv horizons as a result of mobilization of organic compounds and their displacement to lower horizons. The C/N ratio of the soil supporting spruce at Jezeří and of the soil supporting beech at Červená Jáma varies from 20 in the O horizon to 18 in the Bv horizons. No statistically significant differences between these two sites have been observed. The $\delta^{15}N$ ratios of organic N in all soil profiles change to more positive values with depth (Figure 3). The fractionation is more efficient at Načetín than at the other sites.

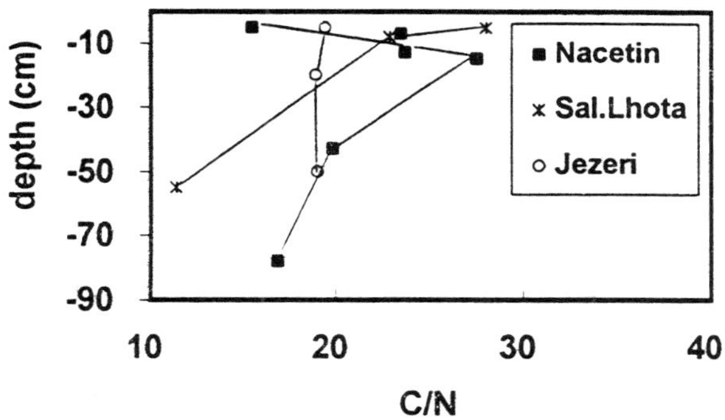

Fig.2. Elemental composition of soil in the profile at Načetín (spruce), Jezeří (beech), and Salačova Lhota (spruce).

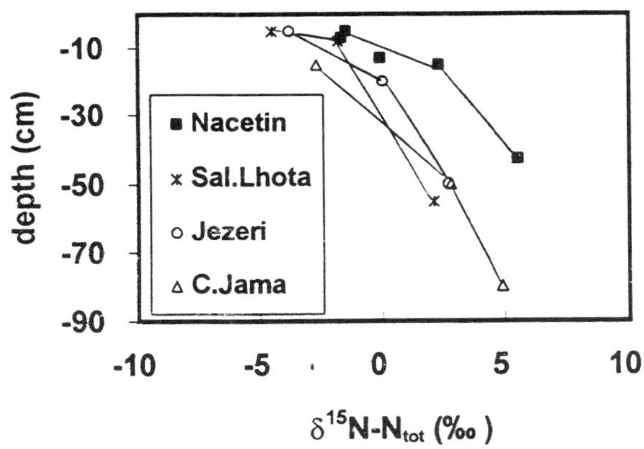

Fig.3. $\delta^{15}N$ ratios of N_{tot} in the soil profiles at the spruce sites at Načetín and Červená Jáma, at the beech site of Jezeří in the acidified region of the Erzgebirge, and in the Salačova Lhota catchment in the less acidified region of Bohemian-Moravian Upland.

3.4. SOIL WATER

Two sets of lysimeters were sampled regularly from spring to winter in 1995/1996 at Načetín. The $\delta^{15}N$ ratios of NO_3^- from lysimeters are compared with the $\delta^{15}N$ ratios of NO_3^- in bulk precipitation and throughfall in Figure 4. The $\delta^{15}N$ ratio in soil water NO_3^- in from the lysimeters is about 2 ‰ more negative than the $\delta^{15}N$ ratio of atmospheric NO_3^-. Black and Waring (1977) observed an opposite trend between NO_3^- added to the soil column and leached NO_3^-. The enrichment was explained by an equilibrium fractionation of adsorbed NO_3^-. However, it is unlikely that NO_3^- may be adsorbed on soil to such an extent to change the $\delta^{15}N$ of leached NO_3^-. The isotopically more negative soil NO_3^- may originate from nitrification. Admixture of isotopically more positive atmogenic NO_3^- to soil-derived NO_3^- could affect measured $\delta^{15}N$ ratios of soil NO_3^-, resulting in a lower apparent fractionation between NH_4^+ and NO_3^-. Extremely positive $\delta^{15}N$ ratios of atmospheric NO_3^- were determined in December 1995 and May 1996. On these dates, according to a calculated ^{15}N balance, the direct contribution of precipitation NO_3^- to soil NO_3^- varied from 15 to 40%.

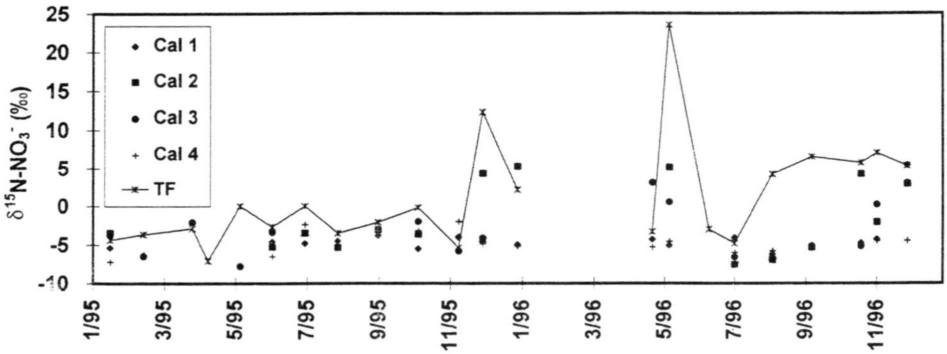

Fig.4. $\delta^{15}N$ ratios of NO_3^- in soil-water from Cal lysimeters and throughfall (TF) from January 1995 to December 1996. Cal lysimeters are numbered according to increasing distance from a tree.

3.5. Mineralized Nitrogen

Mineralized (inorganic) N in the soil sorption complex is present as NH_4^+ and NO_3^- ions. The relations between the exchangeable NH_4^+ and NO_3^- pools in different horizons of the soil profile may clarify the partitioning of mineralized N between uptake and losses in acidified forest soils (Mariotti *et al.*, 1980). A portion of the soil NH_4^+ is immobilized as nonexchangeable NH_4^+ fixed in silicate minerals. This portion is not available for plant uptake or for nitrification. The major portion of soil NH_4^+ is immobilized as exchangeable NH_4^+, which is available for nitrification and partly also for plant uptake.

In the O and A soil horizons, the minimum $\delta^{15}N$ ratio of exchangeable NH_4^+ was similar to the $\delta^{15}N$ ratio of N_{org} (-2.2 ‰). It was shifted to more positive values by nitrification, reaching up to 25.9 ‰ for minimum NH_4^+ content. The difference between $\delta^{15}N$-NO_3^- and $\delta^{15}N$-NH_4^+ ratios (enrichment factor, ε) in A, Bv and C horizons varied between -10 to -24 ‰ in spring and late summer. Some differencies existed due to admixing of atmospheric NO_3^- to soil NO_3^- in the O horizon. The very positive $\delta^{15}N$ ratio in atmospheric NO_3^- (May 1996) made the NO_3^--N isotopically much heavier.

The effect of dilution of soil NH_4^+ by atmogenic N is well observable in the $\delta^{15}N$ - NH_4^+ versus c_{NH4+} plot in Figure 5. A shift of $\delta^{15}N$-NH_4^+ ratios to more positive values during consumption of NH_4^+ is moderated by a decrease to less positive (or more negative) values due to the addition of more negative $\delta^{15}N\text{-}NH_4^+$ coming from the atmosphere. As the NH_4^+ content in soil decreases, the whole process of the replacement is more efficient and more visible on the $\delta^{15}N\text{-}NH_4^+$ plot. The dilution of soil NH_4^+ with atmospheric NH_4^+ takes place mostly in B (15-30 cm) and B/C (40-85 cm) horizons.

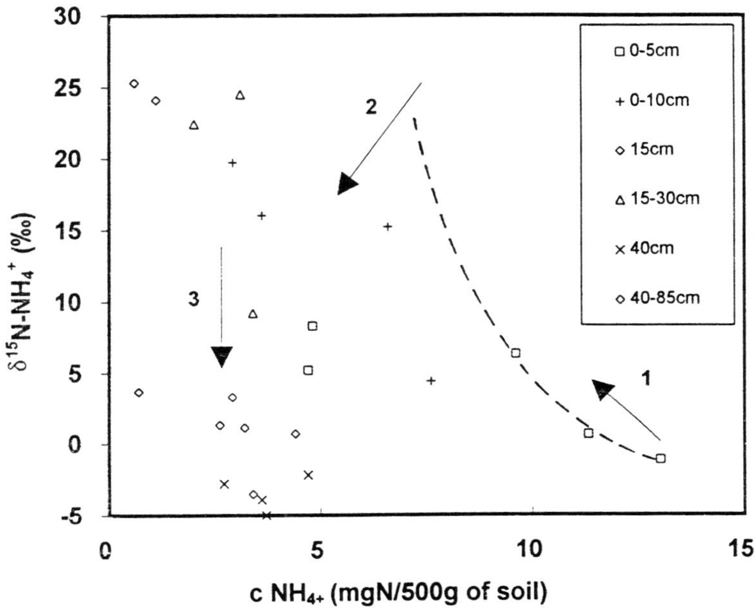

Fig.5. $\delta^{15}N\text{-}NH_4^+$ vs. NH_4^+ content in soil profile at Načetín. Path 1 represents calculated nitrification effect, path 2 represents gradual dilution with isotopically light ($\delta^{15}N$ from +3 to -12 ‰) atmospheric NH_4^+, and path 3 represents adsorption-desorption replacement of soil NH_4^+ with atmospheric NH_4^+.

The mineralized N pool ($N_{min} = N_{NH4+} + N_{NO3-}$) in soil can be approximated by a simple open reaction system. In such a system, ^{15}N fractionation between substrate (NH_4^+) and product (NO_3^-) is described by Eq.1:

$$\delta^{15}N\text{-}NO_3^- = \delta^{15}N\text{-}NH_4^+ + \varepsilon_{NO3/NH4} \quad (1)$$

where $\varepsilon_{NO3/NH4}$ is the enrichment factor. That means that the given system has a constant difference between the $\delta^{15}N$ of instantaneous residual NH_4^+ and produced NO_3^-. This difference is typical for the particular conditions under which the partitioning takes place (Mariotti *et al.*, 1981). Ammonium isotopic composition can be described by the Rayleigh equation:

$$\varepsilon_{NO3/NH4} = (\delta^{15}N\text{-}NH_4^+ - \delta^{15}N\text{-}NH_4^+{}_0)/\ln f \quad (2)$$

where $\delta^{15}N\text{-}NH_4^+{}_0$ and $\delta^{15}N\text{-}NH_4^+$ are the isotopic compositions $\delta^{15}N$ of NH_4^+ at the beginning of the reaction and after the reaction has progressed. The variable f is a molal fraction of the residual NH_4^+ (Fritz and Fontes, 1981).

The total ^{15}N fractionation of mineralized N pool is more complex. The produced NO_3^- is removed continuously from the open system. The residual NH_4^+ ions are affected by fractionation and their δ^{15}N ratios change to more positive values. The content of N_{min} decreases and its average δ^{15}N changes to more positive values (Figure 6). An apparent fractionation of N_{min} depends mainly on NO_3^- production and removal rates. If there is a consumption of both NH_4^+ and NO_3^- inside the system (forest uptake), the plot of δ^{15}N-N_{min} versus N_{min} concentration moves to lower N_{min} concentration and the slope on the plot of c N_{min} vs. δ^{15}N-N_{min} will be less steep. The more NO_3^- produced and transported out of the system with infiltrating water, the lower N_{min} content is in the soil and the more positive the δ^{15}N-N_{min} ratio becomes. Admixing of atmogenic NH_4^+ and NO_3^- moves the curve closer to their δ^{15}N values (the system is isotopically "diluted"). The following processes are shown in Figure 6.: 1) moderate NO_3^- production in areas with low atmospheric input, 2) nearly complete N consumption in a deciduous forest, 3) leaching of produced NO_3^- ions in an acidified spruce forest, and 4) the effect of higher atmospheric input. The data from our sites are presented in Figure 7.

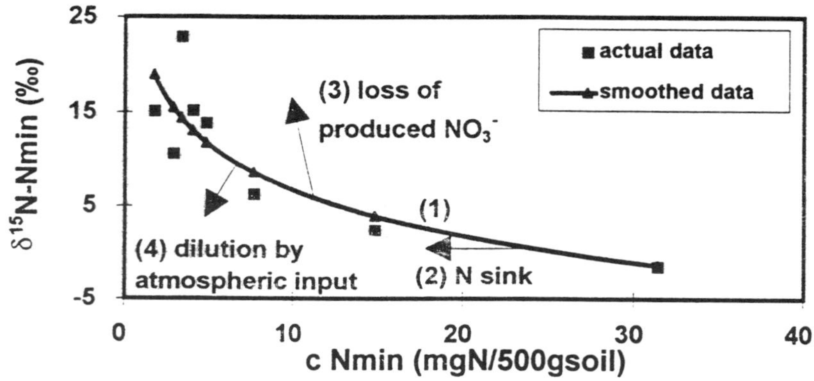

Fig. 6. Fractionation of δ^{15}N-N_{min} with decreasing N_{min} contcentration. The plot corresponds to logarithmic correlation of data from Salačova Lhota. Numbers in parenthesis are shifts explained in the text.

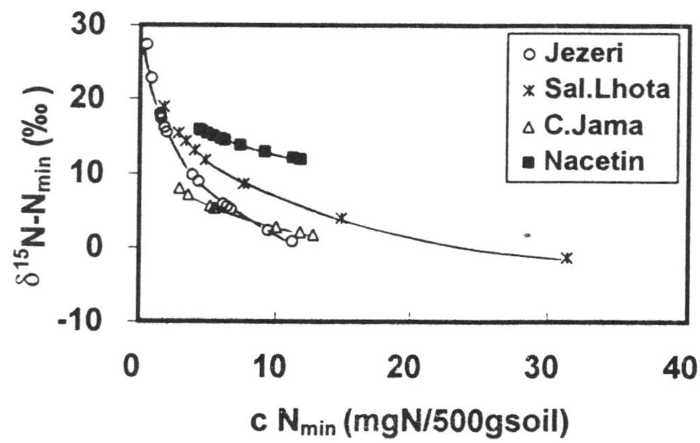

Fig. 7. δ^{15}N-N_{min} vs. N_{min} plot for all monitored sites. Weighted average values were correlated by applying logarithmic regression. The curves are discussed in the text.

The $\delta^{15}N$-N_{min} vs. cN_{min} plots represent a logarithmic regression according to fractionation equation (Eq. 2). The relatively low NO_3^- production in the spruce forested soil at the Salačova Lhota was taken as the “base curve“ in the plot. The plot for Načetín is above the base curve, indicating that the acidified soil loses isotopically lighter NO_3^- than the less acidified site. The plot for the damaged spruce at Červená Jáma is below the base curve. This is due to higher atmospheric contribution of N_{min}, which dilutes the soil-produced N_{min}. The curve representing the soil supporting the beech stand at Jezeří is also below the base curve. This shift is the result of more efficient uptake of N_{min} by undamaged beech.

4. Summary

Soil profiles under three spruce stands and one beech stand were sampled to study the effect of atmospheric acid deposition on ^{15}N fractionation of mineralized N. Atmospheric deposition varies from 25.8 to 35 kg of S ha^{-1} yr^{-1} and from 10.9 to 14.5 kg of N ha^{-1} yr^{-1}. Processes influencing mineralized N in soil can be identified by $\delta^{15}N$-N_{min} vs. cN_{min} plots.

Mineralized N generated in soils is affected by atmospheric input of NO_3^- and NH_4^+ in different ways. Exchangeable NH_4^+ is immobilized by adsorption and a new chemical and isotopic equilibrium is attained after every new NH_4^+ input to the studied soil horizon. Replacement of adsorbed NH_4^+ by new atmospheric input is more efficient at lower NH_4^+ concentrations in soil because, proportionally, more ions are replaced according to an equilibrium fractionation factor. Consequently, the “dilution“ effect is more visible on $\delta^{15}N$ of low concentrations of exchangeable NH_4^+. Moreover, the desorbed isotopically light NH_4^+ ions are entering another subsystem (a lower horizon) to attain a new equilibrium there. It results in a stepwise dilution from upper to lower soil horizons with $\delta^{15}N$ of exchangeable NH_4^+ more negative than corresponds to the residuum after nitrification.

Nitrate ions in soil are mobile and exported in soil solution without any adsorption in the horizon of their origin. Nitrate coming from the atmosphere mixes with NO_3^- produced in soil to form NO_3^- that is consumed partially by trees. The rest of the NO_3^- is removed by percolating soil solution. Frequency and abundance of precipitation events control the NO_3^- distribution in the soil profile. As the $\delta^{15}N$ ratio of atmospheric NO_3^- differs from the $\delta^{15}N$ ratio produced in soil by only 2 ‰ on average, only extreme $\delta^{15}N$ ratios of rain can be directly identified in the soil solutions.

Acknowledgments

This work was supported by the project No. ERB IC20 CT960024 of EEC and by Grant No.205/95/085 of the Czech Grant Agency.

References

Black, A.S. and Waring, S.A.: 1977, *Aust. J. Soil Res.* **15**, 51.
Černý, J., Havel, M. and Krejčí, R.: 1998, *Lesnictví* (in press).

Fottová, D.: 1996, *Calculation of Critical Loads based on Monitoring of Mass Balances in a Network of the Small Catchments, Czech Republic*, Final Report of Grant No. 1832/93, MS Czech Geological Survey, Prague (in Czech).
Fritz, P. and Fontes, J.C.: (eds.): 1980, *Handbook of Environmental Isotope Geochemistry, vol. 1a, Terrestrial Environment*, Elsevier, Amsterdam.
Gebauer, G. and Schulze, E.-D.: 1991, *Oecologia* **87**, 198.
Gebauer, G. and Dietrich, P: 1993, *Isotopenpraxis Environ.Health Stud.* **29**, 35.
Gillet, R.W. and Ayres, G.P.: 1991, *Atmos. Environ.* **25**, 2677.
Hübner, H.: 1986, "Isotope Effects of Nitrogen in the Soil and Biosphere', in P.Fritz and J.C. Fontes (eds.), *Handbook of Environmental Isotope Geochemistry, vol. 2b, The Terrestrial Environment*, Elsevier, Amsterdam, pp. 361-425.
Mariotti, A., Pierre, D., Vedy, J.C., Bruckert, S. and Guillemot, J.: 1980, *Catena* **7**, 293.
Mariotti, A., Germon, J.C. , Hubert, P. et al.: 1981, *Plant and Soil* **62**, 413.
Nadelhoffer, K.J. and Fry, B.: 1994, 'Nitrogen Isotope Studies in Forest Ecosystems', in K. Lajtha and R.H. Michener (eds.), *Stable Isotopes in Ecology and Environmental Science*, Blackwell Scientific Publications, London, pp. 22-42.
Pačes, T.: 1985, *Nature*, **315**, 31.
Schulze, E.-D.: 1996, *NIPHYS - Nitrogen Physiology of Forest Plants and Soils*, Final Report for EEC project EV5V-CT92-0433, Bayreuth, Germany.

CANOPY AND SOIL RETENTION OF NITROGEN DEPOSITION IN A MIXED BOREAL FOREST IN EASTERN FINLAND

S. PIIRAINEN[1], L. FINÉR[1] and M. STARR[2]

[1]*The Finnish Forest Research Institute, Joensuu Research Station, P.O.Box 68, FIN-80101 Joensuu, Finland*
[2]*The Finnish Forest Research Institute, Vantaa Research Centre, P.O.Box 18, FIN-01301 Vantaa, Finland*

Abstract. Nitrogen deposition, leaching, and retention were monitored in a mature spruce (*Picea abies* Karsten) dominated mixed boreal forest in eastern Finland. Bulk precipitation, throughfall, stemflow, and percolation through the podzolic soil profile were monitored from 1993 to 1996. Mean annual bulk deposition of total N was 3.83 kg ha^{-1}, of which 33 % was NH_4^+-N, 26 % was NO_3^--N, and 41 % was organic N. Throughfall+stemflow flux of total N was 2.93 kg ha^{-1} yr^{-1}. Sixty-four % of NH_4^+-N and 38 % of NO_3^--N in bulk precipitation was retained by the tree canopy. Organic N was released (0.27 kg ha^{-1} yr^{-1}) from the tree canopy. Nitrate-N was retained and organic N was leached as the water passed through the ground vegetation and soil O-horizon. Ammonium-N and organic N were retained mainly in the E-horizon. The output of total N from the E-horizon was only 5 % of the total N deposition in the forest stand during the study period and it was mainly as organic N. The output of inorganic N forms from under B-horizon was seasonal and occurred mainly at spring snowmelt.

Keywords: canopy, forest, nitrogen leaching, nitrogen retention

1. Introduction

Nitrogen (N), often the growth limiting nutrient in boreal coniferous forests, is tightly cycled with minor leaching (Tamm, 1991). The deposition of N may result in the acidification and eutrophication of N-limited freshwater ecosystems because the critical load for the most sensitive ecosystems (e.g., softwater lakes) is 5-10 kg N ha^{-1} yr^{-1} (Bobbink and Roelofs, 1995). In Finland, the annual N deposition ranges from 10.6 kg ha^{-1} in the south to 0.9 kg ha^{-1} in the north (Järvinen and Vänni, 1994). Because of retention in the canopy, amounts of throughfall N below boreal coniferous forest canopies are typically less than in bulk precipitation. Studies in Finnish boreal forests have shown that canopy retention of N may be 10-40 % of total N deposition (Hyvärinen,1990; Helmisaari, 1995; Starr, 1995). In areas with high N deposition (>10 kg ha^{-1} yr^{-1}), NO_3^- can leach in considerable amounts to groundwater or surface water (Gundersen, 1995). In areas with low N deposition (< 3 kg ha^{-1} yr^{-1}), < 5-10 % of N deposition leaches below the surface layers in podzolic soils, and then mainly during snowmelt when there is little or no uptake (Helmisaari, 1995). Disturbances to forest ecosystems, such as forest harvesting, decrease canopy and root uptake and can increase the leaching of N to surface water (Lepistö *et al.*, 1995).

In this paper, we present data on the fate of NH_4^+-N, NO_3^--N, and organic N in bulk deposition as it passes through the canopy and soil of a coniferous stand. The extent of N retention by the forest canopy and soil and the amount of N leaching from the soil are determined and evaluated. These baseline data are a part of an ongoing project that evaluates the effects of forest harvesting on the leaching of nutrients to surface waters (Finér *et al.*, 1997).

Water, Air, and Soil Pollution **105:** 165–174, 1998.

2. Material and Methods

2.1. Study Area

Data were collected from three 50 m × 50 m sample plots located in a mature mixed boreal forest in eastern Finland (63° 51' N, 28° 58' E, 220 m a.s.l.). Norway spruce (*Picea abies* Karsten) dominates the stand, but Scots pine (*Pinus sylvestris* L.), white and silver birch (*Betula pubescens* Ehrh. and *Betula pendula* Roth.) and European aspen (*Populus tremula* L.) also are present. Norway spruce accounts for 50 %, Scots pine 34 %, and the deciduous tree species 16 % of the total volume of 266 m^3 ha^{-1} in the tree stand. The canopy coverage in the stand is 54 %. The forest is uneven-aged; the oldest trees are 170 years old.

The site type is *Vaccinium-Myrtillus* type according to the classification of Cajander (1949). The field layer is dominated by dwarf shrubs (*Vaccinium vitis-ideaea* L. and *V. myrtillus* L.) and the bottom layer by feather mosses (*Pleurozium schreberi* Brid. and *Hylocomnium splendens* (Hedw.) B.S.&G.).

The soil, derived from sandy till, is a weakly developed iron-podzol (according to the classification of Kubiëna, 1953), with a clay content of <2 %. The bedrock is composed of granodiorite. The average thickness of the mor humus layer is 3 cm. Mean annual air temperature recorded at 2 m for the period 1993-1995 averaged +1.0 °C and the temperature sum was 908 °C (sum of daily mean temperatures >+5 °C). A more detailed description of the study area is presented by Finér *et al.* (1997).

2.2. Deposition and Soil Water Sampling

Bulk precipitation and throughfall were monitored from 1993 to 1996 and stemflow from 1993 to 1995. Bulk precipitation was collected in an open, treeless site within 500 m of the sample plots. Five plastic collectors (each 130.7 cm^2) were used during the snow-free period and three plastic bucket snow collectors were used during the winter (each 298.7 cm^2 for winter 1992/1993 and 1146.1 cm^2 thereafter). The collectors were placed 0.6 m above the ground during the snow-free period and 1.5 m above the ground in winter. Throughfall was measured in each sample plot using 16 collectors during the snow-free period and eight snow collectors during winter. The collectors were of the same type as used for bulk precipitation and they were placed systematically around the sides of the plots. Stemflow was collected at two of the sample plots. At each plot, five white or silver birches, Scots pines and Norway spruces representing different diameter classes (>6 cm) were fitted with spiral type stemflow collectors made of silicon tubing. The bulk precipitation and throughfall collectors were emptied once a week in the snow-free period and once a month in winter. On each sampling occasion, the samples from all collectors were bulked for analysis, by plot in the case of bulk deposition. Stemflow was collected weekly and only during the snow-free period. The samples were combined by species and plot for chemical analyses.

Soil water was sampled in each sample plot with nine systematically placed zero-tension lysimeters installed at three depths: under the humus layer (O-horizon), under the eluvial horizon (E-horizon), and 23 cm below the boundary of eluvial and illuvial horizons (B-horizon). The lysimeters under the O-horizon were of the Jordan-type (Jordan, 1968) and made of a plexiglass with a collecting area of 420 cm^2. The lysimeters installed under the E- and B-horizons were made of a polythene plastic funnel with a collecting area of 299 cm^2 and fitted to a 2 L bottle. The lysimeters were emptied on the same day as the precipitation collectors and the volume collected was recorded. Soil water samples were taken irregularly from October to April. The soil water samples from each lysimeter were analyzed separately.

2.3. Laboratory Analyses and Calculations

Samples were kept cold after sampling and were transported to laboratory the day after sampling. Bulk precipitation, throughfall, and stemflow samples were filtered through a Schleiche & Schuell no. 5891 filter and the soil percolate samples through a Schleiche & Schuell no. GF 52 glass wool filter; filtrates were stored in a freezer before analyzing. Ammonium and total N were determined by flow injection analysis (Tecator FIA-Star 5020) and NO_3^- by ion chromatography (Dionex series 100 or 300 or 500). Organic N was calculated by subtracting the sum of NH_4^+-N and NO_3^--N concentrations from the total N concentration. Deposition and leaching values were calculated by multiplying the concentrations by the amount of water collected. Pearson product-moment correlations and the paired *t*-test with a probability of Bonferroni's inequality were used to evaluate the data statistically (Snedecor and Cochran, 1980) using SAS for Windows 6.11.

3. Results

3.1. Bulk Precipitation, Throughfall, and Stemflow

The mean annual bulk precipitation for 1993-1996, was 502 mm (±52 mm); monthly bulk precipitation ranged from 11 to 116 mm. Precipitation fell mainly as snow from November to April, and accounted for 45 % of the annual bulk precipitation. Throughfall was 86-100 % of annual bulk precipitation and stemflow was almost negligible, only 1.3-1.5 % of annual bulk precipitation. Monthly total throughfall (throughfall+stemflow) ranged from 9 to 114 mm.

The N in bulk precipitation was dominated by inorganic forms, whereas in throughfall and stemflow, organic N concentrations were higher (Table I). The NH_4^+-N and NO_3^--N concentrations in throughfall were significantly lower than those in bulk precipitation, and organic N concentration in stemflow was significantly higher than that in throughfall and bulk deposition.

Over the study period, deposition of total N in bulk precipitation averaged 3.83 kg ha^{-1} yr^{-1} (Figure 1), 59 % being inorganic N. The annual bulk deposition of NH_4^+-N (1.27 kg ha^{-1}) was slightly higher than that of NO_3^--N (1.01 kg ha^{-1}), contributing 33 % and 26 % to the annual total N deposition, respectively. Deposition of NO_3^--N was greater during the dormant season than during the growing season; the opposite pattern was observed for NH_4^+-N (Figure 2).

TABLE I

Water volume-weighted mean annual concentrations (mg L^{-1}; SD in parentheses) of different N compounds during 1993-1996. Values within columns followed by the same letter do not differ significantly (p = 0.0125) from each other.

	Total N	NH_4^+-N	NO_3^--N	Organic N
Bulk precipitation	0.76 (0.22) ef	0.25 (0.04) a	0.21 (0.02) a	0.32 (0.20) a
Throughfall	0.60 (0.05) abe	0.11 (0.03) b	0.15 (0.02) b	0.35 (0.03) a
Stemflow	1.35 (0.12) cdf	0.07 (0.06) ab	0.04 (0.01) bc	1.21 (0.11) b
O-horizon	1.06 (0.34) acef	0.15 (0.09) ab	0.04 (0.02) abc	0.85 (0.23) b
E-horizon	0.94 (0.26) ac	0.17 (0.08) ab	0.06 (0.06) abc	0.71 (0.18) b
B-horizon	0.35 (0.20) bd	0.06 (0.03) b	0.04 (0.04) c	0.25 (0.15) a

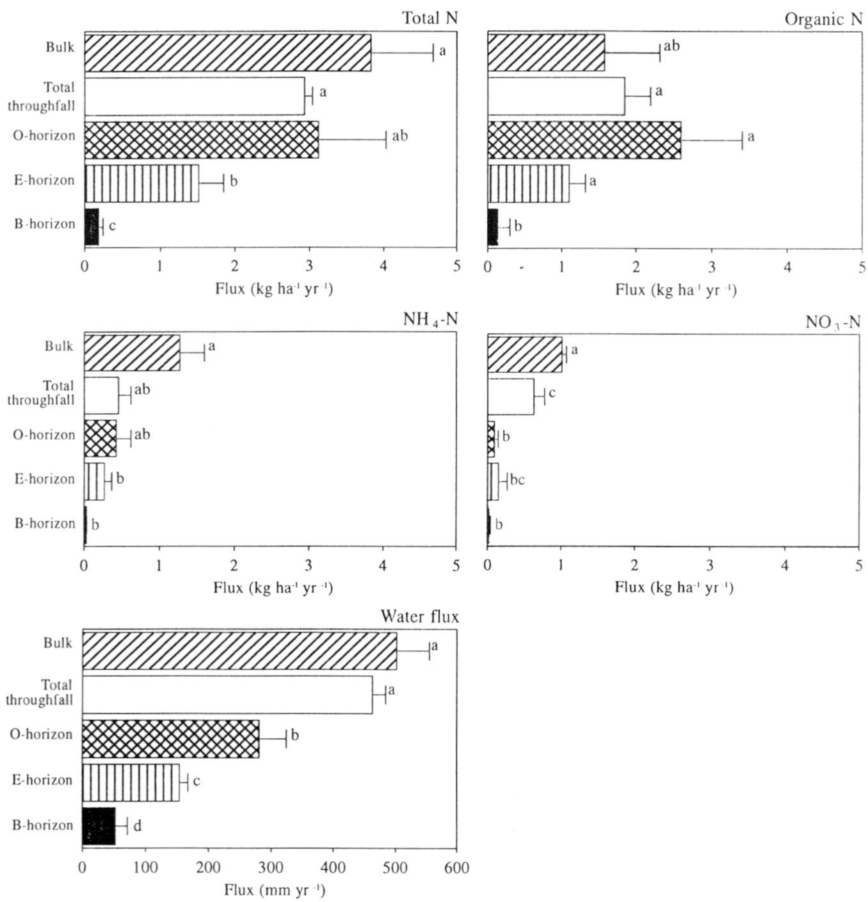

Fig. 1. Mean annual fluxes (±SD) of different N compounds and water in bulk deposition, total throughfall, and percolates from below the O-, E- and B-horizons; 1993-1996. Values with the same letter do not differ significantly (p = 0.0125) from each other.

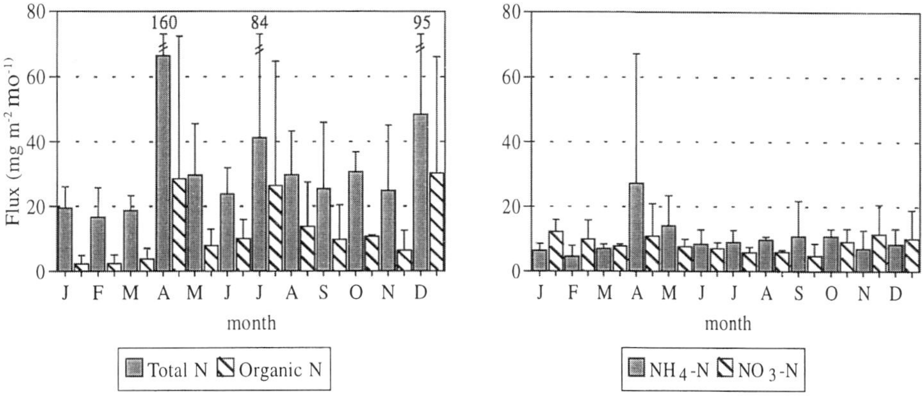

Fig. 2. Monthly mean bulk deposition (±SD) of different N compounds; 1993-1996.

Annual N deposition in total throughfall averaged 2.93 kg ha^{-1}, i.e. 77 % of that in bulk deposition (Figure 1). Total N that reached the forest floor via stemflow was only 2-3 % of total N deposition in total throughfall during the study period. Annual deposition of NH_4^+-N and NO_3^--N in total throughfall was 0.46 and 0.63 kg ha^{-1}, respectively (Figure 1). However, the difference between bulk precipitation and total throughfall (net throughfall) shows that the forest canopy was a sink for inorganic N (Figure 3). The forest canopy retained 38 % of the NO_3^--N bulk deposition and 64 % of the NH_4^+-N bulk deposition, corresponding to a net uptake of 1.19 kg N ha^{-1} yr^{-1}. Canopy retention of NH_4^+-N averaged 0.81 kg ha^{-1} yr^{-1} and of NO_3^--N, 0.38 kg ha^{-1} yr^{-1} and was the least during the dormant season. During the growing season, 76 % of NH_4^+-N and 61 % of NO_3^--N deposition were retained by the canopy (Figure 3). Monthly canopy uptake of NH_4^+-N and NO_3^--N was correlated with the monthly bulk deposition of NH_4^+-N ($r = 0.98$, $n = 44$, $p < 0.01$) and of NO_3^--N ($r = 0.72$, $n = 38$, $p < 0.01$).

The annual deposition of organic N in total throughfall averaged 1.85 kg ha^{-1}, corresponding to 56 % of the total N in total throughfall (Figure 1). Organic N was released from the canopy in spring and autumn, but was retained in the canopy during the growing season (Figure 3). The annual release of organic N from the canopy averaged 0.27 kg ha^{-1}.

3.2. Soil Percolates

During the study period, 61, 33 and 11 % of the total annual throughfall, on average, passed through the O-, E- and B-horizons, respectively (Figure 1). The average monthly soil percolate from below the O-horizon ranged from 0 to 59 mm. Usually the soil was frozen from December to March and no percolation occurred. At snowmelt, and occasionally also after heavy rainfall in summer and autumn, soil water percolation was at its greatest. In the summer months, the average monthly amount of soil water percolate from below the B-horizon varied from 1 to 3 mm.

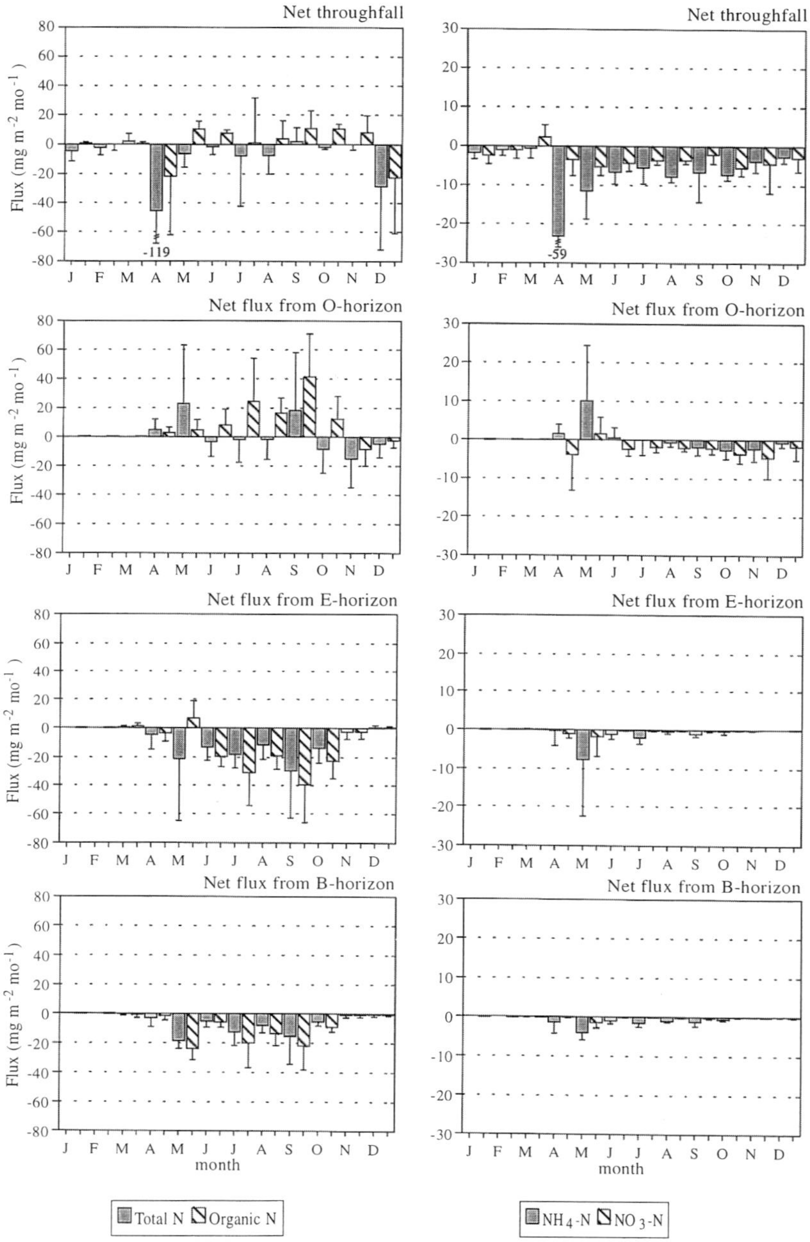

Fig. 3. Monthly mean (±SD) net flux (input-output) of different N compounds from forest and under different soil horizons in 1993-1996; negative values indicate retention and positive values indicates leaching.

Mean annual NO_3^--N concentrations in the soil percolate from the O-horizon were lower than concentrations in throughfall, and remained almost constant deeper in the soil (Table I). Mean annual NH_4^+-N concentration of the soil percolate was similar at all depths. Mean annual organic N concentration in the soil percolate increased as it passed through the O-horizon, but decreased below the B-horizon.

Some leaching of NO_3^--N from the O-horizon also was observed at spring snowmelt, but most of the NO_3^--N was retained (Figure 3). The retention of NO_3^--N in the O-horizon was considerable; 84 % (0.53 kg ha^{-1} yr^{-1}) of the annual average total throughfall deposition. Monthly total throughfall deposition of NO_3^--N and net retention of NO_3^--N by the O-horizon were correlated for the period when soil water percolation occurred ($p < 0.01$, $r = 0.91$, $n = 30$). The retention of NO_3^--N in the mineral soil layers also was high, especially in the B-horizon; output from below the B-horizon was only 2 % of the input to the forest stand via bulk deposition.

Some NH_4^+-N leaching from the O-horizon occurred at spring snowmelt (Figure 3). During the summer and autumn, NH_4^+-N was retained in the O-horizon. Monthly total throughfall deposition of NH_4^+-N and net retention of NH_4^+-N in the O-horizon during growing period also were correlated ($p < 0.01$, $r = 0.47$, $n = 29$). On an annual basis, NH_4^+-N output from below the O-horizon was balanced by the total throughfall deposition (Figure 1). The mineral soil horizons retained NH_4^+-N during the growing period; the output from below the B-horizon being 2 % of the bulk precipitation flux. The monthly flux of NH_4^+-N to the E- and B-horizons was correlated with the retention of NH_4^+-N by these horizons ($r = 0.94$, $p < 0.01$, $n = 30$ and $r = 0.99$, $p < 0.01$, $n = 28$, respectively). On an annual basis, 93 % of the NH_4^+-N input to the mineral soil horizons from O-horizon was retained.

Retention of inorganic N in total throughfall deposition by the soil was 0.43 kg ha^{-1} yr^{-1} for NH_4^+-N and 0.61 kg ha^{-1} yr^{-1} for NO_3^--N. The organic horizon retained most of the NO_3^--N (87 %) and the B-horizon retained most of the NH_4^+-N (56 %).

On an annual basis, 0.75 kg ha^{-1} yr^{-1} of organic N was leached from the O-horizon, but it was subsequently retained in mineral soil horizons (Figure 3). The retention of the organic N leached from O-horizon was 1.50 kg ha^{-1} yr^{-1} in the E-horizon and 0.96 kg ha^{-1} yr^{-1} in the B-horizon. Output from below the B-horizon was only 9 % of the bulk precipitation input of organic N.

The output of total N from below the B-horizon (0.19 kg ha^{-1} yr^{-1}) was 5 % of bulk precipitation total N input (Figure 1). Almost 74 % of total N in the percolate from below the B-horizon was organic. In total, 2.46 kg of organic N was retained jointly by the forest canopy and soil, with the E-horizon retaining the most (61 %).

4. Discussion

4.1. Retention by Canopy

The critical load of N to coniferous forests has been estimated as 10-20 kg ha^{-1} yr^{-1}, (Bobbink and Roelofs, 1995). In Finland, only 65 % of the inorganic N input in bulk precipitation is deposited as wet precipitation (Tuovinen *et al.*, 1990). Dry deposition of N is dominated by HNO_3, NO_2^--N and NH_3-N; particulate nitrate and particulate ammonium sulfate are also notable components (Tuovinen *et al.*, 1990). On the basis of the values given by Tuovinen *et al.* (1990), we estimated that the inorganic dry deposition of N to the study area is 1.23 kg ha^{-1} yr^{-1}, and the total deposition of N is about 5.0 kg N ha^{-1} yr^{-1}.

The N in total throughfall represents the balance between leaching, canopy uptake and dry deposition (Parker, 1983). Total throughfall contained 52 % of the inorganic N present in bulk precipitation; a similar percentage has been observed in other coniferous forests (Rosén and Lundmark-Thelin, 1985; Helmisaari and Mälkönen, 1989; Hyvärinen, 1990; Starr and Ukonmaanaho, 1995). The retention of inorganic N in the canopy may be due to foliar uptake (Carlisle *et al.*, 1966; Tukey, 1970; Reiners and Olson, 1984), uptake by epiphytic lichens (Lang *et al.*, 1976; Reiners and Olson, 1984), and in the case of NH_4^+, by cation exchange (Tukey, 1970). Spruce canopies have been shown to be the strongest sinks for inorganic N and also to have the highest biomass of epiphytic lichens (Lovett, 1992). Of the mean annual retention of inorganic N by the forest canopy (1.19 kg ha^{-1}), most was NH_4^+-N. If the estimate for dry deposition is considered, the total N retention would be as much as 2.42 kg ha^{-1} yr^{-1}.

The bulk deposition of NH_4^+-N was 26 % greater than that of NO_3^--N, but the canopy retention of NH_4^+-N was 113 % greater than that of NO_3^--N. Norway spruce utilizes NH_4^+ directly via the needles more readily than NO_3^- (Bowden *et al.*, 1989; Eilers *et al.*, 1992; Lumme, 1994). Increasing inorganic N uptake with increasing inorganic N deposition may be a general characteristic of spruce canopies (Eilers *et al.,* 1992; Lovett, 1992).

The demand for N by coniferous forests is 30-70 kg ha^{-1} yr^{-1} (Nihlgård, 1972; Mälkönen, 1975; Finér, 1989; Helmisaari, 1995). On this basis, Likens *et al.* (1977) and Helmisaari (1995) concluded that N deposition supplies about 10 % of the annual N demand; the rest being supplied by recycling. According to our results, the bulk deposition of total N (including dry deposition) supplied about 3-8 % of the annual N requirement of the stand.

4.2. Leaching of N from Soil

The mean annual NO_3^--N concentration in the percolate from below the organic horizon was lower than in total throughfall. The uptake by roots and microorganisms is probably efficient; only 16 % of the NO_3^--N deposition in total throughfall passed through the O-horizon (cf. Brinkmark, 1980; Helmisaari, 1995). Some leaching of NO_3^--N occurred in spring, during snowmelt, when the biological uptake is small and water flux is high. The deposition of NO_3^--N during winter months was higher than during growing season. As a result, the water equivalent of the snowpack contains more NO_3^- than a corresponding amount of rainfall deposition (Overrein *et al.*, 1980). Nitrate leaching from soil during snowmelt has also been reported by Helmisaari and Mälkönen (1989) and Pardo *et al.* (1995). Stream water NO_3^--N concentrations also are typically higher during the spring meltwater than during other seasons in boreal forest ecosystems (Overrein *et al.*, 1980).

As also observed by Helmisaari and Mälkönen (1989), NH_4^+-N was leached from the O-horizon in spring, during snowmelt, and retained during the summer and autumn. Bringmark (1980) suggested that the higher retention of NH_4^+-N in autumn compared to other times of the year corresponds to uptake resulting from a larger population of fungi and bacteria at this time. On an annual basis, however, the input and output of NH_4^+-N for the O-horizon were in balance. Helmisaari (1995) reported a net annual leaching of NH_4^+-

N from the organic horizon. However, her study site was a 100-year-old Scots pine stand. Norway spruce roots have been shown to prefer NH_4^+ over NH_3- (Ek, 1993; Lumme, 1994). Nevertheless, the NH_4^+-N percolating from the O-horizon during spring snowmelt was effectively retained in the mineral soil (93 % of the annual input). This retention of NH_4^+-N may not have been entirely due to uptake by roots and micro-organisms, however, as NH_4^+ ions can be retained by cation exchange and some may have been converted to NO_3^- through nitrification. Persson and Wirén (1995) have reported nitrification at a depth of 10-20 cm and deeper in acid mineral soil, although they found none in the humus layer.

In conclusion, the spruce canopy acted as a sink for N (retaining almost 50 % of total deposition) and there was little leaching of N beyond the B-horizon (7 % of the N in throughfall deposition). Because the amount of NO_3^--N leaching from the B-horizon was low (<0.1 kg ha^{-1} yr^{-1}) and represented such a small fraction of the amount arriving to the O-horizon (<2 % of NO_3^--N and <1 % of inorganic N), the ecosystem clearly can not be considered N saturated (cf., Tietema *et al.* 1994). In the future, we intend to examine the effect of harvesting on NO_3^--N leaching.

Acknowledgments

The authors wish to thank Prof. Hannu Mannerkoski for comments to the manuscript. Ms. Seija Repo and Ms. Anita Pussinen for analyzing the water samples, Ms. Leena Karvinen for preparing the figures and tables, and Mr. Timo Pelkonen, Mr. Esko Kokkonen and Mr. Markku Tiainen for field work. We also wish to acknowledge the National Forest and Park Service for making the site available.

References

Bowden, R., Geballe, G. and Bowden, W.: 1989, *Can. J. For. Res.* **19**, 382.

Bobbink, R. and Roelofs, J.G.M.: 1995, *Water, Air, Soil Pollut.* **85,** 2413.

Brinkmark, L.: 1980, 'Ion Leaching through a Podsol in a Scots Pine Stand', in T. Persson (ed.), *Structure and Function of Northern Coniferous Forest - An Ecosystem Study, Ecol. Bull. 32,* Stockholm, pp. 341-361.

Cajander, A.K.: 1949, *Acta For. Fenn.*, **56**, 1.

Carlisle, A., Brown, A.H.F. and White, J.: 1966, *J. Ecol.* **54**, 87.

Eilers, G., Brumme, R. and Matzner, E.: 1992, *For. Ecol. Manage.*, **51**, 239.

Ek, H.: 1993, *Doctoral Thesis, University of Lund, Sweden*

Finér, L.: 1989, *Acta For. Fenn.* **208**, 43 pp.

Finér, L., Ahtiainen, M., Mannerkoski, H., Möttönen, V., Piirainen, S., Seuna, P. and Starr, M.: 1997, *Finnish Forest Research Institute, Research Papers* **648**, 38 pp.

Gundersen, P.: 1995, *Water, Ai,r Soil Pollut.* **85**, 1179.

Helmisaari, H-S. and Mälkönen, E.: 1989, *Scand. J. For. Res.* **4**, 13.

Helmisaari, H-S.: 1995, *Plant and Soil* **168-169**, 327.

Hyvärinen, A.: 1990, 'Deposition on Forest Soils - Effects of Tree Canopy on Throughfall', in P. Kauppi, P. Anttila and K. Kenttämies. (eds.), *Acidification in Finland*, Springer-Verlag, New York, pp. 199-213.

Jordan, C.F.: 1968, *Soil Sci.* **106**, 81.
Järvinen, O. and Vänni,T.: 1994, *Vesi- ja ympäristöhallituksen monistesarja* **579**, 68 pp. (in Finnish)
Kubiëna, W.L.: 1953. *The Soils of Europe*, 1st ed., Thomas Murby and Co., Madrid, Spain.
Lang, G.E., Reiners, W.A. and Heier, R.K.: 1976, *Oecologia* **25**, 229.
Lepistö, A., Andersson, L., Arheimer, B. and Sundblad, K.: 1995, *Water, Air, Soil Pollut.* **84,** 81.
Likens, G.E., Bormann, F.H., Pierce, R.S., Eaton, J.S. and Johnson, N.M.: 1977, *Biogeochemistry of a Forested Ecosystem,* Springer-Verlag, New York.
Lovett, G.M.: 1992, 'Atmospheric Deposition and Forest Nutrient Cycling', in D.W. Johnson and S.E. Lindberg (eds.), *Ecol. Studies* **91**, Springer-Verlag, pp. 152-166.
Lumme, I.: 1994, *For. Ecol. Manage.* **63**, 87.
Mälkönen, E.: 1975, *Comm. Inst. For. Fenn.* 84.5..
Nihlgård, B.: 1972, *Oikos* **23**, 69.
Overrein, L.N., Seip, H.M. and Tollan, A.: 1980, *Acid Precipitation - Effects on Forest and Fish,* Final report of the SFNS-project, Sur Nedbors Virkning på Skog og Fisk, Research Report Fagrapport FR.
Pardo, L.H., Driscoll, C.T. and Likens, G.E.: 1995, *Water, Air, Soil Pollut.* **8**, 1659.
Parker, G.G.: 1983, *Adv. Ecol. Res.* **13**, 58.
Persson, T. and Wirén, A.: 1995, *Plant and Soil* **168-169**, 55.
Reiners, W.A. and Olson, R.K.: 1984, *Oecologia* **63**, 320.
Rosén, K. and Lundmark-Thelin, A.: 1985, *Reports in Forest Ecology and Forest Soils* **51**, Swedish University of Agricultural Science, Department of Forest Soils, 1-15.
Snedecor, G.W. and Cochran, W.G.: 1980, *Statistical Methods, 7th ed.*, The Iowa State University Press, Ames, Iowa.
Starr, M.: 1995, 'Quantity of Throughfall and Stemflow', in I. Bergström, K., Mäkelä and M. Starr (eds.), *Integrated Monitoring in Finland,* Ministry of the Environment, Environment Policy Department, Report **1**, pp. 59-60.
Starr, M. and Ukonmaanaho, L.: 1995, 'Stemflow Chemistry', in I. Bergström, K., Mäkelä and M. Starr (eds.), *Integrated Monitoring in Finland*, Ministry of the Environment, Environment Policy Department, Report **1**, pp. 61-62.
Tamm, C. O.: 1991, *Ecological Studies* **81**, Springer-Verlag, New York.
Tietema, A., Riemer, L., Verstraten, J.M., van deer Maas, M.P., van Vijk A.J. and van Voorthuyzen, I.: 1994, *For. Ecol. Manage.* **57**, 20.
Tukey, H.B.J.: 1970, *Ann. Rev. Plant Physiol.* **21**, 305.
Tuovinen, J.-P., Kangas, L. and Norlund, G.: 1990, 'Model Calculations of Sulphur and Nitrogen Deposition in Finland', in P. Kauppi, P. Anttila and K. Kenttämies (eds.), *Acidification in Finland*, Springer-Verlag, New York, pp. 167-197.

EFFECTS OF CHRONIC NITROGEN AMENDMENTS ON PRODUCTION OF DISSOLVED ORGANIC CARBON AND NITROGEN IN FOREST SOILS

WILLIAM H. MCDOWELL[1], WILLIAM S. CURRIE [2], JOHN D. ABER[3] and YURIKO YANO[1]

[1]*Department of Natural Resources, University of New Hampshire, Durham, NH 03824, USA*
[2]*Ecosystems Center, Marine Biological Laboratory, Woods Hole, MA 02541, USA*
[3]*Complex Systems Research Center, University of New Hampshire, Durham, NH 03824, USA*

Abstract. Chronic N deposition has been hypothesized to affect DOC production in forest soils due to the carbon demand exerted by microbial immobilization of inorganic N. We tested this hypothesis in field experiments at the Harvard Forest, Petersham, Massachusetts, USA. During four years of sampling soil solution collected beneath the forest floor in zero-tension lysimeters, we observed little change in DOC concentrations (10-30 % increase, not statistically significant) associated with elevated N inputs, but did observe significant increases in DON concentrations. Both DOC and DON varied seasonally with highest concentrations in summer and autumn. Mean DON concentrations increased 200-300 % with the highest rate of inorganic N fertilization, and concentrations of DON were highest in samples with high inorganic N concentrations. We conclude that the organic chemistry of soil solution undergoes qualitative changes as a result of long-term N amendment at this site, with small changes in DOC, large increases in DON, and a decline in the C:N ratio of dissolved organic matter.

Keywords: dissolved organic carbon, dissolved organic nitrogen, Harvard Forest, nitrogen fertilization

1. Introduction

Assessing the long-term impacts of anthropogenic N deposition is an important goal for forest ecosystem management. The impacts are potentially severe, and include decline in forest vigor as well as degradation of water quality (Aber *et al.*, 1989). The ability of a forest ecosystem to retain N and thus forestall the "N saturation" that can occur with continued high N inputs varies considerably from site to site or stand to stand (e.g., Kahl *et al.*, 1993; Magill *et al.*, 1997; Gundersen *et al.*, 1997). Understanding the mechanisms of N retention in forested ecosystems is essential to accurately predict the long-term consequences of N saturation, as well as the likelihood that a given site will undergo N saturation within a specific time frame.

Nearly complete retention of N has been observed in the Harvard Forest hardwood plots from 1988-1993 (Magill *et al.*, 1997), despite addition of 826 kg ha^{-1} N in the high-N treatments. This added N appears to have been retained in a non-exchangeable form in soil organic matter. The mechanism of N incorporation is unknown, but a plausible mechanism is immobilization by soil microbes (Aber *et al.*, 1993; Magill *et al.*, 1997). If microbial immobilization is responsible for this N retention, then decreases in the concentration and bioavailability of dissolved organic carbon (DOC) in soil solution were predicted to occur due to the increased energy demand associated with immobilization (Aber, 1992).

To test this hypothesis, we examined changes in soil solution chemistry associated with chronic N fertilization of hardwood and pine plots at the Harvard Forest. Concentration and flux of DOC and dissolved organic nitrogen (DON) were measured to assess the effects of increased N deposition on soil solution organic chemistry beneath the forest floor.

Water, Air, and Soil Pollution **105:** 175–182, 1998.

2. Methods and Materials

2.1. Site Description

The study site is the Harvard Forest, Petersham, Massachusetts, USA (42° 30' N, 72° 10' W). Climate is mildly continental, with average temperatures of -7 °C in January and 19 °C in July. Precipitation averages 110 cm and is evenly distributed throughout the year (Currie *et al.*, 1996). Snow typically covers the ground for several months during winter, but mid-winter thaws and rain storms are common. At the Harvard Forest, we studied soil solution in the chronic N plots (Aber *et al.*, 1993, Magill *et al.*, 1997). Plots (one per combination of vegetation and N deposition) are 30 x 30 m, with either hardwood (predominantly *Quercus velutina* and *Q. alba*) or pine (*Pinus resinosa*). The hardwoods are the result of regeneration at the site after clearcutting approximately 50 years ago; the pine was planted in 1926 after agricultural abandonment. Soils are derived from glacial till, and are rocky and well-drained Typic Dystrocrepts with C:N in the forest floor of 22-26 prior to manipulation. Soils of the pine plot were plowed prior to abandonment; those of the hardwood plots do not have an Ap horizon. The plots have received N additions since 1988. Current rates of addition are 150 kg N ha^{-1} yr^{-1} (high-N), 50 kg N ha^{-1} yr^{-1} (low-N), and control (no added N). Atmospheric deposition of N at the site is approximately 8 kg ha^{-1} yr^{-1} inorganic N in wet and dry deposition combined (Aber *et al.*, 1993). Further descriptions of the study site, methods of N application, and results to date are given by Aber *et al.* (1993), Currie *et al.* (1996), and Magill *et al.* (1997).

2.2. Sample Collection and Analysis

Soil solution was collected from polyethylene zero tension lysimeters (5 in each of the 6 plots) installed beneath the forest floor in August 1992 (Currie *et al.*, 1996). Lysimeters were sampled after major rain events; bottles were kept in place in the field at all times. Solution collected from the 5 lysimeters was combined by plot prior to analysis, yielding one sample per plot per event. Over the course of the study, approximately 400 samples were collected, corresponding to approximately 65 events in each of the 6 plots. Samples were transported on ice to the University of New Hampshire, where they were filtered through pre-combusted Whatman GF/F glass fiber filters within 36 hr of collection. After filtration, samples were frozen prior to analysis.

Dissolved organic C was analyzed as total C minus inorganic C using a Shimadzu TOC 5000 Total Organic Carbon analyzer with autosampler (Pt-catalyzed combustion at 680 °C). Total dissolved N (TDN) was measured by alkaline persulfate digestion (Solorzano and Sharp, 1980) from October 1992 to September 1994, and high temperature Pt-catalyzed combustion (Merriam *et al.*, 1996) from October 1994 to November 1996. Samples collected in 1994 were used to compare the two TDN methods; similar results were obtained with each (Merriam *et al.*, 1996). Nitrate and NH_4^+ were measured with automated colorimetry (Technicon methods 782-86T and 780-86T, respectively); NO_2^- was negligible (Currie *et al.*, 1996). Dissolved organic N was estimated as the difference between concentrations of TDN and dissolved inorganic N (DIN; the sum of NH_4^+-N and NO_3^--N).

To assess the effects of vegetation type and rate of N application on DOC and DON concentrations, the results were analyzed with ANOVA as a 2x3 factorial design without replication (factors = vegetation and N application, with 2 levels for vegetation and 3 levels for N treatment). To assess differences in solution chemistry associated with season, the results were analyzed with ANOVA as a 3x3 factorial design without replication (factors = N application and season, with 3 levels each). Data were grouped into the following seasons: spring (March-May), summer (June-August), and autumn (September-November). Where significant differences were found, Tukey's test was used to make comparisons among means. Significance was established at $p < 0.05$; all results are reported on that basis. Fluxes were calculated using the method described in Currie *et al.*, (1996) on a calendar year basis; the method involves estimating monthly water losses from the forest floor using PnET II (Aber and Federer, 1992) and multiplying seasonal water flux by seasonal mean concentrations in soil solution. For winter, solution concentrations were estimated from the average of the last autumn sample and the first spring sample for the appropriate year.

3. Results

Concentrations of DOC in forest floor leachate varied significantly by season, with lowest concentrations (as low as 10 mg L^{-1}) in spring and highest concentrations in summer and autumn (up to 300 mg L^{-1}; Figure 1). Mean concentrations in summer and autumn were significantly higher than those for spring. Concentrations of DOC were elevated by approximately 8 mg L^{-1} (15-25 %) in high-N plots relative to the controls (Table I), but this difference was not statistically significant. The highest DOC concentrations were observed in the N-amended pine plots in the first year after lysimeter installation, when concentrations consistently exceeded 100 mg L^{-1} (Figure 1).

Concentrations of DON in forest floor leachate showed significant seasonal variation, following patterns similar to those of DOC (higher in summer and autumn than spring; Table I). Concentrations varied significantly with N treatment; high-N plots were significantly higher in DON concentration than controls, with a doubling or tripling of average concentrations (Table I). Concentrations of DON reached levels as high as 30 mg L^{-1}, but were typically no higher than 5-10 mg L^{-1} (Figure 1). Overall, there was a strong positive relationship between concentrations of DON and DIN (DON=0.14DIN - 0.04, r^2= 0.57, $p < 0.0001$).

Changes in the relative proportions of DOC and DON in solution resulted in a significant decrease in the C:N ratio (mass:mass) of dissolved organic matter with N treatment (Table I). Although the C:N ratio declined with N fertilization, it was not significantly affected by vegetation type. The C:N ratio showed significant seasonal variation, with summer and autumn different from spring.

Fluxes of DOC and DON in soil solution followed the patterns seen in concentrations, with the highest fluxes typically found in high-N plots. Fluxes of DOC ranged from 27 to 75 g m^{-2} yr^{-1}, with highest values for high-N pine plots (Table II). Fluxes of DON ranged from 0.7 to 3.6 g m^{-2} yr^{-1}, and values were highest in the high-N treatments from both stands.

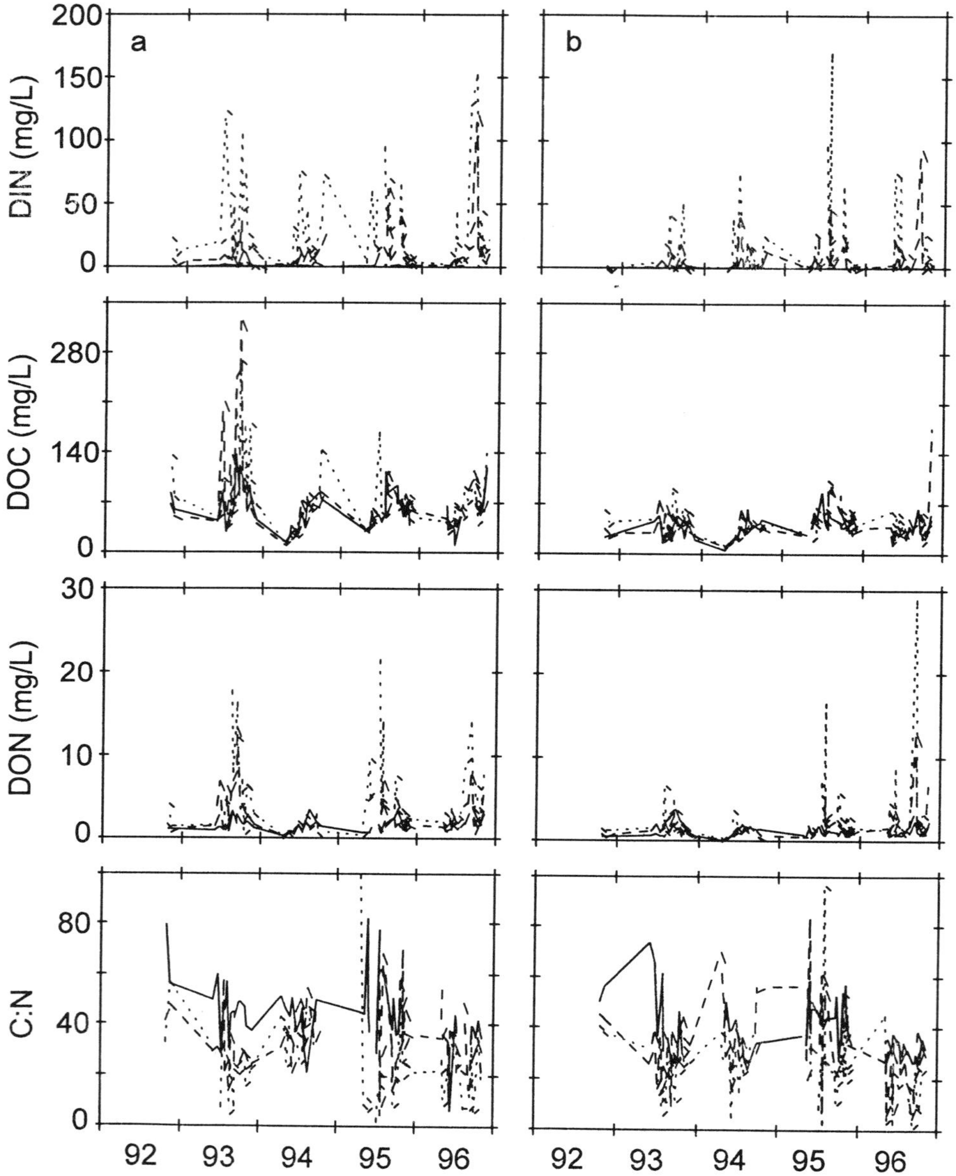

Fig. 1. Concentrations of inorganic N (DIN), dissolved organic carbon (DOC), dissolved organic nitrogen (DON), and the ratio DOC:DON (C:N; mass:mass) in soil solution collected from zero-tension lysimeters (forest floor leachate) in the pine (a) and hardwood (b) stand from 1992-1996. Solid line indicates control, dashed line the low-N treatment (50 kg ha^{-1} yr^{-1}), and dotted line the high-N treatment (150 kg ha^{-1} yr^{-1}).

TABLE I

Mean concentration of DOC and DON (mg L^{-1}) and mean DOC:DON (mass:mass) in forest floor solution collected from pine (P) and hardwood (H) plots at the Harvard Forest with zero tension lysimeters from 1992-1996. Since 1988, N has been applied to the plots at rates of 0 (control), 50 (Low-N), and 150 (High-N) kg ha^{-1} yr^{-1} (Magill *et al.*, 1997).

		Season			N-treatment		
	Stand	Spring	Summer	Autumn	Control	Low-N	High-N
DOC	Pine	36.5	69.3	86.3	60.8	75.5	76.6
	Hardwood	31.4	43.5	43.9	35.7	42.7	46.1
DON	Pine	1.50	3.19	3.21	1.46	2.68	458
	Hardwood	1.05	2.02	2.07	0.98	1.67	3.02
DOC:DON	Pine	39.6	32.9	37.6	43.6	36.4	28.7
	Hardwood	42.4	30.6	31.4	40.1	32.8	25.6

TABLE II

Flux of DOC and DON (g m^{-2} yr^{-1}) by N treatment and stand type from 1993-1996. Fluxes are calculated as the seasonal average concentration×seasonal water flow estimated by PnET II (Aber and Federer, 1992) for each year.

	Year							
	1993		1994		1995		1996	
Plot Treatment	DOC	DON	DOC	DON	DOC	DON	DOC	DON
Pine								
Reference	54	1.2	46	1.2	45	1.0	45	1.4
Low N	72	2.6	44	1.2	50	1.6	52	1.7
High N	75	3.6	56	1.3	54	3.2	45	2.9
Hardwood								
Reference	33	0.7	28	0.8	33	0.8	27	0.8
Low N	31	1.1	28	0.8	37	1.1	35	1.8
High N	42	1.8	31	1.0	40	1.9	36	2.9

4. Discussion

We originally had hypothesized that N saturation would be accompanied by a decrease in DOC concentration and flux due to the immobilization of added N by free-living soil microbes and the subsequent depletion of C available to fuel the immobilization process (Aber, 1992). Our results directly contradict that hypothesis; we observed a significant effect of N treatment on DOC concentrations in the hardwood plot, but the response was a moderate (and statistically insignificant) increase in DOC concentrations with high N application rather than the predicted decrease (Table I). Based on one year (1993-1994) of the four years of data presented here, Currie *et al.*, (1996) did not observe a statistically significant change in DOC or DON concentrations, although they did observe a significant change in DOC:DON. Our longer record shows there also has been a long-term change in DON concentrations associated with N treatment, but no change in DOC concentrations. In a comparison of 5 sites (NITREX project), Gundersen *et al.*, (1997) also concluded that DOC flux was not clearly related to N status at their control sites, but found that N addition did not consistently affect DON production. Rustad *et al.*, (1996) found no statistically significant changes in DOC concentrations with 4 yr of high N treatment of a spruce stand in Maine, USA. Guggenberger (1994) found that DOC fluxes in soil solution are highest in a site receiving the highest inputs of N in acid deposition, and suggested that this high DOC flux was the result of higher microbial activity due to increased N availability. In summary, neither our data nor data from other field experiments support the initial hypothesis of Aber (1992) that DOC concentrations will decline under conditions of N saturation.

Laboratory experiments, however, do show that DOC production by forest soils can be sensitive to N status. Gödde *et al.* (1996) found that in a comparison of six spruce stands, DOC produced during laboratory incubations of both cores and sieved soils was significantly lower in stands with high total N and low C:N in soil organic matter. At Harvard Forest, similar results have been obtained. DOC production by N-saturated soils (pine plot, high-N treatment collected in 1991) was less than half that of control and low-N samples when sieved soils were incubated for 35 days in laboratory microlysimeters (Benjamin, 1996). This discrepancy between field and laboratory results may be most easily explained by the lack of vegetation in the microlysimeters. In the lab, in the absence of vegetation or fully functioning tree roots, sources of DOC in high-N soils were depleted rapidly (Gödde *et al.*, 1996). We infer that in the field, this decline in DOC production by forest soil might be masked by a corresponding increase in DOC production by plant processes such as decomposition of recently fallen plant litter, root exudation, sloughing and decomposition of fine roots, or decomposition of mycorrhizal fungi.

The relative importance of various sources of DOC in soil solution is largely unknown. Guggenberger *et al.*, (1994) found that both oxidative degradation of plant-derived organic matter and production of microbial metabolites contribute to DOC in soil solution. Isotopic analysis (^{14}C) of DOC suggests that much of it is of recent origin, the result of photosynthetic fixation during the past 40 years (Schiff *et al.*, 1990). The striking disagreement between field and laboratory data for Harvard Forest points out the potentially large significance of recent plant photosynthate in providing DOC to soil solution, and warrants further investigation.

Our results do not provide conclusive evidence regarding the mechanism of N retention in forest soils experiencing heavy N loading. The original hypothesis that DOC should decline under conditions of N saturation was based on the premise that soil solution reflects the status of soil organic matter dynamics, and that soil organic matter would be depleted of labile materials with the increased C demand posed by N immobilization by soil microbes. One or both parts of our initial premise might be wrong. Rather than reflecting soil organic matter dynamics, DOC in soil solution might be more responsive to C balance in vegetation than previously thought. Alternately, N immobilization might be an abiotic process or due to immobilization of added N by mycorhizzal fungi rather than by free-living soil microorganisms.

The marked change in DOC:DON ratio (from above 40 to 26-29) which we observed under heavy N loading (Table I) shows that large qualitative shifts in the organic chemistry of soil solution have occurred. Other studies have shown that this ratio typically ranges from 40 to 12 (Qualls and Haines, 1991; Sollins and McCorison, 1981; Gundersen *et al.*, 1997), and thus the mean values that we report are within the range of those previously published. Fertilization has resulted in a 25-67 % increase in foliar N content on the high-N treatments (Magill *et al.*, 1997). This change in litter chemistry might have contributed to the change in soil solution chemistry by altering the soluble products of leaf decomposition, but such linkages are largely speculative.

The average concentrations of DON that we have measured in the control plots (1.0-1.5 mg L^{-1}) at Harvard Forest are similar to those observed at other sites. The average concentration of DON collected beneath the forest floor over multiple years and sites (hardwood and conifer) by the Integrated Forest Study in the United States was 0.4 to 2.1 mg L^{-1} (Johnson and Lindberg 1992); an identical range in concentration was obtained for coniferous forests in Sweden by Stuanes *et al.*, (1995). Flux values that we report in the control plots also are comparable to those reported by the Integrated Forest Study (Johnson and Lindberg, 1992) and NITREX (Gundersen *et al.*, 1997).

Neither the chemical structure nor the density of identifiable functional groups is known for the DON produced in response to inorganic N fertilization at Harvard Forest. Without such information, it is difficult to assess the ecological significance of this additional DON. If the added DON is comprised of compounds such as amines, then it is potentially available to plant roots or microbes growing in the mineral soil (Kaye and Hart, 1997). If it contains more refractory compounds, then long-term accumulation of this DON in the soil profile could occur. No net increase in DON concentration or flux from the mineral soil occurs under conditions of N saturation (Currie *et al.*, 1996), suggesting that either biotic uptake or abiotic adsorption (cf. McDowell and Wood, 1984) maintains a constant DON flux from the solum despite the increased flux from the forest floor.

Acknowledgments

Financial support was provided by the USDA National Research Initiative Competitive Grants program (award#93-37101-8700), the National Science Foundation LTER program, a NASA Graduate Training Grant Fellowship award, and the Agricultural Experiment Station of the University of New Hampshire. Contribution 1977 of the Agricultural Experiment Station of the University of New Hampshire. We thank J. Merriam, A. Magill, T. Eklund, and C. Goodale for technical assistance.

References

Aber, J.D.: 1992, *TREE* **7**, 220.

Aber, J.D. and Federer, C. A.:1992, *Oecol.* **92**, 463.

Aber, J.D., Magill, A., Boone, R., Melillo, J.M., Steudler, P. and Bowden, R.: 1993, *Ecol. Applic.* **3**, 156.

Aber, J.D., Nadelhoffer, K.J., Steudler, P. and Melillo, J.M.: 1989, *BioScience* **39**, 378.
Benjamin, S.S.: 1996, *Effects of Nitrogen Status on Dissolved Organic Carbon Dynamics in Forest Soils*, Thesis, University of New Hampshire, Durham.
Currie, W.S., Aber, J.D., McDowell, W.H., Boone, R.D. and Magill, A.H.: 1996, *Biogeochemistry* **35**, 471.
Gödde, M., David, M.B., Christ, M.J., Kaupenjohann, M. and Vance, G.F.: 1996, *Soil Biol. Biochem.* **28**, 1181.
Guggenberger, G.: 1994, *Environ. Internat.* **20**, 31.
Guggenberger, G., Zech, W. and Schulten, H.: 1994, *Org. Geochem.* **21**, 51.
Gundersen, P., Emmett, B.A., Kjønaas, O.J., Koopmans, C.J. and Tietema, A.: 1997, *Forest Ecol. Manag.*, in press.
Johnson, D.W. and Lindberg, S.E.: 1992, *Atmospheric Deposition and Forest Nutrient Cycling*, Springer-Verlag, New York.
Kahl, J.S., Norton, S.A., Fernandez, I.J., Nadelhoffer, K.J., Driscoll, C.T. and Aber, J.D.: 1993, *Environ. Sci. Techol.* **27**, 565.
Kaye, J.P. and Hart, S.C.: 1997, *TREE* **12**, 139.
Magill, A.H., Aber, J.D., Hendricks, J.J., Bowden, R.D., Melillo, J.M. and Steudler, P.A.: 1997, *Ecol. Applic.* **7**, 402.
McDowell, W.H. and Wood, T.: 1984, *Soil Sci.* **137**, 23.
Merriam, J., McDowell, W.H. and Currie, W.S.: 1996, *Soil Sci. Soc. Am. J.* **60**, 1050.
Qualls, R G. and Haines, B.L.: 1991, *Soil Sci. Soc. Am. J.* **55**, 1112.
Rustad, L.E., David, M.B., Mitchell, M.J., Nadelhoffer, K.J. and Fuller, R.B.: 1996, *Soil Sci. Soc. Am. J.* **60**, 1933.
Schiff, S.L., Aravena, R., Trumbore, S.E. and Dillon, P.J.: 1990, *Water Resourc. Res.* **26**, 2949.
Sollins, P. and McCorison, F.M.: 1981, *Water Resourc. Res.* **17**, 1409.
Solorzano, L. and Sharp, J.H.: 1980, *Limnol. Oceanogr.* **25**, 751.
Stuanes, A.O., Kjonaas, O.J. and Van Miegroet, H.: 1995, *For. Ecol. Manage.* **71**, 99.

EFFECT OF CHRONIC NITROGEN ADDITIONS ON SOIL NITROGEN FRACTIONS IN RED SPRUCE STANDS

M.B. DAVID[1], A.M. CUPPLES[1], G.B. LAWRENCE[2], G. SHI[1], K. VOGT[3] and P.M. WARGO[4]

[1]*University of Illinois, Dept. of Natural Resources & Environmental Sciences, Urbana, Illinois 61801, USA*
[2]*U.S. Geological Survey, Troy, New York 12180, USA*
[3]*Yale University, School of Forestry and Environmental Studies, New Haven, Connecticut 06511, USA*
[4]*USDA Forest Service, Northeastern Forest Experiment Station, Hamden, Connecticut 06514, USA*

Abstract. The responses of temperate and boreal forest ecosystems to increased nitrogen (N) inputs have been varied, and the responses of soil N pools have been difficult to measure. In this study, fractions and pool sizes of N were determined in the forest floor of red spruce stands at four sites in the northeastern U.S. to evaluate the effect of increased N inputs on forest floor N. Two of the stands received 100 kg N $ha^{-1}yr^{-1}$ for three years, one stand received 34 kg N ha^{-1} yr^{-1} for six years, and the remaining stand received only ambient N inputs. No differences in total N content or N fractions were measured in samples of the Oie and Oa horizons between treated and control plots in the three sites that received N amendments. The predominant N fraction in these samples was amino acid N (31-45 % of total N), followed by hydrolyzable unidentified N (16-31 % of total N), acid-soluble N (18-22 % of total N), and NH_4^+-N (9-13 % of total N). Rates of atmospheric deposition varied greatly among the four stands. Ammonium N and amino acid N concentrations in the Oie horizon were positively related to wet N deposition, with respective r^2 values of 0.92 and 0.94 (n = 4, $p < 0.05$). These relationhips were somewhat stronger than that observed between atmospheric wet N deposition and total N content of the forest floor, suggesting that these pools retain atmospherically deposited N. The NH_4^+-N pool may represent atmospherically deposited N that is incorporated into organic matter, whereas the amino acid N pool could result from microbial immobilization of atmospheric N inputs. The response of forest floor N pools to applications of N may be masked, possibly by the large soil N pool, which has been increased by the long-term input of N from atmospheric deposition, thereby overwhelming the short-term treatments.

Keywords: forest soil chemistry, forest nitrogen cycling, nitrogen deposition, forest soil nitrogen fractions

1. Introduction

Recent studies have examined the response of forest ecosystems to inputs of N to determine whether N saturation (Aber *et al.*, 1989) has occurred, and whether it has affected tree health (e.g., Sullivan, 1993; McNulty *et al.*, 1996). Nitrogen cycling in spruce-fir stands in the northeastern United States has been studied extensively because: 1) many of these stands receive high rates of atmospheric deposition, especially at high elevations; 2) decomposition rates in these stands are low; and 3) these trees have low N requirements relative to hardwood species in this region (McNulty *et al.*, 1996). Studies of both northern hardwood and spruce-fir forests in this region have found no changes in soil N pools in response to experimental N inputs (Christ *et al.*, 1995; Magill *et al.*, 1996).

Difficulty in measuring the responses of forest ecosystems to experimental N inputs also may be due to high rates of N inputs from atmospheric deposition. For example, at the Fernow Experimental Forest in West Virginia, few changes in the N status of soils were observed from applications of 54 kg N ha^{-1} yr^{-1} for 4 years (Gilliam *et al.*, 1996). Both the treated and reference watersheds at this site receive about 20 kg N ha^{-1} yr^{-1} from atmospheric deposition.

Water, Air, and Soil Pollution **105:** 183–192, 1998.

Soil N fractions at sites that have received experimental additions of N have not been examined, other than to determine total N, despite the large soil N pools relative to N additions. Therefore, the objective of this study was to determine N fractions and pool sizes in the forest floor of three sites that received experimental N additions, and one site that received only ambient inputs, to evaluate the effects of N inputs on forest floor N.

2. Methods

Forest floor (Oie, Oa horizons) and Bh horizons were collected from three red spruce (*Picea rubens* Sarg.) stands in the northeastern United States that have received experimental applications of N (Hubbard Brook, New Hampshire; Big Moose, New York; Bear Brook, Maine), and from one site that received ambient N inputs (Howland, Maine). Site characteristics are summarized in Table I. Each site was dominated by red spruce, supported by Spodosols with well-developed forest floors (12 to 24 cm thickness). Additional details on soils and soil solutions at these sites are given by David and Lawrence (1996), as are data on atmospheric deposition inputs (Lawrence *et al.*, 1997).

TABLE I

Characteristics of the northeastern red spruce stands used in this study.

Site	Code	Elevation (m)	Wet N Deposition (mmol m^{-2} yr^{-1})	Forest Floor Thickness (cm) Oie	Oa	Forest Floor Mass (Mg ha^{-1}) Oie	Oa
Hubbard Brook	HB	720	51	4.3	9.8	52	125
Big Moose	BM	550	52	5.2	18.7	52	243
Bear Brook	BB	395	30	4.1	7.9	41	130
Howland	HW	80	18	5.5	7.1	55	92

The Hubbard Brook and Big Moose sites received 100 kg N ha^{-1} yr^{-1} of NH_4NO_3 fertilizer by hand application (in three increments during growing season) during 1992-94; soil samples were collected at these sites in June 1995. At each of these sites three replicate control and treatment plots (30 m by 30 m) were sampled. Three samples from each horizon within each plot were collected. Samples from each plot were combined and mixed by horizon for analysis. The top 10 cm of the B horizon (which generally consisted of Bh and Bs horizon soil at these sites) was also collected, except at Hubbard Brook, where no mineral soil was present within the experimental plots.

The Bear Brook site received 34 kg N ha^{-1} yr^{-1} of $(NH_4)_2SO_4$ from November 1989 through October 1995 by helicopter in six increments per year (Norton *et al.*, 1994). Samples were collected from Oa and Bh horizons of nine individual soil pits (three in each of three subplots) in red spruce stands of the treated and control plots (both at the same elevation). Soils were then combined as for the other sites, by subplot and horizon. Samples were collected in October 1992, June and October 1993, August 1994, and October 1995. The 1995 sampling included the Oie horizon. Additional data were provided by a separate study of red spruce soils at Howland, Maine, in which three soil profiles were sampled in 1992 for N analysis (Shi, 1995; Dai *et al.*, 1996). Sampling at the Howland site entailed collection of individual samples from Oe, Oa, and Bh horizons from three large soil pits.

All soil samples were air-dried and sieved (2 and 4 mm screen size for respective Bh and O horizons). All samples were analyzed for loss-on-ignition (combustion at 450 °C for 18 h), and a subset of samples was analyzed for total C concentrations with an elemental C analyzer (Nelson and Sommers, 1982). A regression equation was developed from loss-on-ignition values and C concentrations to estimate C concentration from loss-on-ignition for all samples (David, 1988). Total N concentration was determined through Kjeldahl digestion (Bremner and Mulvaney, 1982) followed by NH_4^+ measurement by a Wescan ammonia analyzer for samples collected before 1994 (Scott *et al.*, 1989), and an automated phenate analysis for the rest (APHA, 1989).

All forest floor samples from the Hubbard Brook, Big Moose and Howland sites, and the 1995 samples from Bear Brook were analyzed for N fractions. Samples underwent acid hydrolysis (6 M HCl), and the N fractions in the hydrolyzate (total hydrolyzable, acid insoluble, NH_4^+-N, amino sugar, and amino acid) were analyzed by the methods of Stevenson (1982). The samples from the Big Moose, Bear Brook, and Hubbard Brook sites were hydrolyzed for 8 hr, whereas those from the Howland site were hydrolyzed for 12 hr. With this exception, all analyses for N fractions were conducted with the same methods, in the same laboratory. Increasing the length of hydrolysis time has been shown to increase the estimated recovery of NH_4^+-N and amino acid pools relative to the other fractions (Bremner, 1967). The amino sugar values were corrected for differences in hydrolysis time by multiplying Hubbard Brook, Big Moose and Bear Brook by values by 1.23, and Howland values by 1.4 (Bremner, 1967). The added amino sugar was assumed to be derived from ammonia, which was reduced by a corresponding amount. Results from Bremner (1949) suggest that the amino acid N concentrations would be increased by about 10 % by increasing hydrolysis time from 8 to 12 hr.

Statistical analyses consisted of a two factor analysis of variance to test concentration differences in total C, total N, C:N ratios, and N fractions among sites and between the control plots and the treatment plots. Where significant differences ($p < 0.05$) among sites were found, Tukey's studentized range test was used to separate means. Regression analysis was used to examine relationships between mean concentrations of N fractions and rates of atmospheric N deposition.

3. Results and Discussion

The experimental applications of N to three of the sites produced no change ($p > 0.05$) in concentrations of total C, total N, N fractions, or C:N ratios, in Oie, Oa or Bh horizons. Therefore, all data are shown as means by site and horizon (Tables II, III). Total C and total N concentrations were typical for Spodosols (David *et al.*, 1995); that is, high in the Oie and Oa horizons, and much lower in the Bh horizon (Table II). The C:N ratios in the Oie and Oa horizons ranged from about 26 to 36 at all sites except Howland, where the ratios were significantly higher (54.1 and 61.9, respectively), in accordance with the low total N concentrations at this site.

TABLE II
Mean total C and N concentrations and C:N ratio in soils from plots of northeastern red spruce stands by horizon. Each value is a mean that represents 18 samples that were collected individually, then combined into 6 samples for analysis, except for Bear Brook where 90 samples were collected and composited into 30 for analysis. For Howland, samples were collected from 3 large pits. Standard errors given in parentheses.

Horizon Site	n	Total C ($g\ kg^{-1}$)	Total N ($g\ kg^{-1}$)	C:N ($mol\ mol^{-1}$)
Oie				
Hubbard Brook	6	531 (1) a[1]	20.2 (0.6) ab	30.8 (1.0) b
Big Moose	6	524 (1) ab	22.8 (0.5) a	27.0 (0.6) b
Bear Brook	6	497 (12) b	18.4 (1.1) b	32.2 (2.6) b
Howland	3	507 (26) ab	11.0 (0.8) c	54.1 (4.2) a
Oa				
Hubbard Brook	6	480 (12) a	18.5 (0.7) a	30.5 (1.0) b
Big Moose	6	494 (8) a	19.3 (0.3) a	29.9 (0.5) b
Bear Brook	30	406 (11) b	16.1 (0.6) a	29.8 (0.6) b
Howland	3	495 (25) a	9.3 (0.1) b	61.9 (2.7) a
Bh				
Big Moose	6	79 (11) a	2.5 (0.3) b	35.7 (1.9) a
Bear Brook	30	91 (4) a	4.3 (0.2) a	25.1 (0.5) b

[1]Within horizons means followed by the same letter(s) are not significantly different ($p = 0.05$).

McNulty *et al.* (1991) found in a previous study of spruce-fir forests, that nitrification in the forest floor was observed only if C:N ratios were less than 31 or total N concentrations were greater than 14 g kg^{-1}, and suggested that these were threshold values. David and Lawrence (1996) reported NO_3^- concentrations in soil solutions of 78 μeq L^{-1} for the Big Moose site and 19 μeq L^{-1} for the Bear Brook site, where total N concentrations exceeded 14 g kg^{-1}, and C:N ratios were less than 31:1. Fernandez *et al.* (1995) also reported mean soil solution NO_3^- concentrations of 1 μeq L^{-1} or less at the Howland site, where soils data suggested that nitrification was not occurring. Total N concentrations at Hubbard Brook suggested that nitrification was occurring, but David and Lawrence (1996) were not able to detect NO_3^- in soil solutions. On the assumption that elevated concentrations of NO_3^- in soil solutions are an indication of nitrification, the relationship developed by McNulty *et al.* (1991) was valid for each of the sites except Hubbard Brook.

Amino acid N was the predominant fraction of total N in Oie and Oa horizons, ranging from 31 to 45 %, followed by hydrolyzable unidentified N that ranged from 16 to 31 %, acid insoluble N that ranged from 18 to 22 %, and NH_4^+-N that ranged from 9 to 13 % (Table III, Figure 1). Results from the untreated Howland site were similar to results from the treated sites, except that the amino acid fraction was a larger percentage of total N than at the other sites, and the hydrolyzable unidentified fraction was a smaller percentage of

TABLE III

Nitrogen fractions (g N kg^{-1}) in soils from northeastern red spruce stands by horizon. Each value is a mean that represents 18 samples, collected individually then combined into 6 samples for analysis, except for Howland where each value is the mean of 3 samples. Standard errors given in parentheses.

Horizon Site	Hydrolyzable N	Amino-acid N	Ammonia N	Amino sugar N	HUN[1]	Insoluble N
Oie						
Hubbard Brook	16.5 (0.5) ab[2]	7.8 (0.4) a	2.3 (0.1) a	1.0 (0.1) ab	5.4 (0.6) a	3.7 (0.4) ab
Big Moose	18.2 (0.5) a	7.6 (0.2) a	2.8 (0.1) a	1.0 (0.1) ab	6.8 (0.7) a	4.6 (0.8) a
Bear Brook	15.2 (0.8) b	6.3 (0.2) b	1.6 (0.2) b	1.5 (0.2) a	5.7 (0.5) a	3.3 (0.3) ab
Howland	8.8 (0.7) c	4.9 (0.3) b	1.3 (0.1) b	0.5 (0.1) b	2.2 (0.4) b	2.1 (0.6) b
Oa						
Hubbard Brook	14.8 (0.3) a	7.1 (0.4) a	2.4 (0.1) a	1.1 (0.2) c	4.0 (0.3) b	3.7 (0.5) a
Big Moose	15.0 (0.2) a	6.0 (0.2) b	2.5 (0.1) a	1.9 (0.1) ab	4.6 (0.3) a	4.3 (0.3) a
Bear Brook	15.4 (0.5) a	6.3 (0.2) ab	2.3 (0.2) a	1.3 (0.1) bc	5.5 (0.2) a	3.8 (0.2) a
Howland	7.5 (0.1) b	4.2 (0.2) c	0.8 (0.2) b	0.7 (0.1) c	1.7 (0.1) c	1.8 (0.1) b

[1]HUN is acid hydrolyzable unknown-N. [2]Within horizons means followed by the same letter(s) are not significantly different at the 0.05 level.

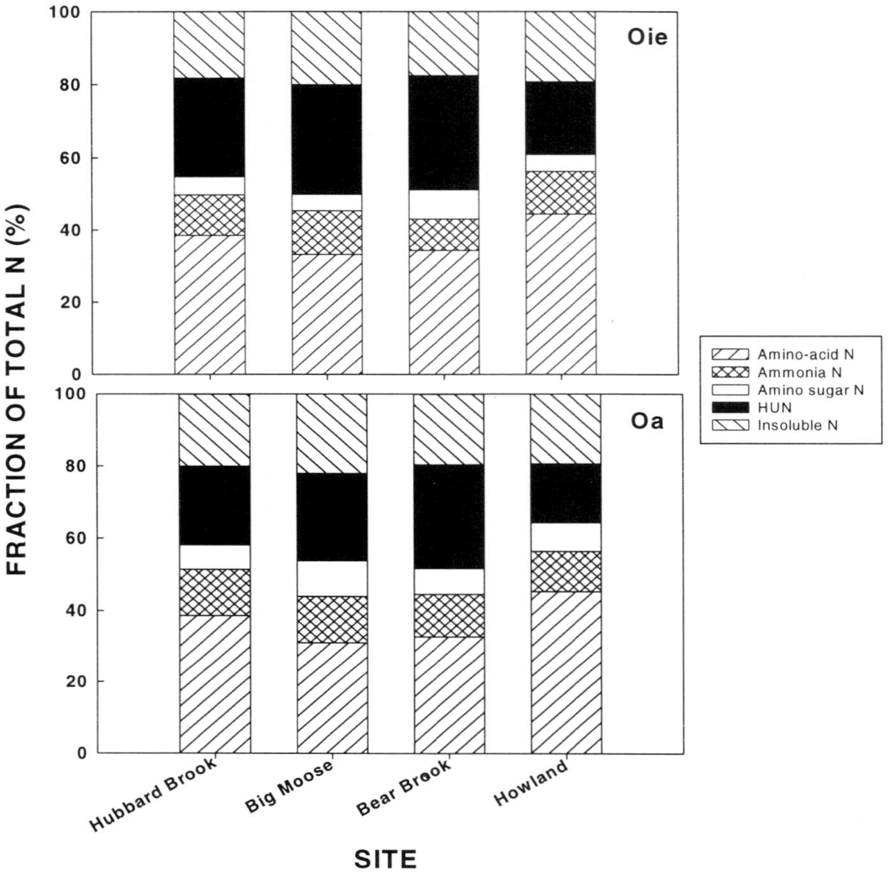

Fig. 1. Mean N fractions as a percent of total N in the Oie and Oa horizons of four red spruce stands in the northeastern U.S. HUN is hydrolyzable unidentified N.

total N than at the other sites. This result may be related to the longer hydrolysis time used to analyze the Howland samples than the samples from the other sites, which allowed for more complete hydrolysis of N, and a larger proportion of amino acid N. In general, however, the N fractions in the forest floors of these sites had similar proportions, suggesting little difference in N processing and storage, despite a considerable range in total N and N fraction concentrations. For example, Howland Oie and Oa horizons had significantly ($p < 0.05$) lower N concentrations for all N fractions compared to the other sites.

Few data are available on N fractions in organic horizons, and little is known about the factors that determine the distribution of N forms in soils (Stevenson, 1994). The proportions of N fractions measured in organic horizons of this study were within the range of proportions of total N summarized by Stevenson (1994) for mineral soils, although the amino acid fraction was at the upper end and the NH_4^+-N at the lower end of these ranges. In most mineral soils, amino acid N, as a proportion of total N, is highest in the upper part of the soil profile, whereas NH_4^+-N shows the reverse trend with depth. The distribution of N in the Oie and Oa horizons in the present study was similar to that in Histosols (cf. Stevenson, 1994), in which the amino acids also were the largest N fraction. Stevenson (1994) attributed these findings to a low rate of microbial turnover of organic N in aquatic sediments that contributed to Histosol formation. The cold, acidic, slowly decomposing forest floors at the four sites in the present study may also have lowed the rate of microbial turnover, which would lead to high proportions of the amino acid fraction.

Microbial cycling of N in the forest floor is an important process, and microbial biomass N was found to range from 2 to 4 % of total N in the forest floor at Howland (Shi, 1995). The turnover of microbial biomass could account for much of the observed amino acid and amino sugar forms of N in the forest floor. Inorganic N added by atmospheric deposition or by experimental applications could be incorporated into organic N pools through microbial assimilation into biomass, as well as from plant uptake and decomposition of organic matter (e.g., leaves, wood) containing N.

As in other manipulation studies, the N applications as a percentage of total N in the forest floor were small–at Big Moose 5.1, at Hubbard Brook 8.8, and at Bear Brook, 6.3, over the entire application period. Forest soils show extremely high spatial variability in nutrient concentrations and content. Therefore, the probability that a response to applications of that magnitude would be detected is small, especially given that much of the added N would be expected to be taken up by the trees or incorporated into the large mass of the mineral horizons. Nadelhoffer *et al.* (1995) used $^{15}NO_3^-$ as a tracer in a hardwood stand at Bear Brook to estimate that from one-fourth to one-third of applied N was retained by above-ground biomass, forest floor, and the upper 5 cm of mineral soil. Half of the retained N was measured in the above-ground biomass, half in the upper soil, presumably a result of microbial assimilation. The small percentage of applied N that was probably retained by the upper soil in the treated plots at the three sites of this study demonstrates the difficulty in detecting treatment effects on the proportions of N fractions. McNulty *et al.* (1996) also examined spruce-fir soils after long-term N applications (either 200 kg NH_4Cl ha^{-1} or 180 kg $NaNO_3$ ha^{-1}, over seven yr) and, similarly, found no response in forest floor N concentrations or in C:N ratios, although the tree-species composition was altered, possibly through induced nutrient imbalances.

One factor that differed among sites was the rate of atmospheric N deposition (Table I, Figure 2). Other work (unpublished data, M.B. David) has shown a significant relationship between wet N deposition and total N concentrations in the Oa horizon for samples collected in 12 spruce-fir stands across the northeastern U.S. (r^2= 0.58), that included the four sites from this study. McNulty *et al.* (1991) obtained a similar correlation in samples from other spruce-fir forests in New England. In this study, we compared N fractions to rates of atmospheric deposition to determine if inputs from atmospheric deposition were related to specific fractions (Figure 2). Comparison of rates of wet deposition N with NH_4^+-N and amino acid N concentrations in the forest floor, gave respective r^2 values of 0.92 and 0.94. These relationships were somewhat stronger than that observed between rates of wet N deposition and total N in the forest floor ($r^2 = 0.85$), and may indicate that NH_4^+-N and amino acid N may be the primary pools in which acidic deposition is stored. Ammonium-N may represent atmospherically deposited N that becomes incorporated into organic matter, whereas amino acid N may represent N that has been microbially assimilated following deposition.

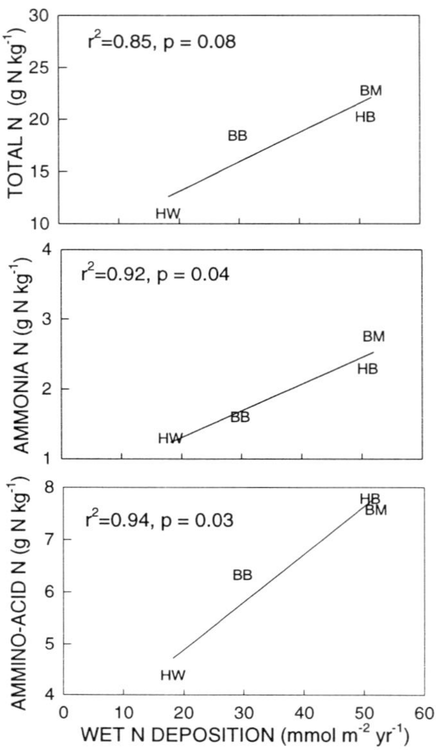

Fig. 2. Wet N deposition as a function of total N, ammonia N and amino acid N in the Oie horizon from four red spruce stands in the northeastern U.S. R^2 values and probability values also shown.

In summary, the response of forest floor N pools to applications of N may be masked, possibly by the large soil N pool, which has been increased by the long-term input of N from atmospheric deposition. Variations in atmospheric deposition to the four sites may have overwhelmed these short-term treatments through effects on soil N pools.

Acknowledgments

We thank Michael Graham for laboratory assistance. Site and logistical support for various research sites is acknowledged from the USDA Forest Service, International Paper, Champion International, and the University of Maine. This project was funded in part by the USDA Forest Service Global Change program, and USDA NRI Competitive Grants program.

References

Aber, J.D., Nadelhoffer, K.J., Steudler, P. and Melillo, J.M.: 1989, *BioScience* **39**, 378.
APHA.: 1989, *Standard Methods for the Examination of Water and Wastewater*. 17th Edition, American Public Health Association, Washington, DC.
Bremner, J.M.: 1949, *J. Agric. Sci.* **39**, 183.
Bremner, J.M.: 1965, 'Total Nitrogen', in C.A. Black (ed.), *Methods of Soil Analysis*, American Society of Agronomy, Madison, Wisconsin, pp. 1148-1178
Bremner, J.M.: 1967, 'Nitrogenous Compounds', in A.D. McLaren and G.H. Peterson (eds.), *Soil Biochemistry*, Marcel Dekker Inc., New York, pp. 19-66.
Bremner, J.M. and Mulvaney, C.S.: 1982, 'Nitrogen-Total', in A.L. Page (ed.), *Methods of Soil Analysis,* American Society of Agronomy, Madison, pp. 595-624.
Christ, M., Zhang, Y., Likens, G.E. and Driscoll, C.T.: 1995, *Ecol. Appl.* **5**, 802.
Dai, K.H., David, M.B. and Vance, G.F.: 1996, *Biogeochemistry* **35**, 339.
David, M.B.: 1988, *Commun. Soil Sci. Plant Anal.* **19**, 1593.
David, M.B. and Lawrence, G.B.: 1996, *Soil Sci.* **161**, 314.
David, M.B., Vance, G.F. and Krzyszowska, A.J.: 1995, 'Carbon Controls on Spodosol Nitrogen, Sulfur, and Phosphorus Cycling', in W.W. McFee and J.M. Kelly (eds.), *Carbon Forms and Functions in Forest Soils*, Soil Science Society of America, Madison, pp. 329-353.
Fernandez, I.J., Lawrence, G.B. and Son, Y.: 1995, *Water, Air Soil Pollut.* **84**, 129.
Gilliam, F.S., Adams, M.B. and Yurish, B.M.: 1996, *Can. J. For. Res.* **26**, 196.
Lawrence, G.B. and David, M.B.: 1997, *Environ. Sci. Technol.* **31**, 825.
Lawrence, G.B., David, M.B., Bailey, S.W. and Shortle, W.C.: 1997, *Biogeochemistry* **38**, 19.
Magill, A.H., Downs, M.R., Nadelhoffer, K.J., Hallett, R.A. and Aber, J.D.: 1996, *For. Ecol. Manage.* **84**, 29.
McNulty, S.G., Aber, J.D. and Boone, R.D.: 1991, *Biogeochemistry* **14**, 13.
McNulty, S.G., Aber, J.D. and Newman, S.D.: 1996, *For. Ecol. Manage.* **84**, 109.
Nadelhoffer, K.J., Downs, M.R., Fry, B., Aber, J.D., Magill, A.H. and Melillo, J.M.: 1995, *Oecologia* **103**, 292.
Nelson, D.W. and Sommers, L.E.: 1982, 'Total Carbon, Organic Carbon, and Organic Matter', in A.L. Page (ed.), *Methods of Soil Analysis,* American Society of Agronomy, Madison, pp. 539-579.
Norton, S.A., Kahl, J.S., Fernandez, I.J., Rustad, L.E., Scofield, J.P. and Haines, T.A.: 1994, *For. Ecol. Manage.* **68**: 61.

Scott, T.J., Mitchell, M.J., Santos, A. and Destaffen, P.: 1989, *Commun. Soil Sci. Plant Anal.* **20**, 1131.

Shi, G.: 1995, *Nitrogen Chemistry and Accumulation in a Low Elevation Spruce-fir Forest Soil, Maine*, M.S. thesis, Univ. of Illinois.

Stevenson, F.J.: 1982, 'Nitrogen-Organic Forms', in A.L. Page (ed.), *Methods of Soil Analysis,* American Society of Agronomy, Madison, pp. 625-641.

Stevenson, F.J.: 1994, *Humus Chemistry: Genesis, Composition, Reactions,* 2nd ed., John Wiley & Sons, New York.

Sullivan, T.J.: 1993, *Environ. Sci. Technol.* **27**, 1482.

AN EMPIRICAL APPROACH FOR ASSESSING THE RELATIONSHIP BETWEEN NITROGEN DEPOSITION AND NITRATE LEACHING FROM UPLAND CATCHMENTS IN THE UNITED KINGDOM USING RUNOFF CHEMISTRY

R. HARRIMAN[1], C. CURTIS[2] and A.C. EDWARDS[3]

[1]*Freshwater Fisheries Laboratory, Pitlochry, Perthshire, Scotland PH16 5LB, UK*
[2]*Environmental Change Research Centre, University College London, London WC1H OAP, UK*
[3]*Macaulay Land Use Research Institute, Craigiebuckler, Aberdeen AB9 2QJ, UK*

Abstract. Samples were collected from 13 upland sites (main inflow and loch outflow) in the UK along an N deposition gradient of 12 - 50 kg $ha^{-1} yr^{-1}$ to determine the relationship between N deposition and NO_3^- concentrations in surface waters. There was no direct correlation between NO_3^- leaching and soluble inorganic N deposition at these sites, but a significant relationship with NO_3^- was found using a deposition function incorporating dissolved organic carbon (DOC) flux from each catchment. A similar but less significant relationship was found between NO_3^- concentration and DOC:DON ratio in runoff water. Catchments showed evidence of N saturation, i.e., when mean NO_3^- concentration exceeded 5 μeq L^{-1}, when the mean DOC:DON ratio fell below an approximate value of 25. Five other large loch sites (LLS) with multiple subcatchments were used to test these relationships and for four of these mostly heathland sites the predicted NO_3^- concentrations closely matched measured values. At the fifth site, where most subcatchments were forested, the deposition function and DOC:DON ratios gave conflicting predictions and both methods generally underestimated measured NO_3^- concentrations. If the capacity of these catchments to retain deposited N is determined by C supply then many upland catchments in the UK may experience increasing NO_3^- concentrations in runoff in the future at current or increased levels of N deposition.

Keywords: catchment, monitoring, nitrogen deposition, nitate leaching, runoff chemistry, stream chemistry

1. Introduction

The signing of the European sulfur protocol in June 1994 committed the United Kingdom government to a 70 % reduction in sulfur (S) emissions by the year 2005. Since then, there has been mounting concern that the perceived improvements to ecosystems will not be achieved because increasing nitrate (NO_3^-) leaching may result if increasing nitrogen (N) emissions and corresponding increasing N deposition move terrestrial ecosystems toward N saturation. The concept of N saturation (and N limitation) has been discussed by numerous authors (e.g., Aber *et al*, 1989; Stoddard, 1994), both in terms of changes in N cycling by vegetation in terrestrial ecosystems and the onset of N runoff into surface and ground water. In the context of this study, N saturation implies that vegetation demand and microbial activity have reached an upper limit resulting in significant NO_3^- concentrations in surface water runoff. In non-agricultural catchments, NO_3^- runoff invariably follows a seasonal pattern when the first signs of N saturation occur during the dormant winter months.

In pristine, low N deposition areas (<2 kg ha^{-1} yr^{-1}), mean NO_3^- concentrations in lakes and streams draining catchments rarely exceed 2-3 μeq L^{-1} (Muniz and Framstad, 1997); therefore, waters with mean concentrations >3 μeq L^{-1} could be considered to be showing signs of N saturation. Most experimental studies of N cycling in Europe have concentrated on the effects of soluble inorganic N additions to or exclusions from forest plots or small forest catchments, e.g., the NITREX project (Wright and van Breemen, 1995), but this approach requires detailed information on biogeochemical pools and fluxes in soil and vegetation, thus making it difficult to develop a regional assessment that would include both forest and non-forested catchments. Therefore, to provide an assessment of the potential effects of increasing N deposition, the freshwater group of the Department of Environment, Critical Loads Advisory Committee set up a network of 13 sites along the United Kingdom N deposition gradient to determine the relationship between NO_3^- runoff and N deposition. Five additional sites in Scotland were used to validate the derived relationships. These sites covered a much smaller range of N deposition. We selected upland sites devoid of agricultural practices, commercial forest plantations and other land management practices to try and ensure that any N inputs were from atmospheric sources. Previous studies (Emmett *et al.,* 1995) have suggested that forest management practices may affect the N status of soils and vegetation; therefore, only one validation site was selected to include forested subcatchments.

Studies by Gundersen and colleagues (pers. comm.) have indicated that differences between C:N ratios in forest soils are correlated with N input. In this study, we assess similar relationships, but using soluble C and N content of surface waters and deriving a NO_3^- leaching function to predict NO_3^- runoff.

2. Materials and Methods

The 13 nitrogen gradient sites (NGS) were selected on the basis of their physical, chemical and deposition attributes (Figure 1, Table I). Each site included a small lake (area ranging from 1 to 25 ha), which received only one major inflow. Monthly dip samples for chemical analysis were collected from each lake inflow and outflow between September 1995 and April 1997 to provide estimates of both catchment retention and in-lake retention of N. These catchments were unforested (mainly heathland or acid grassland) reflecting the major UK upland land type, covering an N-deposition range of about 12-50 kg ha^{-1} yr^{-1}. The large loch sites (LLS), used to validate the NGS relationships, were located in Scotland (Figure 1, Table I), and dip samples from the outflows of the main subcatchments where they entered the lake were collected every three months. The areas of these large lakes ranged from 1.1 to 13.3 km^2 while those of the subcatchments encompassed a much larger range than the NGS sites (0.2 - 129 km^2). The LLS sites receive N deposition ranging from 16 to 27 kg ha^{-1} yr^{-1} and only the Loch Grannoch catchments have substantial forest cover where three of the six subcatchments have about 80 % mature forest cover. It was anticipated that the LLS sites would provide a better estimate of within-site variation in NO_3^- runoff over a narrow N deposition range.

Nitrogen deposition and runoff values were extracted from the UK deposition database, which provides an average value of total deposition (wet plus dry plus cloud deposition enhanced for altitude and vegetation type) in each 20 km^2 grid for the period 1989 - 91 (CLAG 1995, 1997). Estimating N deposition at the catchment scale is extremely difficult, especially for those catchments with a large altitude range (Reynolds *et al.,* 1997); therefore, the above data were considered to be the best available in the absence of any 1995-97 deposition data. Ion chromatography was used to determine major anions and cations, including NH_4^+, NO_3^- and Ca^{2+} concentrations. Total N was determined using the alkaline persulfate digestion method of Williams *et al.* (1995). Dissolved organic carbon (DOC) was determined on samples filtered through 0.45μm polycarbonate membranes and dissolved organic nitrogen (DON) was calculated as the difference between total N and inorganic N (mainly NO_3^-) on similarly filtered samples.

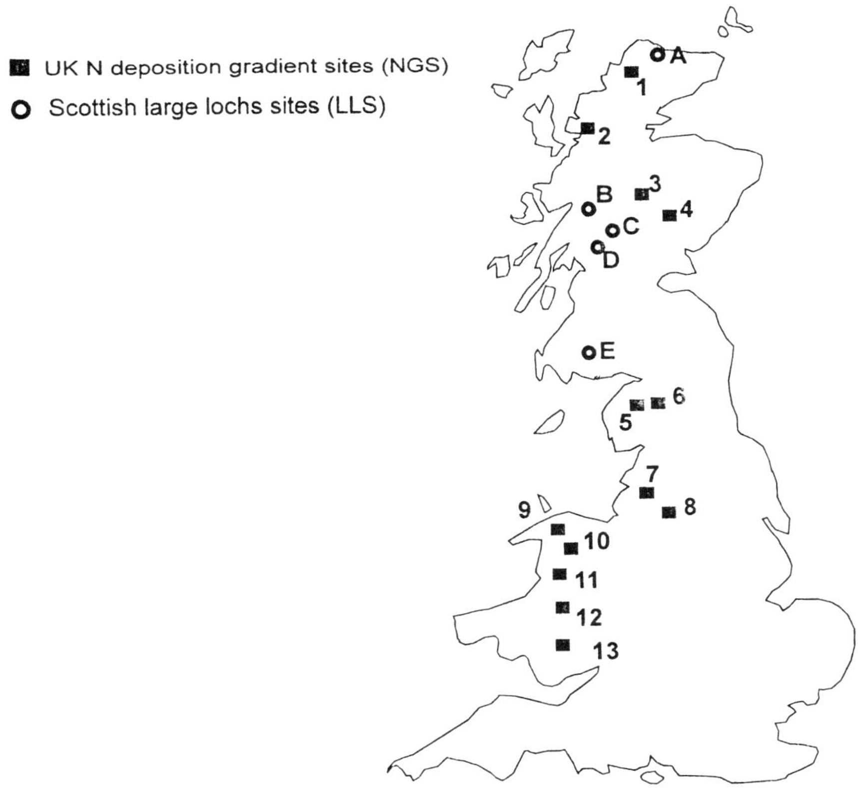

Fig.1. Sampling sites for nitrogen gradient studies (NGS) and large loch validation studies (LLS).

3. Results and Discussion

3.1. Empirical Relationships

At the NGS sites inflow and outflow pH ranged from 3.78 to 7.47, mean Ca^{2+} concentration ranged from 40 to 726 μeq L^{-1} and mean NO_3^- from 0.8 to 68 μeq L^{-1}. Nitrate concentrations showed significant negative correlations ($p < 0.01$) with pH, alkalinity, Na^+ and Cl^- although the r values were all less than 0.4. This reflects the general north to south decline in sea-salt deposition compared to the north to south increase in N (and S) deposition. Significant positive correlations ($p < 0.01$) were found with NH_4^+, K^+, Mg^{2+} and SO_4^{2-} concentrations, but no significant ($p < 0.01$) correlation was found with conductivity or

Ca^{2+} concentrations. The most significant correlation ($r = -0.49, p < 0.001$) was with DOC (Figure 2) although the relationship was not linear. Over 90 % of samples with high NO_3^- concentrations (>25 μeq L^{-1}) had a DOC concentration less than 5 mg L^{-1}, while the large variation in NO_3^- within this DOC band was due to seasonal differences at individual sites. Surface water DOC concentration generally is used as an indicator of catchment peat cover and, by implication, a measure of the C pool. Mulder *et al.* (1997) showed that most of

TABLE I
Site details of NGS and LLS catchments.

	Sites		N Deposition kg ha^{-1} yr^{-1}	NO_3^--N Flux kg ha^{-1} yr^{-1}	Mean outflow or inflow NO_3^- concentration μeq L^{-1}
1.*	Loch Nan Eun	Inflow	17	0.4	1.4
		Outflow	17	0.2	0.8
2	Lochan Dubha	Inflow	13	0.5	1.7
		Outflow	13	0.4	1.2
3	Loch Caoldair	Inflow	12	0.2	1.3
		Outflow	12	0.4	2.4
4	Loch Beanie	Inflow	16	0.5	3.2
		Outflow	16	0.3	2.4
5	Small Water	Inflow	34	7.4	20.5
		Outflow	35	11.2	31.1
6	Easedale Tarn	Inflow	51	9.1	22.4
		Outflow	50	9.7	24.6
7	Smithills Reservoir	Inflow	24	1.4	13.1
		Outflow	24	1.7	16.5
8	Chew Reservoir	Inflow	30	3.4	24.5
		Outflow	30	9.4	67.8
9	Llyn Nadroedd	Inflow	16	16.6	36.3
		Outflow	16	14.3	31.5
10	Llyn Cwm Dulyn	Inflow	17	6.8	27.2
		Outflow	17	6	25.2
11	Llyn Arran	Inflow	21	8.3	27.1
		Outflow	21	5.7	18.6
12	Llyn Gwngu	Inflow	26	1	4.3
		Outflow	26	1.1	4.9
13	Ystradfellte	Inflow	35	2.9	11.8
		Outflow	35	3.2	13.1
LLS Sites		Number of sub-catchments			
A.*	Loch Hope	10	16	0.5	2
B.	Loch Tulla	4	22	0.9	3.9
C.	Loch Voil	7	16	1.7	7
D.	Loch Katrine	9	25	1.7	6.8
E.	Loch Grannoch				
	Mature Forest (> 40 yr)	3	27	6.7	34
	Moorland/Young Forest (< 15 yr)	3	24	0.9	9.6

* Numbers and Letters indicate site location in Figure 1.

the N deposition to a forested ecosystem in southwestern Norway was immobilized in the C-rich soils; therefore, we evaluated the relationship between N deposition and NO_3^- concentration in runoff both directly and by incorporating mean DOC into a modified deposition function (f), where

$$(f) = \frac{\text{N deposition (kg ha}^{-1}\text{ yr}^{-1})}{\text{DOC flux in inflow or outflow (kg ha}^{-1}\text{ yr}^{-1})}$$

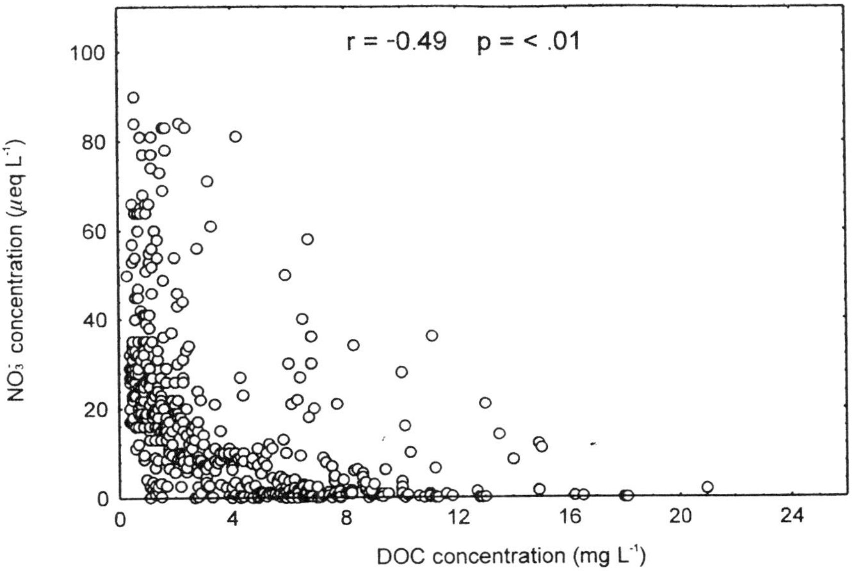

Fig. 2. Relationship between inflow and outflow NO_3^- concentrations and DOC for individual samples at NGS sites.

In the first case (Figure 3a), the direct relationship was not significant, although sites with negligible NO_3^- leaching also coincided with low N deposition. In the second case, incorporating DOC, a significant linear relationship was obtained (Figure 3b). From this relationship and using 95 % confidence intervals, NO_3^- concentrations were calculated for each 0.2 band of the deposition function range, viz., the 0-0.2 band covers a NO_3^- runoff range of 0-4 ±4 μeq L^{-1}; the 0.2-0.4 band covers a range of 4-8 ±4 μeq L^{-1}; the 0.4-0.6 band covers a range of 8-14 ±3 μeq L^{-1}; and the 0.6-0.8 band covers a range of 14-20 ±3 μeq L^{-1}, etc. These values were then used to predict and compare with NO_3^- concentrations at the LLS sites. For comparative purposes we also used the approach by Gunderson *et al.* (pers. comm.), who showed a significant relationship between C:N ratios in forest floor and N status of the site and suggested that a C:N ratio of about 25 separated N limited sites (ratio > 25) from N saturated sites (ratio < 25). Using individual data from the NGS sites a significant relationship was found between DOC and DON in runoff water (Figure 4). Consequently, when mean DOC:DON ratios were plotted against mean concentrations, a separation of N limited sites (NO_3^- < 5 μeq L^{-1}) from N saturated sites at a DOC:DON

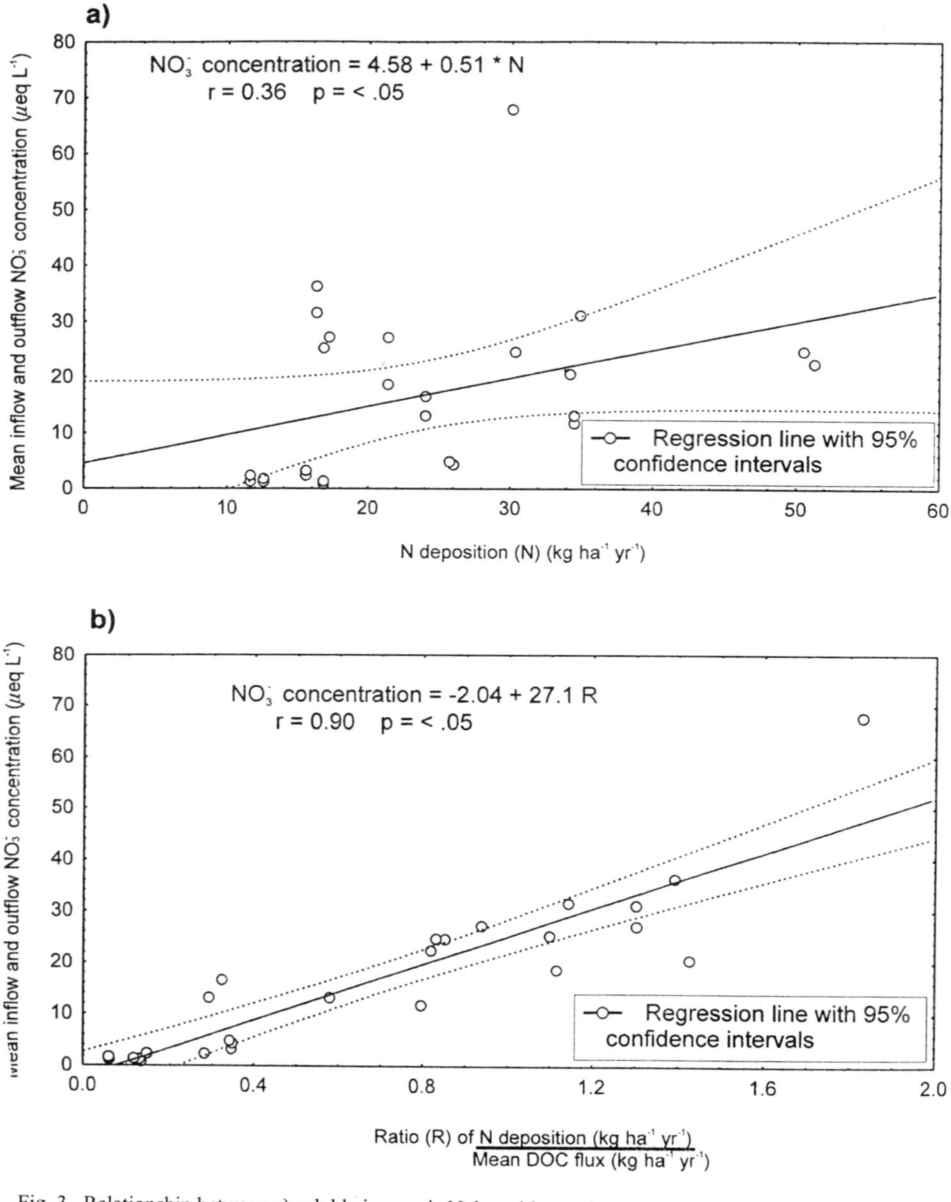

Fig. 3. Relationship between a) soluble inorganic N deposition and mean NO_3^- concentration at NGS sites and b) mean NO_3^- concentration and a modified N deposition function.

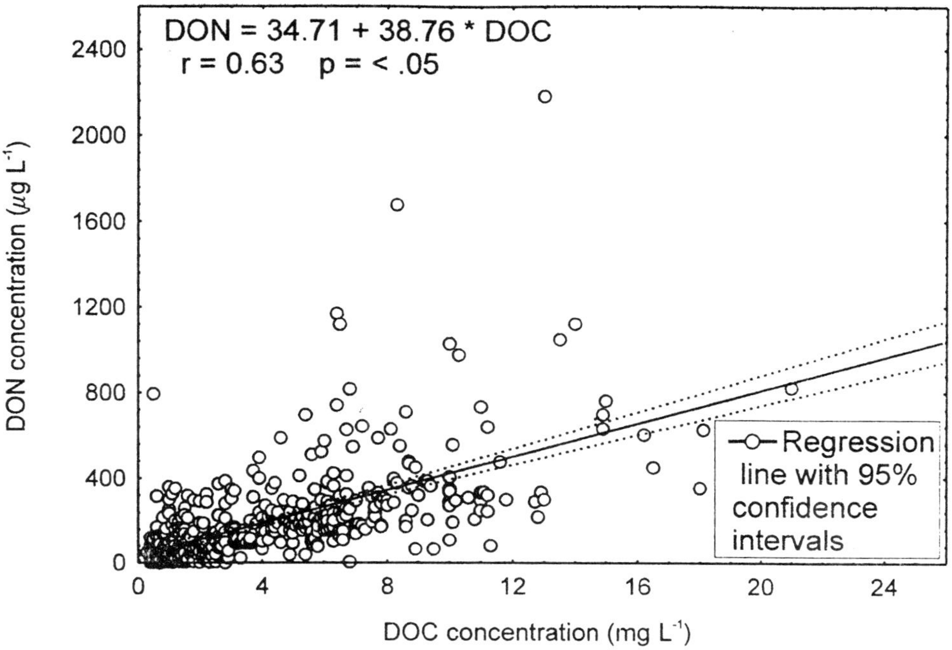

Fig. 4. Relationship between inflow and outflow concentrations of DON and DOC at NGS sites.

ratio around 25 was apparent (Figure 5). Despite obtaining a significant relationship between DOC:DON ratios and mean NO_3^- concentration for sites with DOC:DON ratios < 25, we could not validate this method at LLS sites because the confidence intervals of the regression were too large. Therefore only values predicted from the modified deposition function were used to compare with measured NO_3^- concentrations at LLS sites (Table 2). Vitousek *et al.* (1997) suggested that, in fully saturated systems, N losses could approach total inorganic N deposition once the retention capacity is reached. Results from the NGS sites indicate a clear relationship between NO_3^- retention and DOC leaching with some sites approaching complete saturation (Table I). The extent of this retention also appears to be linked to DOC:DON ratios, which may provide a measure of the current capacity of a catchment to immobilize N into the soil N pool. NGS sites with mean NO_3^- leaching < 5 μeq L^{-1} had a DOC:DON ratio >30 while NO_3^- leaching gradually increased as DOC:DON ratios declined below 25 . These findings, which suggest a DOC:DON threshold of 25-30 in catchment runoff for separating N-limited from N-saturated sites, are very similar to those reported for soils in plot and small catchment studies in the NITREX project (Dise and Wright, 1995).

The extent to which NO_3^- is retained in the NGS lakes was estimated as the difference between the inflow (catchment retention) and outflow fluxes (catchment plus in-lake retention) given in Table I. For the majority of sites in-lake retention was very low (<5%) compared to N deposition even allowing for large errors in these estimates, especially N deposition. This finding is not so surprising as most of these upland sites appear to be unproductive with only a small proportion of the N deposition reaching each lake.

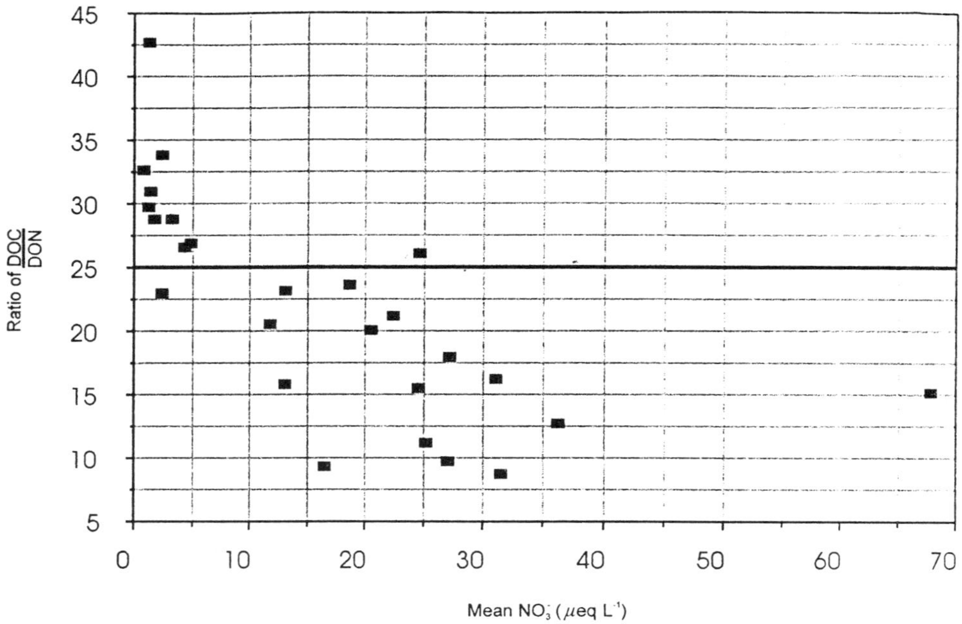

Fig. 5. Relationship between mean DOC:DON ratios and mean NO_3^- concentrations at NGS sites.

3.2. Role of NH_4^+ and Organic N

At the NGS sites, mean ourflow and inflow NH_4^+ concentrations of greater than 2 μeq L^{-1} were evident only where N deposition exceeded 25 kg ha^{-1} yr^{-1}, presumably reflecting an excess of NH_4^+ supply over retention/ transformations by cation exchange, microbial uptake and nitrification. The extent to which the NH_4^+ component of deposition (~ 50:50 NH_4^+-N to NO_3^--N in the UK) contributes to NO_3^- leaching is unclear, however N addition experiments, especially those where N additions are solely in the NH_4^+ form (see Norton *et al.*, 1998, this issue), suggest that NO_3^- release from the soil N pool may be an indirect but significant consequence of NH_4^+ deposition. For example, at Llyn Nadroedd, NO_3^- runoff could only be accounted for if total inorganic N deposition was used rather than NO_3^- deposition. Nevertheless, many of the NGS catchments appear to be currently immobilizing around 10-15 kg ha^{-1} yr^{-1} of inorganic N, some of which is probably lost as DON or is denitrified.

At a few of the NGS catchments, leaching of DON represents the largest N removal process from these catchments. The greatest loss of DON occurred at Loch Dubha (6.5 kg ha^{-1} yr^{-1}), which is equivalent to 50 % of the total N deposition. Whether DON leaching can be viewed as a delaying mechanism in the context of N saturation is still a matter of conjecture, but based on the evidence of declining ratios of DOC:DON at NGS sites, it does appear that more DON is lost per unit release of DOC as N deposition increases.

3.3. Testing the Empirical Relationships

The relationships between the N deposition function, DOC:DON ratios and runoff NO_3^- concentrations provide a general basis for assessing the current N status (i.e., the extent of NO_3^- leaching) of any catchment in the UK. In selecting the LLS sites to test these relationships, we recognized that no forest sites had been used in the original evaluation. Therefore, we included the Loch Grannoch site, which has three adjacent sub-catchments with young forest (< 15 yr) and heathland plus three with mature forest (> 40 yr) (Table I). For each of the LLS sites, the average of the measured mean NO_3^- concentrations in the subcatchment outflows was compared with predicted values derived from the mean deposition functions and mean DOC:DON ratios. Apart from the Loch Grannoch site, there was little variation between subcatchments at each site in terms of mean and standard deviation (SD) of NO_3^- concentrations (Table II). For Loch Hope, the deposition function values were less than 0.2 for all subcatchments and outflow DOC:DON ratios were within the range of 35-55. All the 10 subcatchments would therefore be predicted to be N-limited (mean NO_3^- concentration less than 5 μeq L^{-1}), which closely mirrors the measured values (Table II). Interestingly, subcatchments with DOC:DON ratios between 35-40 have the lowest DOC flux and mean NO_3^- concentrations between 3-4 μeq L^{-1}, while sites with DOC:DON ratios between 40-55 exhibited concentrations between 0.7 - 1.4 μeq L^{-1}. This suggests that the subcatchments with the lowest DOC flux at Loch Hope may be showing the first signs of N deposition responses. Subcatchments at Loch Voil and Loch Tulla, with similar N deposition (16-22 kg ha^{-1} yr^{-1}), showed slightly higher NO_3^- concentrations due to lower DOC flux, and this is again reflected in lower DOC:DON ratios. At Loch Katrine, where N deposition is higher than the above sites, the predicted and measured values again compared quite favorably, but in this instance one of the subcatchments exhibited elevated summer NO_3^- concentrations (> 7 μeq L^{-1}) with a DOC:DON ratio of 22. This was the only subcatchment at Loch Katrine with a summer minimum NO_3^- of greater than 1 μeq L^{-1}. At the Loch Grannoch site, measured NO_3^- concentrations from the mature forest subcatchments were significantly higher than concentrations from the young forest or heathland catchments. These differences were predicted using the deposition function; however, the measured values for mature forest sites were more than double the predicted values while the young forest and heathland catchments gave similar measured and predicted concentrations (Table II). Estimated NO_3^- concentrations from DOC:DON ratios in the mature forest subcatchments were less than 5 μeq L^{-1}, compared with mean measured values of 34 μeq L^{-1}. These forest sites had low DOC fluxes, but DOC:DON ratios greater than 40, indicating an unusually low ratio for such a high N deposition site. Site management effects, such as ploughing and ditching prior to planting, may have disturbed the N cycle in these forested subcatchments causing elevated concentrations of DOC and NO_3^-. Our final observation concerns the high aluminum (Al^{3+}) values (up to 1152 μg L^{-1}) recorded at the Grannoch forest sites and some NGS sites and the question as to whether N deposition is potentially more damaging to biota than S deposition due to increased Al^{3+} leaching from mineral soil horizons where N transformations occur (Gundersen and Ramussen, 1990). Further studies are required to include more forest catchments in the network to provide a more detailed assessment of the relationship between NO_3^- and Al^{3+} concentrations in runoff. In addition, the efficacy of the deposition function should be tested in other countries with similar heathland catchments.

TABLE II
Predicted NO_3^- concentrations at LLS sites.

LLS Sites	(1) Ratio of N Deposition: DOC Flux (units as kg ha^{-1} yr^{-1})	Ratio of DOC: DON (units as $\mu g\ L^{-1}$)	Predicted NO_3^- concentration using deposition function (1) ($\mu eq\ L^{-1}$ ± 95 % CL)	Measured mean catchment outflow NO_3^- concentration ($\mu eq\ L^{-1}$ ±SD)
A. Loch Hope	0.14	48	0 - 4 ± 4	2.0 ± 1.3
B. Loch Tulla	0.36	35	4 - 8 ± 4	3.9 ± 1.2
C. Loch Voil	0.27	35	4 - 8 ± 4	7.0 ± 1.5
D. Loch Katrine	0.29	31	4 - 8 ± 4	6.8 ± 2.3
E. Loch Grannoch				
Mature Forest	0.47	45	8 - 14 ± 3	34.0 ± 14.0
Moorland/Young Forest	0.22	35	4 - 8 ± 4	9.6 ± 3.2

In summary, we have shown how important the C and N status of catchments is in determining NO_3^- runoff from catchments. By applying a deposition function linked to DOC:DON ratios in runoff we have provided a simple empirical technique that predicts NO_3^- leaching from catchments. This method requires only the analysis of water samples for a few key parameters plus current information on N deposition at the catchment scale and can be used on a regional basis in the UK. Accurate information on N deposition for catchments may be the most difficult data to obtain on a regional basis, while further comparative studies of heathland and managed commercial forest catchments would be helpful in clarifying the effects of management practices on C and N cycling in soils. These studies only give a snapshot of the current N status of UK catchments, however some of the catchments with small C pools already appear to be close to N saturation with respect to N deposition. If NO_3^- and NH_4^+ retention in catchments with larger C pools does decline in the future then NO_3^-, and possibly NH_4^+, concentrations will inevitably increase with associated consequences for freshwater acidification and eutrophication.

4. References

Aber, J.D., Nadelhoffer, K.J., Steudler, P. and Melillo, J.M.: 1989, *BioScience* **39**, 378.
CLAG.: 1995, *Critical Loads of Acid Deposition for United Kingdom Freshwaters,* Department of Environment.
CLAG.: 1997, *Deposition Fluxes of Acidifying Compounds in the United Kingdom,* Department of Environment.
Dise, N.B. and Wright, R.F.: 1995, *For. Ecol. Manage.* **71**, 153.
Emmett, B.A., Brittain, S.A., Hughes, S. and Kennedy, V.: 1995, *For. Ecol. Manage.* **71,** 61.
Gundersen, P. and Ramussen, L.: 1990. *Rev. Environ. Contamin. Toxicol.* **13**, 1.

Gundersen, P., Emmett, B.A., Kjonaas, O.J., Koopmans, C.J. and Tietma, A.: 1997, *For. Ecol. Manage.,* (in press).
Mulder, J., Nilsen, P., Stuanes, A.O. and Huse, M.: 1997, *Ambio* **26**, 273.
Norton, S.A., Kahl, J.S. and Fernandes, I.J.: 1998, *Water Air, Soil Pollut.*, (this issue).
Reynolds,B., Fowler, D., Smith, R.I. and Hall, J.R.: 1997, *J. Hydrol.* **194**, 305.
Stoddard, J.L.: 1994, *Environmental Chemistry of Lakes and Reservoirs,* American Chemical Society.
Muniz, I.P. and Framstad, E.: 1997, *Hydrobiologia,* **348**, 49.
Williams, B.L., Shand, C.A., Hill, M., O'Hara, C., Smith, S. and Young, M.E.: 1995, *Commun. Soil Sci. Plant Anal.* **26**, 91.
Vitousek, P.M., Aber, J.D. Howarth, R.W., Likens, G.E., Matson, P.A., Schindler, D.W., Schlesinger, W.H. and Tilman, D.G.: 1997, *Ecol. Applic.* **7**, 737.
Wright, R.F., and van Breemen, N.: 1995, *For. Ecol. Manage.* **71**, 1.

THE PREDICTION OF NITRATE LEACHING WITH THE FIRST-ORDER ACIDITY BALANCE (FAB) MODEL FOR UPLAND CATCHMENT IN GREAT BRITAIN

C.J. CURTIS[1], T.E.H. ALLOTT[1], B. REYNOLDS[2] and R. HARRIMAN[3]

[1]*Environmental Change Research Centre, University College London,London, WC1H 0AP, UK*
[2]*Institute of Terrestrial Ecology, Bangor Research Unit, Deiniol Road, Bangor, Gwynedd, LL57 2UP, UK*
[3]*Freshwater Fisheries Laboratory, Faskally, Pitlochry, Perthshire, PH16 5LB, UK*

Abstract. The relative contribution of N deposition to the acidification of freshwaters in Great Britain has increased over the last few years as S deposition has fallen in line with reduced emissions. In certain high deposition areas of Great Britain, NO_3^--based acidity can equal or exceed the contribution of SO_4^{2-}-based acidity in some upland waters. Here we apply the first-order acidity balance model (FAB) to predict the maximum N leaching from 13 study catchments at future steady state. Using mean water chemistry and catchment soils data, along with long-term default values for N sink processes, we predict NO_3-N leaching at much higher rates than currently are being measured in surface waters, with a mean increase of 10.5 kg ha^{-1} yr^{-1}. As a result, mean acid neutralizing capacity would decline to less than 0 meq L^{-1} at 4 sites. While there are uncertainties associated with model parameterization relating to the short-term storage of N within catchment soils and vegetation, model outputs do indicate much greater leaching of N at some time in the future as steady-state is achieved.

Keywords: acidification, catchment, first-order acidity model, nitrate leaching, predictive model

1. Introduction

The effects of acid deposition on terrestrial ecosystems and surface waters are well documented in the United Kingdom (e.g., CLAG, 1994) and there is a vast international literature on the subject regarding the effects of S deposition. However, only in the last few years has the contribution of N species to acidification of surface waters been widely reported (e.g., Dillon and Molot, 1990; Grennfelt and Thörnelöf, 1992; Stoddard, 1994; Allott *et al.*, 1995).

While S emissions and deposition continue to decline in Great Britain, emissions of N species are slowly increasing. NO_x emissions increased by about 20 % between 1970 and 1990, mainly due to greater car usage (INDITE, 1994). Ammonia emissions are more difficult to quantify than NO_x emissions because of the diversity of diffuse sources, but are thought to have increased with agricultural intensification (INDITE, 1994). Although there is regional variability, S and H^+ wet deposition have declined on a national basis between 1986 and 1995, while NO_3^- and NH_4^+ deposition show no significant overall change, except in some areas that show a slight increase (Vincent *et al.*, 1997). As a result, both the absolute and relative contributions of N species to acid deposition have increased in recent years, and are at least partially offsetting the beneficial effects of reduced S deposition.

In Europe the critical loads approach has been adopted as a policy tool in order to enable emissions reductions to be targeted most efficiently (Bull, 1991), and this requires the modeling of changes in water chemistry in response to changes in acid deposition load. Two empirical models, the Steady-State Water Chemistry model (Henriksen *et al.*, 1992), and the diatom model (Battarbee *et al.*, 1996) were developed to link SO_4^{2-} deposition, assuming SO_4^{2-} is a relatively mobile anion at steady-state in catchments (Seip, 1980), to biological change through acidification. However, these models are unable to predict critical loads under different N

Water, Air, and Soil Pollution **105:** 205–215, 1998.

deposition scenarios because they take no account of the N processes within catchments that determine the rate of N leaching. To address this problem, a process-based model that allowed for some quantification of long-term N fluxes through catchments was required. One such steady-state model has been proposed for critical loads mapping in Europe, based on a simple mass- and charge-balance (Posch *et al.*, 1993, 1995, 1997). This model previously has been referred to as the first-order acidity balance (FAB) model (Posch, 1995).

To assess the speciation and significance of N in British upland surface waters and to improve modeling of N leaching, a monitoring network of lake sites and their major inflow streams was created in June 1995. This paper uses the results to date from this network for a catchment based application of the FAB model to provide estimates of long-term N sinks in Great Britain and to predict future NO_3^- leaching under current deposition loads. Only dynamic models can predict the timing of acidification processes, but in general, the steady-state models employed for defining critical loads assume a long-term steady-state situation which may only be achieved over a timescale of decades.

2. Methods

2.1. Site Selection, Sampling and Data Sources

The 13 lake and inflow stream monitoring sites were selected along gradients of N deposition and lake-water Ca^{2+} concentration (as a measure of sensitivity to acidification) and represent non-forested catchments with minimal human disturbance other than atmospheric deposition and low level extensive sheep grazing. The presence of a major inflow stream for monitoring was a prerequisite for site selection. Low N deposition sites generally are located in northern Scotland, intermediate N deposition sites are found in Wales, and sites experiencing the highest N deposition loads occur in the Pennines and Lake District of northern England. Monthly water dip samples (0.5 L) were taken from both the major inflow stream and the lake outflow stream, and refrigerated in the dark prior to analysis within three days according to the methods described in Harriman *et al.* (1990). The data reported here are mean values for the period June 1995 to March 1997. Discharge and deposition were not measured directly at these sites. Modeled deposition estimates on a 20 km grid basis were obtained from the modified total (wet + dry) deposition fields derived as mean annual values for the period 1992-94. Catchment weighted runoff was derived from the 1 km grid mean modeled runoff data set for 1992-94 using catchment outlines, digitized from 1:25,000 maps.

2.2. The First-order Acidity Balance Model

Critical load definitions require that under steady-state conditions, no acidification of the surface waters below a pre-specified acid neutralizing capacity (ANC) is permitted. The FAB model assumes that sources of alkalinity must therefore balance the input and production of acidity less the permitted critical load (Posch *et al.*, 1997). Short-term dynamic interactions are ignored because critical loads are set relative to long-term steady-state. The charge balance for the lake catchment, incorporating major sources and sinks of S and N, is:

$$\mathbf{N_{dep} + S_{dep} = \mathit{f}N_{upt} + (1\text{-}\mathit{r})(N_{imm} + N_{den}) + \mathit{r}N_{ret} + \mathit{r}S_{ret} + BC_l - ANC_l,}$$

N_{dep}	=	total N deposition
S_{dep}	=	total S deposition
N_{upt}	=	net growth uptake of N by forest vegetation (removed by harvesting)
N_{imm}	=	long-term immobilization of N in catchment soils
N_{den}	=	N lost through denitrification in catchment soils
N_{ret}	=	in-lake N retention
S_{ret}	=	in-lake S retention
BC_l	=	base cation leaching from catchment
ANC_l	=	ANC leaching from catchment
f	=	fraction of forested area in the catchment
r	=	lake:catchment area ratio

(all units, except f and r, are equivalents per unit area and time).

The acidity balance model relies on four key assumptions (Posch *et al.*, 1997): 1) long-term sinks of SO_4^{2-} in catchment soils are negligible, 2) there are no agricultural N inputs to the catchment (such catchments will generally not be vulnerable to acidification), 3) the effect of internal N cycling within the ecosystem can be ignored, and 4) NH_4^+ leaching is negligible, implying complete uptake, adsorption or nitrification.

In national mapping exercises it is impossible to measure all the model parameters at each mapping site, so a library of default values has been developed for general mapping within Europe (Hornung *et al.*, 1995). The major recognized sinks for N are described below, along with suggested long-term rate values.

2.2.1. *N_{upt} (net growth forest uptake)*

Because none of the network catchments are afforested, this parameter is set to zero. However, in grazed or burned catchments there may be small net losses of N from other types of catchment vegetation in terms of weight gain of removed livestock and losses in smoke. Suggested working ranges for N losses have been 0.5-1.0 kg ha^{-1} yr^{-1} for grazed acid grassland, and 0-1 kg ha^{-1} yr^{-1} from burning, i.e., a maximum of 2 kg ha^{-1} yr^{-1} in unafforested catchments (Hornung *et al.*, 1995). These losses are, however, assumed to be negligible in this model application because there are no data on grazing and burning practices at N network catchments.

2.2.2. *N_{imm} (long term net immobilization of N in catchment soils)*

A suggested long-term range of net N immobilization rates (including N-fixation) is 2-5 kg ha^{-1} yr^{-1} (Posch *et al.*, 1995), although it is noted that under present conditions with high growth due to elevated N deposition this range may extend to a much higher figure in the short-term (see Section 3.3). For UK soils, a working range of 1-3 kg ha^{-1} yr^{-1}, depending on soil type, has been adopted (Hall *et al.*, 1997). This term incorporates fixation and the export of organic N from the catchment (Posch *et al.*, 1997).

2.2.3. N_{den} *(denitrification in catchment soils)*

Denitrification is primarily a function of soil moisture, and a range of 0-4 kg ha^{-1} yr^{-1} has been suggested for common soil types in Europe (Hornung *et al.*, 1995). For British soils the suggested working range is 1-4 kg ha^{-1} yr^{-1} (Hall *et al.*, 1997). Catchment weighted values for both N_{imm} and N_{den} have been obtained from estimates of the proportion of major soil types (from digitized 1:250,000 soils maps) and default values associated with each soil type (Hall *et al.*, 1997).

2.2.4. rN_{ret} *(in-lake N retention)*

While long-term, sustainable maximum rates of immobilization, denitrification and growth uptake are assumed to be independent of the atmospheric N load, in-lake N retention is assumed to be proportional to the net N input to the lake (Posch *et al.*, 1997):

$$rN_{ret} = \rho_N\,[N_{dep} - fN_{upt} - (1-r)(N_{imm} + N_{den})].$$

The retention factor ρ_N is modeled by a kinetic equation;

$$\rho_N = \frac{S_N}{S_N + Q/r}$$

(Kelly *et al.*, 1987), where Q is the runoff and S_N is the net mass transfer coefficient for N, set here to the default value 5 m yr^{-1} derived for 12 Canadian lakes (Dillon and Molot, 1990), where reported in-lake retention of total inorganic N varied from 24 % to 82 % depending on mean depth and retention time.

3. Results and Discussion

3.1. Current Catchment Budgets

Catchment budgets for N and S at both lake sites and inflow streams have been calculated using the data sets described above (Table I). While deposition rates of total N and non-marine SO_4^{2-} are very similar for a given site, the mean leaching output of S and N from the catchments is very different. Non-marine SO_4^{2-} leaching is closely linked to deposition with a near 1:1 relationship (Figure 1a) indicating that SO_4^{2-} is a highly mobile anion in these catchments, as also assumed by the steady-state water chemistry model described above. The small net export of SO_4^{2-} calculated for many catchments could be due to an underestimate of S deposition at these sites, because most of these catchments lie in the uplands and the use of 20 × 20 km resolution deposition data may lead to systematic underestimates of wet deposition for sites lying above the mean altitude for a grid square. One outlier (site code CZSD61 and inflow) is thought to lie in an area of geological SO_4^{2-} sources. The mean export figure is still, however, only around 11 % greater than inputs of non-marine SO_4^{2-} (Table I).

In nonforested catchments with no net removal of N in vegetation, biological processes such as immobilization and denitrification contribute to the much greater retention of N within the catchment soils (Figure 1b). In the current context, N retention refers to that proportion of N inputs to a catchment that is not exported in the runoff. The denitrification process does not lead to physical retention of N within catchment soils because gaseous N products are exported to the atmosphere, but the N is prevented from contributing to surface water acidification. For total inorganic nitrogen (TIN), exports are much lower than deposition inputs, with a mean catchment retention of 76 %. Only sites with total N deposition in excess of 10 kg ha^{-1} yr^{-1} show significant N leaching losses, and this deposition threshold has been observed in other studies (Gundersen, 1995; Dise and Wright, 1995). There is a major difference between reduced and oxidized N species (Table I), with NH_4^+-N retention in excess of 95 % as recorded by other authors (e.g., Dillon and Molot, 1990) while NO_3^--N retention is much lower, at a mean value of only 43 %. This may be due to the preferential biological uptake of NH_4^+ over NO_3^-, reducing its relative mobility in catchments (Stoddard, 1994), or to the adsorption of NH_4^+ on the exchange complex within catchment soils.

TABLE I

Estimated N and S budgets for the 13 lake and inflow stream sites in the monitoring network. * denotes the non-marine component. Inflow stream sites are denoted by the site code suffix _IF

		Deposition (kg ha^{-1} yr^{-1})				Leaching output (kg ha^{-1} yr^{-1})				Overall retention (kg ha^{-1} yr^{-1})			
Sitecode	Region	NH_4-N	NO_3-N	Total N	SO_4*-S	NH_4-N	NO_3-N	TIN	SO_4*-S	NH_4-N	NO_3-N	Total N	SO_4*-S
CZNN68	Central Scotland	2.5	3.2	5.8	7.7	0.1	0.4	0.5	7.7	2.4	2.9	5.3	0.0
CZNN68_IF	Central Scotland	2.5	3.3	5.8	7.8	0.1	0.2	0.3	8.8	2.5	3.0	5.5	-1.1
CZNG83	NW Scotland	2.9	4.6	7.6	10.6	0.2	0.2	0.5	2.5	2.7	4.4	7.1	8.1
CZNG83_IF	NW Scotland	2.9	4.6	7.6	10.6	0.4	0.4	0.8	2.5	2.5	4.3	6.8	8.1
NEUN	NW Scotland	3.5	6.0	9.5	11.0	0.1	0.2	0.3	3.4	3.4	5.8	9.2	7.6
NEUN_IF	NW Scotland	3.5	6.0	9.5	11.0	0.2	0.4	0.6	3.5	3.3	5.6	8.9	7.5
CZNO16	Central Scotland	5.6	5.6	11.2	11.5	0.2	0.3	0.6	11.8	5.4	5.3	10.6	-0.3
CZNO16_IF	Central Scotland	5.6	5.6	11.2	11.5	0.1	0.4	0.5	12.3	5.5	5.2	10.7	-0.8
CZSH44	North Wales	11.8	5.9	17.6	13.9	0.1	6.6	6.6	15.7	11.7	-0.7	11.0	-1.8
CZSH44_IF	North Wales	11.8	5.9	17.6	13.9	0.2	7.4	7.6	17.6	11.6	-1.5	10.0	-3.7
VSH7105	North Wales	11.5	6.6	18.1	16.0	0.4	6.7	7.1	28.1	11.1	-0.1	11.0	-12.1
VSH7105_IF	North Wales	11.5	6.6	18.1	16.0	0.5	9.8	10.4	26.0	10.9	-3.2	7.7	-10.0
VSH5502	North Wales	12.8	6.9	19.7	16.3	0.5	13.3	13.8	24.1	12.3	-6.4	5.9	-7.9
VSH5502_IF	North Wales	13.1	7.2	20.4	17.1	0.2	15.6	15.8	26.2	12.9	-8.3	4.6	-9.1
CZSN87	Mid Wales	13.7	8.0	21.7	20.2	0.2	1.3	1.4	22.4	13.5	6.7	20.3	-2.2
CZSN87_IF	Mid Wales	13.7	8.0	21.7	20.2	0.3	1.1	1.4	24.2	13.4	6.9	20.3	-4.0
CZSD61	Pennines, N.England	16.5	9.8	26.3	29.3	2.9	2.1	5.0	45.3	13.6	7.7	21.3	-16.0
CZSD61_IF	Pennines, N.England	16.5	9.8	26.3	29.3	1.8	1.7	3.5	47.6	14.7	8.1	22.8	-18.3
CZSN91	South Wales	16.6	11.1	27.6	25.9	0.3	3.4	3.8	26.6	16.2	7.6	23.9	-0.8
CZSN91_IF	South Wales	16.3	10.5	26.7	24.6	0.4	3.1	3.5	30.0	15.9	7.4	23.3	-5.3
VNY4101	Lake District, N.England	18.6	11.8	30.4	29.4	0.5	11.4	12.0	36.8	18.1	0.3	18.4	-7.4
VNY4101_IF	Lake District, N.England	18.6	11.8	30.4	29.4	0.6	7.3	7.9	46.4	18.0	4.5	22.5	-17.0
CZSE00	Pennines, N.England	21.3	12.3	33.6	39.2	0.5	11.3	11.8	65.6	20.8	1.0	21.8	-26.4
CZSE00_IF	Pennines, N.England	21.3	12.3	33.6	39.2	1.7	4.1	5.8	42.8	19.6	8.2	27.8	-3.6
CZNY30	Lake District, N.England	22.8	14.4	37.2	35.8	0.5	9.4	10.0	25.9	22.3	5.0	27.3	9.9
CZNY30_IF	Lake District, N.England	22.8	14.4	37.2	35.8	0.8	8.7	9.5	27.9	22.0	5.7	27.7	7.9
Mean value:		12.32	8.16	20.48	20.51	0.53	4.88	5.41	24.30	11.79	3.28	15.07	-3.8

While catchment retention of N is evidently much greater than for SO_4^{2-}, there are certain sites in areas of high total N deposition where NO_3^- concentrations approach or even exceed SO_4^{2-} concentrations for part of the year (e.g., site VSH5502; CLAG, unpublished data). Any future increases in NO_3^- fluxes could lead both to a greater number of sites where the contribution of NO_3^- to total acidity will exceed that of SO_4^{2-} on a seasonal basis, and to longer periods at some sites during which this occurs.

Changes in the seasonal pattern of N leaching have been attributed to the progression of catchment ecosystems through stages of "N saturation" (Aber *et al.*, 1989; Stoddard, 1994), whereby the supply of N exceeds biological demand leading to enhanced leaching patterns. The retention of large proportions of elevated N inputs implies that these study catchments must to some degree be undergoing this process, and the very high leaching rates at some of the catchments does indicate a degree of saturation. This inevitably has a resultant effect on the acidity and, possibly, nutrient status of the surface waters, with adverse biological implications.

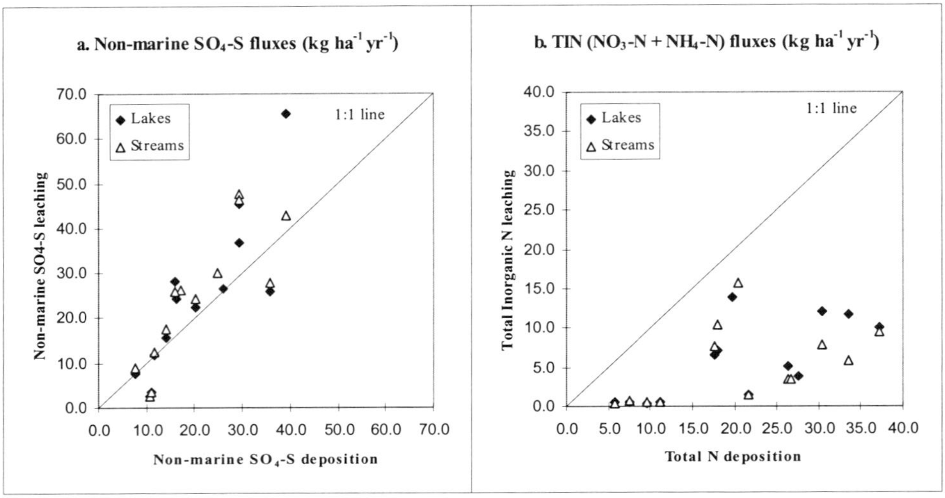

Fig. 1 Comparison of input and output fluxes of S and N species at network sites

3.2. Increases in NO_3^- Leaching Predicted by the FAB Model

It is accepted that increased rates of all N sink processes may occur temporarily under increased N deposition loads (Posch *et al.*, 1995; Emmett and Reynolds, 1996). However, for planning and policy-making, knowledge of the long-term steady-state condition of the system is required. An important output of the FAB model is the quantification of the long-term, sustainable, catchment retention rates for N. The literature-based default values described above therefore have been used for the application of the FAB model at the study sites to calculate these outputs. Although other European studies have identified soil pH and N deposition as important determinants of N leaching (Dise and Wright, 1995; Gundersen, 1995) they have only considered forest catchments and have not attempted to quantify future N leaching.

The catchment weighted N sink terms and the predicted changes in NO_3^- leaching are presented in Table II. In every case, there is a predicted increase in NO_3^- leaching under current total N deposition loads. The clear implication here is that current levels of NO_3^- retention are not sustainable in the long term. While current estimates of total catchment N sink values give a range of 4.5-27.8 kg ha^{-1} yr^{-1} (mean = 15.1 kg ha^{-1} yr^{-1}), those predicted by FAB using default values are in the range 2.0-9.5 kg ha^{-1} yr^{-1} (mean = 4.6 kg ha^{-1} yr^{-1}). The mean increase in TIN leaching predicted by FAB across all sites is 10.5 kg ha^{-1} yr^{-1} (Table II). There are appreciable levels of NH_4^+ leaching at a few sites (Table I), but in general the leaching of NO_3^- is much higher. The mean contribution of NH_4^+ to total inorganic N leaching is less than 10 %. In terms of the FAB model, it is assumed that under future steady-state conditions this NH_4^+ will be nitrified to NO_3^-, thereby contributing to a decline in ANC and pH. Under this scenario, the predicted mean increase in NO_3^--N leaching for all study sites is 11.0 kg ha^{-1} yr^{-1}, which equates to an increase in mean concentration from 16.8 to 64.7 meq L^{-1} under current mean runoff levels. This will have a significant effect on annual mean ANC concentrations at some sites (Figure 2).

TABLE II

Comparison of current and predicted N budgets using FAB (*averaged for lake sites only; not applicable [n/a] to inflow streams).

Sitecode	Lake: catchment ratio (r)	Net denitrification kg ha^{-1} yr^{-1}	Net N immobilisation kg ha^{-1} yr^{-1}	In-lake N retention kg ha^{-1} yr^{-1}	Measured N sink kg ha^{-1} yr^{-1}	FAB N sink kg ha^{-1} yr^{-1}	Measured TIN leaching kg ha^{-1} yr^{-1}	FAB predicted NO_3-N leaching kg ha^{-1} yr^{-1}
CZNN68	0.079	0.9	1.9	0.7	5.3	3.6	0.5	2.1
CZNN68_IF	n/a	1.0	2.4	n/a	5.5	3.4	0.3	2.3
CZNG83	0.055	3.8	0.9	0.5	7.1	5.2	0.5	2.4
CZNG83_IF	n/a	4.0	1.0	n/a	6.8	5.0	0.8	2.6
NEUN	0.105	3.6	0.9	1.0	9.2	5.5	0.3	4.0
NEUN_IF	n/a	4.0	1.0	n/a	8.9	5.0	0.6	4.5
CZNO16	0.070	2.0	2.0	1.8	10.6	5.9	0.6	5.3
CZNO16_IF	n/a	2.8	1.8	n/a	10.7	4.6	0.5	6.6
CZSH44	0.055	0.9	2.8	1.8	11.0	5.6	6.6	12.1
CZSH44_IF	n/a	1.0	3.0	n/a	10.0	4.0	7.6	13.6
VSH7105	0.023	1.4	2.1	0.6	11.0	4.1	7.1	14.0
VSH7105_IF	n/a	1.0	2.1	n/a	7.7	3.1	10.4	14.9
VSH5502	0.034	1.3	1.0	0.9	5.8	3.2	13.8	16.4
VSH5502_IF	n/a	1.0	1.0	n/a	4.5	2.0	15.8	18.3
CZSN87	0.018	1.0	2.9	0.8	20.3	4.8	1.4	16.9
CZSN87_IF	n/a	1.0	3.0	n/a	20.3	4.0	1.4	17.7
CZSD61	0.025	1.0	2.9	2.7	21.3	6.6	5.0	19.8
CZSD61_IF	n/a	1.0	3.0	n/a	22.8	4.0	3.5	22.3
CZSN91	0.026	2.6	2.9	1.4	23.8	7.0	3.8	20.6
CZSN91_IF	n/a	3.0	3.0	n/a	23.3	6.0	3.5	20.8
VNY4101	0.047	1.0	1.1	2.3	18.4	4.4	12.0	26.0
VNY4101_IF	n/a	1.0	1.0	n/a	22.5	2.0	7.9	28.4
CZSE00	0.056	0.9	2.8	5.7	21.8	9.5	11.8	24.1
CZSE00_IF	n/a	1.0	3.0	n/a	27.8	4.0	5.8	29.6
CZNY30	0.043	1.0	1.0	2.6	27.3	4.5	10.0	32.7
CZNY30_IF	n/a	1.0	1.0	n/a	27.7	2.0	9.5	35.2
Mean value:	**0.05***	**1.7**	**2.0**	**1.8***	**15.1**	**4.6**	**5.4**	**15.9**

The effect of predicted increases in future leaching rates of NO_3^- is most marked for the two Pennine sites (CZSD61 and inflow, CZSE00_IF) where the ANC decline would be more than 150 meq L^{-1}. However, these sites are already so highly acidified and damaged that further biological effects would be limited. The most significant effects of increased NO_3^- concentrations are found at those sites where current ANC values are less than 100 meq L^{-1} (Figure 2). The predicted increase in NO_3^- concentrations would reduce ANC at these sites to near or less than 0 meq L^{-1}. This is widely recognized as an important biological threshold, for example for the health of brown trout populations (Lien *et al.*, 1992). Furthermore, with a mean ANC of zero, the seasonality of NO_3^- leaching would lead to periods in which ANC values fall well below zero. The sites at which this transition from positive to negative ANC is predicted (VSH5502 and inflow, CZSH44 and inflow, VSH7105_IF, CZNY30 and inflow) are all found in North Wales and the Lake District of England. The Scottish study sites tend to have higher current ANC values and lower deposition loads, so that predicted ANC values are still quite high. The ecological impacts are likely to be less important at these sites.

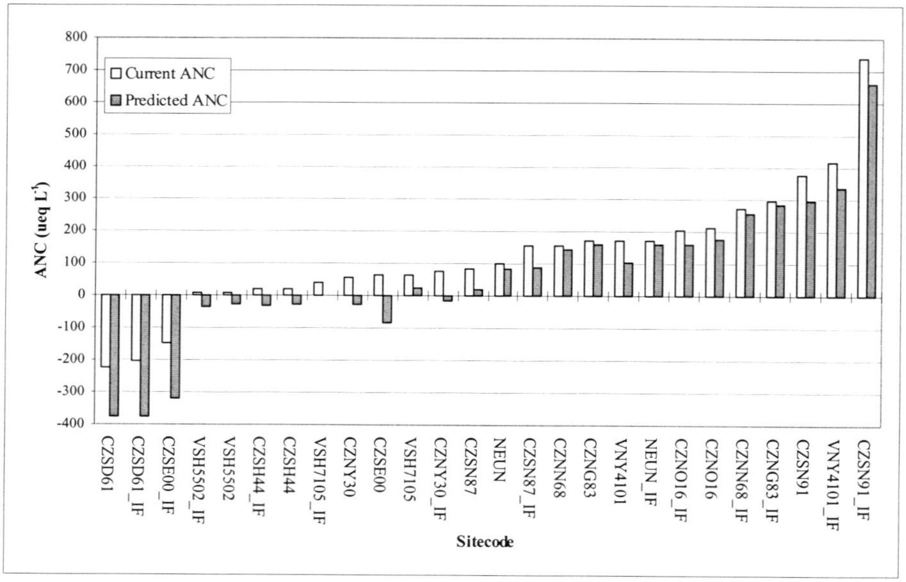

Fig. 2. Reduction in ANC due to increased NO_3^- leaching predicted by the FAB model for nitrogen network sites, assuming no change in current non-marine SO_4^{2-} concentrations.

3.3. Uncertainties in Budgets and Model Predictions

While these model predictions have been calculated using best available data, there are inevitably uncertainties associated with several aspects of the modeling. The calculation of current catchment budgets relies on modeled runoff and deposition data, both of which will have associated errors. Uncertainty in the output of the UK deposition models used has been estimated at ±38 % (Smith *et al.*, 1995), although this could be reduced with the use of 1 km^2 resolution data instead of 20 × 20 km estimates. Uncertainties in the runoff data are unknown. The near 1:1 relationship for non-marine SO_4^{2-} deposition and leaching at many sites does, however, imply that the compound error from these data sets is unlikely to be large. Mean water chemistry data for deriving leaching outputs are not flow-weighted, because discharge has not been measured at the study sites, introducing another potential source of error in budget calculations. The annual mean data are based on 22 months of monitoring so there will be a small error associated with the missing two months of data due to seasonal variability.

Another source of uncertainty in the application of the model lies in the default values used for the N sink terms. The in-lake retention component of the model is based on Canadian studies of a small number of lakes (Dillon and Molot, 1990), and will at best only approximate the true retention of N for the British study sites. The data in Table II show that when the current catchment sinks are estimated, the values for stream inflow sub-catchments where N retention within the water body is zero (mean = 15.3 kg ha^{-1} yr^{-1}) are actually slightly higher than for their respective lake catchments (mean = 14.8 kg ha^{-1} yr^{-1}). There is therefore no evidence from this study that in-lake retention is currently a major sink for N inputs.

The other sink terms in this application of the FAB model are net denitrification (range 0.9-4.0, mean = 1.7 kg ha^{-1} yr^{-1}) and net N immobilization (range 0.9-3.0, mean = 2.0 kg ha^{-1} yr^{-1}). The mean total *terrestrial* sink for N therefore is estimated as 3.7 kg ha^{-1} yr^{-1}, compared with the observed value of 15.3 kg ha^{-1} yr^{-1} for stream sites. The uncertainty associated with these long-term sink estimates is unknown, but even taking the maximum default value (Hall *et al.*, 1997) for each parameter in Britain (N_{upt}=2, N_{den}=3, and N_{imm}=4 kg ha^{-1} yr^{-1}), the total available long-term terrestrial N sink is 9 kg ha^{-1} yr^{-1}, i.e., still about 6 kg ha^{-1} yr^{-1} less than the current observed mean sinks. It is recognized that N immobilization rates in soils might considerably exceed the suggested default range in the short-term. However, for the purposes of critical load modeling it is the longer timescale (decades) that is important and it is not known whether the high N immobilization rates are sustainable over this period. It has been suggested that high soil percent N can lead to N leaching (Dise and Wright, 1995) implying that there is indeed a finite capacity of soils to immobilize N at high rates. This study demonstrates the need for further data on the long-term, sustainable rates at which soils can immobilize N.

It is possible that there are other processes in these catchments omitted from the model that operate as effective N sinks, for example in-stream N retention and retention or enhanced denitrification in riparian wetland zones. These potential sinks are as yet largely unquantified. However, if the cited literature values do truly represent the top end of the steady-state range for the known major N sinks, then even taking into account the uncertainties in the model, all the evidence from this study indicates that current rates of NO_3^- retention are not sustainable. It follows that NO_3^- leaching under constant deposition will increase to varying degrees for each catchment at some point in the future.

4. Conclusions: A "Nitrogen Time-bomb" or a "Missing Sink"?

It could be hypothesized that the current high rates of N retention reflect an adjustment of the catchment ecosystems (vegetation and microbial communities in both terrestrial and aquatic habitats) to anthropogenic inputs of atmospheric N, as implied by Posch *et al.* (1995) and by Emmett and Reynolds (1996). The question is then whether the enhanced levels of denitrification, uptake (removal), and/or immobilization in the soil microbial biomass are transient or sustainable over the long-term. For example, permanent changes in N cycling within catchments might sustain higher levels of immobilization within soils. Changes in vegetation cover could occur due to differences in the ability of plant species to utilize excess N. A transition from heather to grass species, attributed to N deposition, has been observed in lowland heathlands in the Netherlands, and changes in other communities have been recorded (Bobbink *et al.*, 1992). If there is at present a "missing sink" (or sinks) in these study catchments, it cannot be accounted for by current knowledge of long-term steady-state processes. It seems possible that long-term changes in catchment ecosystem structure and function might be induced by both the acidifying and nutrient effects of N (and S) deposition. This could in turn lead to changes in cycling and leaching of N. However, if the steady-state rates of net N uptake, immobilization and denitrification for any modified ecosystem still fall within the known ranges cited above, then at some point in the future there must be a significant increase in NO_3^- leaching from these catchments under current deposition loads, leading ultimately to a decreasing ANC.

Acknowledgments

The authors would like to thank the Critical Loads Advisory Group (CLAG) freshwaters subgroup for the provision of water chemistry data, David Fowler at the Institute of Terrestrial Ecology for deposition data, and David G. Morris at the Institute of Hydrology for provision of rainfall and runoff data. Any misinterpretation of their data is in no way attributable to them. Thanks are also due to Martin Kernan and Dan Bird of the Environmental Change Research Centre, London for assistance with fieldwork and catchment digitizing. And finally, we are grateful to the UK Department of the Environment for funding this work.

References

Aber, J.D., Nadelhoffer, K.D., Steudler, P. and Melillo, J.M.: 1989, *BioScience* **39**, 378.

Allott, T.E.H., Curtis, C.J., Hall, J., Harriman, R. and Battarbee, R.W.: 1995, *Water, Air, Soil Pollut.* **85**, 297.

Battarbee, R.W., Allott, T.E.H., Juggins, S., Kreiser, A.M., Curtis, C. and Harriman, R.: 1996, *Ambio* **25**, 366.

Bobbink, R., Boxman, D., Fremstad, E., Heil, G., Houdjik, A. and Roelofs, J.:1992, in P. Grennfelt and E. Thörnelöf, *Critical Loads for Nitrogen - A Workshop Report*, Nord 1992:41, Copenhagen, pp. 111-160.

Bull, K.R.: 1991, *Environ. Pollut.* **69**, 105.

Critical Loads Advisory Group (CLAG): 1994, *Critical Loads of Acidity in the United Kingdom*, Institute of Terrestrial Ecology, Penicuik, UK, 61pp.

Dillon, P.J. and Molot, L.A.: 1990, *Biogeochemistry* **11**, 23.

Dise, N.B. and Wright, R.F.: 1995, *For. Ecol. Manag.* **71**, 153.

Downing, C.E.H., Vincent, K.J., Campbell, G.W., Fowler, D. and Smith, R.I.: 1995, *Water, Air, Soil Pollut.* **85**, 659.

Emmett, B.A. and Reynolds, B.: 1996, *Forestry* **69**, 205.

Grennfelt, P. and Thörnelöf, E.: 1992, *Critical Loads for Nitrogen - A Workshop Report*, Nord 1992:41, Copenhagen.

Gundersen, P.: 1995, *Water, Air, Soil Pollut.* **85**, 1179.

Hall, J., Hornung, M., Freer-Smith, P., Loveland, P., Bradley, I., Langan, S., Dyke, H., Gascoigne, J. and Bull, K.: 1997, *Current Status of UK Critical Loads Data - December 1996,* Institute of Terrestrial Ecology, Monks Wood, UK.

Harriman, R., Gillespie, E., King, D., Watt, A.W., Christie, A.E.G., Cowan, A.A. and Edwards, T.: 1990, *J. Hydrol.* **116**, 267.

Henriksen, A., Kämäri, J., Posch, M. and Wilander, A.: 1992, *Ambio* **21**, 356.

Hornung, M., Sutton, M. and Wilson, R.B. (eds.): 1995, *Mapping and Modeling of Critical Loads for Nitrogen: A Workshop Report.* Proceedings of the Grange-Over-Sands Workshop, 24-26 October 1994, Institute of Terrestrial Ecology, Penicuik, UK.

INDITE: 1994, *Impacts of Nitrogen Deposition in Terrestrial Ecosystems. Report of the United Kingdom Review Group on Impacts of Atmospheric Nitrogen.* UK Department of the Environment, London.

Kelly, C.A., Rudd., J.W.M., Hesslein, R.H., Schindler, D.W., Dillon, P.J., Driscoll, C.T., Gherini, S.A. and Hecky, R.E.: 1987, *Biogeochemistry* **3**, 129.

Lein, L., Raddum, G.G. and Fjellheim, A.: 1992, *Critical Loads of Acidity to Freshwater - Fish and Invertebrates.* Fagrapport No. 23, NIVA, Oslo, Norway, 36pp.

Patrick, S., Monteith, D.T. and Jenkins, A.: 1995, *UK Acid Waters Monitoring Network: The First Five Years.* ENSIS Publishing, London, 320 pp.

Posch, M., Forsius, M. and Kämäri, J.: 1992, *Water, Air, Soil Pollut.* **66**, 173.

Posch, M., deVries, W. and Hettelingh, J.-P.: 1995, in M. Posch, P.A.M. de Smet, J.-P. Hettelingh and R.J. Downing (eds.), *Calculation and Mapping of Critical Thresholds in Europe,* CCE, RIVM, Bildhoven, the Netherlands, pp. 31-42.

Posch, M., Kämäri, J., Forsius, M., Henriksen, A. and Wilander, A.: 1997, *Environ. Manag.* **21**, 291.

Posch, M.: 1995, in M. Hornung, M.A. Sutton and R.B. Wilson (eds.), *Mapping and Modeling of Critical Loads for Nitrogen - A Workshop Report*, Institute of Terrestrial Ecology, Penicuik, UK, pp. 63-68.

Reynolds, B., Emmett, B.A. and Woods, C.: 1992, *J. Hydrol.* **136**, 155.

Reynolds, B. and Edwards, A.: 1995, *Agric. Water Manag.* **27**, 181.

Sansom, L.E., Briscombe, C. and Page, H.M.: 1997, *Acid Deposition Monitoring in the United Kingdom: The First Ten Years*, Report AEAT/20017001/REMA-304, AEA Technology, Culham, UK.

Seip, H.M.: 1980, in D. Drabløs and A. Tollan (eds.), *Ecological Impact of Acid Precipitation, Proceedings of an International Conference*, Sandefjord, Norway, March 11-14, 1980, pp. 358-366.

Smith, R.I., Hall, J.R. and Howard, D.C.: 1995, *Water, Air, Soil Pollut.* **85**, 2503.

Stoddard, J.L.: 1994, in L.A. Baker (ed.), *Environmental Chemistry of Lakes and Reservoirs*, ACS Advances in Chemistry Series No. 237, American Chemical Society, Washington, D.C., pp. 223-284.

Vincent, K.J., Campbell, G.W., Downing, C.E.H., Hasler, S.E., Davies, M., Stedman, J.R.,

NITROGEN AND SULFATE EXPORT FROM HIGH ELEVATION CATCHMENTS OF THE SIERRA NEVADA, CALIFORNIA

JAMES O. SICKMAN and JOHN M. MELACK

Institute for Computational Earth System Science, University of California, Santa Barbara, California 93106, USA

Abstract. Based on studies of high-elevation, Sierra Nevada catchments during the period from 1983 through 1996, we describe temporal variations in the concentrations of NO_3^- and SO_4^{2-} in surface waters. During snowmelt, some catchments had a pattern of NO_3^- increase to a plateau between the start of snowmelt and some weeks before runoff peaked, and a decline as runoff increased to its maximum. In other catchments, NO_3^- concentrations peaked during the autumn and winter. Long-term trends in surface water chemistry were evident in only two catchments: an increase in SO_4^{2-} concentrations in surface waters of the Ruby Lake basin, and a lowering of annual maxima and minima of NO_3^- concentrations at Emerald Lake. From October 1987 through April 1994, SO_4^{2-} concentrations increased from about 6 μeq L^{-1} to about 12 μeq L^{-1} in Ruby Lake, and in Emerald Lake, NO_3^- maxima declined by 25-50 %.

Keywords: California, monitoring, nitrate export, Sierra Nevada, stream water chemistry, sulfate export, watershed

1. Introduction

Investigations into the dynamics of nitrogen (N) and sulfur (S) in high elevation catchments in the Sierra Nevada can further our understanding of how biogeochemical processes are influenced by atmospheric deposition and climatic anomalies such as drought. In the Sierra Nevada, both elements are deposited in the form of strong acid anions by wet and dry deposition or released, in the case of S, as weathering products from sulfide minerals. Both N and S are involved in the generation and consumption of H^+ (Stoddard, 1994; Williams *et al.*, 1993), and evidence suggests that S dynamics are influenced primarily by processes such as mineral weathering and abiotic sulfate adsorption and desorption in soils (Clow *et al.*, 1996; Williams and Melack, 1997). Nitrogen dynamics are controlled largely by biological processes in catchments, and N supply often determines the productivity of terrestrial and aquatic ecosystems.

Atmospheric deposition of inorganic N to high elevations in the Sierra Nevada is usually less than 30 % of N loading in the eastern United States or the Rocky Mountains (Melack and Stoddard, 1991; Williams *et al.*, 1996a; Melack *et al.*, 1997a), however, the oligotrophic nature of aquatic ecosystems in the Sierra Nevada makes them sensitive to nutrient inputs. Fertilization experiments conducted in mesocosms in Emerald Lake have shown that phytoplankton biomass and species composition are altered by varying inputs of N and P (Sickman, 1991; Sickman and Melack, 1992). In another Sierran lake, Lake Tahoe, atmospheric deposition of N has resulted in a shift from co-limitation of phytoplankton by N and P to persistent P limitation (Jassby *et al.*, 1994). Nitrogen deposition in excess of uptake capacity has resulted in N saturation of terrestrial and aquatic ecosystems in forested watersheds in North America (Stoddard, 1994; Sullivan *et al.*, 1997; Aber *et al.*, 1995), and evidence is growing that this process may be occurring in high elevation watersheds in the Colorado Front Range of the Rocky Mountains (Williams *et al.*, 1996a).

Water, Air, and Soil Pollution **105:** 217–226, 1998.

In light of these findings, an evaluation of the nutrient status of Sierran watersheds is needed.

In this paper we present temporal variations in the concentrations of NO_3^- and SO_4^{2-} in surface waters of high elevation catchments in the Sierra Nevada from 1983 through 1996. These catchments vary considerably with respect to solute chemistry, and climatic conditions during the study had periods of above normal precipitation as well as prolonged drought. Our objectives are to evaluate the status of N saturation in these catchments and to describe interannual variations in surface water chemistry caused by climatic disturbances such as drought.

2. Methods

The seven headwater catchments in our study are located in the alpine and subalpine zones of the Sierra Nevada of California, and their geographic locations span most of the north-south extent of the range. Catchments range in size from 25 ha to 441 ha and their outlet elevations vary from 2,475 m to 3,390 m. For complete descriptions of the sites, see Melack and Sickman (1995).

Lakes were sampled bimonthly with the exception of Emerald Lake where samples were collected biweekly during the summers of 1983 through 1987. Outflows were sampled concurrently with the lakes and on a biweekly to weekly basis during snowmelt (i.e., April through July).

From 1986 through 1996, NO_3^- and SO_4^{2-} concentrations were measured by ion chromatography on a Dionex model 2010i instrument. From 1983 through 1985, NO_3^- concentrations were determined colorimetricaly using cadmium reduction (Strickland and Parsons, 1972), and SO_4^{2-} was measured with ion chromatography.

Stream stage was monitored continuously using pressure transducers installed in the stream bed and attached to solid-state data loggers. Calibration of the transducers to discharge was done using salt dilution estimates of flow. V-notch weirs were installed in the outlets to Emerald and Spuller lakes during 1990.

3. Results

The majority of efflux of solutes and water from the catchments occurred during the period of snowmelt, and solute concentrations changed rapidly during this period (Figures 1 and 2). During snowmelt, some catchments (Emerald, Pear, Ruby and Spuller) typically had a pattern where NO_3^- concentrations increased between the start of snowmelt and some weeks before runoff peaked and later declined as runoff increased to its maximum (Figure 1). In other catchments, i.e., Crystal and Lost lakes, a small increase in NO_3^- concentration was evident during snowmelt, or concentrations remained at or below the detection limit throughout the spring and summer (Figure 1). In contrast, at Topaz Lake, NO_3^- concentrations during the autumn and winter were sometimes 40 to 175 μeq L^{-1} (Figure 3). At Topaz Lake, NO_3^- concentrations often declined during snowmelt, reaching levels below detection during the early summer. As discharge fell to low levels, NO_3^- concentrations in the outflow to Topaz Lake increased to a few μeq L^{-1} to tens of μeq L^{-1} in the late summer.

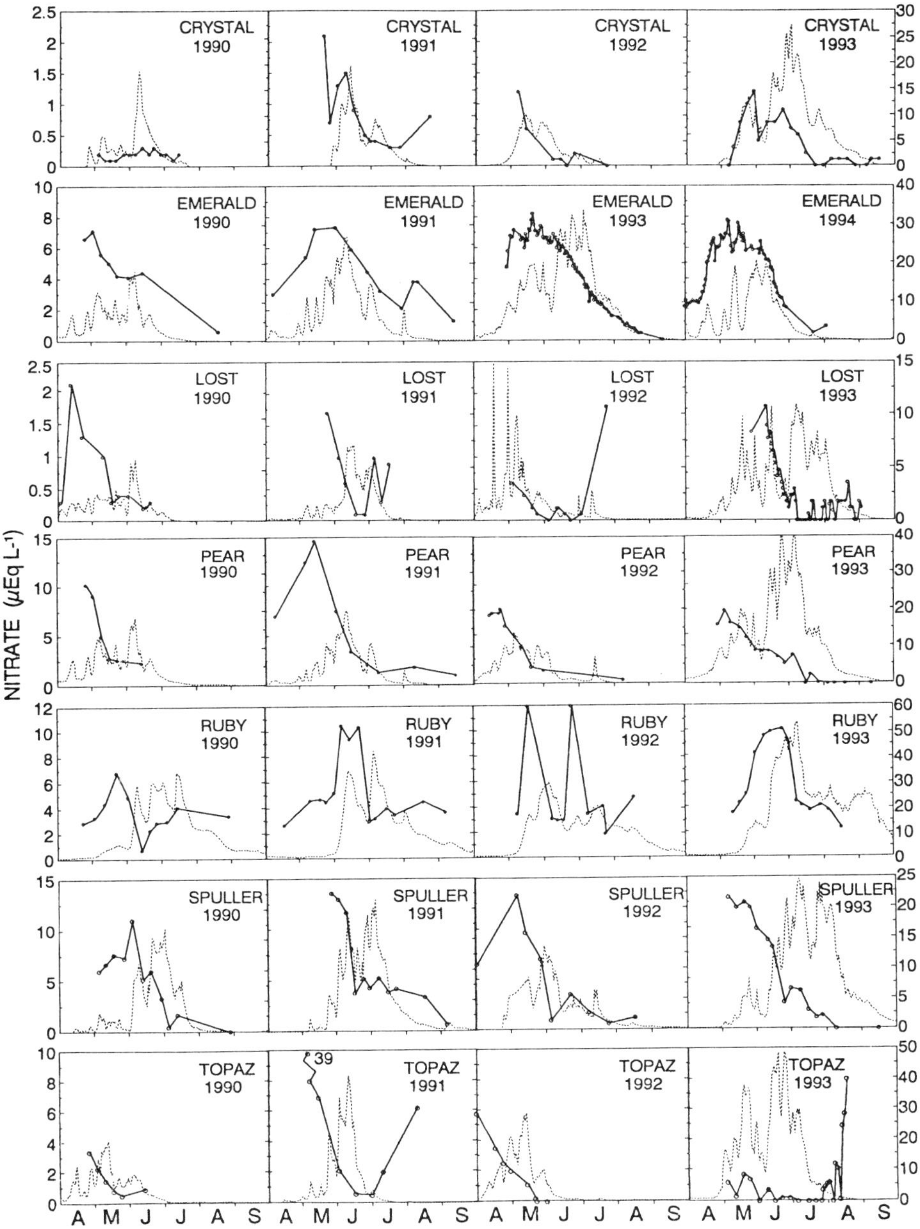

Fig. 1. Time series of NO_3^- and daily outflow discharge at Emerald, Pear, Topaz, Ruby, Crystal, Spuller and Lost Lakes from 1990 through 1993; for Emerald Lake, data from 1994 are substituted for 1992. Solid line with circles (samples) are NO_3^- concentrations and the dashed line without symbols is daily discharge.

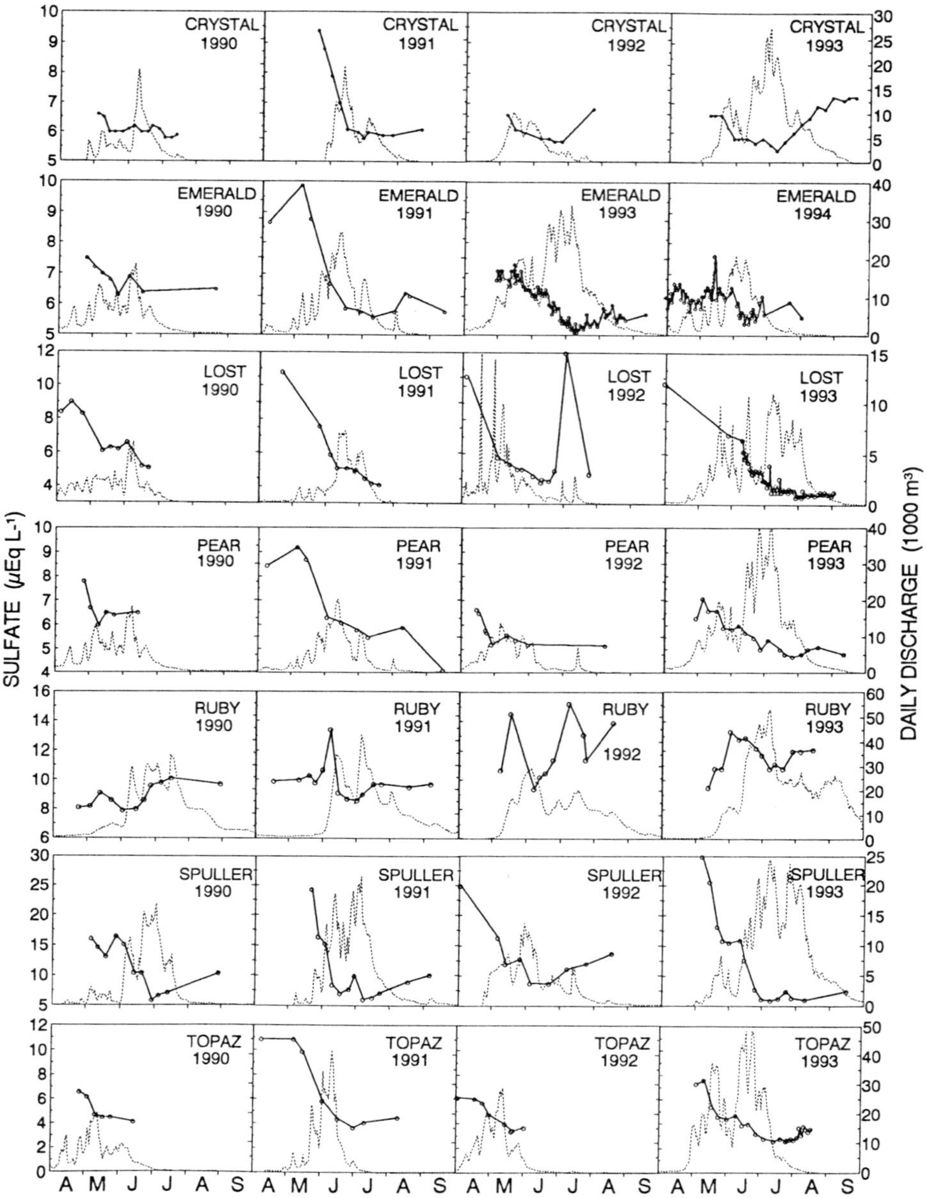

Fig. 2. Time series of SO_4^{2-} and daily outflow discharge at Emerald, Pear, Topaz, Ruby, Crystal, Spuller and Lost Lakes from 1990 through 1993; for Emerald Lake, data from 1994 are substituted for 1992. Solid line with circles (samples) are SO_4^{2-} concentrations and the dashed line without symbols is daily discharge.

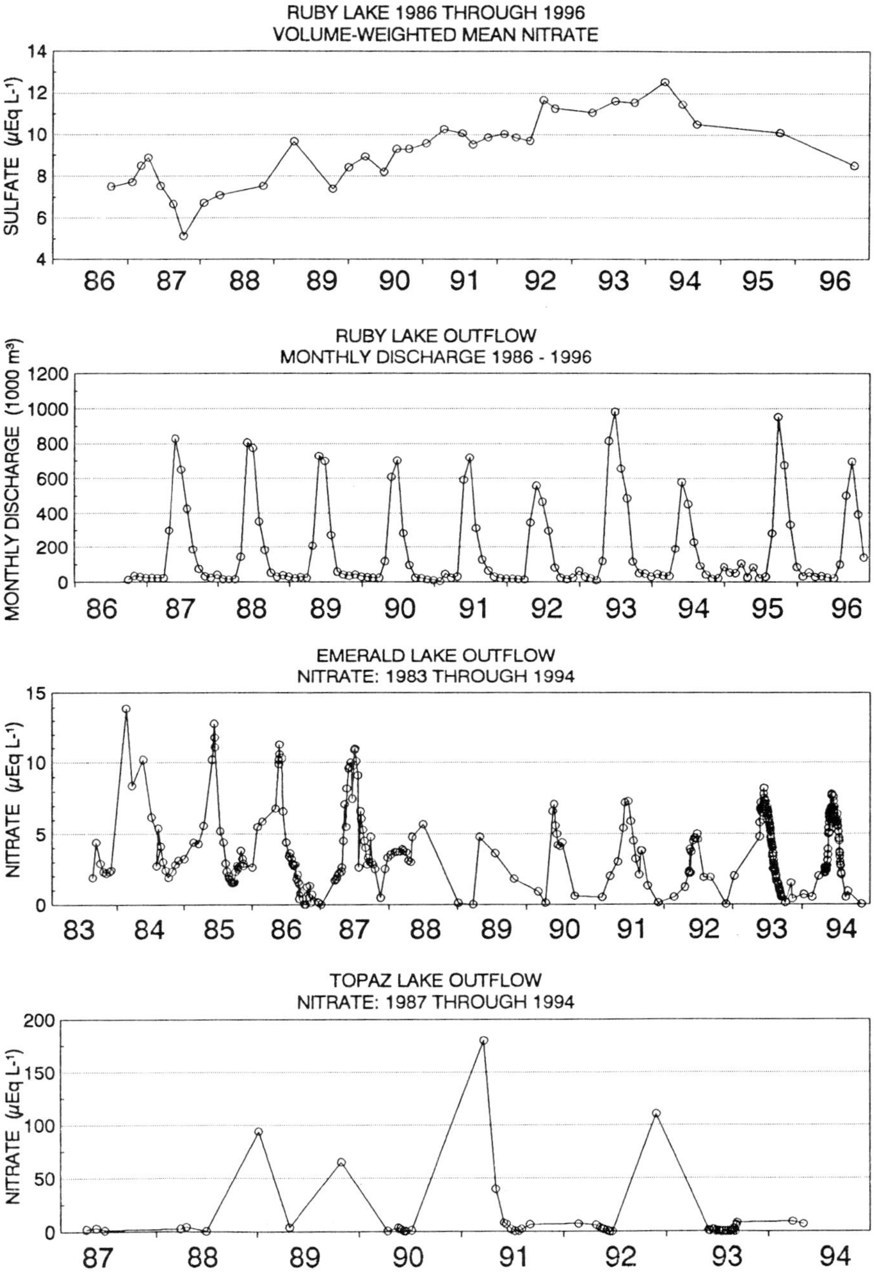

Fig. 3. First panel: volume-weighted mean SO_4^{2-} concentration in Ruby Lake; 1986-1996. Second panel: monthly outflow discharge from the Ruby Lake watershed; 1986-1996. Third panel: NO_3^- concentration in the Emerald Lake outflow; 1983-1994. Fourth panel: NO_3^- concentration in the Topaz Lake outflow; 1983-1994.

Peak NO_3^- concentrations in outflows varied by as much as ten-fold among the catchments. Low concentrations were observed at Lost and Crystal Lakes where peak values ranged from about 0.5 to 2.0 μeq L^{-1}, despite inflows to the lakes having NO_3^- concentrations greater than 5 μeq L^{-1} (Melack *et al.*, 1997b). The Emerald drainage had peak NO_3^- concentrations between 11 and 14 μeq L^{-1} during 1984 to 1987 and between 6 and 8 μeq L^{-1} during 1990 through 1994 (Figures 1 and 3).

In some catchments, SO_4^{2-} concentrations varied in a pulse/dilution pattern similar to NO_3^-, though changes in SO_4^{2-} concentration during snowmelt were smaller (Figure 2). During early snowmelt, SO_4^{2-} concentration increased by 1 to 3 μeq L^{-1} when pulses were observed, the best examples being Ruby 1991 and 1993 and Emerald 1991 and 1994 (Figure 2). In other cases, only dilution occurred, as at Spuller where SO_4^{2-} concentrations declined from 15 to 30 μeq L^{-1} before melt to about 5 to 10 μeq L^{-1} at peak runoff. In some catchments, SO_4^{2-} concentrations remained fairly stable, declining by only 1 or 2 μeq L^{-1}, despite large changes in discharge (e.g., Crystal 1992, 1993 and Emerald 1993, 1994).

Long-term trends (i.e., > 5 year) in surface water chemistry were evident in only two catchments: an increase in SO_4^{2-} concentrations in surface waters of Ruby Lake (Figure 3), and a lowering of annual NO_3^- maxima and minima at Emerald Lake during the period from 1983 through 1994 (Figure 3). During the period from October 1987 through April 1994, SO_4^{2-} concentrations increased from about 6 μeq L^{-1} to about 12 μeq L^{-1} in Ruby Lake and its outflow (Figure 3). Similar trends were seen in only one of the three inflows to the lake routinely sampled; this inflow drains the major portion of the catchment including an area of rock glaciers. The upward trend in SO_4^{2-} concentrations ended early in 1994.

The other trend in surface water chemistry was the decline in NO_3^- concentrations in the Emerald catchment. From 1984 through 1987, peak snowmelt concentrations in the lake and outflow were above 10 μeq L^{-1} (Figure 3). Peak concentrations during water years 1990 through 1994 were less than 5 μeq L^{-1} in the lake and 8-9 μeq L^{-1} in the outflow. Prior to 1986, NO_3^- concentrations levels did not fall below 1 μeq L^{-1} in the outflow or below 0.5 μeq L^{-1} in the lake; in later years, NO_3^- concentrations regularly declined to below the detection limit in both the lake and outflow stream. The change from high to low NO_3^- concentrations appears to have been fairly abrupt, and occurred during 1988. However, during much of 1988 and 1989, the sampling frequency was low and the time-series was not well defined.

4. Discussion

High NO_3^- concentrations at Topaz Lake during the period of late summer through the onset of snowmelt were not associated with NO_3^--rich precipitation events or melting snow. The large annual range of NO_3^- concentrations at Topaz suggests biological processes are largely responsible for the accumulation and release of NO_3^- during periods of low runoff, i.e., late autumn through onset of snowmelt. The shallow lake and extensive, seasonally flooded meadow at Topaz Lake accentuated the influence of biological processes. High NO_3^- concentrations during the autumn and winter seasons are likely the result of low N uptake by plants and relatively high rates of N mineralization and nitrification. While we have no direct measurements of these processes, studies in the Colorado Rockies have found that N mineralization and nitrification occur in snow-covered soils (Williams *et al.*, 1996b;

Brooks *et al.*, 1996).

At Topaz Lake, NO_3^- export occurred during the latter portion of the growing season during some years (i.e., 1991 and 1993), however, NO_3^- concentrations were typically low or below detection during late spring or early summer. Thus, Topaz Lake falls somewhere between Stage 1 of N saturation (Stoddard 1994; chronic N export during non-growing season) and Stage 2 (chronic N export during growing season). However this classification may be inappropriate, because the N exported during the growing season was likely derived from catchment sources and not atmospheric deposition. At other lakes, during the period of 1990 through 1994, NO_3^- was exported only during non-growing seasons (primarily snowmelt) and hence would be classified as at Stage 1 of N saturation. Given the short growing season and sparse vegetation in the High Sierra and the possibility that much of the NO_3^- exported during snowmelt derives from catchment sources (Melack *et al.*, 1997b; Kendall *et al.*, 1995), N export during non-growing seasons probably represent a natural, undisturbed condition for most of these watersheds and suggests that the N saturation concept should not be strictly applied to all Sierran watersheds.

Prior to 1988, Emerald Lake seemed to be experiencing Stage 2 symptoms of N saturation as NO_3^- export occurred year-round, but the catchment was a net sink for N deposition (Melack *et al.*, 1997b). After 1988, conditions at Emerald Lake shifted to Stage 1 of N saturation. There are several possible explanations for the decline in NO_3^- concentrations levels observed at Emerald. One option is that the shift is an artifact of changes in analytical techniques. Nitrate was determined by ion chromatography during most years, but a colorimetric technique was employed during 1983, 1984 and 1985. The switch to ion chromatography was made prior to the snowmelt season of 1986, thus it is unlikely that a change in analytical technique explains the reduction of NO_3^- maxima. In addition, NO_3^- minima,\ were in the range of 1-2 $\mu eq\ L^{-1}$, well above the detection limit of 0.2 $\mu eq\ L^{-1}$ for ion chromatography. Hence, changes in analytical methods can be ruled out as a possible explanation for the shift. In the early years of the project (1983 through 1987), NO_3^- determinations routinely were done within a few days to weeks of collection. In later years, delays of weeks to months were the norm. However, sample storage tests conducted on Emerald Lake samples indicate that analytical delays in the 1990's are not responsible for the decline in NO_3^- concentration. Thus, the shift in NO_3^- concentrations is genuine and requires some other explanation.

Results from nutrient limitation studies corroborate the NO_3^- decrease seen at Emerald Lake. From 1983 to 1985, bioassays were conducted to determine the nutrients limiting phytoplankton primary productivity and biomass in Emerald Lake (Sickman, 1991). In five out of a total of seven experiments, P was the sole nutrient controlling phytoplankton biomass; in the other experiments either no limitation was found or both N and P were limiting. During these experiments, NO_3^- concentrations in the lake were above 2 $\mu eq\ L^{-1}$. Another bioassay experiment was conducted in Emerald Lake during the autumn of 1992 (Sickman and Melack, unpublished). Nitrogen was found to be the limiting nutrient in this experiment. Phytoplankton in Emerald Lake had shifted from P limitation in 1983-85 to N limitation in 1992, which supports the conclusion that the decline in surface water NO_3^- concentrations observed during the past 12 years is real.

A shift to N limitation could be the result of increased P loading or declining N supply to the lake. Given that there was no identifiable trend in atmospheric P or N loading to the Emerald Lake basin during the study period (Melack *et al.*, 1997a), biogeochemical processes within the lake or catchment are probably the cause of the observed change in trophic status. Comparison of inflow and outflow chemistry at Emerald Lake shows that the lake itself exerts little influence on NO_3^- concentrations during the spring and summer seasons (i.e., April through September, Melack *et al.,* 1997b). This observation suggests that lower NO_3^- concentrations in the 1990's are a result of greater retention of N in the terrestrial portions of the catchment. While we have no definitive explanation for increased N retention, recent studies in forested watersheds in the northeastern U.S. (Mitchell *et al.*, 1996) and high elevation catchments of the Alps (Sommaruga-Wögrath *et al.*, 1997) suggest that climate change may effect N cycling within catchments. These studies found that increases in air temperature stimulated biological activity and demand for N, thereby decreasing NO_3^- concentrations in surface waters and lowering N losses from catchments. Another possible explanation is that the 1987-1992 drought may have increased N uptake in the Emerald Lake catchment by lengthening the growing season; snowpacks were shallower and melted faster than in the relatively wet years of 1983 through 1986. In addition, the shallower snowpacks tended to form later in the year, which may have affected rates of N mineralization and nitrification based on findings from microbial studies in soils underlying Rocky Mountain snowpacks (Brooks *et al.*, 1995). Assuming that some NO_3^- exported from the Emerald Lake watershed was derived from catchment sources (Melack *et al.*, 1997b), less NO_3^- was produced within the catchment, hence less was available for export during snowmelt.

Interestingly, Williams and Melack (1997) also measured a decline in NO_3^- concentration in Log Creek ,which was contemporaneous with changes seen in Emerald Lake. This stream drains a small (50 ha) mixed conifer catchment (elevation 2,067 m) located within 15 km of the Emerald Lake basin. Prior to 1990, NO_3^- concentrations ranged from the detection limit to 1.0 mM; after 1990, concentrations were nearly always below the detection limit. The correspondence of temporal changes in NO_3^- concentrations between Emerald Lake and Log Creek supports the hypothesis that N dynamics in the Sierra Nevada are susceptible to climate forcings.

The cause of increased SO_4^{2-} concentrations in surface waters in the Ruby Lake basin is not resolved, but several alternatives can be discounted. No increase in SO_4^{2-} deposition to the Ruby Lake basin occurred during the period in question (Melack *et al.*, 1997a). Analytical error can be ruled out because: (1) no increasing trend was observed for SO_4^{2-} concentrations in other catchments or in two of the major inflows to Ruby Lake, and (2) no changes in analytical or field methods occurred during the period of study. The upward trend in SO_4^{2-} concentrations at Ruby Lake occurred during a long drought. Below average precipitation was recorded during water years 1987 through 1992 in the Sierra Nevada. The drought ended in water year 1993, and a downward trend in SO_4^{2-} concentrations began in early 1994. Although low precipitation may have contributed to the increase in SO_4^{2-} concentration in surface waters of the Ruby Lake basin, the lack of similar changes in the other six catchments requires identifying special circumstances at Ruby Lake. Several morphological characteristics do set the Ruby basin apart including its large size and high elevation, substantial storage and release of groundwater and the presence of rock-glaciers. In addition, much of the Ruby Lake basin in underlain by granites that are high in sulfide minerals. The upward trend in SO_4^{2-} concentrations was not detected until the year after the start of the drought (1988), and initiation of the downward trend did not begin until a year after the end of the drought in 1993. Possible explanations for the hydrochemical inertia include a sizable groundwater reservoir in the basin and the large volume of the lake. Ground water influence on surface water chemistry might increase as precipitation quantity and snowmelt runoff decline, resulting in higher solute concentrations in streams and lake water. Monthly runoff from the Ruby Lake basin from 1987 through 1994 gradually declined (Figure 3). As precipitation and snowmelt decreased, the relative influence of alternate sources of water to streams increased, which could have resulted in changes in stream chemistry. Also, sulfide mineral weathering in the catchment may have increased due to the shorter duration of snow cover during the drought and/or increased temperature.

5. Conclusions

Research conducted in high elevation watersheds of the Sierra Nevada during the last two decades indicates these ecosystems are potentially sensitive to increased nutrient loading and climatic perturbations. Drought conditions in the Sierra Nevada probably were responsible for increasing the proportion of runoff derived from shallow groundwater in the Ruby Lake basin as evidenced by increased SO_4^{2-} concentrations. Drought also may be partially responsible for increased retention of N in the Emerald Lake catchment. In the Emerald Lake basin, long-term monitoring has revealed a decrease in annual NO_3^- maxima and minima with a concomitant shift of the lake's phytoplankton community from P limitation toward N limitation. These findings support recent evidence that N uptake in alpine catchments may increase due to climate warming and runs counter to the recent shift of Lake Tahoe to P limitation of phytoplankton. At Emerald Lake, a decrease in NO_3^- concentrstions represents a shift from Stage 2 to Stage 1 of N saturation and is a counter-example to lakes and streams in the Front Range of the Rocky Mountains.

Acknowledgments

We thank all the dedicated field assistants who helped sample streams and lakes, and thank F. Setaro, D. Lucero and S. Sippel for performing chemical analyses. Permission to conduct our studies within Sequoia National Park was granted by the National Park Service, and permission to work in National Forests was provided by the U.S. Forest Service. The California Air Resources Board and the National Aeronautics and Space Administration funded our research. The manuscript was improved by suggestions from A. Leydecker, T. Meixner and two anonymous reviewers.

References

Aber, J.D., Magill, A., McNulty, S.G., Boone, R.D., Nadelhoffer, K.J., Downs, M. and Hallet, R.: 1995, *Water, Air, Soil Pollut.* **85**, 665.

Brooks, P.D., Williams, M.W. and Schmidt, S.K.: 1995, 'Snowpack Controls on Soil Nitrogen Dynamics in the Colorado Alpine,' in K.A. Tonnessen, M.W. Williams and M.Tranter (eds.), *Biogeochemistry of Seasonally Snow-Covered Catchments*, IAHS Publication no. 228, pp. 283-292.

Brooks, P.D., Williams, M.W. and Schmidt, S.K.: 1996. *Biogeochemistry* **32**, 93.

Clow, D.W., Mast, M.A. and Campbell, D.H.: 1996, *Hydrol. Process.* **10**, 727.

Jassby, A.D., Reuter, J.E., Axler, R.P., Goldman, C.R. and Hackley, S.H.: 1994, *Water Resourc. Res.* **30**, 2207.

Kendall, C., Campbell, D.H., Burns, D.A., Shanely, J.B., Silva, S.R. and Chang, C.C.Y.: 1995, 'Tracing Sources of Nitrate in Snowmelt Runoff using the Oxygen and Nitrogen Isotopic Compositions of Nitrate,' in K.A. Tonnessen, M.W. Williams and M.Tranter (eds.), *Biogeochemistry of Seasonally Snow-Covered Catchments*, IAHS Publication no. 228, pp. 339-347.

Melack, J.M. and Stoddard, J.L.: 1991, 'Sierra Nevada,' D.F. Charles (ed.), *Acidic Deposition and Aquatic Ecosystems*, Springer-Verlag, New York, in pp. 503-530.

Melack, J.M. and Sickman, J.O.: 1995, 'Snowmelt Induced Changes in Seven Streams in the Sierra Nevada, California,' in K.A. Tonnessen, M.W. Williams and M.Tranter (eds.), *Biogeochemistry of Seasonally Snow-Covered Catchments*, IAHS Publication no. 228, pp. 221-234.

Melack, J.M., Sickman, J.O., Setaro, F. and Dawson, D.: 1997a. *Monitoring of Wet Deposition in Alpine Areas in the Sierra Nevada.* Final Report, Contract A932-081. California Air Resources Board.

Melack, J.M., Sickman, J.O., Leydecker, A. and Marrett, D.: 1997b. *Comparative Analyses of High-elevation Lakes and Catchments in the Sierra Nevada: Susceptibility to Acidification.* Final Report, Contract A032-188. California Air Resources Board.

Mitchell, M.J., Driscoll, C.T., Kahl, J.S., Likens, G.E., Murdoch, P.S. and Pardo, L.H.: 1996, *Environ. Sci. Technol.* **30**, 2609.

Sickman, J.O. 1991. *Planktonic Primary Productivity and Responses of Phytoplankton to Acid and Nutrient Additions in Emerald Lake, Sierra Nevada, California.* M.A. Thesis, University of California, Santa Barbara.

Sickman, J.O. and Melack, J.M.: 1992, *Hydrobiologia* **230**, 37.

Sommaruga-Wögrath, S., Koinig, K.A., Schmidt, R., Sommaruga, R., Tessadri, R. and Psenner, R.: 1997, *Nature* **387**, 64.

Stoddard, J.L.: 1994, 'Long-term Changes in Watershed Retention of Nitrogen: Its Causes and Aquatic Consequences', in Baker L.A. (ed.) *Environmental Chemistry of Lake and Reservoirs*, Advances in Chemistry Series No. 237, American Chemical Society, Washington D.C., pp. 223-284.

Strickland, J.D. and Parsons, T.: 1972, *A Practical Handbook of Seawater Analysis*, Bulletin of the Fisheries Research Board of Canada.

Sullivan, T.J., Eilers, J.M., Cosby, B.J. and Vaché, K.B.: 1997, *Water, Air, Soil Pollut.* **95**, 313.

Williams, M.W., Brown, A.D. and Melack, J.M.: 1993, *Limnol. Oceanogr.* **38**, 775.

Williams, M.W., Bales, R.C., Brown, A.D. and Melack, J.M.: 1995, *Biogeochemistry* **28**, 1.

Williams, M.W., Baron, J.S., Caine, N., Sommerfeld, R. and Sanford Jr., R.: 1996a, *Environ. Sci. Technol.* **30**, 640.

Williams, M.W., Brooks, P.D., Mosier, A. and Tonnessen, K.S.: 1996b, *Water Resourc. Res.* **32**, 3161.

Williams, M.R. and Melack, J.M.: 1997, *Biogeochemistry* **37**, 111.

LINKS BETWEEN RUNOFF GENERATION, CLIMATE AND NITRATE-N LEACHING FROM FORESTED CATCHMENTS

LOTTA ANDERSSON[1] and AHTI LEPISTÖ[2]

[1]*Department of Water and Environmental Studies, Linköping University, S-581 83 Linköping, Sweden*
[2]*Finnish Environment Institute, P.O.Box 140, FIN-00251 Helsinki, Finland*

Abstract. To assess links between hydroclimatological factors and NO_3^--N concentrations in streamflow from boreal forests with shallow soils, data from two catchments were analyzed. TOPMODEL was used to calculate the surface runoff fraction, daily dynamics of soil moisture, groundwater levels, and extensions of saturated areas. The stable isotope ^{18}O was used for isotopic hydrograph separation (IHS) during one snowmelt season. Air-temperature and flow increase were the dominating factors explaining annual NO_3^--N dynamics. Correlation also was found between NO_3^--N concentrations and the surface runoff fraction. Increased concentrations during times of shallow groundwater were found both during cold and warm periods in one catchment. In the other, shallow groundwater was correlated to decreasing concentrations during cold periods, and increasing concentrations during warm periods. A two component model of event and pre-event water fractions and corresponding NO_3^--N concentrations was set up for the snowmelt season. Model predictions mirrored NO_3^--N concentration s during the first five days of the snowmelt. After that, the model overestimated NO_3^--N concentrations, which indicates retention of NO_3^--N in the event water fraction, originating from the snowmelt. The highest concentrations occurred during the initiations of flow increase, which indicates flushing of surficial NO_3^--N.

Keywords: boreal forest, hydrograph separation, stable oxygen isotopes, nitrate leaching, TOPMODEL

1. Introduction

The ability to predict dynamics of N flow is important, as it leads to the potential to predict impacts of imposed environmental changes, such as increased N deposition (e.g., Lepistö, 1995), climatic change (e.g., Wright and Schindler, 1995) and forest management (e.g., Swank and Johnson, 1994). In a riparian forested wetland area, water sources have been found to have distinct and evolving nitrate (NO_3^-) signatures, which can vary both vertically and laterally, depending upon their specific flowpaths (Hill, 1990; Cirmo and McDonnell, 1997). Water pathway is a key concept for the transport of substances to streams in at least three ways. First, water with short transit times is more likely to have a composition that is close to that of precipitation. Constituents in precipitation, e.g. of NO_3^--N fallen on saturated areas, may be transported to the stream with limited soil interaction, which might increase the concentrations in the streams if concentrations in precipitation are high compared with those of streamwater (Löfgren, 1991; Lepistö, 1994). Secondly, because biological transformations mainly are confined to the upper soil horizons, the availability of NO_3^--N in the soil often decreases with depth (Burt and Arkell, 1986). Third, NO_3-N may be flushed from areas close to the stream in the initial phase of a flow increase, during horizontal extension of saturated areas or during vertical extension of saturated soil layers (Andersson and Sundblad, 1992; Stoddard, 1994; Arheimer *et al.* 1996; Creed *et al.*, 1996).

Within a particular region or forest stand, mineralization and nitrification rates vary considerably in response to temperature and moisture (e.g., Vitousek, 1977; van Miegroet and Johnson, 1993). Temperature is a critical factor for terrestrial assimilation, i.e., uptake and metabolic use of N by plants and soil microbes, which is a major form of N removal

Water, Air, and Soil Pollution **105:** 227–237, 1998.

from watersheds (Vitousek, 1977). Climate will thus both influence the available amount of NO_3^--N that can be leached from the soil and the water flow that will transport it to the streams (e.g., Arheimer *et al.*, 1996). Because a large proportion of the dynamics is governed by climatological factors, these should be considered before any further assumptions of changes over time are made, and when interpolating concentrations to estimate monthly or annual export.

In an isotopic hydrographic separation (IHS), water with short transit times corresponds to rainfall or snowmelt that reached the stream during an event. Water with longer transit times corresponds to the pre-event water. It should be noted, however, that the event water fraction could include water that has passed the soil surface, although with short transit times, before it enters the stream. A separation of components can also be made with an hydrological model that separates the surface runoff from the subsurface.

In this study, the stable isotope ^{18}O was used for isotopic hydrograph separation (IHS) into event and pre-event water during a snowmelt period, and in the longer term. TOPMODEL (Beven and Kirkby, 1979) was applied for calculation of the daily surface runoff fraction and dynamics of soil moisture and groundwater depths. The study was based on data from two small forested catchments, Rudbäck in Finland and Stubbetorp in Sweden. The following questions were addressed, concerning streamwater concentrations of NO_3^--N: to what extent can air temperature or other factors explain seasonal variations, to what extent can the proportion of event-water or surface runoff explain variations, and to what extent can processes related to flushing of N accumulated close to the stream, or in soil layers that become saturated, explain variations?

2. Studied Catchments

2.1. Catchment Characteristics

The Rudbäck catchment is located in Siuntio, southern Finland, about 40 km west of Helsinki. A subcatchment of 0.18 km^2 was used in this study. The Stubbetorp catchment (0.87 km^2) is situated in southeastern Sweden, about 150 km south of Stockholm. Soils are shallow in both catchments, with depths varying between 0 and 5 m (Mäkinen, unpublished data; Fleetwood and Olofsson, 1994). The vegetation cover is mainly coniferous forest; no lakes or agricultural areas are included. Detailed descriptions of the catchments are given by Kämäri *et al.* (1992) and Robinson (1993).

2.2. Hydrological and Hydrochemical Monitoring

Monitoring of hydrology and hydrochemistry has taken place since 1991 in Rudbäck (Kämäri *et al.*, 1992; Lepistö, 1994) and since 1985 in Stubbetorp (Robinson, 1993). For this study, data from 1995-1996 were used for Stubbetorp, and from 1991-1995 for Rudbäck. From these periods, frequent time series of $\delta^{18}O$ and NO_3^--N concentrations were available. Runoff was continuously monitored and daily precipitation was recorded. The $\delta^{18}O$ of water samples was measured by mass spectrometry at the University of Copenhagen; NO_3^--N was analyzed using the cadmium reduction method.

3. Methods

3.1. HYDROLOGICAL SIMULATIONS

The semi-distributed TOPMODEL (Beven and Kirkby, 1979) was used with daily time steps for simulation of runoff, divided into a surface and a subsurface component. Simulations of the water equivalent of the snowpack, snowmelt, groundwater tables and soil moisture contents also were used in the study. The model uses a distributed topographical index to give all grid squares with the same index identical groundwater levels at each time step. Saturated excess overland flow is assumed to occur from grids in which the groundwater level is at the soil surface. We included a simple snow subroutine, based on a degree-day approach.

3.2. ISOTOPIC HYDROGRAPH SEPARATION

From the mass and tracer balance in steady state, the fraction of event water contribution in the stream is (e.g., Fritz *et al.*, 1976):

$$F_{event} = 1 - [(\delta_s - \delta_p)/(\delta_g - \delta_p)] \quad (1)$$

where: δ_s, δ_p, and δ_g are the tracer composition of streamwater, precipitation and groundwater. Hydrograph separation was carried out for the snowmelt period that occurred in Stubbetorp during April 1996. The water equivalent of the snowpack (SWE) was modeled to 43 mm before the initiation of the melt period. The $\delta^{18}O$ ratio of groundwater was estimated from initial streamwater $\delta^{18}O$ ratios (-11.6 ‰) and was assumed to be stable during the melt period, as it returned to a similar level after the melt period. The $\delta^{18}O$ ratio of meltwater was not known in detail because the collected meltwater amount only corresponded to a third of the snowpack in the catchment. Therefore, the volume weighted average $\delta^{18}O$ ratio (-15.9 ‰) of all the snowfall events (n=31) during the snow accumulation period (16 Nov. 1995 - 31 March 1996) was used as the snowmelt $\delta^{18}O$ ratio, which was close to that measured from meltwater samples (-16.4 ‰). Samples used for analyses of $\delta^{18}O$ sometimes were lumped to represent a period of increasing or decreasing flow. When using lumped samples of $\delta^{18}O$, the same lumping was used for the corresponding NO_3^--N water sample.

3.3. ANALYSES OF LINKS TO NO_3^--N CONCENTRATIONS

Linear regression models were used to explain daily variations of NO_3^--N concentrations. The significance of various factors was assessed both for the total time series, and separately for warm (air temperature > +10 °C) and cold (air temperature < +5 °C) periods. The following variables were included: measured runoff, absolute and relative runoff increase since the preceding day, daily air temperature, 60- and 120-day moving averages of air temperature, and TOPMODEL simulations of groundwater depths, soil moisture deficits, surface runoff fractions, total volumes of surface runoff, and extension of the saturated area. The variables were selected stepwise, using a probability of obtaining a higher F value of 0.05 on entry and of 0.10 on removal.

In addition, a simple two component model was set up for the snowmelt event in Stubbetorp (Eq. 2).

$$NO_3^- \text{-}N_{str} = (F_{pre\text{-}event} \times NO_3^- \text{-}N_{ground}) + (F_{event} \times NO_3^- \text{-}N_{melt}) \quad (2$$

where: $F_{pre\text{-}event}$ and F_{event} are fractions of pre-event and event water, determined by isotopic separation. NO_3^--N_{str} is the concentration in the stream prior to the event, and NO_3^--N_{melt} is the volume weighted average concentration during the snow accumulation period.

4. Streamflow Dynamics, Climate and NO_3^--N Concentrations

4.1. Seasonal Variations

The NO_3^--N concentrations at different temperature intervals were similar for the two catchments (Figure 1), ranging from below detection limit to 0.5 mg L^{-1}, which is very low compared to forested sites in mid-Europe where over one order of magnitude higher concentration are common (Hauhs *et al.*, 1989). If no separation into seasons was made, a negative correlation to temperature was the most significant variable in regression models of NO_3^--N dynamics. For Rudbäck, the best simulation was obtained with the moving average air temperature, using a time lag of 120 days, whereas a lag of 60 days gave the best fits for Stubbetorp. For both catchments, the relative flow increase factor was also included in the best model. The regression model for Rudbäck (1991-1995, $R^2 = 0.33$) is shown in Figure 2. The model for Stubbetorp (1995-1996) obtained an R^2 of 0.55.

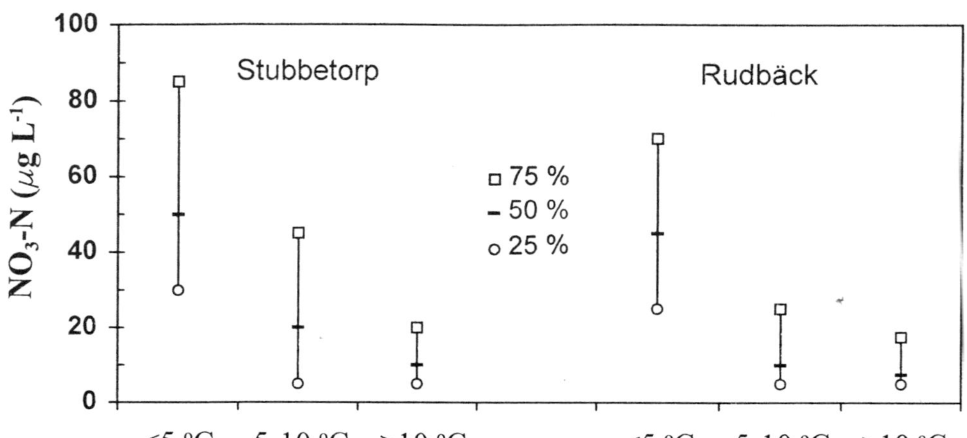

Fig. 1. Median, 25 and 75 percentiles of NO_3^--N concentrations in Stubbetorp (1995-1996) and in Rudbäck (1991-1995) for different air temperature intervals.

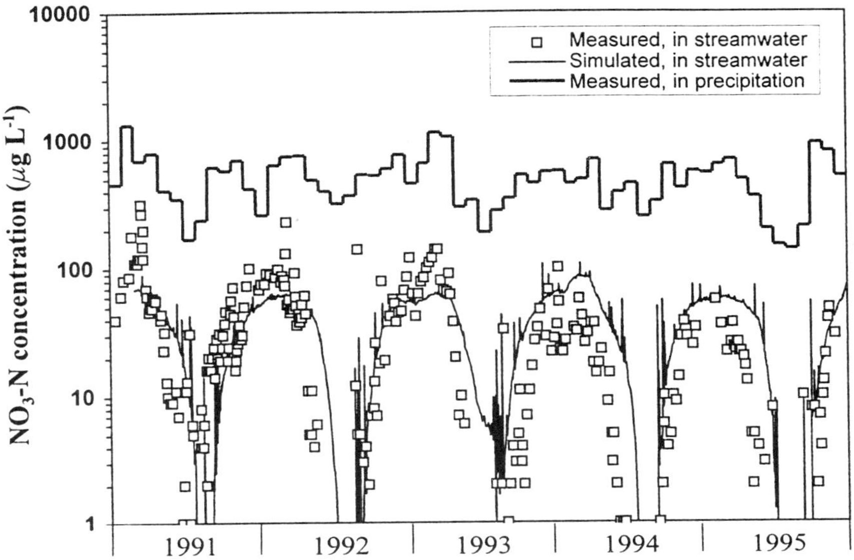

Fig. 2. NO_3^--N concentrations in precipitation and streamwater at Rudbäck (1991-1995). Measured and modeled streamwater concentrations are compared. Observe that a logarithmic scale is used.

Models also were set up for the upper and lower temperature intervals, defined in Figure 1. During the warm season (Figure 3a), the groundwater depth explained a significant variability in NO_3^--N concentrations, with higher concentrations when groundwater depths were closer to the soil surface. In Rudbäck, groundwater depth and a positive correlation to the relative flow increase explained 34 % of the variation. In Stubbetorp, only 24 % of the variation was explained by groundwater depth, whereas flow increase not was selected as a significant variable. The highest NO_3^--N concentrations occurred, however, during events with relatively high groundwater levels or a flow increase.

During the cold season (Figure 3b), a positive correlation to the flow increase factor seemed to be the most significant explaining variable in Stubbetorp, explaining 41 % of the variation. In addition, groundwater depth was included in the selected model, with lower NO_3^--N concentrations when the groundwater was shallow, the model explaining 55 % of the variation. In Rudbäck, concentrations were significantly higher when the groundwater depth decreased also during the cold season. Groundwater depths and flow increase explained 28 % of the variation. The highest concentrations in Rudbäck occurred during the start of the snowmelt in March 1991, with the maximum during the beginning of the flow increase.

In Stubbetorp, a 4.6 % surface runoff fraction was calculated with TOPMODEL, compared with 7.3 % for Rudbäck. Between the surface runoff fraction and NO_3^--N concentrations, significant ($p < 0.05$) although weak correlation was found in both catchments. When only considering flow increase during cold periods, the correlation was higher for Stubbetorp ($R^2 = 0.32$, $p < 0.01$). For Rudbäck, the highest NO_3^--N concentrations were linked to relatively high (20-30 %), but not the highest, surface runoff fractions (50-60 %).

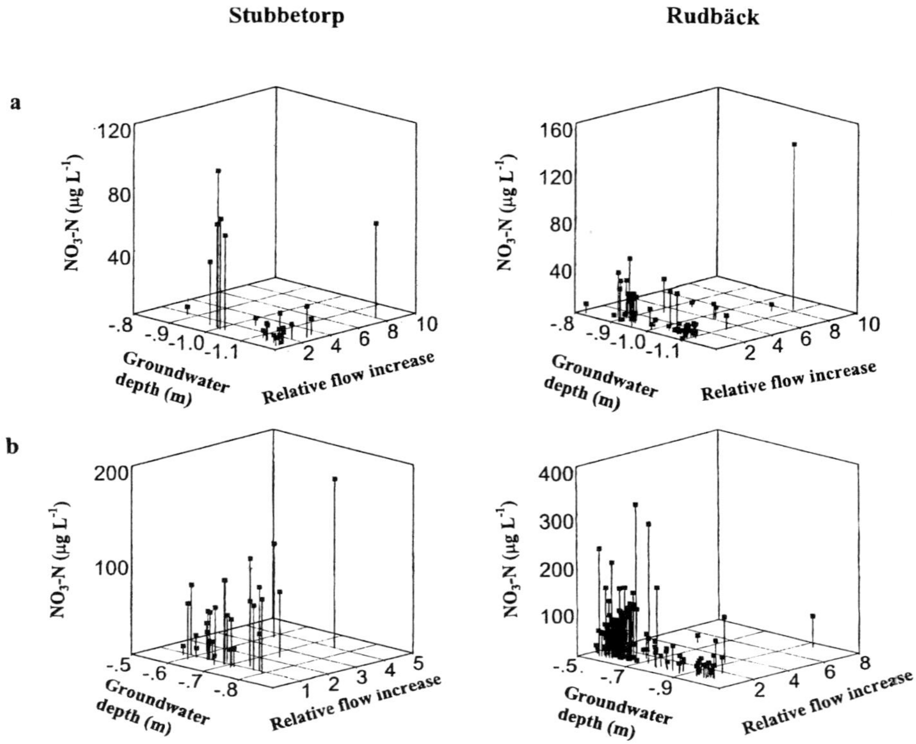

Fig. 3. NO_3^--N concentrations vs. groundwater depths and relative flow increase in Rudbäck and Stubbetorp: a) air temperature above +10 °C; b) air temperature below +5 °C .

4.2. A Snowmelt Event

During flow increase, the $\delta^{18}O$ ratio of streamwater decreased by not more than 0.5 ‰ (Figure 4a). Thereafter, it returned to the pre-event baseflow ratio within two weeks, with considerable fluctuations within and between days. The total runoff during the intensively monitored period was 57 mm. The event water fraction varied between 0 and 17 %, with an average event water fraction of only 7 % (Figure 4b). The fraction was highest in the beginning of the snowmelt period (Figure 4c).

The hypothesis that the event water fraction would provide the runoff with a NO_3^--N concentration that corresponds to its part of the total runoff was tested by setting up a two component model. The event water fraction was given a NO_3^--N concentration equal to the average snowpack NO_3^--N concentration, using monthly observed concentrations in precipitation during the snow accumulation period (610 mg L^{-1}). For the pre-event fraction, a concentration of 30 mg L^{-1} was used, which corresponded to the lowest 25th percentile of cold period concentrations (Figure 1), and to the concentration monitored two weeks before, and directly before the start of the snowmelt. The model predictions mirrored two of the characteristics of the dynamics (Figure 4d): the general decrease of NO_3^--N concentrations during the snowmelt, and the increase of NO_3^--N concentration during the start of the flow increase during April 4-8. After that, however, increased amounts of event water were not reflected by a corresponding increase in NO_3^--N concentration, as expected by the model. The model did not explain the three peaks in NO_3^--N concentrations (Figure 4d). The first took place during the first flush of meltwater in late March, the second in the very beginning of the flow peak on April 8 and the third and highest concentration during the flow peak on April 19, which was just after the modeled maximum extension of the saturated area during the snowmelt was reached.

5. Discussion

A negative correlation to temperature was undoubtedly the most important factor for annual dynamics of NO_3^--N in streamwater. Owing to nutrient uptake processes, affected by temperature, NO_3^--N concentrations are depressed during the growing season, giving an amplified seasonal pattern (Stoddard, 1994). In addition to temperature, the flow increase factor was significant for explaining annual dynamics, indicating the importance of flushing (Murdoch and Stoddard, 1992; Stoddard, 1994).

Significant positive correlation between modeled surface runoff and concentrations was only found when considering periods of flow increases during cold periods. Flushing is characterized by peak concentrations occurring prior to the peak flow (Hornberger *et al.*, 1994). During cold periods, availability of N is high, but during flow decrease it is possible that available N is already washed out. Low NO_3^--N concentrations sometimes were observed when surface runoff fractions were remarkably high, indicating that it was not the direct contribution from the N-deposition in the precipitation that caused the concentrations to rise, but flushing of N.

The study of the snowmelt event in Stubbetorp, however, showed that the rise in NO_3^--N concentrations during the snowmelt followed the rise of the event water fraction separated by isotopes, although the two-component not could explain the highest concentrations. The first two of these peaks occurred during the initiation of flow increase, and the third when the maximum extension of the saturated area was reached. Possible explanations could thus be flushing of accumulated N from areas close to the stream in the initiation phase, and from areas in the extended saturated area later on.

Another sign of the possible importance of NO_3^--N concentrations in the event water was the fact that seasons with high concentrations in precipitation during the snowmelt accumulation periods in Rudbäck were mirrored by high concentrations in the stream during the following snowmelt periods. Considering groundwater depths, it could be hypothesized that the availability of NO_3^--N could be decreasing with depth, because biological transformations are confined to the upper soil horizons (Burt and Arkell, 1986). No data were, however, at hand concerning N-availability in the soil. To understand connections between groundwater depths and NO_3^--N concentrations it also has to be considered that high groundwater levels are linked to large volumes of water being transported through the soil, which may cause dilution. Fröberg (1997) showed that in Stubbetorp, during the winter months there was a negative correlation ($R^2 = 0.72$) between flow and NO_3^--N concentrations monitored in a spring in the catchment, which contributes with approximately 2 5% of the total discharge.

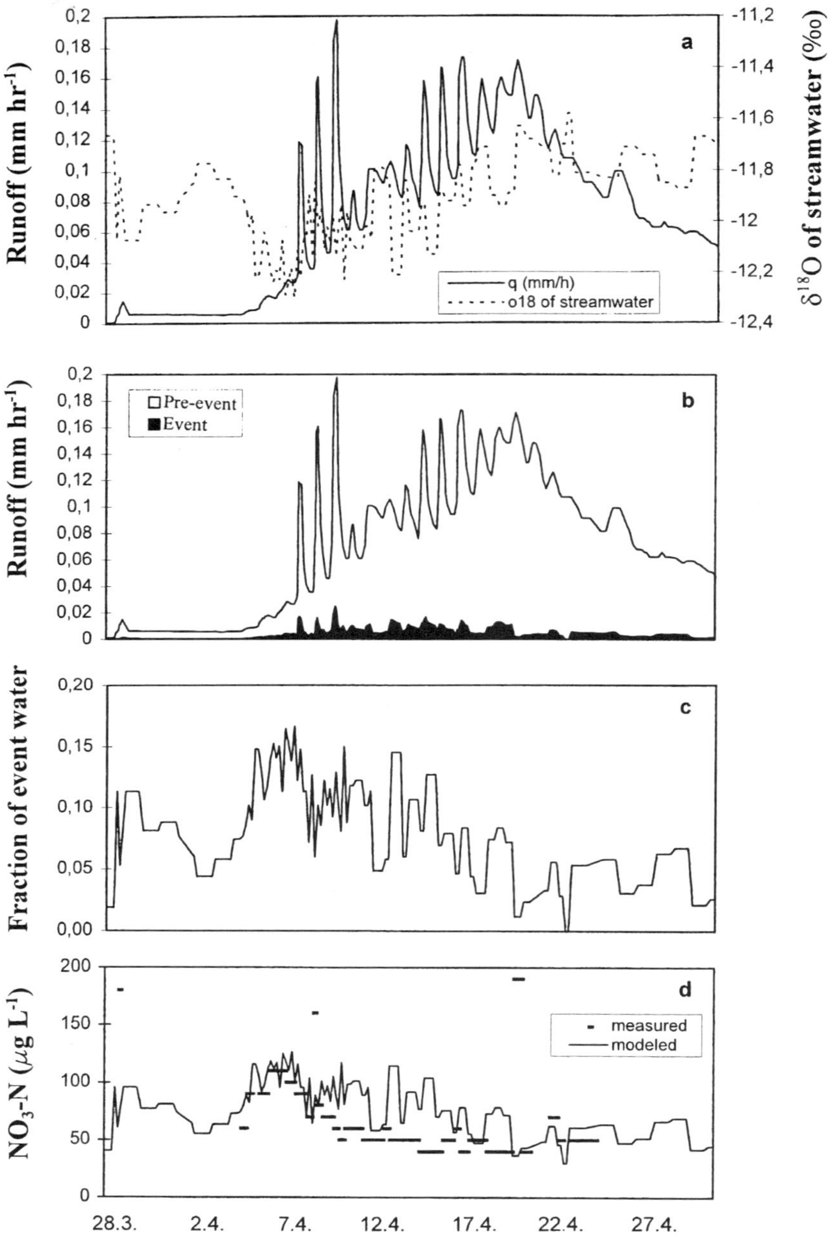

Fig. 4. Snowmelt period, Stubbetorp, March 28 - April 30, 1996: a) δ^{18}O ratio of streamwater; b) hourly event and pre-event fractions, separated by IHS; c) fraction of event water; d) measured and modeled NO_3^--N concentrations, using a two-component model. In some cases, samples of δ^{18}O and NO_3^--N concentrations are accumulated to represent a period of flow increase or decrease.

Positive correlation to groundwater depths was found during warm periods in both catchments, when groundwater depths are lower and the flow is small. In Rudbäck, positive correlation also was found during warm periods, whereas the correlation during cold periods was negative in Stubbetorp, indicating that the dilution effect was overriding.

6. Conclusion

Air temperature had an overriding importance for concentration dynamics. Concerning the importance of the flushing effect vs. the proportion of event water or surface runoff, it was difficult to separate the effects from the two processes, as both are related to periods of flow increase. The acceptable fit in Stubbetorp, obtained with the two-component model during the snowmelt event, and the link between high NO_3^--N concentrations in deposition and in the following snowmelt in Rudbäck, indicate that the proportion and composition of the event water is of importance for streamwater concentrations of NO_3^--N during snowmelt. On the other hand, because a correlation between concentrations and the modeled surface runoff fraction only was found for periods with a combination of low temperatures and flow increase, and because the sometimes very high proportions of surface runoff in Rudbäck did not correspond to high concentrations, it seems that there had to be N available to be flushed out to obtain high concentrations. The fact that high surface runoff fractions did not correspoed to high concentrations, with the above exceptions, indicates that N in the event water is retained in the near-stream area.

Acknowledgments

The Swedish Geological Survey (SGU) and Finnish Environment Research, Helsinki financed the study. Prof. Keith Beven, Lancaster University, is acknowledged for providing us with the TOPMODEL code.

References

Andersson, L. and Sundblad, K.: 1992, In: Lundteigen-Fossdal (ed.), *NHP Workshop, Water and Mass Balances in Small Catchments*, Norwegian National Committee for Hydrology, Oslo.

Arheimer, B., Andersson, L. and Lepistö, A.: 1996, *J. Hydrol.* **179**, 281.

Beven, K., and Kirkby, M.J.: 1979, *Hydrol. Sci. Bull.* **24,** 43.

Burt, T.P., and Arkell, B.P.: In: Lerner, D. (ed), *Monitoring to Detect Changes in Water Quality Series*, Proc. of Budapest Symposium, July 1986, IAHS Publ. No. 157, 155.

Cirmo, C.P., and McDonnell, J.J.: 1977, *J. Hydrol.* **199**, 88.

Creed, I.F., Band, L.E., Foster, N.W, Morrison, I.K., Nicolson, J.A., Semkin, R.S., Fleetwood, Å. and Olofsson, B.: 1994, *VLF, A General Survey from the Stubbetorp Catchment*, Report of the Department of Land and Water Resources, KTH, Stockholm, in Swedish.

Fritz, P., Cherry, J.A., Weyer, K.V. and Sklash, M.G.: 1976, in *Interpretation of Environmental Isotope and Hydrochemical Data in Groundwater Hydrology*, IAEA Vienna, p. 111.

Fröberg, A.: 1997, *Seasonal and Spatial Variations of Nitrogen Concentrations in a Small Forested Catchment in Kolmården*, Master Thesis, Division of Biology, Linköping University, Sweden (in Swedish).

Hauhs, M., Rost-Siebert, K., Raben, G., Pačes, T. and Vigerust, B.: 1989, 'Summary of European Data', in J.L. Malanchuk and J. Nilsson (eds.), *The Role of Nitrogen in the Acidification of Soils and Surface Waters*, Nordic Council of Ministers, Copenhagen, Nord 92: 5-1, 5-37.

Hill, A.R.; 1990, *Hydrobiologica* **206**, 29.

Kämäri, J., Seuna, P., Ahlberg, T., Holmberg, M., Kivinen, Y., Lepistö, A., Roos, J. and Soveri, J.: 1992, in M. Kanninen and P. Anttila (eds.), *The Effect of Climate Change on the Hydrology and Material Fluxes of Forested Catchments*, The Finnish Research Program on Climate Change, Progress Report, Publ. of the Academy of Finland. p. 93.

Lepistö, A.: 1994, *Aqua Fenn.* **24**, 103.

Lepistö, A.: 1995, *J. Hydrol.* **171**, 103.

Löfgren, S. (ed.): 1991, *Influence of Atmospheric Nitrogen Deposition on Nitrogen Leakage to Surface Water in Forest Ecosystems*, Nord 1991:32, Nordic Council of Ministers, Copenhagen.

van Miegroet, H. and Johnson, D.W.: 1993, 'Nitrate Dynamics in Forest Soils', in T.P. Burt, A.L. Heathwaite and S.T. Trudgill (eds.), *Nitrate: Processes, Patterns and Management,* John Wiley & Sons, Chichester, p. 75.

Murdoch, P.S. and Stoddard, J.L.: 1992, *Water Resourc. Res.* **28**, 2707.

Robinson, M. (Ed.): 1993, *FRIEND, Vol. III, Inventory of Streamflow Generation Studies*, Institute of Hydrology, Wallingford, UK.

Stoddard, J.L.: 1994, in L.A. Baker (ed.), *Environmental Chemistry of Lakes and Reservoirs,* American Chemical Society, Washington D.C., Advances in Chemistry Series No. 237, p. 237-231.

Swank, W.T. and Johnson, C.E.: 1994, in B. Moldan and J. Černý (ed.), *Biogeochemistry of Small Catchments: A Tool for Environmental Research,* SCOPE 51, p. 419.

Wright, R.F. and Schindler, D.W.: 1995, *Water, Air, Soil Pollut.* **85**, 89.

LEACHING OF NUTRIENTS, ORGANIC CARBON AND IRON FROM FINNISH FORESTRY LAND

P. KORTELAINEN[1] and S. SAUKKONEN[2]

[1]*Finnish Environment Institute, P.O. Box 140, 00251 Helsinki, Finland*
[2]*Päijänne Nature Center, Tallukantie 1, 17200 Vääksy, Finland*

Abstract. This study provides an assessment of the spatial variability of the long-term leaching of nutrients, total organic carbon (TOC) and iron (Fe_{tot}) from 22 forested catchments (0.69-56 km^2), distributed across all but the northenmost areas of Finland. The natural, unmanaged Kruunuoja catchment is located in a national park, while the other catchments represent Finnish forestry land. The average leaching of P_{tot} (4.2 kg km^{-2} yr^{-1}) from the Kruunuoja catchment was small compared to the catchments representing forestry land (on average 10 kg km^{-2} yr^{-1}). Moreover, P fertilization was the most important predictor for the spatial variation in P_{tot} leaching (r^2 = 0.45). Leaching of TOC, Fe_{tot} and N compounds was not closely related to forestry practices. Median C/N ratio in the study streams was high (range 34-66). The average inorganic N proportion and leaching of NO_3^--N were lowest in the Kruunuoja catchment (7.3 % and 2.8 kg km^{-2} yr^{-1}, respectively) and highest in the southernmost Teeressuonoja catchment (54 % and 100 kg km^{-2} yr^{-1}, respectively) located in the highest N deposition area. The most important forestry practices since the 1960's have affected about 2.4 % of the area of study sites per year (cf. 2 % in the entire country in 1991). Moreover, the mean annual runoff from the catchments (230-430 mm yr^{-1}) agrees with the mean annual runoff from Finland (301 mm yr^{-1}). Consequently, the results of the study catchments can be used to estimate average total annual leaching from Finnish forestry land: 2,700 t of P_{tot}, 48,000 t of N_{tot}, 110,000 t of Fe_{tot} and 1.5 million t of TOC.

Key words: Finland, forested catchments, iron, leaching, nitrogen, nutrients, organic carbon, peatland, phosphorus

1. Introduction

The cold climate and flat topography in Finland provide favorable conditions for organic matter accumulation. Peatlands covered originally one third of the total land area, a higher percentage than in any other country. Ditching and fertilization of peatlands has been intensive, peaking in the late 1960's when it reached 300,000 ha yr^{-1}. Presently, half of the peatlands have been ditched mostly for forestry.

About 86 % of the total land area of Finland (78 % of the total area) is classified as forestry land (Finnish Statistical Yearbook of Forestry, 1997). Coniferous forests are dominant and each year about 2-3 % of all forestry land in Finland is subjected to forestry practices. The impacts of forestry are most pronounced in small lakes and brooks, where single forestry practices might concern most of the catchment area. Surface waters in Finland are typically colored due to a high concentrations of dissolved organic carbon and Fe. The relative importance of non-point source pollution as a cause of eutrophication in Finnish surface waters recently has increased due to improved point source control.

This study provides an assessment of long-term leaching of P_{tot}, N_{tot}, NO_3^--N, NH_4^+-N, TOC and Fe_{tot} from 22 forested catchments distributed across all but the northernmost areas of Finland. The catchments represent typical Finnish forestry land including variable proportions of ditched peatlands. Impacts of catchment characteristics, hydrological conditions, forestry practices and deposition on spatial and temporal variability are considered.

Water, Air, and Soil Pollution **105:** 239–250, 1998.

2. Material and Methods

2.1. Study Catchments

The 22 catchments (0.69-56 km^2) are distributed across all but the northernmost areas of Finland (Figure 1). Forestry and atmospheric deposition can be considered the only significant human impacts; less than 6 % of the catchments is covered by agricultural land (Table I). The Kruunuoja catchment is located in a national park and has been in a natural state for the past 30 years. The other catchments represent ordinary Finnish forestry land where a number of forestry practices have occurred. No data base with continuous information of forest management is available. Some catchments were divided between numerous land-owners and the forestry practices since the 1960's were collected from several authorities. The most important forestry practices during the past 30 years have been described thoroughly in Saukkonen and Kortelainen (1995) and are summarized in Table I.

The proportion of the catchments covered by peatlands ranges from 10 to 87 %. Ditching was most intensive in the late 1960's and in the early 1970's. In Sydänmaanoja and Kesselinpuro, ditching began in the early 1930's. In Joutenpuro and Heinästönluoma, small-scale ditching occurred into the early 1990's. Clear-cutting occurred in 87 % of the Joutenpuro catchment area in the late 1960's; 48 % of the Kirsioja catchment was clear-cut from the late 1960's to the early 1980's. In other catchments, clear-cutting and scarification (soil tilling, a side preparation method connected with forest regeneration) have occurred on less than 20 % and 15 % of the catchment area, respectively (Table I).

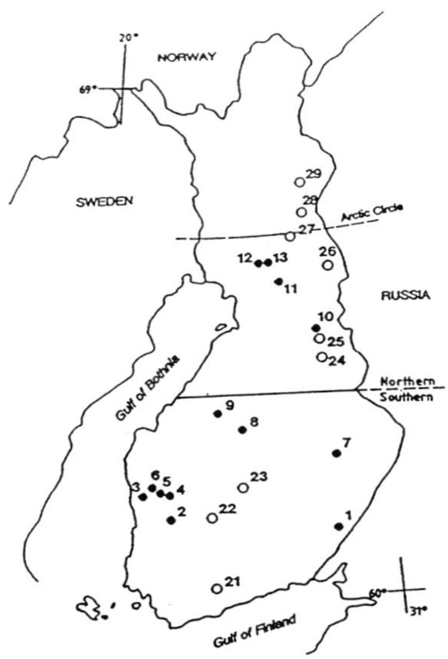

Fig. 1. Location of study catchments in southern and northern Finland (see Table I).
● = Peatland percentage < 35 %; ○ = peatland percentage >35 %.

TABLE I

Characteristics of the study catchments, showing the sum of all forestry practices carried out since the 1960's, the mean annual precipitation (1971-1992) and the mean annual runoff (1960-1992), The fertilization described the fertilization since the 1960's, while the mean annual inorganic N and P deposition describe the average annual values.

Catchment number and name (cf. Fig. 1)	Area (km^2)	Field (%)	Peatland (%)	Ditching (%)	Clearcutting (%)	Scarification (%)	N fertilization ($kg\ km^{-2}$)	P fertilization ($kg\ km^{-2}$)	N deposition ($kg\ km^{-2}\ yr^{-1}$)	P deposition ($kg\ km^{-2}\ yr^{-1}$)	Precipitation ($mm\ yr^{-1}$)	Runoff ($mm\ yr^{-1}$)
1 Huhtisuonoja[a]	5.0	0	47	85	4	4	740	2900	400	14	630	230
2 Katajaluoma[a]	11	2.7	37	100	-	-	0	600	520	11	700	320
3 Heinästönluoma[a]	16	1.4	59	57	9	9	790	400	520	11	640	290
4 Sydänmaanoja[a]	3.8	0	52	20	3	1	1500	53	520	11	670	280
5 Kruunoja[a]	10	0	87	0	0	0	0	0	520	11	710	280
6 Töllinoja[a]	3.4	0	38	100	18	15	1600	880	520	11	600	280
7 Kesselinpuro[a]	21	3.0	37	94	17	9	1000	440	440	13	620	260
8 Vertailuelue[a]	2.5	0	65	61	-	-	-	-	350	9.5	570	320
9 Pahkaoja[a]	21	2.1	41	49	13	-	86	48	350	9.5	580	320
10 Joutenpuro	3.5	0	46	100	87	3	710	2000	470	8.6	620	320
11 Kirsioja	23	0	59	54	49	14	39	130	240	5.8	600	370
12 Kotioja	18	1.0	54	49	-	-	-	-	240	5.8	680	370
13 Ylijoki	56	3.0	59	50	-	-	-	-	240	5.8	680	400
Mean	***15***	***1.0***	***52***	***63***					***410***	***9.8***	***640***	***310***
21 Teeressuonoja[a]	0.69	0	12	0	0	0	0	0	710	9.5	620	270
22 Paunulanpuro[a]	1.4	2.0	10	100	18	-	500	57	350	14	680	290
23 Heinäjoki[a]	9.5	5.1	10	73	5	-	630	190	350	14	640	300
24 Kellojoki	1.9	0	34	100	8	5	0	2200	470	8.6	630	330
25 Myllypuro	11	1.0	21	100	12	13	2600	480	470	8.6	560	380
26 Vääräjoki	19	0.1	34	5	17	15	110	0	300	5.5	540	420
27 Vähä-Askanjoki	16	0	10	96	-	-	-	-	170	3.6	520	430
28 Kuusivaaranpuro	28	2.4	31	95	-	-	46	140	170	3.6	470	280
29 Myllyoja	29	0.3	12	78	-	9	21	21	170	3.6	550	390
Mean	***13***	***1.0***	***19***	***72***					***350***	***7.9***	***580***	***340***
Southern Finland	8.8	1.4	41	62					460	12	640	290
Northern Finland	21	0.8	36	73					300	6.0	580	370

Upper group: peatland percentage > 35; lower group: peatland percentage < 35. Within the groups, the number of catchments increases from south to north. The dashed spaces in the clearcutting, scarification, and N fertilization columns indicate that all the forestry practices have not been completely documented; [a]Southern Finland.

Fourteen of the catchments have been fertilized with N and fifteen with P (Table I). In Töllinoja, all the N fertilization (1,600 kg km^{-2}) occurred in 1974. In Katajaluoma, all the P fertilization (600 kg km^{-2}) occurred in 1978, in Kellojoki (2,200 kg km^{-2}) in 1973 and in Töllinoja (880 kg km^{-2}) in 1974. In other catchments with the high N or P fertilization doses, fertilization was spread out over several years.

2.2. Sampling and Analytical Methods

Daily runoff was recorded in 14 catchments by a V-notch weir and a water level recorder; in the remaining eight catchments, discharge calculations were based on runoff data from nearby catchments (Table II). Manual stream water quality monitoring was started in some catchments as early as 1962. However, all catchments have not been monitored continuously. The total study period for the 22 catchments ranges from 8 to 31 years. The study period of NO_3^--N, NH_4^+-N, and TOC was restricted to the years 1970-1992 due to many missing values in the 1960's. The sampling frequency was approximately 12 samples per year. Prior to 1981, samples were collected once a month. In 1981, the sampling strategy was changed; sampling was concentrated during spring and autumn high flow periods.

Concentrations of P_{tot}, N_{tot} and Fe_{tot} were measured colorimetrically following oxidation with $K_2S_2O_8$ (National Board of Waters 1981). Concentrations of TOC were measured by UV-persulfate oxidation followed by IR gas measurements or using high-temperature oxidation. Ammonium-N was measured by a spectrophotometric method with hypochlorite and phenol, and the sum of NO_2^--N and NO_3^--N by the cadmium reduction method (Erkomaa *et al.*, 1977). Organic N fractions were calculated as the difference between total N and inorganic N $[(N_{tot} - (NO_3^-\text{-N} + NH_4^+\text{-N})]$. The C/N ratio was calculated by dividing TOC (mol L^{-1}) by organic nitrogen (mol L^{-1}).

2.3. Leaching Calculations and Multiple Regression Analyses

The annual leaching of nutrients, TOC, and Fe_{tot} was calculated by multiplying concentration values with the discharge period around the sampling (method 2 in Rekolainen *et al.* 1991). The spring period was determined graphically for each study year in all catchments. Spring was defined to begin when runoff started to increase due to snowmelt; spring was considered to be completed when runoff decreased to less than the mean annual runoff or when the recession runoff was over. Spring runoff represented 50 % of the annual runoff, despite the spring period representing only of about 10-15 % of the whole year.

The impact of catchment characteristics, climatic factors, deposition and forestry practices on leaching was studied by correlation analysis and by developing stepwise multiple regression models. The average annual deposition, precipitation and runoff data, as well as the sum of all the forestry practices since the 1960's, were used to describe the average conditions (see Table I). Monthly monitoring of bulk deposition of NO_3^--N and NH_4^+-N in the nearest deposition stations was used as an index of total N deposition.

TABLE II

The leaching of P_{tot}, N_{tot}, NO_3^--N, NH_4^+-N, TOC and Fe_{tot} from the study catchments, the proportion of inorganic N $[(NO_3^-\text{-N} + NH_4^+\text{-N})/N_{tot}]$ and C/N ratio (TOC/Organic N; both in mol L^{-1}) in the stream water. The leaching of NO_3^--N, NH_4^+-N and TOC was calculated since the 1970's due to many missing values in the 1960's.

Catchment number and name (cf. Fig. 1)	P_{tot} leaching (kg km^{-2} yr^{-1})	N_{tot} leaching (kg km^{-2} yr^{-1})	NO_3^--N leaching (kg km^{-2} yr^{-1})	NH_4^+-N leaching (kg km^{-2} yr^{-1})	Inorg. N leaching (kg km^{-2} yr^{-1})	TOC leaching (kg km^{-2} yr^{-1})	Fe_{tot} leaching (kg km^{-2} yr^{-1})	C/N ratio	Sampling years
1 Huhtisuonoja[a,b]	14	170	27	25	32	4200	640	50	1963-1992
2 Katajaluoma[a,b]	13	290	45	82	41	6600	870	42	1962-1977, 1992-1992
3 Heinästönluoma[a,c]	10	250	11	37	20	8699	430	54	1971-1977, 1991-1992
4 Sydänmaanoja[a,c]	10	250	19	51	27	7400	410	57	1970-1976, 1991-1992
5 Kruunoja[a,c]	4.2	130	2.8	16	7.3	6300	450	66	1971-1978, 1991-1992
6 Töllinoja[a,c]	8.4	180	25	27	22	4300	170	48	1971-1978, 1991-1992
7 Kesselinpuro[a,b]	12	220	22	7.5	13	7400	330	49	1962-1992
8 Vertailuelue[a,c]	9.4	270	13	55	21	8800	980	53	1970-1978, 1991-1992
9 Pahkaoja[a,b]	10	210	21	40	17	6300	450	50	1962-1978, 1980-1989, 1991-1992
10 Joutenpuro[c]	11	160	4.1	9.4	9.1	6600	400	66	1971-1977, 1991-1992
11 Kirsioja[c]	12	140	11	6.1	12	5900	650	53	1972-1997, 1991-1992
12 Kotioja[b]	9.3	200	32	6.0	21	5900	500	44	1976-1992
13 Ylijoki[b]	12	240	33	19	25	5300	740	35	1976-1992
Mean	***10***	***240***	***19***	***29***	***21***	***6400***	***540***	***51***	
21 Teeressuonoja[a,b]	6.4	230	100	14	54	3700	170	40	1962-1986, 1980-1992
22 Paunulanpuro[a,b]	10	180	41	6.7	19	4800	270	38	1962-1992
23 Heinäjoki[a,b]	10	200	36	13	21	5800	250	45	1962-1992
24 Kellojoki[c]	20	140	5.7	17	17	7100	250	65	1970-1977, 1991-1992
25 Myllypuro[b]	11	170	8.3	12	17	5900	240	60	1962-1992
26 Vääräjoki[b]	6.8	180	14	5.0	12	3700	220	34	1962-1978, 1991-1992
27 Vähä-Askanjoki[b]	11	150	14	11	16	4000	420	47	1962-1975, 1977-1992
28 Kuusivaaranpuro[b]	5.8	130	12	8.6	29	3500	250	41	1962-1975, 1977-1980, 1992-1992
29 Myllyoja[b]	6.3	100	7.8	7.8	18	2600	140	45	1962-1975, 1977-1980, 1991-1992
Mean	***9.7***	***160***	***27***	***11***	***23***	***4600***	***250***	***46***	
Southern Finland	9.8	210	30	31	25	6200	450	49	
Northern Finland	11	160	14	14	18	5100	380	49	

[a]Southern Finland; [b]Daily runoff available in the study catchment; [c]Runoff data from a nearby catchment was used for leaching calculations

3. Results

3.1. Concentrations and Leaching of Nutrients, TOC and Fe_{TOT}

Median P_{tot} concentration in the study catchments ranged from 9 to 40 μg L^{-1}. Concentrations were higher in catchments with high peatland proportion (> 35 %) than in those with a low peatland proportion (28 vs. 19 μg L^{-1}) and were slightly higher in southern Finland than in northern Finland (27 vs. 22 μg L^{-1}). In the Kruunuoja catchment, both the average leaching of P_{tot} (4.2 kg km^{-2} yr^{-1}) and the interannual variation in leaching (Figure 2) were lower compared to the catchments representing forestry land (on average 10 kg km^{-2} yr^{-1}).

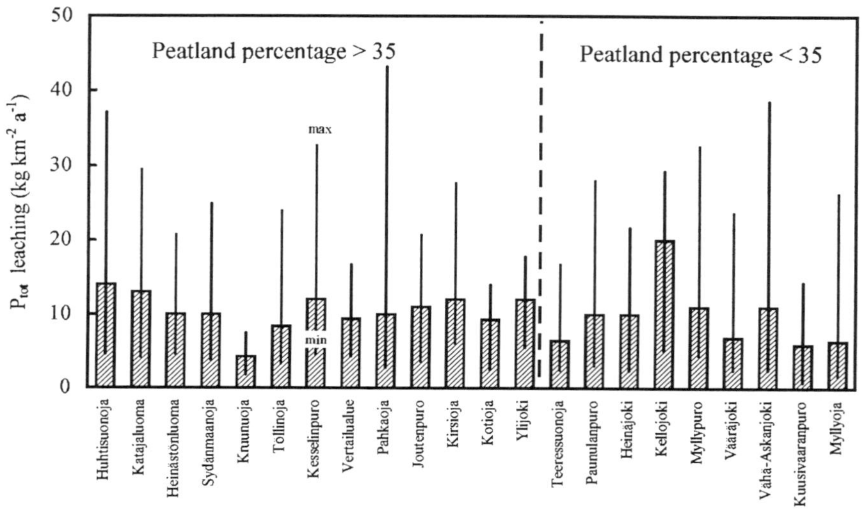

Fig. 2. Mean annual leaching of P_{tot} (min=minumum annual P_{tot} leaching, max=maximum annual P_{tot} leaching). Within the two groups the catchments are presented from south to north.

Median TOC and Fe_{tot} concentration in the 22 streams ranged from 5 to 31 mg L^{-1} and from 230 to 3,300 μg L^{-1}, respectively. Both TOC and Fe_{tot} concentrations were higher in the catchments with a high peatland proportion than in those with a low peatland proportion (20 vs. 12 mg L^{-1} and 1,900 vs. 660 μg L^{-1}, respectively) and also were higher in southern Finland than in northern Finland (20 vs. 13 mg L^{-1} and 1,600 vs. 1,100 μg L^{-1}, respectively).

The major part of the N transported from the catchments was organic N. Inorganic N proportion was lowest (7.3 %) in the Kruunuoja catchment and highest (54 %) in the southernmost Teeressuonoja catchment (Table II). Median C/N ratio of the study streams was high, ranging from 34 to 66. Nitrate leaching varied between 2.8 (Kruunuoja) and 100 kg km^{-2} yr^{-1} (Teeressuonoja) and was higher in southern Finland than in northern Finland. Retention of NO_3^--N deposition, calculated as (input-output)/input, was high in all catchments ranging from 0.67 (Teeressuonoja) to 0.99 (Kruunuoja). Nitrogen concentrations and leaching from the study catchments have been described more thoroughly in Kortelainen *et al.* (1997).

Leaching of N_{tot}, NH_4^+, TOC and Fe_{tot} was higher in the catchments with a high peatland proportion than in those with a low peatland proportion and also was higher in southern Finland than in northern Finland (Table II). About half of the annual leaching occurred in spring.

3.2. Factors Associated with Leaching

There were few statistically significant relationships between forestry practices and the leaching. P_{tot} leaching was positively related to P fertilization ($p < 0.01$) and with the proportion of ditched peatlands in the catchment ($p < 0.05$). The stepwise multiple regression model selected only P fertilization, which explained 45 % of the variation in P_{tot} leaching [P_{tot} leaching (kg km^{-2} yr^{-1}) = 8.36 + 0.00283 × P fertilization, n=18, $p < 0.01$]. For Fe_{tot} leaching, the model selected only the proportion of peatlands in the catchment, which explained 34 % of the variation in Fe_{tot} leaching [Fe_{tot} leaching (kg km^{-2} yr^{-1}) = 167 + 6.52 × Peatland %, n=22, $p < 0.01$).

The stepwise multiple regression models for N and TOC leaching (since the 1970's) have been presented in Kortelainen *et al.* (1997). Nitrate-N leaching was negatively related to C/N ratio ($p < 0.01$) and to latitude ($p < 0.05$). The most significant two-variable regression model combined C/N ratio with N deposition and explained 72 % of the variation in NO_3^--N leaching. For organic N, the stepwise multiple regression model combined TOC leaching with C/N ratio, which together explained 84 % of the variation. Ammonium-N leaching was negatively related to latitude ($p < 0.05$) and positively related to TOC leaching ($p < 0.05$). N_{tot} leaching was positively related to TOC leaching ($p < 0.01$), precipitation ($p < 0.05$), and N deposition ($p < 0.05$) and was negatively related to latitude ($p < 0.01$). TOC leaching was positively related to peatland proportion in the catchment area ($p < 0.01$), C/N ratio in stream water ($p < 0.01$), and precipitation ($p < 0.05$) and was negatively related to stream water pH ($p < 0.0001$).

Median pH was lower in catchments with a high proportion of peatlands than in catchments with a low proportion of peatlands (5.8 vs. 6.4). Lowest median pH (4.5) was found in the natural, unmanaged Kruunuoja catchment. There were some statistically significant relationships between nutrients, TOC and Fe_{tot}. Leaching of Fe_{tot} was positively related to the leaching of TOC ($p < 0.05$) and of N_{tot} ($p < 0.01$). N_{tot} leaching also was positively related to TOC leaching ($p < 0.05$).

4. Discussion

4.1. Spatial and Interannual Variation in Leaching

In most catchments the interannual variation in the leaching of nutrients, TOC and Fe_{tot} was large probably partly due to variation in natural climatic and hydrological conditions. It is probable that in many catchments the natural interannual variation in runoff and leaching is superimposed on the changes due to forestry practices. To reduce the effect of temporal variability of climatic and hydrological factors, average values over a long period were used for leaching calculations.

The spatial variation in the leaching was highest for NO_3^--N, followed by NH_4^+-N, Fe_{tot}

and P_{tot}. Considering the differences in catchment size and location, as well as different forestry practices, the spatial variation in the average long-term leaching of N_{tot} and TOC was small (Table II). The regional variation was reduced due to the fact that concentrations were generally lower in northernmost catchments while runoff from the study catchments increased in a northernly direction (runoff vs. latitude: $r^2 = 0.61$). Although precipitation is higher in southern Finland, the lower precipitation in northern Finland is compensated by lower evapotranspiration in the colder climatic conditions resulting in higher runoff. Spatial variation in Fe_{tot} and TOC leaching was related to the proportion of peatlands in the catchment, while leaching of P_{tot} was related to P fertilization (see sections 3.2 and 4.2).

The C/N ratios in the study streams were high and much of the N transported from the catchments was bound to organic C. TOC was the most important predictor for organic N leaching. In general, nitrification is favored by low soil C/N ratios, high temperature and moisture, as well as high pH. Soil C/N ratio of the study catchments was not available. The stream water C/N ratio, which can be assumed to be closely related to soil C/N ratio, was, however, the most important single predictor of NO_3^--N leaching (negative relationship), followed by latitude (negative relationship), and pH (positive relationship).

Retention of inorganic N deposition was high in all catchments (see Kortelainen *et al.*, 1997). Most northern forests efficiently retain added N and only in the most fertile forest types NO_3^--N leaching has increased significantly after major disturbances (Vitousek *et al.*, 1979). A recent analysis of an input-output database covering 65 forest ecosystem studies (plots and catchments) across Europe showed that NO_3^--N leaching was close to zero below a deposition threshold of about 10 kg N ha^{-1} yr^{-1} and was highly variable at intermediate deposition levels of 10-25 kg N ha^{-1} yr^{-1}, whereas a considerable portion of the input (10-35 kg N ha^{-1} yr^{-1}) was leached in all systems with deposition inputs above 25 kg N ha^{-1} yr^{-1} (Dise and Wright 1995). In southernmost Finland, the N deposition is close to the deposition threshold of 10 kg N ha^{-1} yr^{-1}; in other parts of the country N deposition is much lower.

4.2. Impact of Land Use vs. Deposition

The leaching values presented in this study represent the combined net effect of deposition, natural leaching, and leaching due to forestry practices. Total P input from fertilization in the catchments during the 30 year study period has ranged from 0 to 2,900 kg km^{-2} (Table I). Mean annual P deposition since the 1970's has ranged from 3.6 to 14 kg km^{-2} yr^{-1}. This deposition level since the 1960's would mean a total P deposition during the 30 year period of about 100-400 kg km^{-2}. Comparison of P fertilization and deposition amounts in the catchments (Table I) suggests that P fertilization has been the main P source in the catchments with the highest fertilization doses. In the catchments with the lowest P fertilization doses, P deposition has been more important.

The total N input from fertilization in the catchments during the 30 year study period has ranged from 0 to 2,600 kg km^{-2} (Table I). Mean annual N deposition since the 1970's has ranged from 170 to 710 kg km^{-2} yr^{-1}. Consequently, deposition can be considered the major long-term N source in the catchments. However, comparison of total long-term deposition and fertilization amounts does not reflect all the impacts of added N in the catchments. Deposition has been a continuous annual N source, while fertilization has been going on occasionally, but includes some high single doses.

The long-term average leaching values based on rather sparse sampling show clear forestry impacts only on P_{tot} leaching and some impact on N leaching. Phosphorus fertilization was the most important single predictor of P_{tot} leaching. Moreover, leaching of P_{tot} from the Kruunuoja catchment was significantly lower than P_{tot} leaching from catchments representing forestry land. Leaching of N_{tot} from the Kruunuoja catchment was comparable to N leaching from the northernmost catchments with a low peatland proportion; NO_3^--N leaching from the Kruunuoja catchment was close to zero (2.8 kg km^{-2} yr^{-1}). Highest NO_3^--N leaching was found from the Teeressuonoja catchment (100 kg km^{-2} yr^{-1}), which is located in the highest N deposition area in southernmost Finland. Besides atmospheric deposition, the only anthropogenic impact in Teeressuonoja since the 1960's has been some excavation of sand and gravel. The catchment-scale effects of these human impacts can be considered small compared to deposition.

Most of the correlation coefficients between the forestry practices and the leaching of nutrients, TOC and Fe_{tot} were not statistically significant, suggesting that a large proportion of the catchment has to be impacted at the same time before clear effects on long-term spatial variability becomes evident. This is in agreement with the study of Lepistö *et al.* (1995) based on 10 Finnish and 12 Swedish forested catchments. In previous studies it has been shown that ditching, scarification and clear cutting increase leaching of nutrients (e.g. Ahtiainen and Huttunen, 1995) and Fe. Moreover, ditching has been found to result in a short-term increase of organic C concentrations (Heikurainen *et al.,* 1978; Moore, 1987). In the long-term, ditching lowers the groundwater level and can result in decreased organic C leaching. Fertilization has been shown to increase P leaching, especially from peatlands, and N leaching, especially from mineral soils.

The present data base with sparse sampling does not allow for an exact evaluation of single forestry practices. The study catchments represent Finnish forestry land where a large number of forestry practices (some of them simultaneously in different parts of the catchments) have been carried out. Some practices typically follow each other: cutting is followed by scarification and ditching by fertilization. Many of the practices in the catchments have affected only small parts of the catchments, and the impact of a single practice is dependent on the distance of the treated area from the stream. Moreover, in many catchments, ditching was going on already in the early 1960's, although these catchments were not monitored until the 1970's.

4.3. Leaching from Finnish Forestry Land

The long-term monitoring and regional representativeness of the catchments enable an assessment on the leaching of nutrients, TOC and Fe_{tot} from Finnish forestry land. Mean annual runoff from the catchments, 230-430 mm yr^{-1} (Table I) agrees with mean annual runoff from Finland, 301 mm yr^{-1}, from 1931 to 1990 (Kuusisto 1992). Moreover, the forestry practices in the study catchments (ditching, clear cutting, scarification and fertilization) have afected annually about 2.4 % of the catchment area (cf. 2.5 % in the entire country in 1980 and 2 % in 1991). When the average leaching from southern and northern Finland are multiplied with the area of forestry land in southern and northern Finland, respectively, the following estimates for total annual leaching from Finnish forestry land (262,760 km^{-2}) are obtained: 2,700 t of P_{tot}, 48,000 t of N_{tot}, 5,700 t of NO_3^--N, 5,800 t of NH_4^+-N, 1.5 million t of TOC, and 110,000 t of Fe_{tot} (Table III).

TABLE III
Leaching of P_{tot}, N_{tot}, NO_3^--N, NH_4^+-N, TOC, and Fe_{tot} from Finnish forestry land (86 % of the total land area).

	Southern Finland		Northern Finland		Finland
Forestry land (km^2)	125,440		137,320		262,760
	Leaching		Leaching		Leaching
	kg km^{-2} yr^{-1}	t yr^{-1}	kg km^{-2} yr^{-1}	t yr^{-1}	t yr^{-1}
Ptot	10	1,200	11	1,500	2,700
N_{tot}	210	26,000	160	22,000	48,000
NO_3^--N	30	3,800	14	1,900	5,700
NH_4^+-N	31	3,900	14	1,900	5,800
TOC	6,200	780,000	5,100	700,000	1,500,000
Fe_{tot}	450	56,000	380	52,000	110,000

Löfgren and Olsson (1990) have calculated the total annual nutrient leaching from Swedish forest land (224,772 km^{-2}, peatlands excluded): 1,729 t of P_{tot}, 36,032 t of N_{tot}, 6,981 t of NO_3^--N, and 954 t of NH_{4+}-N. The results for N_{tot}, and NO_3^--N are similare to the results of the present study, although the calculation approach is different. The present study is based on small catchments, while the results by Löfgren and Olsson (1990) are based on large river basins. Moreover, Finnish forestry land includes ditched peatlands, while peatlands are excluded from the results by Löfgren and Olsson (1990).

The estimated total annual P_{tot} and N_{tot} leaching from Finnish forestry land (2,700 t and 48,000 t, respectively) also similar to the background leaching (including contributions of forestry, deposition and natural leaching, but excluding agriculture, industry and municipalities) calculated by Pitkänen (1994) to Finnish coastal waters. In his study, Finnish coastal waters (which receive about 70 % of the annual runoff from Finnish territory) were estimated to receive annually on average 4,800 t of P_{tot} and 79,000 t of N_{tot}, the contribution of background leaching was estimated to be 2,300 t of P_{tot} and 41,000 t of N_{tot}.

Rekolainen *et al.* (1995) suggested that nutrients leached from small agricultural areas mostly enter coastal waters with negligible retention in river channels. This was suggested to be due to the fact that most of the nutrient losses occur in spring, autumn or early winter associated with high water flows, short residence times of water and low productivity. The results from agricultural areas are supported by the present study from forested catchments with average spring leaching contributing about 50 % and the total annual P_{tot} and N_{tot} leaching, comparable to the riverine background P and N input to Finnish coastal waters. Bearing in mind that agricultural land in Finland is concentrated in southern and western coastal areas, Rekolainen (1989) estimated that the total annual leaching from agriculture in Finland would contribute 2,000-4,000 t of P_{tot} and 20,000-40,000 t of N_{tot}. This suggests that although the leaching from forested areas is much lower per unit surface area, the total leaching from Finnish forestry land is comparable to the total leaching from agricultural land.

5. Conclusions

P fertilization was the major P source in most catchments; spatial variation in the leaching of P_{tot} was closely related to P fertilization. The average leaching of P_{tot} and the interannual variation in the leaching of P_{tot} were lowest in the Kruunuoja catchment located in the national park.

Leaching of TOC, Fe_{tot} and N was not closely related to forestry practices, although both N_{tot} and NO_3^--N leaching (only 2.8 kg km^{-2} yr^{-1}) from the Kruunuoja catchment were small. The spatial variation in the leaching of TOC and Fe_{tot} was related to the proportion of peatlands within a catchment. The results suggest that intensive forestry at a catchment scale is needed before clear long-term impacts on spatial variability can be detected.

Nitrogen deposition was the major long-term N source in the study catchments. The major part of the N transported from the catchments consisted of organic N. Both the average inorganic N proportion and NO_3^--N leaching were lowest in the Kruunuoja catchment and highest in the Teeressuonoja catchment located in the highest N deposition area. The C/N ratio in the study streams was high and it was the most important predictor for NO_3^--N leaching.

Long-term leaching of N_{tot}, P_{tot}, TOC and Fe_{tot} from the study catchments, representing the combined net effect of deposition, natural leaching, and leaching due to forestry practices, provide a good indication of average long-term leaching from Finnish forestry land. Forested catchments in Finland, dominantly covered by coniferous forests and including variable proportions of ditched peatlands, are an important C and Fe source to Finnish coastal waters. Total P_{tot} and N_{tot} leaching from Finnish forestry land (86 % of the total land area) is comparable to the nutrient leaching from agricultural land in Finland, although leaching per unit surface area from forestry land is much lower than leaching per area from agricultural land.

Acknowledgments

Joint Research Project on the Adverse Effects of Forest Management on the Aquatic Environment and Their Abatement (METVE) is gratefully acknowledged for financial support of this study.

References

Finnish Statistical Yearbook of Forestry: 1997, *SVT Agriculture and Forestry, 4*, Finnish Forest Research Institute, Helsinki.

Ahtiainen, M. and Huttunen, P.: 1995, 'Long Term Impacts of Forestry Practices on the Stream Water Quality and the Leaching', (in Finnish), in S. Saukkonen and K. Kenttämies (eds.), *Joint Research Project on the Adverse Effects of Forest Management on the Aquatic Environment and their Abatement. Final Report. Finnish Environment No. 2,* Finnish Environment Institute, Helsinki, pp. 33- 50.

Dise, N.B. and Wright, R.F.: 1995, *For. Ecol. Manag.* **71**, 153.

Erkomaa, K., Mäkinen, I. and Sandman, O.: 1977, *Analytical Methods used by Authorized and Water Authority Laboratories. Report 121,*(in Finnish), National Board of Waters, Helsinki.

Heikurainen, L., Kenttämies, K. and Laine, J.: 1978, *Suo* **29**, 49.
Kortelainen, P., Saukkonen, S. and Mattsson, T.: 1997, *Global Biogeochem. Cycles* **11**, 627.
Kuusisto. E.: 1992, *Aqua Fennica* **22**, 9.
Lepistö, A., Andersson, L., Arheimer, B. and Sundblad, K.: 1995, *Water, Air, Soil Pollut.* **84**, 81.
Löfgren, S.and Olsson, H.: 1990, *Nitrogen and Phosphorus Input to Surface Waters in Sweden* (in Swedish), Naturvårdsverket, Report 3692, Solna.
Moore, T.: 1987, *Water Resourc. Bull.* **23**, 785.
National Board of Waters: 1981, *Methods of Water Analyses Employed by the Water Administration". Report 213,* (in Finnish), National Board of Waters, Helsinki.
Pitkänen, H.: 1994, *Eutrophication of the Finnish Coastal Waters: Origin, Fate and Effects of Riverine Nutrient Fluxes,* Publications of the Water and Environment Research Institute. No. 18, Helsinki, p. 45.
Rekolainen, S.: 1989, *Aqua Fennica* **19**, 95.
Rekolainen, S., Pitkänen, H., Bleeker, A. and Felix, S.: 1995, *Nordic Hydrol.* **26**, 55.
Rekolainen, S., Posch, M., Kämäri, J. and Ekholm, P.: 1991, *J. Hydrol.* **128**, 237.
Saukkonen, S. and Kortelainen, P.: 1995, 'Impact of Forestry Practices on the Leaching of Nutrients and Organic Matter" (in Finnish), in S. Saukkonen and K. Kenttämies (eds.), *Joint Research Project on the Adverse Effects of Forest Management on the Aquatic Environment and their Abatement. Final Report. Finnish Environment No. 2,* Finnish Environment Institute, Helsinki, pp. 15-32.
Vitousek, P.M., Gosz, J.R., Grier, C.G., Melillo, J.M., Reiners, W.A. and Todd, R.L.: 1979, *Science* **204**, 469.

PRECIPITATION CHANGE AND SOIL LEACHING: FIELD RESULTS AND SIMULATIONS FROM WALKER BRANCH WATERSHED, TENNESSEE

D.W. JOHNSON[1], P.J. HANSON[2], D.E. TODD, JR[2], R.B. SUSFALK[1] and C.F. TRETTIN[3]

[1]*Desert Research Institute and University of Nevada, Reno, Nevada, USA*
[2]*Oak Ridge National Laboratory, Oak Ridge, Tennessee, USA*
[3]*US Forest Service, Charleston, South Carolina, USA*

Abstract. To investigate the potential effects of changing precipitation on a deciduous forest ecosystem, an experiment was established on Walker Branch Watershed, Tennessee that modified the amount of throughfall at -33 %, ambient (no change), and +33 % using a system of rain gutters and sprinklers. We hypothesized that the drier treatments would cause: 1) disproportionate changes in soil water flux, 2) increased total ionic concentrations in soil solution that would in turn cause 3) decreased SO_4^{2-}/Cl^- ratios, 4) decreased HCO_3^- concentrations, and 5) increased ratios of Al to $(Ca^{2+} + Mg^{2+})$ and of $(Ca^{2+} + Mg^{2+})$ to K^+. Hypothesis 1 was supported by simulation results. Hypotheses 2 and 3 were supported in part by field results, although interpretation of these was complicated by pre-treatment biases. Hypotheses 4 and 5 were not supported by the field results. Comparisons of field data and Nutrient Cycling Model (NuCM) simulations were favorable for most ions except Cl^- and K^+. The disparities may be due to underestimation of soil buffering in the case of Cl^- and overestimation of soil buffering in the case of K^+ in the model. Long-term simulations with NuCM suggest that reducing water inputs will slow the rate of soil acidification and P loss, but will not materially affect growth or ecosystem N status.

Keywords: leaching, Nutrient Cycling Model, rainfall, soil solution chemistry, Walker Branch Watershed

1. Introduction

Climate change could affect the cycling of nutrients and productivity of forest ecosystems in a number of ways. To date, most studies have emphasized the effects of temperature or elevated CO_2, but changes in precipitation may have equal or greater effects (Kirshbaum *et al.*, 1992; Melillo *et al.*, 1996). Specifically, changes in precipitation could cause: 1) changes in productivity in water-limited ecosystems, 2) changes in water yield, and 3) changes in water quality and soil leaching rates. In the Solling forest ecosystem in Germany, Ulrich (1983) noted pulses of NO_3^- and Al during warm, dry years. He hypothesized that drought intensifies N mineralization and nitrification during summer, resulting in NO_3^- pulses during re-wetting periods. In acidic soils, the NO_3^- pulse is accompanied by a pulse in soil solution Al concentrations. Lamersdorf *et al.* (1994) describe a field research project designed to test Ulrich's hypothesis on a large scale. The EXperimental MANipulation of Forest Ecosystems (EXMAN) project involves the imposition of drought through roofs in the forest subcanopy to five forest ecosystems in Europe. Initial results showed that this artificially imposed drought had no effects upon nitrification, but suggested that natural periods of drought could produce such pulses, as observed by Ulrich (1983).

In addition to affecting N mineralization and nitrification, drought can be expected to cause increased ionic concentrations in soil solutions, especially for conservative ions such as Cl^- and Na^+. Changes in the concentrations of other ions may be either buffered by soil chemical processes or controlled by mechanisms that are not sensitive to drought. In the cases of $H_2PO_4^-$ and SO_4^{2-}, for example, adsorption to Fe and Al hydrous oxides may mitigate concentration increases due to drought. In the case of HCO_3^-, concentrations are

Water, Air, and Soil Pollution **105:** 251–262, 1998.

controlled by the partial pressure of CO_2 (pCO_2) in soil and soil solution pH, and could be affected either positively or negatively by drought. If drought causes reduced root and microbial respiration, pCO_2 could decrease. In addition, reduced effective soil CO_2 diffusivity with lower soil water concentrations would tend to result in lower pCO_2 (Wesseling, 1963; Johnson *et al.*, 1994). On the other hand, drought may cause increased HCO_3^- concentrations by causing increased pH. The effects of drought on cations in soil solution should in theory be a function of changes in total ionic concentration. As noted by Reuss (1983), increased total ionic concentrations cause trivalent cations to increase disproportionately to divalent and monovalent cations (to the 3/2 and 3rd power, respectively) and divalent cations to increase as the square of monovalent cations.

To investigate the potential effects of changing precipitation on forest ecosystems, including nutrient cycling, the Throughfall Displacement Experiment (TDE) was established on Walker Branch Watershed, Tennessee in 1994. This experiment modified the amount of throughfall at -33 %, ambient (no change), and +33 % using a system of rain gutters and sprinklers. In this paper, we report the initial results of the treatment effects upon soil solution chemistry, compare those results to simulation output from the Nutrient Cycling Model (NuCM), and present a long-term forecast of the treatment effects from the NuCM model. In both the field experiment and the model simulations, the changes in water flux also cause proportional changes in ion flux to the system, and thus it may at first seem that soil solution concentrations could remain unchanged. However, in both cases, we expected to observe significant treatment effects on soil solution concentrations because the changes in throughfall or precipitation water flux should cause disproportionate changes in soil water flux. Luxmoore and Huff (1989) noted that, over a period of years, streamflow was much more closely related precipitation than was apparent evapotranspiration (ET; the difference between precipitation inputs and streamflow outputs). This suggests that ET was relatively constant and changes in precipitation caused disproportionately large changes in soil water flux and streamflow. Thus, we hypothesized that the drier treatments would cause: 1) disproportionate reductions in soil water flux, 2) increased total ionic concentrations in soil solution ,which would in turn cause 3) decreased SO_4^{2-}/Cl^- ratios, 4) decreased HCO_3^- concentrations, and 5) increased ratios of Al to (Ca^{2+} + Mg^{2+}) and of (Ca^{2+} + Mg^{2+}) to K^+.

2. Site and Methods

2.1. Site Description

The throughfall displacement system for the experiment is located on the Walker Branch Watershed (35°58' N and 84°17' W), a part of the U.S. Department of Energy's (DOE's) National Environmental Research Park near Oak Ridge, Tennessee (Johnson and Van Hook, 1989). Long-term mean annual precipitation is 1,358 mm and mean temperature is 14.2 °C. The acidic forest soils (pH 3.5 to 4.6) are Typic Paleudults (Fullerton series). Depth to bedrock at this location is approximately 30 m. The site was chosen because of its uniform slope, consistent soils, and a reasonably uniform distribution of vegetation. The site is dominated by *Quercus alba* L., *Quercus prinus* L. and *Acer rubrum* L., but it contains 16 other tree species (Hanson *et al.*, 1995). Stand basal area averages 20 to 25

m^2 ha^{-1}. The experimental area was located at the upper divide of the watershed so that lateral flow of water into the soils at the top of the plots would not confound attempts to create a reduced soil water treatment. The site was also chosen to have a southern aspect so that the impacts of the reduced moisture treatment would be increased.

2.2. Experimental Design

The experimental design and its performance were described in detail by Hanson *et al.*, (1995). Briefly, the manipulations of throughfall reaching the forest floor are made with a system designed to passively transfer precipitation from one experimental plot to another. There are three plots in the TDE; one wet, one dry and one ambient. Each 80 × 80 m plot is divided into 100 8 × 8 m subplots that serve as the locations for repetitive, nondestructive measurements of soil and plant characteristics. Throughfall is intercepted in about 2,000 subcanopy troughs (0.3 × 5 m) suspended above the forest floor of the dry plot (~33 % of the ground area is covered). The intercepted throughfall is then transferred by gravity flow across an ambient plot and is distributed onto the wet treatment plot through paired drip holes spaced approximately 1 m apart. The troughs are arranged in 21 rows of 80 to 90 troughs. Reductions in soil moisture anticipated from the experimental removal of 33 % of the throughfall will be comparable to the driest growing season of the 1980's drought (Cook *et al.*, 1988), which resulted in sapling mortality and reduced growth of some vegetation (Jones *et al.*, 1993).

2.3. Soil Solution Collections

Soil solutions were collected with ceramic cup tension lysimeters (Soil Moisture Equipment Corp., Santa Barbara, California). The lysimeters were installed prior to treatment in a 3 × 3 array in each subwatershed at a spacing of 7.9 m. Prior to collection of samples, tensions of 40 kPa were established in each lysimeter. Soil water content (%, v/v) also was measured in each subwatershed with a time domain reflectometer. Data from those measurements are not used in this analysis; the reader is referred to Hanson *et al.* (1995) for details of methodology and results.

3. The NuCM Model

The NuCM model has been described in detail elsewhere (Liu *et al.*, 1991; Johnson *et al.*, 1993), and only a few relevant details are repeated. NuCM depicts nutrient cycling at a stand level, where the ecosystem is represented as a series of vegetation and soil components. The model provides for one generic conifer and one generic deciduous species of specified biomass and nutrient concentration (foliage, branch, bole, roots). The model also provides for an overstory that can be divided into canopy, bole, and roots. Tree growth in the model is a function of user-defined stand developmental stage and the availability of nutrients and moisture. The soil includes multiple layers (up to 10), and each layer can have different physical and chemical characteristics. The model routes precipitation through the canopy and soil layers, and simulates evapotranspiration, deep seepage, and lateral flow. The movement of water through the system is simulated using the continuity equation, Darcy's equation for permeable media flow, and Manning's equation for free surface flow. Percolation occurs between layers as a function of layer permeabilities and differences in moisture content. In these simulations, meteorological data from the period January 1992 through December 1994 were used to generate hydrologic fluxes. These input files were repeated to produce longer-term simulations. Nutrient pools associated with soil solution, the ion exchange complex, minerals, and soil organic matter are all tracked explicitly. The processes that govern interactions among these pools include user-specified rates for decay, nitrification, anion adsorption, cation exchange and mineral weathering.

The model simulates the noncompetitive adsorption of sulfate, phosphate, and organic acids. Sulfate adsorption can be simulated in NuCM using either linear or Langmuir adsorption isotherms. The Langmuir isotherm was used in these simulations. Phosphate adsorption in the model is represented by a linear isotherm. Cation exchange is represented by the Gapon equation. Mineral weathering reactions in the model use rate expressions that depend upon the mass of mineral present and solution-phase H^+ concentration raised to a fractional power.

4. Field Results

4.1. Pre-treatment Soil Solutions

Unfortunately, there were some pre-treatment differences in soil solution concentrations that complicate interpretations of post-treatment effects. Table I shows some soil solution concentration data from the 70 cm depth lysimeters prior to treatment in May and December, 1992 (nine replicates per treatment and date). There were no pre-treatment differences in pH on either date, but there were statistically significant (ANOVA, $p < 0.05$) pre-treatment differences in Ca^{2+}, K^+, Mg^{2+}, and Na^+ concentrations on 13 May and differences in SO_4^{2-} concentration on both 13 May and 12 December. None of the differences in cation concentration were significant on the 12 December date. The direction of these differences in base cations and SO_4^{2-} was indeed unfortunate: the site fated for the Dry treatment had higher concentrations of all four base cations than the sites fated for either the Ambient or Wet treatments. There were also statistically significant pre-treatment differences in Cl^- concentrations and in SO_4^{2-}/Cl^- ratio between the Wet and Ambient treatments on 12 December (but not between the Ambient and Dry treatments; Student's *t*-test). In these cases, however, the pre-treatment differences were in the opposite direction of hypothesized post-treatment changes (lower pre-treatment Cl^- concentrations and higher SO_4^{2-}/Cl^- ratios in the Ambient than in the Wet treatment). Nitrate and NH_4^+ concentrations in soil solution were very low (< 2 $\mu mol_c\ L^{-1}$) and there were no statistically significant pre-treatment differences. There were no significant differences in Ca^{2+}/K^+ or Mg^{2+}/K^+ ratios; Al concentrations were near trace levels.

4.2. Post-treatment Soil Solutions

The Dry treatment had consistently higher electrical conductivity, SO_4^{2-}, Cl^-, K^+, Ca^{2+}, Mg^{2+}, and Na^+ concentrations than either the Ambient or Wet treatments within the fourth year of treatment (Figure 1). As in the pre-treatment collections, there were no treatment effects on pH, NH_4^+, NO_3^-, Ca^{2+}/K^+ or Mg^{2+}/K^+ ratios, and Al concentrations were near trace levels (not shown). In the cases of all ions except Cl^-, these differences were present prior to treatment and therefore cannot necessarily be attributed to the treatments. In the case of Cl^-, however, the pre-treatment effects were in the opposite direction (Cl^- was lower in the Dry treatment site), and therefore it can be concluded that there was a treatment effect. Also, the post-treatment decreases in SO_4^{2-}/Cl^- ratios in the Dry treatment can be considered a treatment effect because pre-treatment differences between the Ambient and Dry treatments were not significant.

TABLE I

Soil solution concentrations ($\mu mol_c\ L^{-1}$; means ± standard errors, n = 9) at two dates prior to treatment (Wet = site to receive +33 % throughfall input; Ambient = no treatment; Dry = site to receive -33 % throughfall).

	Treatment		
	Wet	Ambient	Dry
13 May 1992			
pH	5.9 ± 0.2	6.2 ± 0.1	5.9 ± 0.2
Ca^{2+}	116 ± 17	191 ± 32	193 ± 20
K^+	40 ± 10	45 ± 7	57 ± 9
Mg^{2+}	57 ± 8	77 ± 5	94 ± 6
Na^+	35 ± 2	35 ± 1	42 ± 1
Cl^-	31 ± 2	28 ± 4	28 ± 3
SO_4^{2-}	360 ± 41	457 ± 50	552 ± 47
SO_4^{2-}/Cl^-	3.9 ± 0.5	5.8 ± 1.0	6.8 ± 1.1
12 Dec. 1992			
pH	5.4 ± 0.1	5.5 ± 0.1	5.4 ± 0.1
Ca^{2+}	138 ± 12	166 ± 25	169 ± 16
K^+	53 ± 9	42 ± 8	52 ± 10
Mg^{2+}	59 ± 8	60 ± 8	79 ± 9
Na^+	35 ± 9	26 ± 2	33 ± 4
Cl^-	40 ± 5	24 ± 6	30 ± 6
SO_4^{2-}	387 ± 23	433 ± 38	497 ± 26
SO_4^{2-}/Cl^-	3.9 ± 0.9	8.2 ± 1.7	7.5 ± 1.5

5. Model Simulations

5.1. Simulated Water Fluxes

Simulated hydrologic fluxes using the NuCM and PROPSER (Luxmoore *et al.*, 1978) models are compared for calendar years 1992 and 1996 in Table II. Simulations for the two models differ most with respect to evapotranspiration (ET) and overland flow (OF).

NuCM simulations suggest that ET increases with water input (from 51 to 81 cm, Dry to Wet) whereas PROSPER suggests that ET is nearly constant. NuCM simulations indicate virtually no overland flow, whereas PROSPER shows a moderate amount of flow that increases with water input (from 1 to 12 cm, Dry to Wet). In the case of soil leaching (SWF), however, the parameter of most interest here, the models produce remarkably similar results. In both cases, SWF increases by approximately 30 cm with each increase in water input. However, in a relative sense, SWF differs more from Dry to Ambient (+74 to 80 %) than from Ambient to Wet (+45 to +47 %), and thus Hypothesis 1 was supported by the simulations.

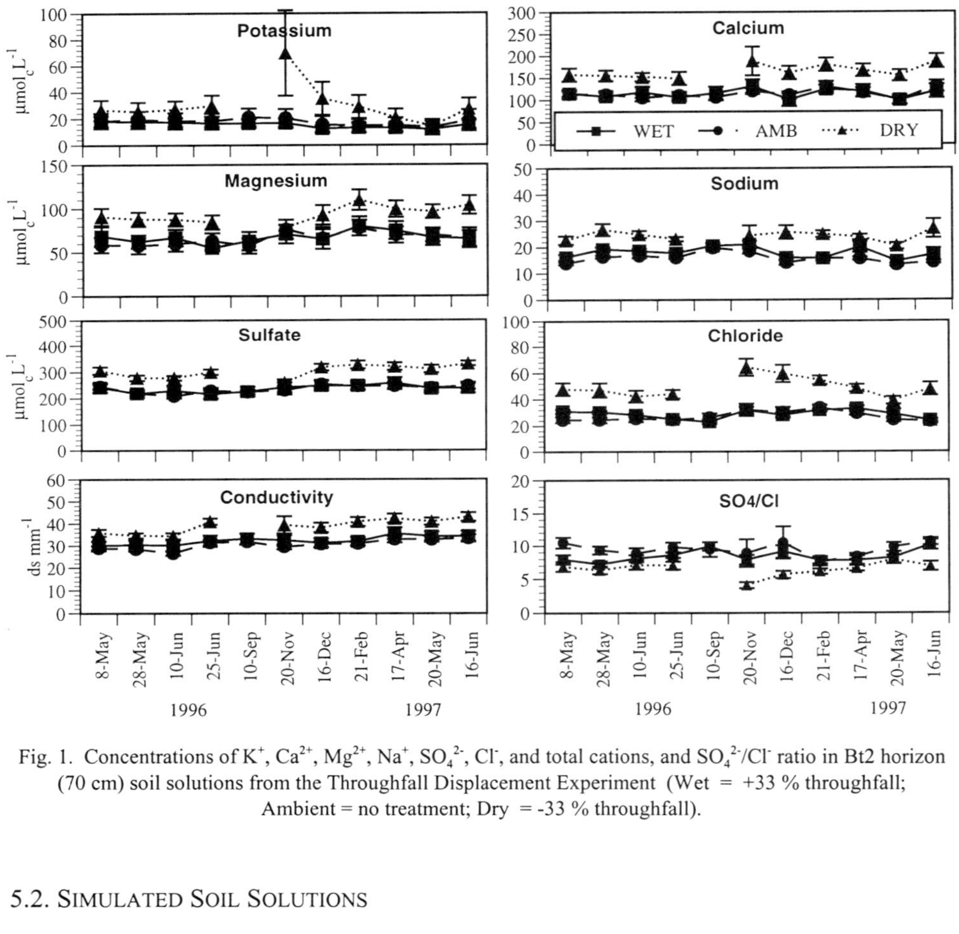

Fig. 1. Concentrations of K^+, Ca^{2+}, Mg^{2+}, Na^+, SO_4^{2-}, Cl^-, and total cations, and SO_4^{2-}/Cl^- ratio in Bt2 horizon (70 cm) soil solutions from the Throughfall Displacement Experiment (Wet = +33 % throughfall; Ambient = no treatment; Dry = -33 % throughfall).

5.2. Simulated Soil Solutions

Simulated soil solution concentrations of total cations, SO_4^{2-}, Cl^-, SO_4^{2-}/Cl^- ratios, K^+, Ca^{2+}, Mg^{2+}, and Na^+ over a four year period are shown in Figure 2. The simulations matched the patterns during year four of treatment fairly well in some cases and poorly in others. In the cases of simulated total cations vs. measured conductivity, SO_4^{2-}, Ca^{2+}, Mg^{2+}, and

Na^+, the simulated patterns during year four followed the patterns in the field to some extent: seasonal variations were minimal and the differences between the Wet and Ambient treatments was less than those between the Dry and Ambient treatments. In the cases of Cl^- and K^+, however, the simulated patterns in year four differed substantially from those in the field. Unlike the field data, simulated Cl^- varied seasonally by a factor of three. In the case of K^+, the simulations indicated very little seasonal variation and very minor treatment effects. In the field, there was a large seasonal variation in K^+ in the Dry treatment and a very large treatment effect during peak concentrations. In contrast to the field results, the NuCM simulations showed only a very slight response of SO_4^{2-}/Cl^- ratios to treatments in the fourth year of treatment. There were larger but inconsistent responses of SO_4^{2-}/Cl^- ratios to treatments in years 1-3 of the simulation.

TABLE II
Simulated water fluxes using the NuCM and PROSPER models (cm). Average of two years (1993 and 1994). (Wet = +33 % precipitation; Ambient = no treatment; Dry = -33 % precipitation).

	Treatment		
	Dry	Ambient	Wet
NuCM			
Precipitation	94	139	186
Evapotranspiration	51	67	81
Overland Flow	<1	<1	<1
Soil Water Flux	42	73	107
PROSPER			
Precipitation	99	141	183
Evapotranspiration	60	56	63
Lateral Flow	2	6	12
Soil Water Flux	41	74	107

5.3. Long-Term Simulations of Nutrient Pools

Increasing water input had only a minor effect (4 to 7 % increase) upon simulated biomass and vegetation nutrient contents after 30 years (Table III). Increasing water had a somewhat greater effect upon forest floor nutrient content (from a 14 % increase for P to a 25 % increase for K). The effect of increasing water upon soil exchangeable pools differed substantially among nutrients: negligible for N, 1 % decrease for SO_4^{2-}, 28 % decrease for P, 62 % decrease for Ca^{2+}, 49 % decrease for K^+, and 35 % decrease for Mg^{2+}. Changes in soil total pools during the simulation were negligible (not shown). None of the decreases in soil exchangeable pools caused a nutrient deficiency to develop during the simulation.

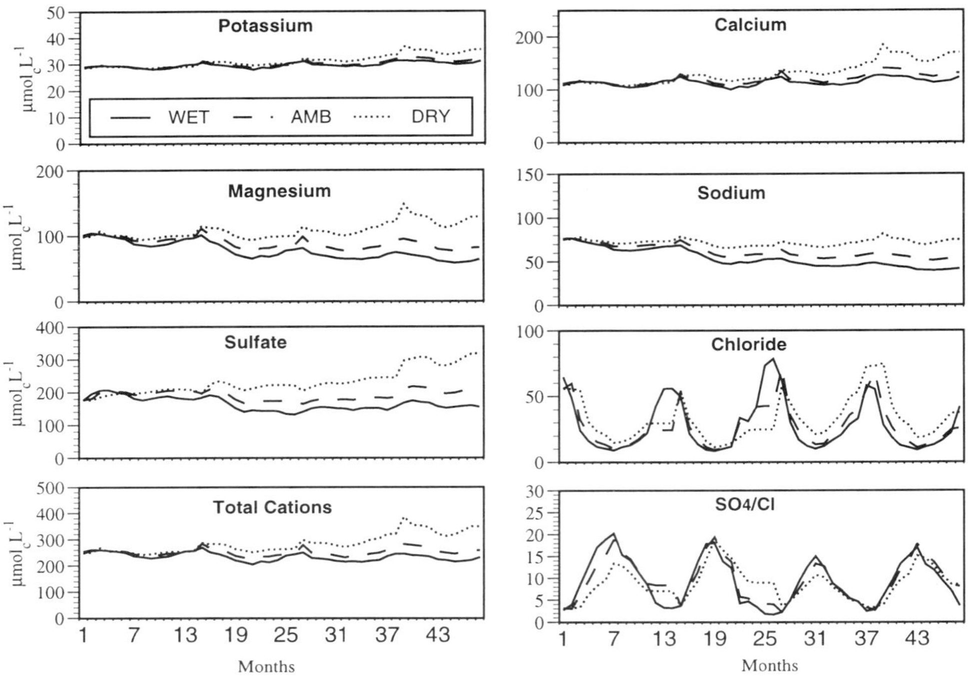

Fig. 2. Simulated concentrations of total cations, SO_4^{2-}, Cl^-, SO_4^{2-}/Cl^- ratios, K^+, Ca^{2+}, Mg^{2+}, and Na^+ in Bt2 horizon (70 cm) soil solutions from the Throughfall Displacement Experiment. (Wet = +33 % throughfall; Ambient = no treatment; Dry = -33 % throughfall).

6. Discussion

Hypothesis 1 (reduced water would lead to disproportionate reduction in soil water flux) was supported by simulation results form both the PROSPER and NuCM models. No field data are available to verify these results, but treatment effects on soil water content have been noted throughout the growing season except during the drought of 1995 (Hanson et al., 1995). Hypothesis 2 (reduced water would result in increased total ionic concentrations) was supported in part by the Cl^- results. The post-treatment increases in Cl^- in the Dry treatment clearly were due to the treatment rather than pre-treatment site bias and suggest that soil solution concentrations other ions also were affected by treatments even though pre-treatment differences existed. The pre-treatment differences in Cl^- between the Wet and Ambient sites (higher in the Wet site) disappeared in the post-treatment collections, also suggesting treatment effects on Cl^- (and perhaps also on total ionic strength).

TABLE III
Simulated ecosystem nutrient contents (kmol ha^{-1}) (Wet = site to receive +33 % throughfall input; Ambient = no treatment; Dry = site to receive -33 % throughfall).

		After 30 years		
	Initial	Wet	Ambient	Dry
Biomass	19.30	24.91	24.66	23.97
Nitrogen				
Vegetation	38.76	48.56	47.85	46.42
Forest Floor	11.59	19.84	18.84	17.59
Soil, Exch.	<0.01	<0.01	<0.01	<0.01
Sufur				
Vegetation	2.57	3.37	3.30	3
Forest Floor	1.41	3.69	3.44	3.13
Soil, Exch.	12.37	13.07	13.12	13.22
Phosphorus				
Vegetation	1.62	2.08	2.05	1.99
Forest Floor	0.29	0.58	0.55	0.51
Soil, Exch.	0.44	0.26	0.29	0.36
Calcium				
Vegetation	25.92	33.10	32.69	31.76
Forest Floor	3.21	10.38	9.65	8.82
Soil, Exch.	22.90	2.33	3.27	6.01
Potassium				
Vegetation	8.95	11.13	10.97	10.61
Forest Floor	2.01	4.29	3.87	3.43
Soil, Exch.	27.65	10.01	14.61	19.63
Magnesium				
Vegetation	4.30	5.52	5.43	5.27
Forest Floor	0.77	2.59	2.40	2.20
Soil, Exch.	4.37	0.70	0.77	1.07

Hypothesis 3 (reduced water would cause decreased SO_4^{2-}/Cl^- ratios) also was supported in part by the results of the field collections. Soil solution SO_4^{2-}/Cl^- ratios were reduced in the Dry compared to the Ambient treatment after treatment whereas there were no significant differences before treatment. The situation between the Wet and Ambient sites for SO_4^{2-}/Cl^- ratios is less clear. The pre-treatment differences in SO_4^{2-}/Cl^- ratios between the Wet and the Ambient sites (Wet was lower) persisted only in the Spring samples of the post-treatment collections. Hypothesis 4 (reduced water would cause decreased HCO_3^- concentrations) was not supported by the limited data on hand. However, the data are not adequate to refute this hypothesis because no pre-treatment data were available, and post-treatment HCO_3^- concentrations were often at trace levels. Hypothesis 5 [reduced water would result in increased ratios of Al to (Ca^{2+} + Mg^{2+}) and of (Ca^{2+} + Mg^{2+}) to K^+] was not supported by the results of the field collections.

The NuCM simulation results offer a few additional insights into factors that might be affecting field responses and additional factors that might be desirable in the model. The simulation results suggest that the differences in concentration between the Dry and Ambient treatments are greater than those between the Wet and Ambient sites because of the greater relative effect on soil water flux in the former than in the latter (Table II; Figure 2). The failure of the NuCM simulations to match observed patterns in soil solution Cl^- suggests that there is some buffering of Cl^- in the field (e.g., adsorption; Johnson and Cole, 1977) that is not accounted for in the model. On the other hand, the failure of NuCM to simulate the observed seasonal variations of K^+ in the field may reflect too much emphasis on buffering by cation exchange. For the purposes of comparison with field data, NuCM results from only the deepest horizon (Bt2) are shown in Figure 2. The simulations also produced output for the surface horizons, and these are shown for K^+ in Figure 3. The NuCM simulations showed considerable seasonal variations and treatment effects in soil solution K^+ in surface horizons, but these variations were buffered by exchange processes in the Bw2 horizon. In the field, the appearance of seasonal variations in the Bw2 horizons may have been due to macropore flow (e.g., Luxmoore *et al.*, 1981), which is not accounted for in the NuCM model.

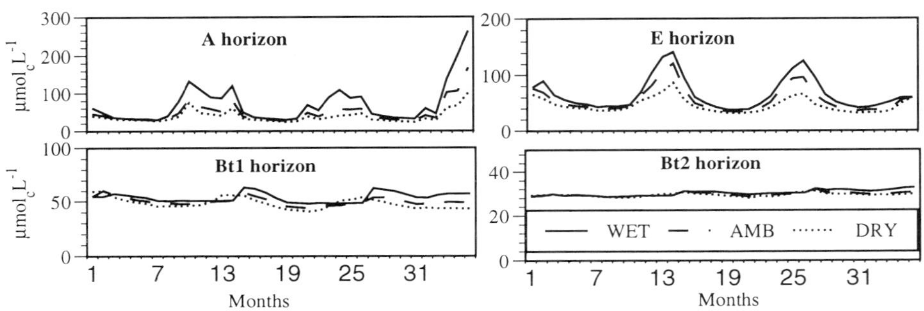

Fig. 3. Simulated concentrations of K^+ in A, E, Bt1 and Bt2 horizon soil solutions from the Throughfall Displacement Experiment (Wet = +33 % throughfall; Ambient = no treatment; Dry = -33 % throughfall.)

There is no way to either verify or disprove the long-term simulation results from the NuCM model. However, it is intuitively obvious that changes in water flux could cause substantial changes in the rate of K cycling because of the fact that so much K is cycled via foliar leaching. Thus, it is to be expected that changing water would have a large effect on forest floor and soil exchangeable K^+ pools (Table III). It is also intuitively obvious that nearly two-fold changes in the soil water flux and, consequently, in the rate of soil leaching would cause significant reductions in adsorbed P exchangeable Ca^{2+} and Mg^{2+} pools. The lack of change in soil SO_4^{2-} pools in the simulations pools suggests that the system is near steady-state with respect to SO_4^{2-} adsorption.

7. Conclusions

Based upon the literature and theoretical considerations, we hypothesized reduced water inputs in the Throughfall Displacement Experiment would cause: 1) disproportionate reductions in soil water flux, 2) increased total ionic concentrations in soil solution, which would in turn cause 3) decreased SO_4^{2-}/Cl^- ratios, 4) decreased HCO_3^- concentrations, and 5) increased ratios of Al to ($Ca^{2+} + Mg^{2+}$) and of ($Ca^{2+} + Mg^{2+}$) to K^+. Hypothesis 1 was supported by simulation results. Hypotheses 2 and 3 were supported in part by field results, although interpretation of these was complicated by pre-treatment biases. Hypotheses 4 and 5 were not supported by the field results. Comparisons of field data and NuCM model simulations were favorable for most ions except Cl^- and K^+. Disparities between simulation outputs and field results for these ions suggest that Cl^- adsorption may be occurring in the field (not accounted for in the model) and that macropore flow (not accounted for in the model) may allow seasonal pulses of K^+ and other ions to penetrate the soil profile. Long-term simulations with the NuCM model suggest that reducing water inputs will slow the rate of soil acidification and P loss, but would not materially affect growth or ecosystem N status.

Acknowledgments

This research was supported by the Department of Energy and the Nevada Agricultural Experiment Station. Publication No. 4733, Environmental Sciences Division, Oak Ridge National Laboratory.

References

Cook, E.R., Kablack, M.A., and Jacoby, G.C.: 1988, *J. Geophy. Res.* **93**, 14257.

Hanson, P.J., Todd, D.E, Jr., Edwards, N.T. and Huston, M.A.: 1995, 'Field Performance of the Walker Branch Throughfall Displacement Experiment', in A. Jenkins, R.D. Ferrier, and C. Kirby (eds.), *Ecosystem Manipulation Experiments: Scientific Approaches, Experimental Design, and Relevant Results*, Ecosystem Research Report #20, Commission of the European Communities, pp. 307-313.

Johnson, D.W. and Cole, D.W.: 1977, *Anion Mobility in Soils: Relevance to Nutrient Transport from Terrestrial to Aquatic Ecosystems.* EPA-600/3-77-068. Ecological Research Series, U.S. Environmental Protection Agency, Corvallis, OR.

Johnson, D.W., Geisinger, D.R., Walker, R.F., Newman, J., Vose, J.M., Elliot, K.J. and Ball, J.T.: 1994, *Plant Soil* **165**, 111.

Johnson, D.W., Swank, W.T. and Vose, J.M.: 1993, *Biogeochemistry* **23**,169.

Johnson, D.W. and Van Hook, R.I. (eds.): 1989, *Analysis of Biogeochemical Cycling Processes in Walker Branch Watershed*, Springer-Verlag, New York.

Jones, E.A., Reed, D.D., Mroz, G.D., Liechty, H.O. and Cattelino, P.J.: 1993, *Can. J. For. Res.* **23**, 229.

Kirschbaum, K.W. and Fischlin, A.: 1992, 'Climate Change Impacts on Forests', in R.T. Watson, M.C. Zinyowera, and R.H. Moss (eds.), *Climate Change 1995 Impacts, Adaptions and Mitigation of Climate Change: Scientific-Technical Analysis*, Cambridge University Press, New York, pp. 95-129.

Lamersdorf, N.P., Bier, C., Blanck, K., Bredemeier, M. Cummins, T., Farrel, E.P., Rasmusson, L. and Ryan, M.: 1995, 'Reactions of soil solution chemistry to drought: results of the EXMAN project', in A. Jenkins, R.D. Ferrier and C. Kirby (eds.), *Ecosystem Manipulation Experiments: Scientific Approaches, Experimental Design, and Relevant Results*, Ecosystem Research Report #20, Commission of the European Communities, pp. 86-95.

Liu, S., Munson, R., Johnson, D.W. Gherini, S., Summers, K., Hudson, R., Wilkinson, K. and Pitelka, L.: 1991, *Tree Physiol.* **9**,173.

Luxmoore, J.R., Grizzard, T. and Patterson, M.R.: 1981, *Soil Sci. Soc. Amer. J.* **45**: 692.

Luxmoore, R.J. and Huff, D.D.: 1989, 'Water', in D.W. Johnson and R.I. Van Hook (eds.), *Analysis of Biogeochemical Cycling Processes in Walker Branch Watershed,* Springer-Verlag, New York, pp. 165-196.

Luxmoore, R.J., Huff, D.D., McConathy, R.K., and Dinger, B.E.: 1978, *For. Sci.* **24**: 327.

Melillo, J.M., Prentice, I.C., Farquhar, G.D., Schulze, E.-D. and Sala, O.E.: 1996, 'Terrestrial Biotic Responses to Environmental Change and Feedbacks to Climate', in J.T. Houghton *et al.* (eds.), *Climate Change 1995: The Science of Climate Change*, Cambridge University Press, New York, pp. 445-481.

Reuss, J.O.: 1983, *J. Environ. Qual.* **12**: 591.

Ulrich, B.: 1983, 'Soil Acidity and its Relation to Acid Deposition', in B. Ulrich and J. Pankrath (eds.), *Effects of Accumulation of Air Pollutants in Ecosystems,* D. Reidel Co., pp

TRACING HYDROLOGIC PATHWAYS USING CHLORIDE AT THE PANOLA MOUNTAIN RESEARCH WATERSHED, GEORGIA, USA

NORMAN E. PETERS[1] and ELIZABETH B. RATCLIFFE[2]

[1]*U.S. Geological Survey, 3039 Amwiler Rd., Suite 130, Atlanta, Georgia, 30360-2824, USA*
[2]*University of Bristol, School of Geographical Sciences, Bristol BS8 1SS, United Kingdom*

Abstract. An analysis of chloride (Cl^-) concentrations and fluxes at the 41 ha Panola Mountain Research Watershed indicates that Cl^- may be used effectively to differentiate "new" and "old" water flow through the hillslope and their respective contributions to streamwater. Rainfall and throughfall, the "new" water inputs, are marked by low Cl^- concentrations (<15 μeq L^{-1}). Stormwater moves rapidly to depth along preferred pathways in a deciduous forest hillslope, as evidenced by low Cl^- concentrations (<20 μeq L^{-1}) in mobile soil water from zero-tension stainless-steel pan lysimeters. "Old" waters, matrix soil waters and groundwater, typically have high Cl^- concentrations (>30 μeq L^{-1}). Timing of soil water transport is not sufficiently rapid to suggest that soil water from this hillslope site (20 m from the stream) contributes to streamwater during individual rainstorms. The source of streamflow, therefore, must be a combination of channel interception, overland flow and soil water from near-channel areas, and runoff from a 3 ha bedrock outcrop in the headwaters. Groundwater contribution to streamflow was estimated using Cl^- concentrations of throughfall and groundwater as the two end members for a two-component hydrograph separation. For the study period, groundwater contributed 79 % of the streamflow and from 1985 to 1995, contributed 75 % of the streamflow. Rainfall was the source of 45 % of the Cl^- flux from the watershed in the long term; the remaining Cl^- is hypothesized to be derived from dry deposition, consistent with the enrichment noted for throughfall. At peak flow during individual rainstorms, "new" water can contribute 95 % of the runoff.

Keywords: chloride, hydrograph separation, Panola Mountain Research Watershed, streamflow, tracer

1. Introduction

Variations in the solute concentrations of event or "new" waters in transit through a forested hillslope are due to the interaction of several processes, leading either to enrichment or depletion of solutes. Processes include transport and mixing of waters along hydrological pathways, biological reactions, and interactions with or release from basin materials. Factors affecting enrichment include atmospheric deposition, leaching from vegetation or soil, release from exchange sites in the soil, biological reactions, or evaporative concentration. Factors affecting depletion include uptake by vegetation, immobilization by soil organisms, adsorption to basin materials, and gaseous loss. Physical conditions of the ecosystem prior to a rainstorm may control the relative importance of these mechanisms. The magnitude of sources or sinks also may be affected by antecedent soil moisture conditions, rainstorm intensity, and seasonality.

Hydrochemical tracers have been used to identify flow paths (Pearce *et al.*, 1986; Sklash *et al.*, 1986; Kleissen *et al.*, 1990; Leaney *et al.*, 1993) and solute sources (Rascher *et al.*, 1987; Peters and Driscoll, 1989). Chloride is highly mobile in most ecosystems because it is not readily adsorbed onto surfaces, nor is it readily incorporated into secondary minerals (Peters, 1991; Reynolds and Pomeroy, 1991). Atmospheric deposition of marine aerosols is the major source of Cl^- in most ecosystems (Eriksson, 1955; Gerritse and George, 1988). Chloride enrichment in throughfall at the Hubbard Brook Experimental

Forest was attributed to washoff of dry deposition (Juang and Johnson, 1967). Enrichment of Cl^- in throughfall and stemflow also has been reported elsewhere (Neal *et al.*, 1990; Rustad *et al.*, 1994, and references therein). Rainfall and washoff of dry deposition are major contributors of Cl^- to most ecosystems, and the magnitude, hence importance, increases with proximity to the ocean (Eriksson, 1955).

Chloride has been used widely to identify hydrological pathways (Pearce *et al.*, 1986; Sklash *et al.*, 1986; Kleissen *et al.*, 1990; Leaney *et al.*, 1993). Groundwater and matrix soil water (i.e., "old" waters) generally have higher Cl^- concentrations than rainfall due to evapotranspiration, and typically have longer residence times (Jenkins *et al.*,1994). Provided a sufficiently large difference exists in Cl^- concentrations of source waters, the Cl^- concentrations of sequentially sampled soil waters during rainstorms can help elucidate timing and mechanisms of solute and water transport in a watershed. Enriched solute concentrations in initial samples of collection sequences may also indicate preferential flow (Shanley and Peters, 1993). In addition, coupling sequential chemical variations with flow rates will help quantify source contributions and elucidate mechanisms of transport and transformation of other solutes.

For this paper, the temporal and spatial variations in Cl^- concentrations are explored for rainfall, throughfall, soil water, and groundwater in a deciduous forest hillslope and for streamwater draining the forested watershed to evaluate the use of Cl^- as a tracer of "new" and "old" water. Sampling occurred at the Panola Mountain Research Watershed (PMRW), Georgia, during selected rainstorms from May 1994 through April 1995 and these short-term dynamics were compared to the longer-term (1985-1995) water and Cl^- fluxes for the watershed.

2. Background

A 20 m × 20 m plot on a deciduous forest hillslope was studied from May 1994 through April 1995 in the 41 ha PMRW (Figure 1), which is in the Panola State Conservation Park, Stockbridge, Georgia, 25 km southeast of Atlanta (84°10'W, 33°37'N). The hillslope plot is at 260 m above sea level on the drainage boundary of a 10 ha gaged subcatchment. Annual rainfall at PMRW averages 1,150 mm, ranging from 760 to 1,580 mm per water year (October to September) from 1985 to present, and annual air temperature averages 15 °C. In summer, rainstorms typically are convective marked by high intensity and short durations, and in winter, rainstorms typically are frontal marked by lower intensity and longer durations than in summer (Shanley and Peters, 1993).

The hillslope plot is underlain by the Panola Granite, a granodiorite in composition. Soils are 60-100 cm deep consisting of Inceptisols and the regolith is greater than 7 m deep. The hillslope plot is dominated by 70-100 year-old Southern red oak (*Quercus falcata*), Northern red oak (*Quercus rubra*), and mockernut hickory (*Carya ovalis*), and younger flowering dogwood (*Cornus florida*). The 41 ha watershed is 93 % forested and contains a 3 ha lichen- and moss-covered bare rock outcrop in the headwaters (Figure 1). The regolith typically is deepest (5-10 m) in the valley bottoms and thins to about 1 m on the ridges. At or near the outlet, however, bedrock is at or near the surface.

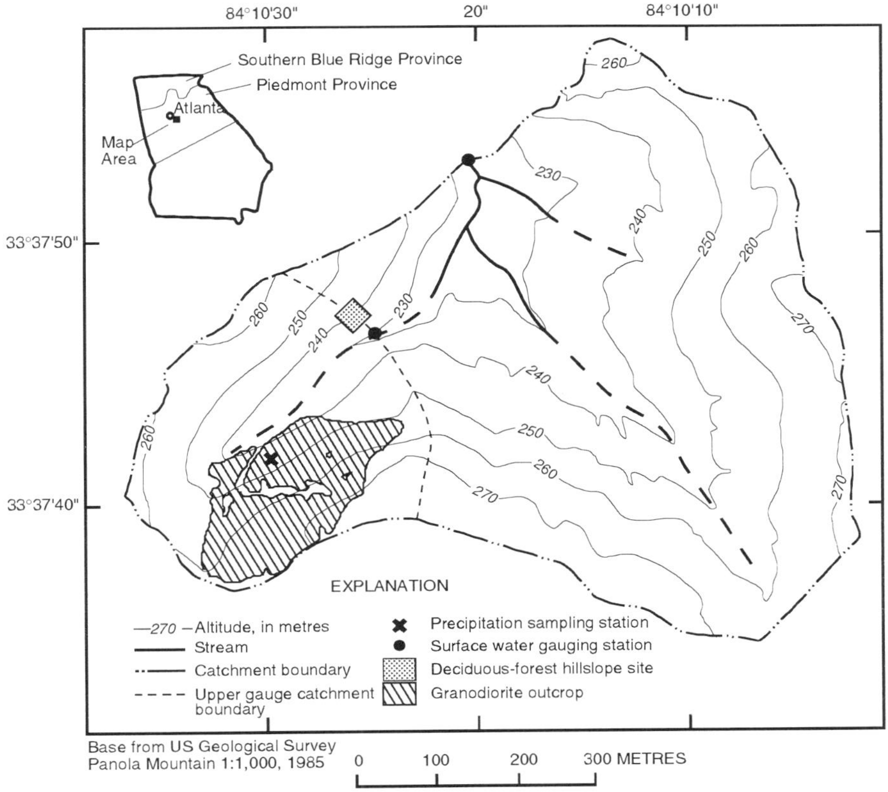

Fig. 1. Map of sampling locations and characteristics of the Panola Mountain Research Watershed near Atlanta, Georgia, USA.

3. Methods

The Cl^- concentrations, and in some cases fluxes, of rainfall, throughfall, soil waters, groundwater and streamwater were evaluated for this paper. Sample collections varied from weekly for rainfall, throughfall, and streamwater to sequentially during select storms for these same components plus soil water and groundwater. Rainfall and throughfall were collected each Tuesday from Aerochem Metrics Model 301 wet/dry rainfall collectors. Sequential (intra-storm) samples of rainfall were collected with a wetfall-dryfall collector, on which the wet-side bucket was replaced with a funnel that drained into a series of sample bottles (Peters, 1994). Sequential samples of throughfall were collected similarly from three 7.62 cm × 1 m long PVC troughs installed at 1.5 m above the forest floor. For the sequential rainfall samples, an adjacent tipping-bucket gage was used to establish timing; for the timing of the throughfall samples, water from the trough drained through a tipping-bucket gage and then into collection bottles.

Soil waters were sampled using both tension and zero-tension lysimeters. Tension lysimeters were installed at 15, 40 and 70 cm depths at each of three sites spaced 2 m apart across the slope. Tension lysimeters were evacuated to 80 cbar before each rainstorm, and were sampled within 24 hr of the end of the rainstorm and during some rainstorms. Total volume of water in each lysimeter was recorded and a sample aliquot taken for chemical determinations. Sequential soil samples were collected from a depth of 50-cm below land surface from a 1 m^2 stainless-steel zero-tension lysimeter (Shanley and Peters, 1993). A 0.08 m^2 polyethylene pan lysimeter similarly was used for sequential sampling of the soil water from the forest floor. Zero-tension water drained from the lysimeters (by polyvinyl chloride (PVC) pipe for the stainless-steel lysimeter and from silastic tubing for the polyethylene lysimeter) through a tipping-bucket gage into a series of sample bottles.

Groundwater samples were collected from 5.1 cm diameter PVC wells screened at various depths below land surface. Four water-quality sampling wells were fitted with 30-cm screens at the bottom having depths below land surface of 150, 260, 400, and 660 cm. One water quality sampling well with a depth of 430 cm below land surface was fitted with a 120 cm screen at the bottom. The wells were evacuated using a peristaltic pump and 1.2 cm diameter silastic tubing (one well volume), and sampled after the well recharged, which was typically within a few minutes.

In addition to the intensive hillslope sampling from May 1994 through April 1995, grab samples of streamwater were collected weekly from 1985 to 1996 at the outlet of the 41 ha watershed. These weekly collections were augmented by automatic sampling of stormflow during 1986-1988 and 1993-1996 using the methods described by Peters (1994). Unfiltered aliquots of the samples from each site were refrigerated and these samples typically were analyzed within a couple of weeks for Cl^- by ion chromatography.

For a Cl^- mass balance, rainfall inputs to the watershed were estimated by summing the product of concentration and rainfall depth per unit area for weekly samples. Streamwater Cl^- fluxes, Φ, were estimated from

$$\Phi = \int (Q(t)C'(Q^*(t) + \varepsilon(t)))dt \qquad (1)$$

where $Q(t)$ is the instantaneous discharge; $C'(Q^*(t) + \varepsilon(t))$ is the estimated concentration from a hyperbolic concentration-discharge regression model (Johnson *et al.*, 1969), which also uses instantaneous discharge and explained 87 % of the variation in Cl^- concentration, and $\varepsilon(t)$ is a piecewise linear function constructed from the model residuals so that the sum of $C'(t)$ and $\varepsilon(t)$ equaled the observed streamwater concentration at every point when it was measured.

4. Results and Discussion

4.1. Spatial Variability of Cl^- Concentrations

The Cl^- concentration of rainfall differs significantly from that of groundwater; Cl^- concentrations of each were the least variable compared to other locations, e.g. throughfall or soil waters (Figure 2). Volume-weighted Cl^- concentration of weekly wet-only rainfall for the study period is 4.2 $\mu eq\ L^{-1}$; median Cl^- concentration of groundwater is 46.5 $\mu eq\ L^{-1}$. In contrast, volume-weighted Cl^- concentration of weekly throughfall in the deciduous canopy is higher than that of rainfall (10.6 μeq L^{-1}), indicating enrichment due to evaporative concentration and washoff of dry deposition and/or leaching from leaves (Juang and Johnson, 1967; Peters, 1989; Cappellato *et al.*, 1993). Only 11-22 % of the rainfall is lost to interception either as retention in the canopy or evaporation (Cappellato *et al.*, 1993; Ratcliffe, 1996). The volume-weighted average rainfall Cl^- concentration due to evaporation in the canopy should only increase by 1.11-1.28. This is far less than the observed 2.5-fold increase, implying that the accumulation and washoff of dry deposition and leaching of Cl^- from leaves contributes as much Cl^- to the watershed as does rainfall.

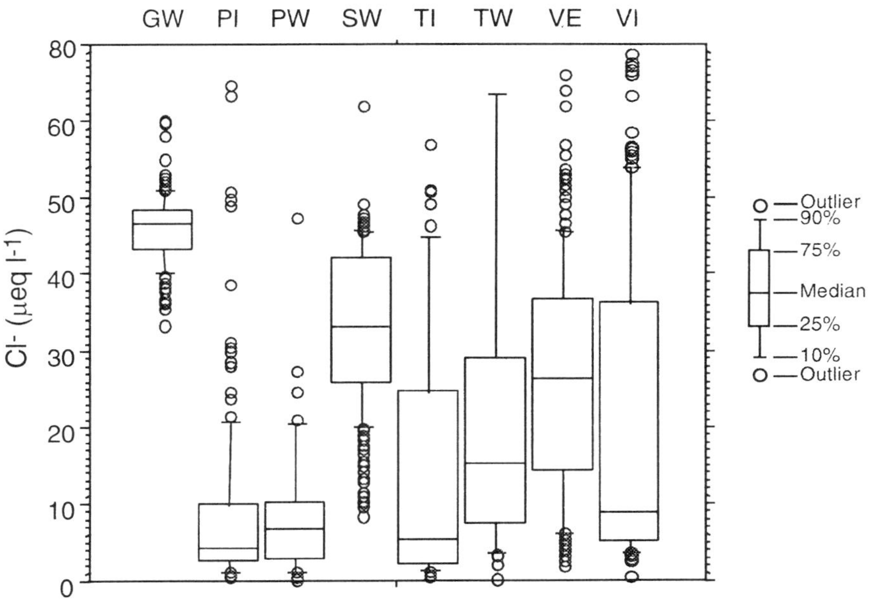

Fig. 2. Variations in Cl^- concentrations of groundwater (GW), rainfall (sequential - PI and weekly - PW), streamwater (SW), deciduous throughfall (sequential - TI and weekly - TW), and soil water (tension lysimeter - VE and zero-tension lysimeter - VI) at the Panola Mountain Research Watershed, May 1994-April 1995.

Streamwater Cl^- concentrations generally ranged between those of rainfall and groundwater. Those of throughfall and soil waters were more variable and sometimes exceeded the median Cl^- concentration of groundwater. The much higher concentrations observed in these samples probably reflect evaporative concentration and rapid dissolution of Cl^- salts from the canopy. The highest Cl^- concentrations in soil water are for samples collected during rainstorms preceded by very dry periods and the highest in throughfall are for low-volume samples (Figure 3).

The Cl^- concentrations of sampling points vertically at the hillslope plot, from rainfall through groundwater show a similar pattern to that above. Waters with low Cl^- concentrations are clearly mobile "new" waters having concentrations more similar to rainfall and throughfall, and are found in both zero-tension (pan) lysimeters and some tension lysimeters. Typically, tension lysimeters collect matrix waters as their design provides a good connection with the soil matrix. However, some of these lysimeters probably intersect macropores as suggested by the low Cl^- concentrations of soil water collected in them. The Cl^- concentrations for soil water from tension lysimeters tend to increase with depth below land surface consistent with the downward transport of Cl^- and dilution of the matrix waters with low Cl^- concentration rainfall and throughfall (Figure 4). For the entire sampling period, groundwater levels from adjacent recording wells also were much lower (2-4 m below land surface) than the depth of the lysimeters, precluding the contribution of groundwater to the lysimeters.

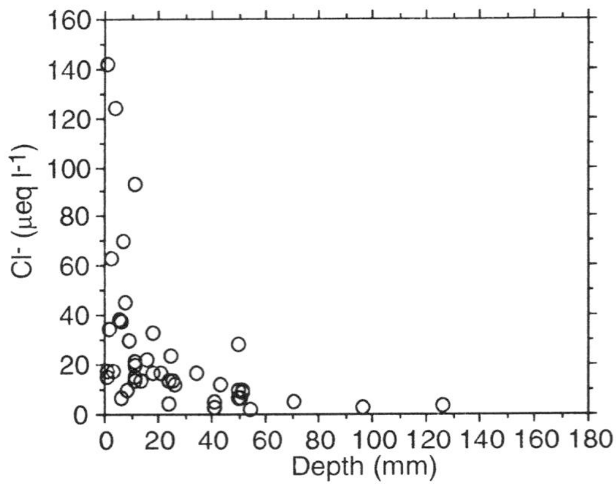

Fig. 3. Relation between Cl^- concentration and depth of weekly deciduous throughfall at the Panola Mountain Research Watershed, May 1994-April 1995.

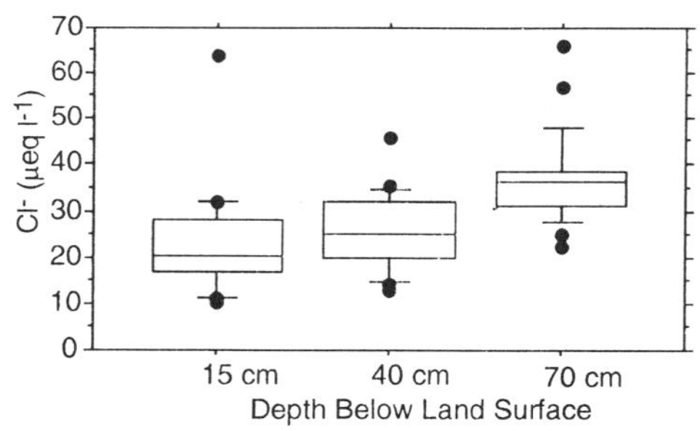

Fig. 4. Relation between Cl^- concentration and depth of soil water tension lysimeters from one site at the Panola Mountain Research Watershed, May 1994-April 1995.

4.2. Temporal Variability of Cl- Concentrations

The Cl^- concentrations of sequential samples of rainfall and throughfall behave similarly (Figure 5). The highest concentrations generally occur at the onset of the rainstorm and decrease thereafter. This pattern has been observed for most solutes in several other studies (Puckett, 1990b; Durana *et al.*, 1992; Hansen *et al.*, 1994; Khare *et al.*, 1996). For rainfall, the initially high concentrations are attributed to washout of particles in the lower atmosphere at the beginning of the rainstorm (Davies, 1983; Durana *et al.*, 1992; Nieto *et al.*, 1994; Wurzler *et al.*, 1995) and for throughfall, dissolution and washoff of dry deposition and leaf exudates increase concentrations over those of the rainfall input (Garten *et al.*, 1988; Puckett, 1990a; Beier *et al.*, 1992; Hansen *et al.*, 1994). However, even the high initial concentrations typically are much lower than those of pre-storm soil water or groundwater (Figures 2 and 5).

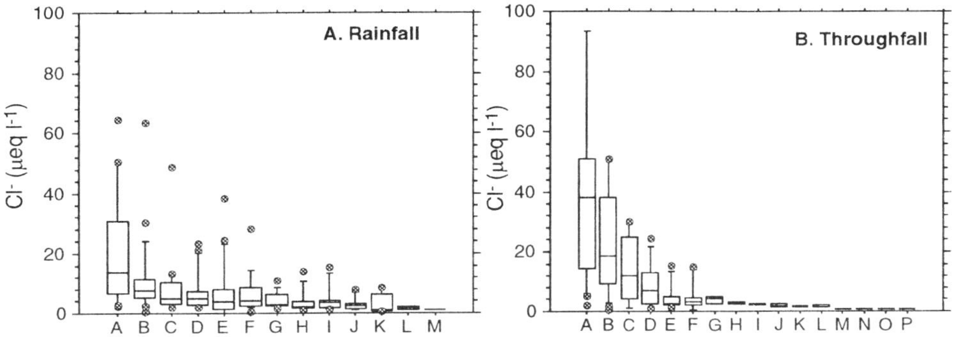

Fig. 5. Relation between Cl^- concentration and time for sequential samples, arranged alphabetically by sample of rainfall (A) and throughfall (B) at the Panola Mountain Research Watershed, May 1994-April 1995.

During the dry summer and fall, Cl^- concentrations of soil-water samples from the deep zero-tension lysimeter are highest in the first sample, decrease and remain relatively constant and then increase slightly during large rainstorms after a lot of water passes through the lysimeter (Figure 6). The Cl^- concentrations also are more similar to those of rainfall (or throughfall), indicating the source to be mobile "new" water. The initially high concentrations correspond with and are partly attributed to rainfall and throughfall. The increase in Cl^- concentrations later in the storm probably is caused by the mixing with matrix waters. However, during the wetter winter period, Cl^- concentrations of samples collected during rainstorms for the same lysimeter are much more variable, and generally are comparable to those of matrix soil water or groundwater. The first sample of zero-tension soil water for the winter typically is more dilute than subsequent samples, possibly indicating the contribution of "new" water to this sample. Likewise, if many samples were collected for a rainstorm, Cl^- concentrations of later samples were lower, which likewise reflects an increasing contribution of "new" water. Although these data indicate that mobile water travels to depth regardless of season, a key factor controlling contributions of "new" water to streamflow is the timing of the mobility.

The timing of water transport in the hillslope qualifies the ability of hillslope water to contribute to streamflow during a rainstorm. To evaluate this, extremes in the short-term behavior of Cl^- concentrations and timing are examined for relatively intense rainstorms, one in the dry summer on 16 August 1994 (Figure 7) and the other during the wet winter on 10 February 1995 (Figure 8). The August rainstorm was an intense 69 mm convective storm and the instantaneous pre-rainstorm runoff, although higher than normal for the summer, was relatively low at 0.7 mm da^{-1}. The February rainstorm had a longer duration and higher magnitude (84 mm) than the August rainstorm and the instantaneous baseflow runoff prior to the rainstorm was 11 mm da^{-1}. The soils were relatively moist throughout the profile for the February rainstorm, but the upper soil horizons were relatively dry prior to the August rainstorm.

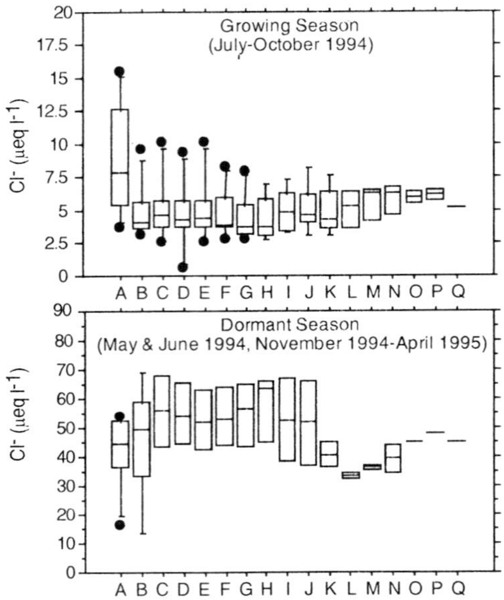

Fig. 6. Relation between Cl^- concentration and time for sequential samples, arranged alphabetically by sample, of zero-tension soil water at 50-cm below land surface for the growing and dormant seasons at the Panola Mountain Research Watershed.

For the August rainstorm (Figure 7), Cl^- concentrations of rainfall, throughfall and soil water from the forest floor and at 50 cm depth decrease as the rainstorm progresses, consistent with the previous discussion. However, although "new" water is traveling to depth in the soil profile, the water arrives much later to the deeper lysimeter than the stormflow response of streamwater, suggesting that water movement through the hillslope does not contribute significantly to stormflow. Streamwater Cl^- concentration decreases rapidly presumably due to dilution by "new" water inputs. Because the timing of water movement in the hillslope is slow relative to that of streamflow, the "new" water is probably derived from the bedrock outcrop, direct contributions of rainfall and throughfall to the channel, and overland flow and soil water from near-channel areas.

For the February rainstorm (Figure 8), rainfall, throughfall and forest-floor soil water behave similarly to that of the August storm, but streamwater Cl^- concentrations do not decrease as rapidly despite the much higher runoff associated with this rainstorm; nor does the deeper soil water reflect the continuous transport of "new" water to depth. The deep soil water, however, responds more rapidly, consistent with the wetter antecedent soil

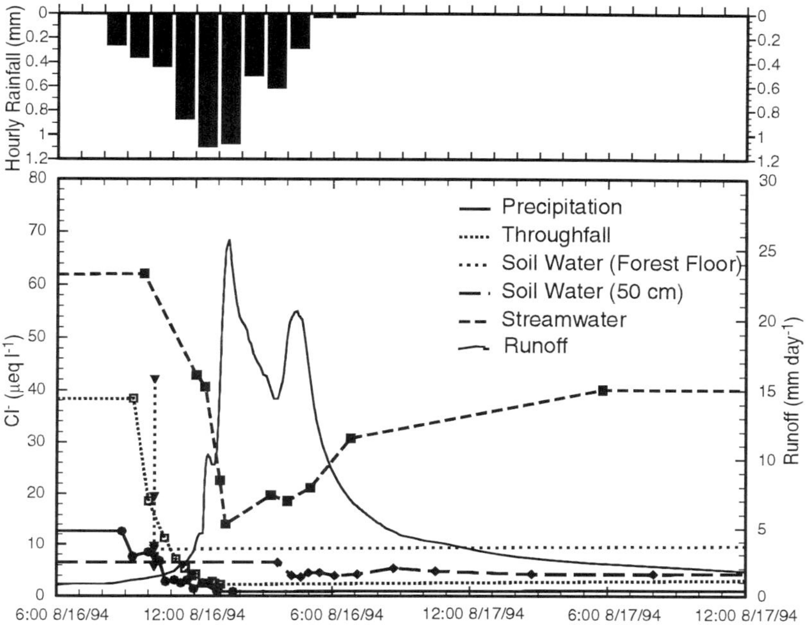

Fig. 7. Temporal variability in Cl⁻ concentrations of rainfall, deciduous throughfall, soil water from the forest floor and at 50-cm depth and streamwater at the Panola Mountain Research Watershed during a rainstorm on 16 August 1994.

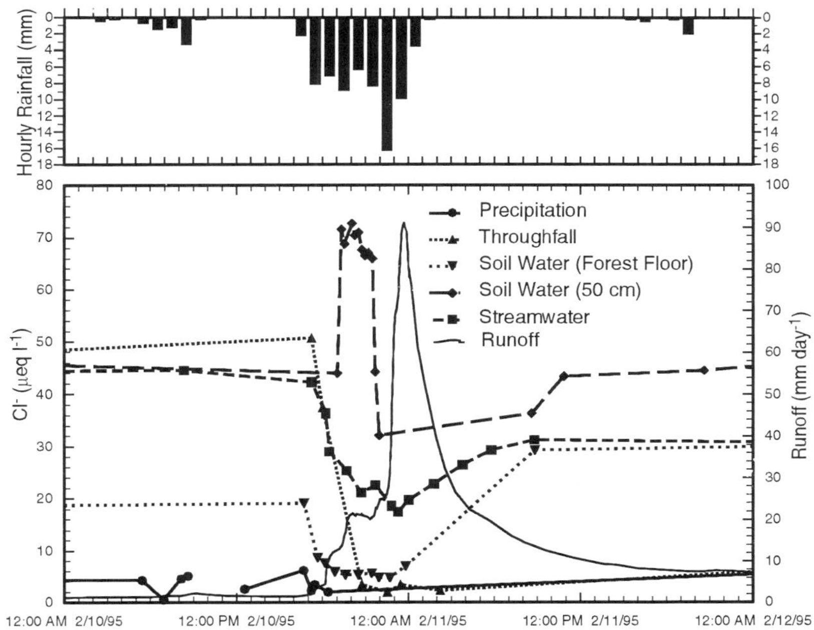

Fig. 8. Temporal variability in Cl⁻ concentrations of rainfall, deciduous throughfall, soil water from the forest floor and at 50-cm depth and streamwater at the Panola Mountain Research Watershed during a rainstorm. on 10-11 February 1995.

moisture conditions, but the high Cl^- concentrations reflect a significantly higher level of interaction with the matrix. Again, the streamwater response does not reflect much contribution of water from the hillslope, i.e. 50 cm soil water, but the smaller response of Cl^- concentrations in the streamwater compared with the August rainstorm suggests a larger contribution of soil matrix water or groundwater.

The similarity of Cl^- concentrations among mobile waters supports the findings of Reynolds and Pomeroy (1991) that Cl^- is relatively inert and is not affected by ion exchange or mineral equilibria processes, i.e., it is highly mobile through the system. However, high Cl^- concentrations in matrix soil waters and groundwater compared with rainfall or throughfall indicate that evaporative concentration has an extremely large impact. Furthermore, evapotranspiration concentrates the Cl^- more in soil waters than in the canopy as indicated by the catchment water yield that on average was 30 % of the total precipitation, which includes the 11 % canopy interception loss. The Cl^- is concentrated in these high Cl^- waters by more than 10-fold compared to the volume-weighted mean concentration of rainfall and by more than 4-fold compared to throughfall.

4.3. Cl^- MASS BALANCE AND CONTRIBITIONS FROM "OLD" AND "NEW" WATER

Atmospheric deposition is the primary source of Cl^- to the watershed. For the study period, rainfall contributed 64 eq ha^{-1}, but the streamwater flux was 243 eq ha^{-1}. The wetter than normal conditions for the study period are directly associated with the excess Cl^- transport. In the long term, annual rainfall input averages 61 eq ha^{-1} and streamwater flux averages 138 eq ha^{-1}. The difference between rainfall input and streamwater flux is similar to the difference in the volume-weighted mean concentrations of rainfall and throughfall, suggesting that the excess transport is input as dry deposition. Interannual variability of fluxes for either rainfall or streamwater is quite large, 2 and >3, respectively, and Cl^- transport is highly correlated with water yield (Figure 9). If more than 50 % of the input to the watershed is derived from dry deposition, i.e. the total of atmospheric inputs should equal the stream transport, then storage of this Cl^- should occur in dry years and excess transport should occur in wet years. The transit time, therefore, will depend on the moisture status of the watershed and the temporal distribution, magnitude and intensity of the rainfall.

The groundwater contribution to streamwater Cl^- transport, i.e. two-component hydrograph separation (Pinder and Jones, 1969), was estimated using the median concentration for groundwater (46.5 μeq L^{-1}), the volume-weighted mean concentration for throughfall (10.6 μeq L^{-1}), and the predicted streamwater concentration from the hyperbolic model (Johnson *et al.*, 1969). For the study period, 79 % of the runoff was derived from "old" water. A similar analysis for the long-term record indicates that "old" water contributes about 75 % to the annual runoff. The runoff at PMRW is dominated by the baseflow component, which typically is most similar to the overall solute composition of groundwater and consequently has the highest Cl^- concentrations. Therefore, in order to have such a high contribution of "new" water to the total runoff annually, large contributions must occur during rainstorms. During the peak discharge for individual events, particularly the peak of record in July, the groundwater contribution decreases to less than 5 %. However, because the Cl^- concentrations of rainfall and throughfall typically decrease as the event progresses, and particularly during extremely intense rainfall, Cl^- concentrations of the "new" water at the time of peak runoff would necessarily be lower, and consequently the contribution of groundwater should be higher than 5 %.

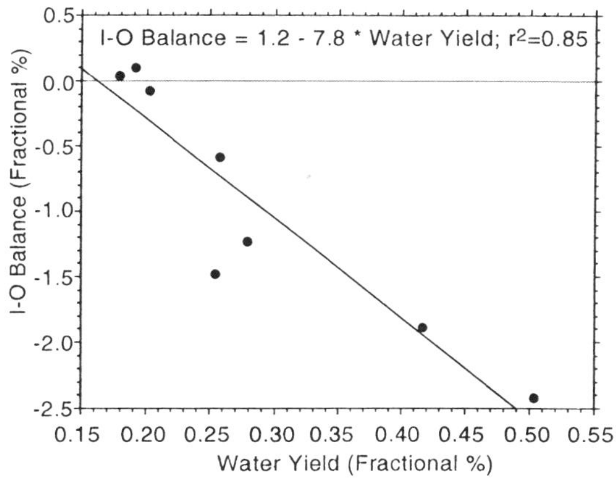

Fig. 9. Relation between annual mass balance expressed as a percentage (rainfall inputs minus streamwater outputs divided by rainfall inputs) and annual water yield at the Panola Mountain Research Watershed for water years 1985 to 1995.

5. Conclusions

Chloride concentration variations in rainfall, throughfall, soil water, groundwater and stream water were evaluated for the short-term response during several rainstorms from May 1994 through April 1995 and for the longer term response from 1985-1995 at the Panola Mountain Research Watershed in the Georgia Piedmont. For rainfall and throughfall, Cl^- concentrations of weekly samples decrease with increasing rainfall amount; Cl^- concentrations of sequential samples decrease with time during individual rainstorms. For rainfall, these observations reflect washout of the lower atmosphere at the onset of a rainstorm; and for throughfall, reflect additional enrichment by washoff of dry deposition and leaching from the leaves. Stormwater moves rapidly to depth along preferred pathways in a deciduous forest hillslope. During the dry growing season, soil water collected from stainless-steel, zero-tension lysimeters and from some of the tension lysimeters indicate that Cl^- moves with rainfall through the canopy and unsaturated zone without changing concentration appreciably. However, during the wet dormant season, matrix water having high Cl^- concentrations also is mobilized and contributes to flow through the unsaturated zone.

Solute concentrations in groundwater and soil water reflect different water types and associated residence times. Concentrations of Cl^- in groundwater and most of the tension lysimeters samples were high and similar, as caused by evaporative concentration. Zero-tension lysimeters only yield water during rainstorms, and for some rainstorms, Cl^- concentrations in zero-tensiometer waters and some of the tension lysimeters were similar to those of rainfall and throughfall. Higher Cl^- concentrations are observed in mobile soil water during the wetter soil-moisture conditions in winter than in summer, which is attributed to more active interaction and mixing of matrix waters with the mobile waters in the winter. The timing of response indicates that flow through the unsaturated zone is too slow to contribute directly to streamflow for an individual rainstorm. For streamwater, Cl^- concentrations are highly correlated with flow having the lowest concentrations at the highest flows, consistent with the rapid transport of "new" waters (rainfall and throughfall) to the stream. Because of the combination of timing and response, the source of this low Cl^- concentration water must be a combination of channel interception, near-channel runoff, and runoff from a 3 ha bedrock outcrop in the headwaters.

For the study period, streamwater Cl^- transport from the watershed exceeds rainfall inputs by more than 3-fold. In the long term (1985-1995), rainfall accounts for 45 % of the streamwater Cl^- flux; the excess is attributed to dry atmospheric deposition, as is consistent with the Cl^- enrichment in throughfall. Also, the excess Cl^- transport is highly correlated with water yield, reflecting relatively rapid mobility of Cl^- provided sufficient water is available for transport. For the study period, groundwater contributes 79 % of the streamflow, and in the long-term, groundwater contributes 75 %. At peak flow during individual rainstorms, groundwater may contribute as little as 5 % of the runoff.

Acknowledgments

This study was conducted in cooperation with the Georgia Department of Natural Resources and the Geography Department of the University of Bristol, United Kingdom. The authors are grateful to the U.S. Geological Survey staff of the Panola Mountain Research Project including database support by B.T. Aulenbach and laboratory support by E.H. Drake, T.P. Pojunas and D.L. Booker. The use of brand names in this report is for information purposes and does not indicate endorsement by the U.S. Geological Survey.

References

Beier, C., Gundersen, P. and Rasmussen, L.: 1992, *Atmos. Environ.* **26**, 1553.

Cappellato, R., Peters, N.E. and Ragsdale, L.H.: 1993, *Can. J. For. Res.* **23**, 1114.

Davies, T.D.: 1983, *Atmos. Environ.* **17,** 797.

Durana, N., Casado, H., Ezcurra, A., Garcia, C., Lacaux, J.P. and Vandinh, P.: 1992, *Atmos. Environ.* **26**, 2437.

Eriksson, E.: 1955, *Tellus* **7**, 243.

Garten, C.T.J., Bondietti, E.A. and Lomax, R.D.: 1988, *Atmos. Environ.* **22**, 1425.

Gerritse, R.G. and George, R.J.: 1988, *J. Hydrol.* **101**, 83.

Hansen, K., Draaijers, G.P.J., Ivens, W.P.M.F., Gundersen, P. and Vanleeuwen, N.F.M.: 1994, *Atmos. Environ.* **28**, 3195.

Jenkins, A., Ferrier, R.C., Harriman, R. and Ogunkoya, Y.O.: 1994, *Hydrol. Proc.* **8**, 335.

Johnson, N., Likens, G., Bormann, F., Fisher, D. and Pierce, R.: 1969, *Water Resour. Res.* **5**, 1353.

Juang, F.H.T. and Johnson, N.M.: 1967, *J. Geophys. Res.* **72**, 5641.

Khare, P., Kapoor, S., Kulshrestha, U.C., Saxena, A., Kumar, N., Kumari, K.M. and Srivastava, S.S.: 1996, *Environ. Technol.* **17**, 637.

Kleissen, F.M., Wheater, H.S., Beck, M.B. and Harriman, R.: 1990, *J. Hydrol.* **116**, 365.

Leaney, F.W., Smettem, K.R.J. and Chittleborough, D.J.: 1993, *J. Hydrol.* **147**, 83.
Neal, C., Smith, C.J., Walls, J., Billingham, P. and Hill, S.: 1990, *Sci. Tot. Environ.* **91**, 1.
Nieto, P.J.G., Garcia, B.A., Diaz, J.M.F. and Brana, M.A.R.: 1994, *Atmos. Environ.* **28**, 2335.
Pearce, A.J., Stewart, M.K. and Sklash, M.G.: 1986, *Water Resour. Res.* **22**, 1263.
Peters, N.E.: 1989, 'Atmospheric Deposition of Sulphur to a Granite Outcrop in the Piedmont of Georgia, USA', in J.W. Delleur (ed.), *Atmospheric Deposition,* IAHS Publ. 179, pp. 173-181.
Peters, N.E.: 1991, *Water Air Soil Pollut.* **59**, 201.
Peters, N.E.: 1994, *J. Hydrol.* **156**, 73.
Peters, N.E. and Driscoll, C.T.: 1987, *Biogeochemistry* **3**, 163.
Peters, N.E. and Driscoll, C.T.: 1989, 'Temporal Variations in Solute Concentrations of Meltwater and Forest Floor Leachate at a Forested Site in the Adirondacks, New York', in *Proceedings of the 46th Annual Eastern Snow Conference*, pp. 45-56.
Pinder, G.F. and Jones, J.F.: 1969, *Water Resour. Res.* **5**, 438.
Puckett, L.J.: 1990a, *Atmos. Environ.* **24A**, 545.
Puckett, L.J.: 1990b, *Can. J. For. Res.* **20**, 1779.
Ratcliffe, E.B.: 1996, *Short-term Hydrological Responses of a Forested Hillslope during Rainstorms, at Panola Mountain Research Watershed, Georgia, USA*, unpublished Ph.D. Thesis, University of Bristol, Bristol, United Kingdom.
Rascher, C.M., Driscoll, C.T. and Peters, N.E.: 1987, *Biogeochemistry* **3**, 209.
Reynolds, B. and Pomeroy, A.B.: 1991, *J. Hydrol.* **99**, 19.
Rustad, L.E., Kahl, J.S., Norton, S.A. and Fernandez, I.J.: 1994, *J. Hydrol.* **162**, 319.
Shanley, J.B. and Peters, N.E.: 1988, *J. Contam. Hydrol.* **3**, 349.
Shanley, J.B. and Peters, N.E.: 1993, *J. Hydrol.* **146**, 361.
Sklash, M.G., Stewart, M.K. and Pearce, A.J.: 1986, *Water Resour. Res.* **22**, 1273.
Wurzler, S., Flossmann, A.I., Pruppacher, H.R. and Schwartz, S.E.: 1995, *J. Atmos. Chem.* **20**, 259.

A COMPARISON OF METHODS FOR DERIVING SOLUTE FLUX RATES USING LONG-TERM DATA FROM STREAMS IN THE MIRROR LAKE WATERSHED

P.A. BUKAVECKAS[1], G.E. LIKENS[2], T.C. WINTER[3] and D.C. BUSO[4]

[1]*University of Louisville, Water Resources Laboratory, Louisville, KY 40292, USA*
[2]*Institute of Ecosystem Studies, Box AB, Millbrook, NY 12545, USA*
[3]*U.S. Geological Survey, Denver Federal Center, Denver, CO 80225, USA*
[4]*Institute of Ecosystem Studies, Box AB, Millbrook, NY 12545, USA*

Abstract. Calculation of chemical flux rates for streams requires integration of continuous measurements of discharge with discrete measurements of solute concentrations. We compared two commonly used methods for interpolating chemistry data (time-averaging and flow-weighting) to determine whether discrepancies between the two methods were large relative to other sources of error in estimating flux rates. Flux rates of dissolved Si and SO_4^{2-} were calculated from 10 years of data (1981-1990) for the NW inlet and outlet of Mirror Lake and for a 40-day period (March 22 to April 30, 1993) during which we augmented our routine (weekly) chemical monitoring with collection of daily samples. The time-averaging method yielded higher estimates of solute flux during high-flow periods if no chemistry samples were collected corresponding to peak discharge. Concentration-discharge relationships should be used to interpolate stream chemistry during changing flow conditions if chemical changes are large. Caution should be used in choosing the appropriate time-scale over which data are pooled to derive the concentration-discharge regressions because the model parameters (slope and intercept) were found to be sensitive to seasonal and inter-annual variation. Both methods approximated solute flux to within 2-10 % for a range of solutes that were monitored during the intensive sampling period. Our results suggest that errors arising from interpolation of stream chemistry data are small compared with other sources of error in developing watershed mass balances.

Keywords: long-term monitoring, Mirror Lake, solute fluxes, stream chemistry, watershed budgets

1. Introduction

Attempts to quantify chemical fluxes for lakes and watersheds have become an important component of ecosystem research (Likens *et al.*, 1977; Driscoll *et al.*, 1989; Likens, 1992; Mitchell *et al.*, 1992; Houle *et al.*, 1995). Analyses of spatial and temporal patterns in the flux of water and chemicals provide insights into the role of hydrologic, geochemical and biotic processes that regulate the transport of materials and the metabolism of aquatic and terrestrial ecosystems (e.g., Vollenweider, 1968; Qualls *et al.*, 1991; Pardo *et al.*, 1995; Clair and Ehrman, 1996; Mitchell *et al.*, 1996). Input-output budgets have been used to quantify the importance of in-lake processes in the context of watershed biogeochemical cycles (Schindler *et al.*, 1976; Chen *et al.*, 1984; Likens *et al.*, 1985; Windolf *et al.*, 1996). Input-output budgets for lakes are derived by measuring or estimating inputs from the atmosphere (deposition directly to the lake surface), inputs from the watershed (lake tributary streams and groundwater), and losses via evapotranspiration, groundwater and the lake outlet. In this paper, we focus on stream inputs and outputs using a long-term data set to test the sensitivity of solute flux data to various methods of estimation.

Flux rates via streamflow are calculated as the product of two measured parameters: stream discharge and the concentration of dissolved or particulate materials. Typically,

Water, Air, and Soil Pollution **105:** 277–293, 1998.

discharge is measured continuously using automated gauging equipment, while chemical measurements are based on discrete samples collected at varying intervals (Likens *et al.*, 1977; Lawrence *et al.*, 1988). Calculation of weekly, monthly or annual fluxes requires interpolation of chemical measurements between sampling dates. Stream chemistry may be inferred from measured values using either a time-averaging approach (hereafter, TIME) or a flow-weighting method (hereafter, FLOW). Time-averaging approaches often use the average of two successive chemistry measurements as an estimate of solute concentrations during the intervening period when no samples were collected (Scheider *et al.*, 1976). During periods of uniform discharge, the average of the two samples is likely to be a strong predictor of stream chemistry. During periods of variable discharge, this method may overestimate or underestimate the flux of solutes whose concentrations are related to discharge. The magnitude of this error will depend on the change in discharge relative to when samples were collected and on the rate of change in solute concentration as a function of discharge.

The FLOW method uses a regression equation derived from paired measurements of concentration and discharge to predict stream chemistry based on measured discharge. The advantage of this method over the time-averaging approach is that the regression model accounts for changes in concentration that may occur during periods of variable discharge (Johnson, 1979). A potential source of error is introduced when regression models are derived from data aggregated at temporal scales that ignore seasonal and interannual variation in the concentration-discharge relationship. The slope and intercept of the concentration-discharge relationship may be affected by antecedent conditions (e.g., soil moisture), vegetative growth cycles, etc., and may result in biased estimates of solute flux.

We compared estimates of solute flux derived by the TIME and FLOW methods to determine whether the discrepancies between the two methods were large relative to other errors associated with flux calculations (e.g., hydrologic measurements and chemical analyses). We were particularly concerned that the variability introduced by the two different computational methods could complicate cross-system comparisons of flux data. Daily, monthly and annual flux rates of dissolved Si and SO_4^{2-} were derived for one of the three tributary streams of Mirror Lake (NW inlet) and for the lake outlet based on 10 years of data (1981-1990). Dissolved Si was chosen to represent solutes that in this and other watersheds (Schindler *et al.*, 1976) exhibit large changes in concentration as a function of discharge. Sulfate was selected as representative of solutes whose concentration exhibits little change in response to varying discharge. To calculate flux estimates using the FLOW method, we derived regression equations from paired measurements of concentration and discharge that were aggregated at a range of temporal scales (<1 to 10 yr). By comparing results derived from the various regressions, we evaluated the sensitivity of flux estimates to seasonal and interannual variation in the concentration-discharge relationship. Finally, we compared the time-averaging and flow-weighted estimates of flux for a 40-da period (March 22 to April 30, 1993) during which we augmented our routine (weekly) chemical monitoring with the collection of daily samples. Results from the intensive sampling period were used to appraise the accuracy of flux estimates derived by interpolation of the weekly chemistry data.

2. Site Description and Methods

Mirror Lake and its watershed are located within the Hubbard Brook valley in north-central New Hampshire, USA. The site has been the focus of long-term study. Detailed descriptions of the chemical, hydrologic and biological characteristics of Mirror Lake are presented in Likens (1985). Mirror Lake is typical of many of the lakes in northern New England and Canada. The lake is small (15 ha) and oligotrophic, with a basin situated largely in glacial till (Winter and Woo, 1990). The lake has a maximum depth of 11 m and a theoretical water residence time of approximately 1 yr. The predominantly forested watershed (85 ha) is comprised of a mixture of deciduous and coniferous species. Annual precipitation averages 135 cm yr^{-1} (Federer *et al.*, 1990) with approximately one-third accumulating as snow.

To measure inflow to Mirror Lake, Parshall flumes were installed in the three tributary streams (W and NW inlets in 1979 and NE inlet in 1981; Winter *et al.*, 1989). Continuous measurements of discharge for the inlet streams and the lake outlet were converted to daily discharge to calculate chemical fluxes. A concrete dam with two spillways is located at the outlet of the lake; horizontal boards regulate the height of the water in the lake. The relationship between lake level and the top of the boards was measured continuously by a water-level recorder and checked weekly by actual measurement. In 1990, a Parshall flume was installed in the stream channel 20 m below the outlet spillway. Regression equations were derived relating lake stage measurements to outflow estimates based on flume data. These relationships were used to calculate lake outflows for stage data collected since 1980.

Chemistry samples were collected weekly (1981-1990) or daily (March 22 to April 30, 1993) and analyzed for all major cations and anions. Chemical analyses were performed by ion chromatography (SO_4^{2-}), inductively-coupled plasma emission spectro-photometry (dissolved Si), flame atomic absorption spectrophotometry (Ca^{2+}, Na^+) or colorimetrically on a Technicon AutoAnalyzer (Cl^-; Likens *et al.*, 1977). On dates when chemistry samples were collected, flux rates were calculated by multiplying the daily discharge by the measured solute concentration. On dates when no chemistry data were available, the TIME and FLOW methods were used to infer solute concentrations. Time-averaged estimates were derived by multiplying the average solute concentration for two successive sampling dates by the daily discharge during the intervening period. FLOW estimates were derived using a concentration-discharge relationship to predict solute concentrations corresponding to each of the daily discharge values during the intervening period. Log(base-10) transformation of the discharge data was found to substantially improve the fit of the linear regression models, whereas transformation of solute concentrations did not. To assess seasonal and interannual variation in the concentration-discharge relationship, regression equations were derived using data aggregated at a range of temporal scales: 1) pooled within years and by seasons (May-November and December-April), 2) pooled within seasons and across all years, and 3) pooled across seasons and years. Annual fluxes were calculated by summing the daily flux estimates for a calendar year.

3. Results

3.1. Dissolved Silica

At the inlet stream, dissolved Si concentrations were inversely correlated with discharge (Table I) indicating that high flows following snowmelt and summer storms had a substantial dilution effect. Regression models derived from data pooled across years and by season (May-November vs. December-April) were significantly different. Baseflow dissolved Si concentrations were significantly greater in May-November than in December-April, and the rate of decrease in concentration with increasing discharge also was significantly different. For the range of discharge observed at the NW inlet (1-100 L sec^{-1}), the corresponding decrease in dissolved Si was from 11.8 to 2.4 mg L^{-1} (May-November) and from 9.7 to 2.9 mg L^{-1} (December-April). At the lake outlet, dissolved Si concentrations were lower and were correlated only weakly with discharge ($R^2 < 0.20$). A small but statistically significant difference was noted in baseflow concentrations, which tended to be higher in December-April than in May-November. Neither winter/spring nor summer/fall high flow events had a substantial dilution effect on outlet dissolved Si (slope < 0.5 mg L^{-1} log[L sec^{-1}]) and slopes were not significantly different for the two regression models. Overall, dissolved Si concentrations in the tributary stream exhibited a wider range of concentration in response to changing discharge and the regression parameters were more sensitive to seasonal variation in comparison to the lake outlet.

We predicted that the time-averaging method would yield higher estimates of dissolved Si inputs to the lake relative to the FLOW method whenever high flow events occurred on non-sampling dates. At the outlet, changes in discharge had little effect on dissolved Si and we expected that the two methods would yield similar estimates of flux. As expected, the TIME method yielded higher annual estimates of influx (for 7 of the 10 years), although the differences between the two methods exceeded 10 % in only two years (1981-1982; Table II). An examination of monthly flux and discharge for 1981-1982 reveals that the

TABLE I

Linear regressions predicting concentrations of dissolved Si and SO_4^{2-} (mg L^{-1}) as a function of daily average flow (log[L sec^{-1}]). Models were derived from data pooled within seasons and across years (1981-1990). Pairwise comparisons (May-November vs. December-April) of model parameters were significantly different ($p < 0.05$) with the exception of the slopes for the outlet dissolved Si regressions.

Solute	Site	Season	Slope	Intercept	R^2	p	N
Dissolved Si	Inlet	May-Nov	-4.68 ± 0.19	11.76 ± 0.14	0.65	0.0001	326
		Dec-Apr	-3.43 ± 0.18	9.70 ± 0.17	0.61	0.0001	239
	Outlet	May-Nov	0.32 ± 0.04	1.89 ± 0.03	0.15	0.0001	305
		Dec-Apr	0.22 ± 0.09	2.31 ± 0.07	0.03	0.0162	223
SO_4^{2-}	Inlet	May-Nov	0.79 ± 0.06	4.94 ± 0.04	0.34	0.0001	326
		Dec-Apr	-0.16 ± 0.07	6.01 ± 0.06	0.02	0.0185	239
	Outlet	May-Nov	0.05 ± 0.04	4.73 ± 0.02	0.01	NS	305
		Dec-Apr	-0.1 ± 0.08	5.13 ± 0.06	0.02	0.0324	223

two methods were in good agreement during periods of high flow in the winter/spring of both years and also during the periods of low flow during the summer/fall of 1982 (Figure 1). However, during high flows in the fall of 1981, the time-averaging method predicted substantially higher fluxes than the flow-weighting method. An examination of daily discharge for this period revealed that three high-flow events (>3,000 m^3 da^{-1}) occurred within two weeks during late-October (Figure 2). The events were of short duration and no chemistry data were collected on the dates when highest discharge occurred. Therefore, the TIME method calculated flux assuming no change in dissolved Si concentration at peak discharge. The FLOW method, on the other hand, estimated flux assuming a decrease in concentration corresponding to the slope of the concentration-discharge relationship (-4.7 mg L^{-1} log[L sec^{-1}]; Table I). As a result, the estimated daily flux was only 50 % of that calculated by the TIME method. These analyses suggest that while annualized estimates for the two methods were generally in good agreement, large discrepancies in daily and monthly flux values did occur.

The lake is a net sink for dissolved Si and therefore flux rates were smaller at the outlet than at the inlet (Table II). Dissolved Si concentrations were correlated only weakly with outlet discharge and therefore we were surprised to find that the two methods yielded substantially different estimates of annual flux. The difference between the two methods exceeded 10 % in 5 of the 10 years and 20 % in 3 of 10 years. There was no consistent pattern to the differences. For example, in 1981, the FLOW method overestimated outflux by 19 % relative to the TIME method, but underestimated flux by 25 % the following year. A scatterplot of daily dissolved Si concentrations revealed that the FLOW method yielded consistently higher estimates in 1981 and lower estimates in 1982 compared to the TIME method (Figure 3).

We recalculated daily concentrations for 1981 and 1982 using concentration-discharge regressions derived from data for a single year and compared these to predicted values derived from the regression based on the pooled (10-yr) data set. Model parameters for the 1981 and 1982 regressions were significantly different. Baseflow dissolved Si concentrations were higher for 1982 than for 1981 (intercept = 2.27 and 1.80 mg L^{-1}, respectively) and the slope of the regression line was greater in 1982 (1.10 mg L^{-1} log[L sec^{-1}]) than in 1981 (0.12 mg L^{-1} log[L sec^{-1}]; not significantly different from zero). Scatterplots comparing the revised estimates with concentrations estimated by the TIME method did not show systematic bias in either year (Figure 4). These results suggest that interannual variation in the slope and intercept of the concentration-discharge relationship is sufficiently large to bias concentration and flux estimates when data are pooled across years.

We compared the revised FLOW estimates against the TIME estimates for a period of rapidly changing discharge. In 1981, the highest flows at the lake outlet occurred during a mid-winter thaw (February 10 - March 5). Routine (weekly) chemistry sampling coincided with two of the three days when peak discharge was observed and, because discharge decreased rapidly after each peak, subsequent samples were collected when discharge was substantially lower (Figure 5). As a result, daily flux estimates derived by the TIME method were weighted by the average concentration from a high flow and low flow sample and followed closely the values calculated using the revised FLOW-method estimates. Results from these analyses suggest that when chemistry samples are collected over a range of discharge, the two methods will yield similar estimates of flux provided that the regression models take into account seasonal and inter-annual variation in the model parameters.

TABLE II

Annual flux of dissolved Si and SO_4^{2-} (kg yr^{-1}) estimated by the time-averaging (TIME) and flow-weighting (FLOW) methods; % denotes the proportional difference between the two methods calculated as (FLOW-TIME)/TIME.

		Inlet			Outlet		
	Year	FLOW	TIME	%	FLOW	TIME	%
Dissolved Si	1981	2084	2387	-12.7	1050	884	18.7
	1982	1367	1533	-10.8	637	852	-25.3
	1983	1577	1553	1.5	285	274	4.1
	1984	1699	1860	-8.7	306	328	-6.8
	1985	1251	1189	5.2	140	115	21.8
	1986	1675	1782	-6.0	652	720	-9.4
	1987	1305	1360	-4.0	413	422	-2.2
	1988	1073	1017	5.6	229	207	10.5
	1989	1511	1556	-2.9	682	555	22.9
	1990	2189	2347	-6.7	842	806	4.4
	Mean	1573	1658	-4.0	524	516	1.6
SO_4^{2-}	1981	1970	2074	-5.0	2214	2423	-8.6
	1982	1195	1238	-3.5	1169	1306	-10.5
	1983	1456	1513	-3.8	606	621	-2.4
	1984	1684	1632	3.2	616	617	-0.1
	1985	1028	1043	-1.5	312	318	-1.8
	1986	1564	1573	-0.6	1246	1257	-0.9
	1987	1115	1100	1.3	785	694	13.1
	1988	814	808	0.7	466	415	12.2
	1989	1451	1427	1.7	1365	1217	12.2
	1990	2259	2129	6.1	1626	1468	10.8
	Mean	1454	1454	0	1040	1034	0.6

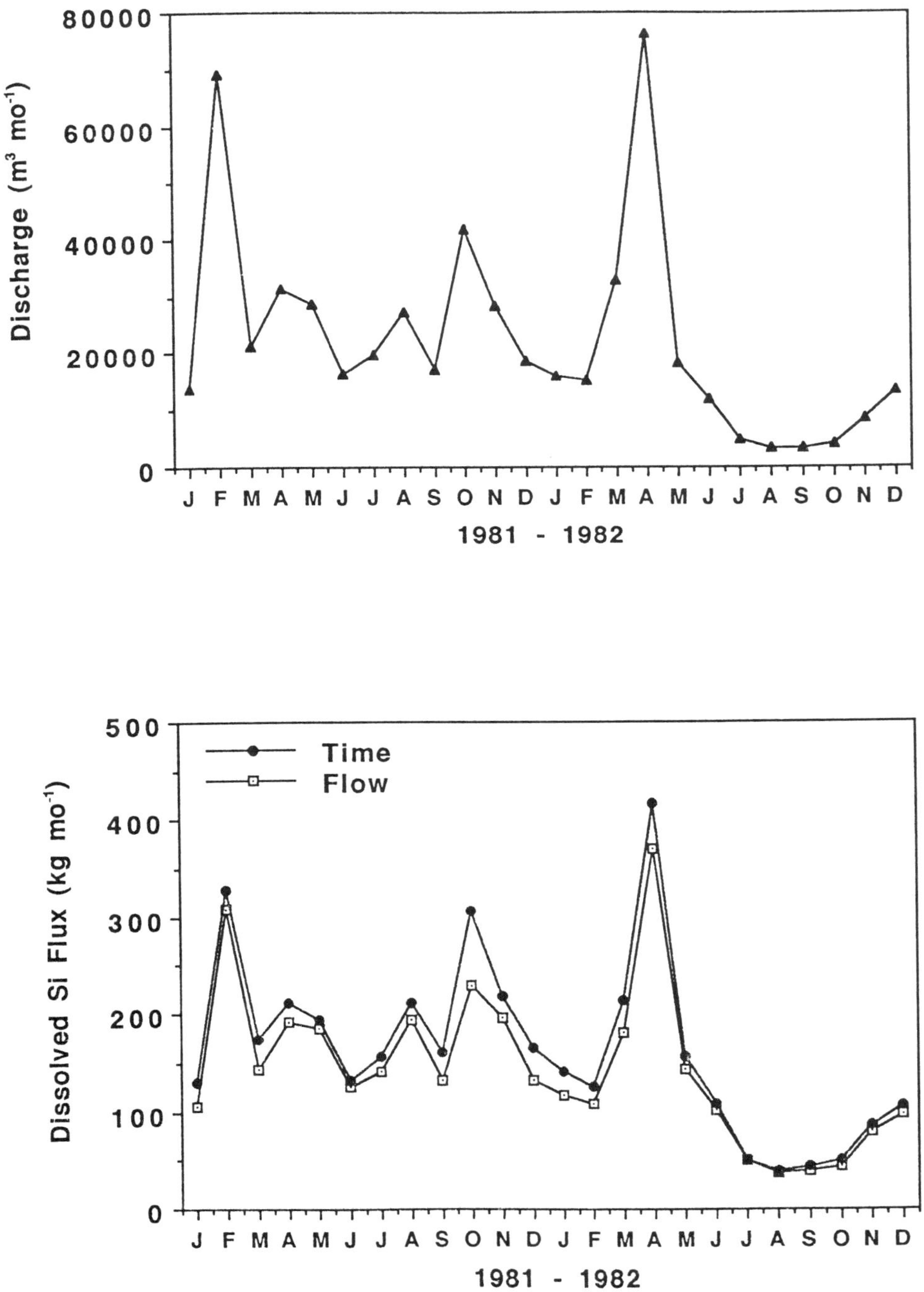

Fig. 1. Monthly discharge and dissolved Si flux for the NW inlet during 1981 and 1982. Flux estimates were calculated using the time-averaging (TIME) and flow-weighting (FLOW) methods (see text).

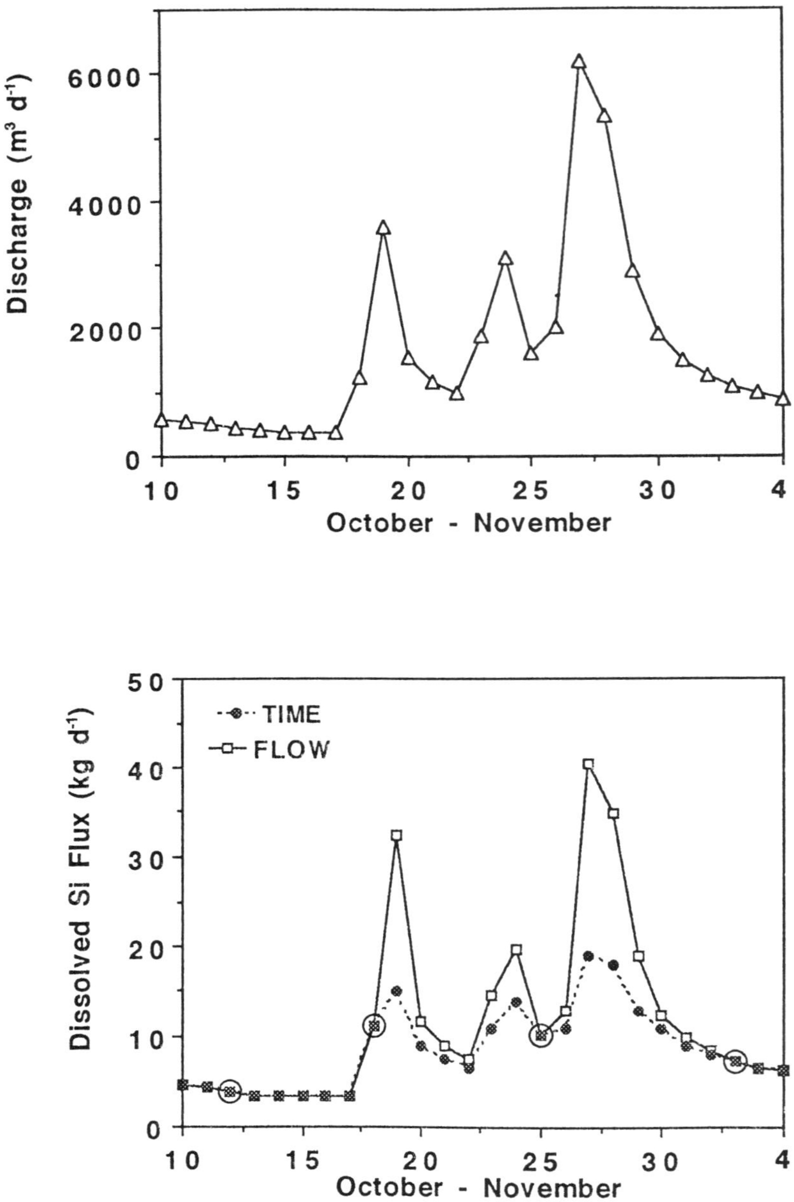

Fig. 2. Daily discharge and dissolved Si flux calculated by the TIME and FLOW methods during a period of variable discharge at the NW inlet in the fall of 1981. Large circles denote dates on which chemistry samples were collected.

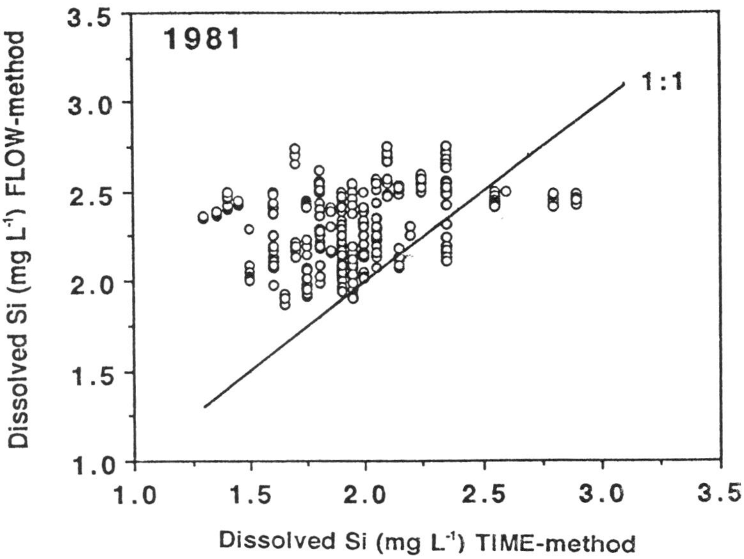

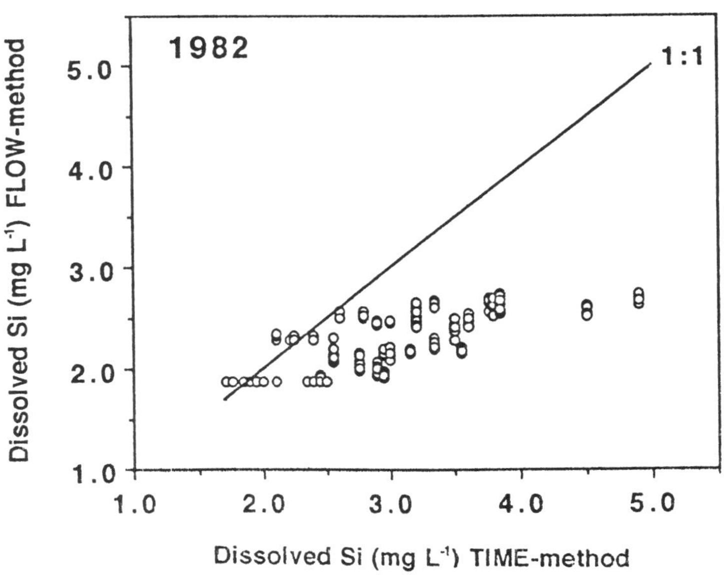

Fig. 3. Daily estimates of outlet dissolved Si concentrations for 1981 and 1982 calculated by the FLOW and TIME methods. Regressions used for FLOW-based estimates were derived from weekly chemistry data pooled by seasons (May-November and December-April) and across years (1981-1990).

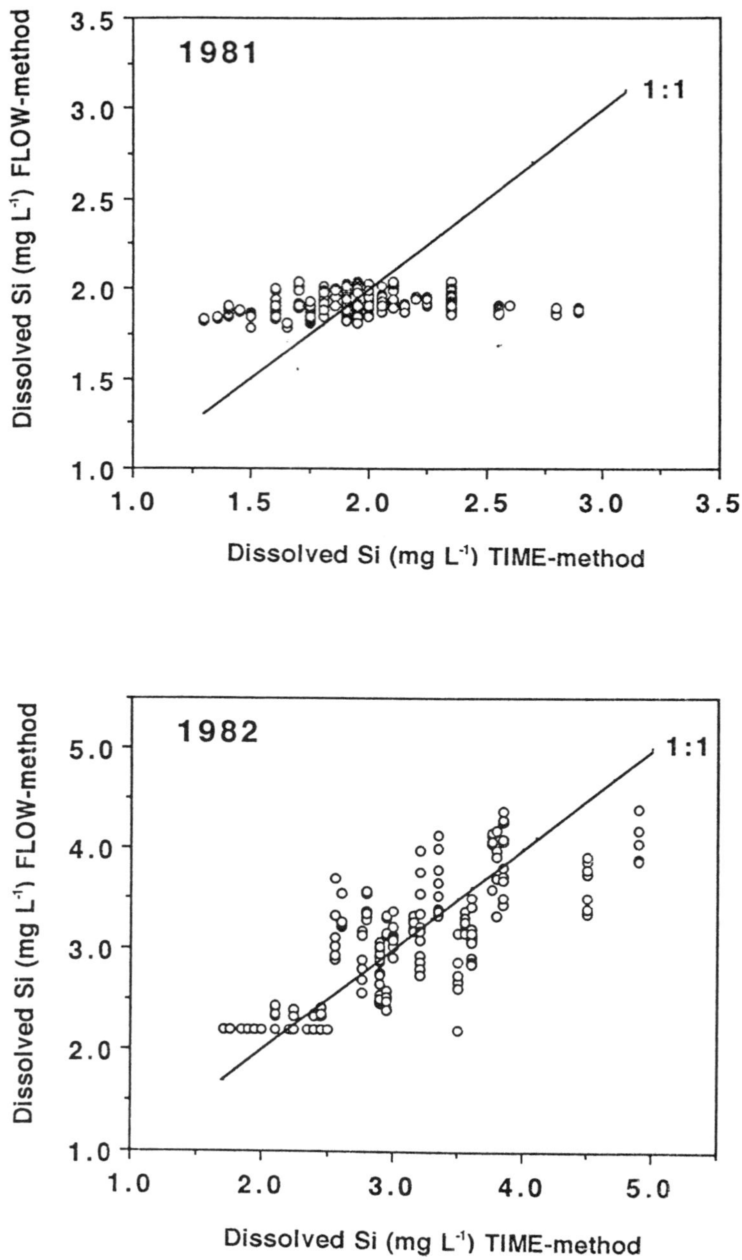

Fig. 4. Daily estimates of outlet dissolved Si concentrations for 1981 and 1982 calculated by the FLOW and TIME methods. Regressions used for FLOW-based estimates were derived from weekly chemistry data for a single year.

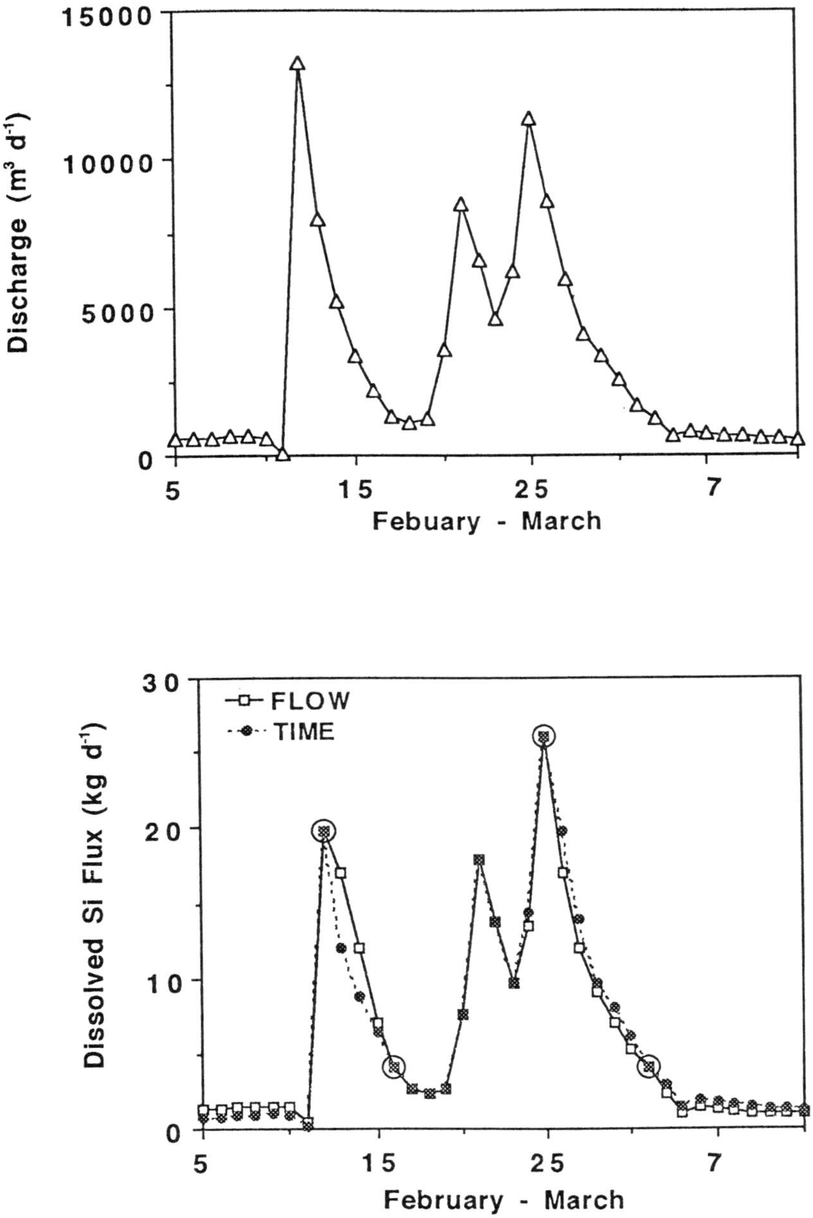

Fig. 5. Daily discharge and dissolved Si flux during a period of variable discharge at the lake outlet in the spring of 1981. Large circles denote dates on which chemistry samples were collected.

3.2. Sulfate

For both the inlet and outlet streams, SO_4^{2-} concentrations were weakly correlated with discharge ($R^2 < 0.40$) and changes in discharge had little effect on concentrations (slope < 1.0 mg L^{-1} log[L sec^{-1}]; Table I). We predicted that the TIME and FLOW methods would yield similar estimates of SO_4^{2-} flux for both the inlet and outlet streams. Annualized flux estimates for the inlet stream were within 10 % in all years and outlet flux estimates were within 10 % in 5 of 10 years. The larger discrepancy between outflux values was due in part to lake retention of SO_4^{2-} resulting in lower concentrations at the outlet. As a result, fluxes at the outlet were lower and the relative error for the two methods of estimation (as a proportion of total flux) was large. Because SO_4^{2-} concentrations were not substantially influenced by changes in discharge, both methods predicted similar flux rates during periods of changing hydrologic conditions even when peak flows were not represented during routine sampling (Figure 6).

3.3. Intensive Sampling Period

Fluxes were calculated for a variety of solutes using data collected during a 40-day period in which we augmented our routine weekly monitoring with daily collection of chemistry samples. Fluxes derived from measured daily concentration and discharge were compared to estimates derived from measured weekly concentration plus time-averaged or flow-weighted estimates of concentration. The flow-weighted estimates were calculated from a concentration-discharge regression based on six months of weekly data (November, 1992 to April, 1993). Both methods approximated fluxes of Ca^{2+}, Na^+, SO_4^{2-}, Cl^- and dissolved Si derived from measured daily concentrations to within 10 % (Figure 7). At the inlet stream, the FLOW method yielded closer estimates to fluxes determined from daily sampling than the TIME method which consistently overestimated for solutes whose concentration was correlated with discharge (Ca^{2+}, Na^+, Cl^-, dissolved Si). At the lake outlet, the TIME method provided better estimates (within 2 % of values derived from daily sampling) than the FLOW method for all solutes.

4. Discussion

Our results show that the time-averaging and flow-weighted methods can yield substantially different estimates of flux for solutes whose concentrations are strongly influenced by changes in discharge. Annualized estimates of dissolved Si flux calculated by the two methods differed by as much as 25 % and daily flux estimates varied by as much as twofold. The magnitude and pattern of the discrepancies were variable from year-to-year and between inlet and outlet streams. Discrepancies between the two methods can be attributed to two factors. First, the TIME method overestimates flux when high flow events are of short duration and the preceding and following chemistry samples are collected during low flow conditions. The data presented for a fall storm event (Figure 2) illustrate the problem of using time-averaged estimates when no chemistry data are collected at peak discharge. In these situations, the TIME method calculates flux assuming no decrease in concentration during the period of increased discharge and predicts flux rates twofold higher

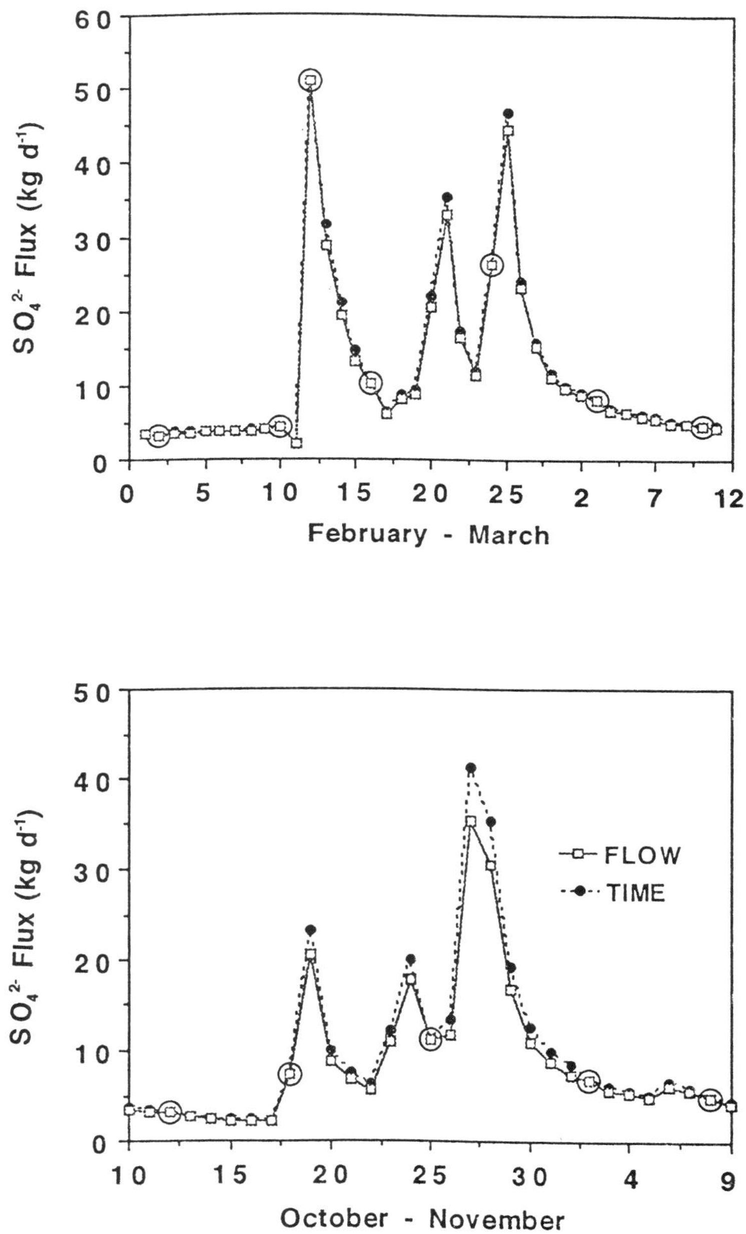

Fig. 6. Daily SO_4^{2-} flux during periods of variable discharge (in 1981) at the NW inlet calculated by the FLOW and TIME methods. Large circles denote dates on which chemistry samples were collected.

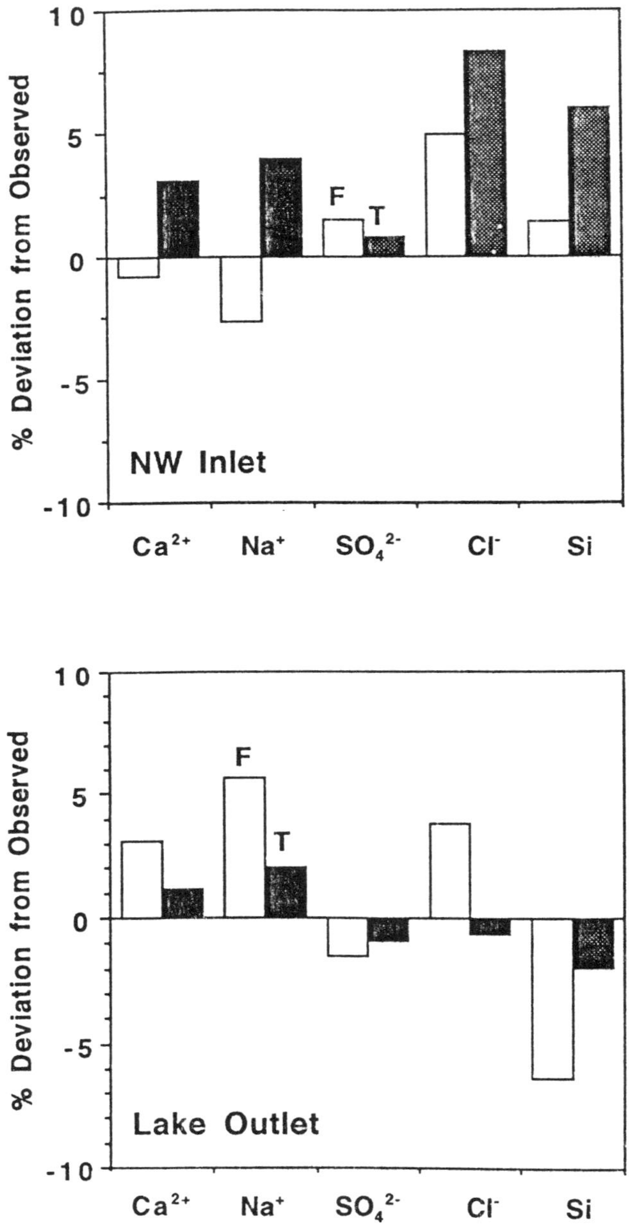

Fig. 7. Deviations (as %) of flux estimates derived by the FLOW (F) and TIME (T) methods from fluxes for a 40-day period during which routine weekly monitoring was augmented by collection of daily chemistry samples.

than those estimated by the flow-weighting approach. The magnitude of this error will be a function of the magnitude and duration of the increase in discharge and the rate of decrease in concentration relative to discharge. Our results suggest that when the slope of the regression relating concentration to discharge is large relative to the intercept (>20 %), the time-averaging approach will likely result in relatively large errors in estimating daily, weekly or monthly flux values.

The effect of this error on annualized estimates of influx/outflux will depend upon the frequency of high flow events that lack corresponding chemistry data. Among lakes in the north temperate region, high flows associated with spring snowmelt and rare storm events account for a substantial fraction of the annual stream discharge. At Mirror Lake, outlet discharges exceeding 2,000 m^3 da^{-1} occurred at an average of 32 dates per year (9 %), but accounted for 63 % of the annual outflow and 67 % of the annual outflux of dissolved Si (Bukaveckas *et al.*, 1993). Our routine weekly monitoring resulted in 4.5 samples per year (9 % of total) corresponding to flows greater than 2,000 m^3 da^{-1}. These results emphasize the importance of obtaining stream chemistry samples during rare high-discharge events.

A second factor that contributes to the discrepancy between the two methods is the error arising from use of regression equations to infer changes in stream chemistry during periods of variable discharge. The inherent assumption in this approach is that the slope and intercept of the concentration-discharge relationship adequately characterize the range of conditions represented by the data set used to derive the regression model. Variation in the concentration-discharge relationship may arise from short-term changes in hydrology (rising-limb vs. falling-limb) as well as seasonal changes due to the differing effects of snowmelt, rainfall, groundwater and evapotranspiration on stream chemistry and flow. For outlet streams, changing in-lake conditions (thermal stratification, ice cover) may contribute to variation in the regression parameters. Watershed disturbance (logging, fires, insect outbreaks) and climatic variability will contribute to inter-annual variation in the concentration-discharge relationship.

When stream chemistry deviates from the concentration-discharge trajectory described by the regression model, this approach will overestimate or underestimate solute concentrations and result in biased estimates of flux. Results of our analyses show that this type of error occurs when chemistry and discharge data are aggregated at temporal scales which ignore seasonal and inter-annual variation in the concentration-discharge relationship. Deviations in stream chemistry from model predictions occurred over seasonal and inter-annual time scales. We found significant differences between regression models derived from summer/fall and winter/spring data collected at the inlet stream. Similar findings were reported by Johnson (1979) who showed that stream phosphorus export estimated from a single (annual) regression was 11 % lower than estimates derived using multiple (monthly) regression equations. A regression model derived from 10 years of weekly chemistry measurements was a poor predictor of lake outlet dissolved Si and SO_4^{2-} concentrations in some years (e.g., 1981, 1982; Figure 3). This may represent a lag imposed by the residence time of Mirror Lake (1 yr) in "resetting" the model parameters to their long-term average. The sources of inter-annual variation in these parameters (e.g., intercept values of dissolved Si concentrations at the outflow) are not known, but may reflect the effects of climatic variability on processes such as Si weathering and biotic uptake. Use of this model resulted in biased estimates of dissolved Si flux relative to the time-averaging method.

We recommend that if a flow-weighting approach is used to calculate flux, regression equations should be adjusted for seasonal and inter-annual variation in the concentration-discharge relationship. This method provides accurate estimates of flux for solutes whose concentrations are discharge-dependent when regressions are derived for relatively short time scales (6-12 months). Problems also can arise when chemistry data are subdivided into progressively smaller groupings (from multi-year to single year to seasonally-aggregated regressions) because the probability that a range of concentration-discharge conditions will be represented within each subset is reduced. This can lead to extrapolation outside the range of measured values used for the regression if, for example, the routine sampling program is not augmented to capture high-flow events. However, our analyses suggest that regressions derived from long-term data sets can yield poor estimates of monthly and annual flux. Choosing the appropriate time scale for aggregating chemistry data is an important consideration for studies that involve planned or unplanned watershed disturbances, as these are likely to influence the trajectory of the concentration-discharge regressions.

In summary, our analyses suggest that caution should be used in pooling multi-year data to define concentration-discharge relationships for streams. Although more highly aggregated data sets (multi-year) can yield more precise estimates of the model parameters (smaller standard error), solute concentrations inferred from chemistry data aggregated at shorter time scales (ca. 6-12 months) were more accurate, particularly for estimating weekly and monthly flux. Our regression models derived from a single year's data yielded flux rates that corresponded closely to measured values. Results from a period of intensive (daily) sample collection show that when regression parameters are adjusted for seasonal and interannual variation, both the FLOW and TIME methods can approximate fluxes to within 2-10 %. This range of error is comparable to those ascribed to chemical analyses (<10 %) and measurements of stream discharge (5 %), whereas other sources of error in developing watershed mass balances (e.g., groundwater fluxes) are likely to be larger (Winter, 1981; LaBaugh and Winter, 1984).

Acknowledgments

This paper is a contribution to the Hubbard Brook Ecosystem Study and the program of the Institute of Ecosystem Studies. We are grateful for use of facilities at the USDA Northeastern Forest Experiment Station. Assistance in the field was provided by L. Derosia, M. Hallett, C. Masson and E. Morency. Funding was provided by the National Science Foundation, the Andrew W. Mellon Foundation and the US Geological Survey.

References

Bukaveckas, P.A., Likens, G.E., Buso, D.C. and Winter, T.: 1993, *Verh. Internat. Verein. Limnol.* **25**, 419.

Chen, C.W., Gherini, S.A., Peters,N.E., Murdock, P.S., Newton, R.M. and Goldstein, R.A.: 1984, *Water Resour. Res.* **20**, 1875.

Clair, T.A. and Ehrman, J.M.: 1996, *Limnol. Oceanogr.* **41**, 921.

Driscoll, C.T., Likens, G.E., Hedin, L.O., Eaton, J.S. and Bormann, F.H.: 1989, *Environ. Sci. Technol.* **23**, 137.
Federer, C.A., Flynn, L.D., Martin, C.W., Hornbeck, J.W. and Pierce, R.S.: 1990, *Thirty years of Hydrometeorologic Data at the Hubbard Brook Experimental Forest, New Hampshire*, USDA Forest Service, General Tech. Report NE-141.
Houle, D., Carignan, R., Lachance, M. and Dupont, J.: 1995, *Limnol. Oceanogr.* **40**, 710.
Johnson, A.: 1979, *Water Resour. Res.* **15**, 1224.
LaBaugh, J.W. and Winter, T.C.: 1984, *Limnol. Oceanogr.* **29**, 322.
Lawrence, G.B., Driscoll,C.T. and Fuller, R.D.: 1988, *Water Resour. Res.* **24**, 659.
Likens, G.E., Bormann, F.H., Pierce, R.S., Eaton, J.S. and Johnson, N.M.: 1977, *Biogeochemistry of a Forested Ecosystem*, Springer-Verlag, New York.
Likens, G.E. (ed.): 1985, *An Ecosystem Approach to Aquatic Ecology: Mirror Lake and its Environment*, Springer-Verlag, New York.
Likens, G.E., Eaton, J.S., Johnson, N.M. and Pierce, R.S.: 1985, 'Flux and Balance of Water and Chemicals', in G.E. Likens (ed.), *An Ecosystem Approach to Aquatic Ecology: Mirror Lake and its Environment*, Springer-Verlag, New York, pp. 135-155.
Likens, G.E.: 1992, *The Ecosystem Approach: Its Use and Abuse. Excellence in Ecology, Vol. 3.* Ecology Institute, Oldendorf-Luhe, Germany.
Mitchell, M.J., Foster, N.W., Shepard, J.P. and Morrison, I.K.: 1992, *Can. J. For. Res.* **22**, 457.
Mitchell, M.J., Raynal,D.J. and Driscoll, C.T.: 1996, *Water, Air, Soil Pollut.* **88**, 355.
Pardo, L.H., Driscoll, C.T. and Likens, G.E.: 1995, *Water, Air, Soil Pollut.* **85**, 1659.
Qualls, R.G., Haines, B.L. and Swank, W.T.: 1991, *Ecology* **72**, 254.
Scheider, W.A., Moss, J.J. and Dillon, P.J.: 1979, 'Measurement and Uses of Hydraulic and Nutrient Budgets', in *Lake Restoration, U.S. Environmental Protection Agency Report 440/5-79-001*, Minneapolis, pp. 77-83.
Schindler, D.W., Newbury, R.W., Beaty, K.G. and Campbell, P.: 1976, *J. Fish. Res. Bd. Can.* **33**, 2526.
Vollenweider, R.A.: 1968, *Scientific Fundamentals of the Eutrophication of Lakes and Flowing Waters, with Particular Reference to Nitrogen and Phosphorus as Factors in Eutrophication*. Water Management Research, Org. Ecol. Coop. Dev. (Paris) Tech. Report DAS/RS1/68.27. 159 pp.
Windolf, J., Jeppesen, E., Jensen, J.P. and Kristensen, P.: 1996. *Biogeochemistry* **33**, 25.
Winter, T.C.: 1981, *Water Resour. Bull.* **17**, 82.
Winter, T.C., Eaton, J.S. and Likens, G.E.: 1989, *Water Resour. Bull.* **25**, 991.
Winter, T.C. and Woo, M.: 1990, 'Hydrology of Lakes and Wetlands', in M.G. Wolman and H.C. Riggs (eds.), *Surface Water Hydrology*, Geological Society of America, Boulder, Colorado, pp. 159-187.

EFFECT OF ALTITUDE AND TREE SPECIES ON $\delta^{34}S$ OF DEPOSITED SULFUR (JEZEŘÍ CATCHMENT, CZECH REPUBLIC)

H. GROSCHEOVÁ, M. NOVÁK, M. HAVEL and J. ČERNÝ

Czech Geological Survey, Geologická 6, 152 00 Prague 5, Czech Republic

Abstract. Precipitation totals, SO_4^{2-} concentrations and $\delta^{34}S$ ratios were measured in bulk (open area) deposition, beech (with some birch and ash) and spruce throughfall at 17 locations in the heavily polluted mountain-slope Jezeří catchment (490-924 m asl) over a 12-month period. Stream water S was analyzed monthly at seven altitudes. The $\delta^{34}S$ ratios decreased in the order: bulk > beech > spruce > stream (1995 annual means across all altitudes were 5.7, 5.1, 4.2 and 4.0 ‰, respectively). The differences were caused by different intensities of scavenging of isotopically distinct pollution-derived SO_2 by tree canopies. The $\delta^{34}S$ ratios of bulk deposition increased and $\delta^{34}S$ ratios of spruce throughfall decreased with increasing altitude. The elevational gradients in $\delta^{34}S$ ratios reflected the changing roles of local vs. distant air pollution sources along the slope.

Keywords: Czech Republic, forest decline, Jezeří catchment, sulfur deposition, stable sulfur isotopes, throughfall

1. Introduction

Sulfur isotopes have been used to fingerprint sources of atmospheric pollution (Mayer *et al.*, 1995; de Caritat *et al.*, 1997) and to study processes of S transformations at the catchment and ecosystem level (Gebauer *et al.*, 1994; Novák *et al.*, 1994; Bottrell and Novák, 1997; Heaton *et al.*, 1997; Ohizumi *et al.*, 1997). In catchment studies, $\delta^{34}S$ ratios are not a conservative tracer due to the occurence of several isotope-selective processes, e.g., atmospheric conversion of SO_2 to SO_4^{2-} (Saltzman *et al.*, 1983), maturation of organic material in the soil (Novák *et al.*, 1996) and anaerobic bacterial dissimilatory sulfate reduction (Krouse and Grinenko, 1991). Still, it may be possible to distinguish between isotope-fractionating processes taking place in the catchment and mixing of S sources with different isotope signatures (Novák, 1998). For this purpose, an extensive study of $\delta^{34}S$ ratios of both inputs/outputs and within-catchment S fluxes is a prerequisite.

There is renewed interest in S cycling in the northern Czech Republic in light of the ever-deteriorating health status of hardwood vegetation in this part of the "Black Triangle" region. In winter 1996, spruce seedlings perished on an area of 1,300 ha. In 1997, 3,000 ha of birch suffered from summertime defoliation, and birch stands completely died back on an area of 500 ha.

In the Jezeří catchment, situated in the northern Czech Republic, S isotope fluxes have been monitored since 1991 (Novák *et al.*, 1995 and unpublished data). Most isotope work in this heavily acidified mountain-slope catchment has focused on its flat upper segment, which compares in altitude and microclimatic conditions to other sites under study in the region. Here we report spatial and temporal trends in $\delta^{34}S$ ratios along the 440-m elevational gradient characterizing the Jezeří catchment. The objectives of the present study were: to assess the influence of altitude on $\delta^{34}S$ ratios of bulk (open area) deposition, beech, and spruce canopy throughfall; to evaluate seasonality in $\delta^{34}S$ ratios of atmospheric inputs and surface runoff across the elevational gradient; and, to calculate the present-day mass- and altitude-weighted $\delta^{34}S$ signature of atmospheric input into the entire catchment.

Water, Air, and Soil Pollution **105:** 295–303, 1998.

2. Methods

2.1. Study Site

The 2.61-ha south-sloping Jezeří catchment (Figure 1) overlooks an industrialized soft-coal basin (elevation of about 250 m asl). Eleven large coal-burning power plants are located in the basin, making it one of the most polluted regions of the world. The catchment reaches the plateau of the Krušné hory Mountains in its upper part (elevation of 924 m asl). Between 1970 and 1975, spruce stands above the elevation of 600-750 m died back due to pollution and were clear-cut. In the lower segment of the catchment, a mature beech stand remains. Four approximately 5-ha patches of mostly young vigorous spruce stands are scattered above the elevation of 770 m. The remaining area (in white in Figure 1) is gradually being overgrown by birch and ash (tree height was < 3 m in 1995). Hydrochemical fluxes at Jezeří have been monitored since 1986 (Černý, 1995). Havel *et al.* (1996) studied the effect of altitude on precipitation chemistry at 17 locations within the catchment.

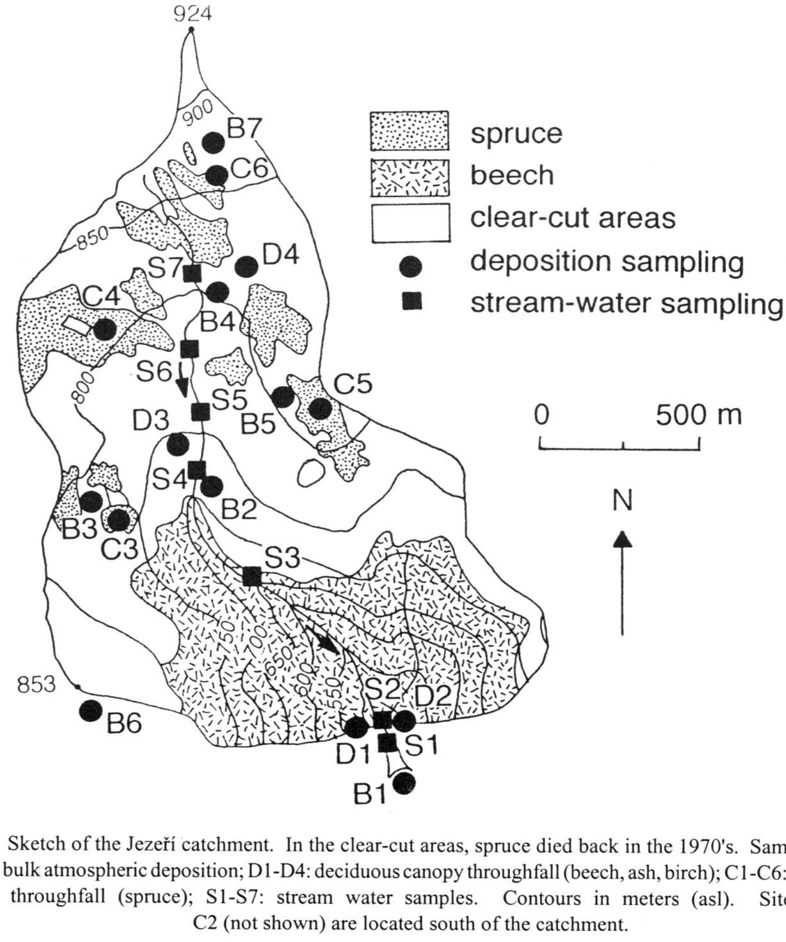

Fig. 1. Sketch of the Jezeří catchment. In the clear-cut areas, spruce died back in the 1970's. Sampling sites: B1-B7: bulk atmospheric deposition; D1-D4: deciduous canopy throughfall (beech, ash, birch); C1-C6: coniferous canopy throughfall (spruce); S1-S7: stream water samples. Contours in meters (asl). Sites C1 and C2 (not shown) are located south of the catchment.

2.2. SAMPLING

Volume-weighted water samples were prepared by combining monthly yields from rainfall collectors installed in nests between an elevation of 365 and 892 m (asl). The nests ("sites"; solid circles in Figure 1) consisted of nine collectors placed in a 10 × 10 m grid for canopy throughfall and two collectors placed 5 m apart for bulk (open area) deposition. Bulk deposition was collected at seven sites (B1-B7; Figure 1), deciduous canopy throughfall at four sites (D1 and D2, beech; D3, birch plus ash; D4, ash), and coniferous canopy throughfall at six sites (C1-C6; all spruce). As a rule, monthly yields within each site were further pooled to represent a bimonthly sampling interval. The sampling started on January 17, 1995 and ended on January 4, 1996. The final day covered by each composite sample was as follows: April 12 (Jan-Mar), June 6 (Apr-May), July 31 (Jun-Jul), September 24 (Aug-Sep), October 23 (Oct), and Jan 4 (Nov-Dec). The time spans for these composite samples thus were: 87, 55, 55, 55, 29 and 73 da, respectively. In addition, 2 L samples of surface runoff (S1-S7; solid squares in Figure 1) were taken monthly over a period of nine months (March 1, April 11, May 9, June 6, July 8, September 24, and October 23, 1995; January 7, 1996). Because the stream draining the catchment was covered with ice in December 1995, we use data from an unpublished pilot study (data were collected in December 1992). Sampling sites for surface runoff were situated between an elevation of 480 and 824 m asl. Water samples were kept in a refrigerator at +2 °C prior to analysis.

2.3. ANALYTICAL TECHNIQUES

Sulfate concentrations were determined on a Shimadzu LC 6A liquid chromatograph. Sulfate for isotope analysis was precipitated as $BaSO_4$ and converted to SO_2 using the method of Yanagisawa and Sakai (1983). Stable isotope measurements were performed on a Finnigan MAT 251 mass spectrometer. The results were expressed in the usual notation as a ‰ deviation of the $^{34}S/^{32}S$ ratio in the sample from the standard (Canyon Diablo Troilite). The reproducibility of $\delta^{34}S$ measurements was 0.2 ‰ (CDT).

3. Results and Discussion

3.1. ISOTOPE SIGNATURES OF INPUTS/OUTPUTS

Bulk atmospheric deposition, deciduous and spruce throughfall, and surface runoff differed in SO_4^{2-} concentrations and S isotope composition. The comparison between individual sample types was based on a total of 154 determinations of SO_4^{2-} concentrations and 148 determinations of $\delta^{34}S$ ratios. When data from all altitudes were combined, the mean annual values of SO_4^{2-} concentrations were 4.6 mg L^{-1} for bulk deposition, 10.8 mg L^{-1} for deciduous (mostly beech) throughfall, 30.7 mg L^{-1} for spruce throughfall, and 46.1 mg L^{-1} for surface runoff. The difference in mean SO_4^{2-} concentrations was statistically significant (one-way, four-level ANOVA, $p < 0.05$). Box plots in Figure 2 illustrate the increasing trend in SO_4^{2-} concentrations in the order bulk < beech < spruce < stream. The mean annual $\delta^{34}S$ ratios were 5.7 ‰ for bulk deposition, 5.1 ‰ for deciduous (mostly beech) throughfall, 4.2 ‰ for spruce throughfall and 4.0 ‰ for surface runoff. Thus, $\delta^{34}S$ ratios showed an opposite trend than SO_4^{2-} concentrations, decreasing in the order bulk > beech > spruce > stream (Figure 2). The difference in mean $\delta^{34}S$ ratios was statistically significant (one-way, four-level ANOVA, $p < 0.05$). Comparisons of each pair (*t*-tests with adjusted *p* level using the JMP package by SAS Institute) indicated a statistically significant difference between bulk and spruce, bulk and beech, bulk and stream, beech and spruce, and beech and stream. The difference was statistically insignificant between spruce and stream.

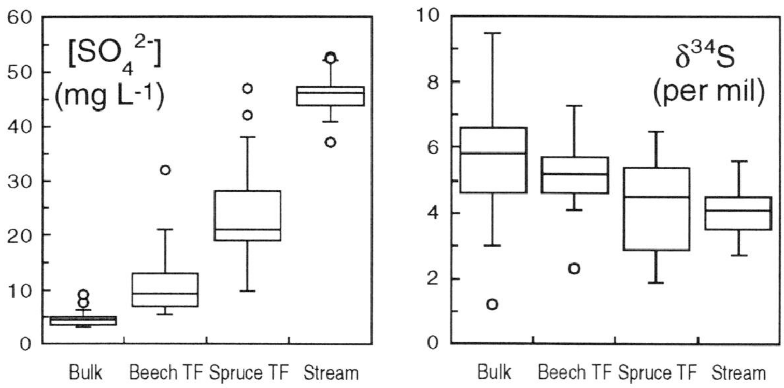

Fig. 2. Comparisons of SO_4^{2-} concentrations and $\delta^{34}S$ ratios between four sample types. Each box encloses 50 % of the data, median is marked with a horizontal line inside the box. Vertical bars show the range of the data, outliers are marked with an open circle. TF - throughfall; "Beech" includes all deciduous TF samples.

Generally, isotope signatures of distant and local emission sources may be different. The $\delta^{34}S$ ratios of coal-burning power plants just south of Jezeří average 2 ‰ (i.e., are relatively low); the more remote German and Polish part of the "Black Triangle" north of Jezeří emits S of unknown isotope composition. As seen in Figure 1, the beech stand is situated in low altitudes (470-700 m), while most surviving spruce are located in high altitudes (800-900 m). Different $\delta^{34}S$ ratios of beech and spruce throughfall could reflect primarily different $\delta^{34}S$ ratios of airborne (local vs. distant) S in each elevation interval. However, our sampling design corrected for stratified vegetation by including small groups of "exotic" species along the slope (spruce in the beech-dominated lower part and decidous trees at high elevation). The differences in $\delta^{34}S$ ratios in Figure 2 rather are explained by differing receptor surfaces. Over 98 % of industrial S emissions are in the form of SO_2. Spruce needles, and to a lesser extent broad-leaves, capture airborne SO_2 in much larger amounts than the surface of an unforested area. This results in higher SO_4^{2-} concentrations in throughfall than in bulk deposition. The larger the leaf surface area, the larger the proportion of dry-deposited, low $\delta^{34}S$ SO_2, and the lower the throughfall $\delta^{34}S$ ratios.

5.2. Elevational Gradient

Sulfate concentrations and $\delta^{34}S$ ratios were averaged for each altitude separately (Figure 3. Means ± standard errors are given for the entire 12-month period (Figure 3a,d), for the summer (samples "April-May" through "October," a total of 194 days; Figure 3b,e), and

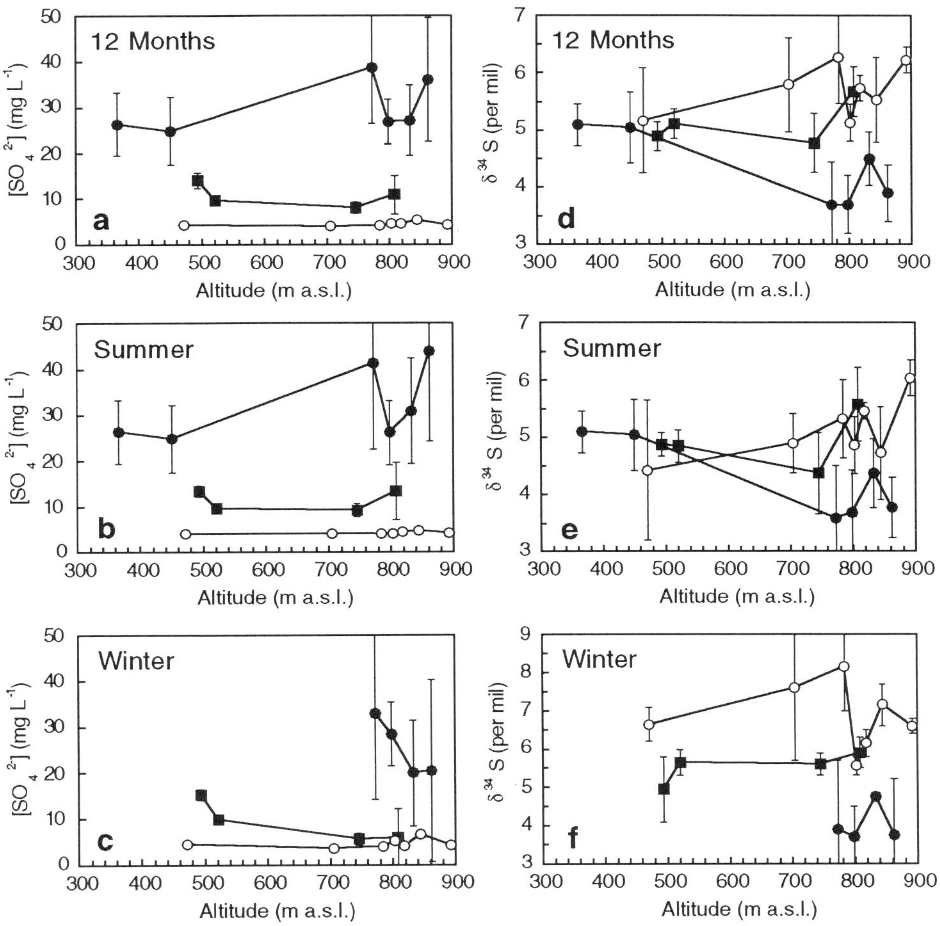

Fig. 3. Elevational gradient in SO_4^{2-} concentrations (a-c) and $\delta^{34}S$ ratios (d-f) of bulk deposition (open circles), spruce throughfall (solid circles) and deciduous throughfall (solid squares); values are means ± standard errors.

for the winter (samples "Jan-Mar" and "Nov-Dec", a total of 160 days; Figure 3c,f). While SO_4^{2-} concentrations of bulk deposition remained virtually constant across the elevation span, there was a clear increase in SO_4^{2-} concentrations of spruce throughfall between 450 and 772 m, followed by a sudden decrease at around 800 m down to values encountered at the foot of the mountain slope. In summer, the highest sampling site (862 m) exhibited again a very high SO_4^{2-} concentration (close to 40 mg L^{-1}). Both SO_4^{2-} concentrations and $\delta^{34}S$ ratios of deciduous throughfall were intermediate between bulk deposition and spruce throughfall, reflecting an intermediate receptor surface area. The effect of altitude on $\delta^{34}S$ ratios of deposited S is summarized in Figure 3d. At the foot of the mountain slope (365-520 m), $\delta^{34}S$ ratios of all three types of atmospheric deposition were identical (close to +5.2

‰). With increasing altitude, between 520 and 780 m, $\delta^{34}S$ of bulk deposition and spruce throughfall formed a nearly symmetrical shape of a "fan" with the difference between the two growing: the $\delta^{34}S$ ratio of bulk deposition increased to 6.2 ‰ and that of spruce throughfall decreased to 3.8 ‰. Interestingly, the symmetry was preserved also at 800 m where the $\delta^{34}S$ ratio of bulk deposition exhibited a valley while the $\delta^{34}S$ ratio of spruce throughfall exhibited a peak. In both cases, this sudden drop/increase approached the $\delta^{34}S$ ratios measured at the base of the slope. In summer, the $\delta^{34}S$ ratio of bulk deposition at the base of the slope was lower than that of spruce throughfall (Figure 3e). The fan-like trends were similar to those in Figure 3d. The convergence of $\delta^{34}S$ ratios of both bulk and spruce throughfall to the horizontal axis of symmetry of the "fan" at the elevation of 800 m existed also in summer. In winter (Figure 3f), the separation of the $\delta^{34}S$ ratios of bulk and spruce throughfall was the largest (4 ‰ at 770 m), however an elevational trend for spruce throughfall could not be evaluated due to missing data at low altitude.

The gradual separation of $\delta^{34}S$ ratios between bulk deposition (higher at higher altitudes) and spruce throughfall (lower at higher altitudes), and higher SO_4^{2-} concentrations under spruce, reflected larger amounts of atmospheric SO_2 captured in the upper parts of the catchment. Enhanced scavenging of low-$\delta^{34}S$ SO_2 explained the dipping spruce throughfall curve in Figures 3d-e. The increase in $\delta^{34}S$ ratios of bulk deposition from 470 to 800 m was not accompanied by higher SO_4^{2-} concentrations and was *per se* harder to explain. Limited exchange between vertically stratified air masses with isotopically distinct S, and S isotope fractionation during the SO_2-to-SO_4^{2-} conversion both could be evoked as possible mechanisms. The convergence of $\delta^{34}S$ ratios of bulk deposition and spruce throughfall to the axis of symmetry of the "fan," and a similar pattern in SO_4^{2-} concentration under spruce (Figures 3a-b, d-f), exactly mark the altitude of the upper limit of frequent winter-time inversions. About 65 % of the annual S input into Jezeří occurs in winter. The inversion cap traps S emitted from the power plants and keeps it above the basin south of the catchment. However, because summer also exhibited localized peaks/valleys in SO_4^{2-} concentrations and $\delta^{34}S$ ratios at 800 m, inversions alone could not explain the trends in Figure 3. Importantly, 78 % of atmospheric S loads at the mountaintop plateau of the Krušné hory Mountains is carried by northwesterly winds, which in this area prevail for most of the year (Czech Hydrometeorological Institute, unpublished data for 1985-1993). The role of S imported from the German/Polish part of the "Black Triangle" will be better understood when its isotope signature is known.

5.3. Across-Altitude Seasonality

Mean $\delta^{34}S$ ratios were evaluated for bulk deposition and spruce throughfall on a bimonthly basis by pooling data from all altitudes (Figure 4). For a short time in early summer ("Jun-Jul"), the normally isotopically lighter spruce throughfall S became isotopically heavier than S in bulk deposition. We suggest that this "reversed" isotope signature in June-July is real and not a random consequence of the complex interplay of elevation gradient with seasonal change in S sources. A 1993-1994 two-year $\delta^{34}S$ time series from a single elevation (B4 and C4 in Figure 1) repeatedly exhibited the same reversal for one month in early summer (Novák *et al.*, 1995). This reversal had been ascribed to more complete SO_2-to-SO_4^{2-} conversion accompanying high precipitation under high humidity relative to other times of the year.

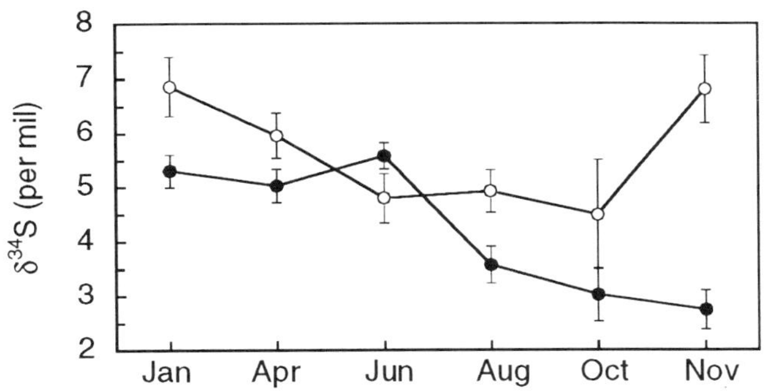

Fig. 4. Seasonality in $\delta^{34}S$ ratios (means ± standard errors) of bulk deposition (open circles) and spruce throughfall (solid circles); for each sampling interval, means were calculated across all altitudes.

Figure 5a depicts a seasonal trend in $\delta^{34}S$ ratios of stream water. The $\delta^{34}S$ ratios systematically increased (by 1.2 ‰) from March to August, possibly reflecting lower contribution of the low-$\delta^{34}S$ dry-deposited S after the heating season ended and power plants started to burn less coal. Figure 5b separates water sampling sites along the longer (west) stream into those draining a forested area and those draining a clearing. We hypothesized that because throughfall generally had lower $\delta^{34}S$ ratios than open area deposition, stream water below a forest would also have lower $\delta^{34}S$ ratios. As seen in Figure 5b, the exact opposite was observed with the difference being statistically significant ($p < 0.05$). The data indicated a more complex nature of streamflow generation, with the direct rainwater component playing only a limited role.

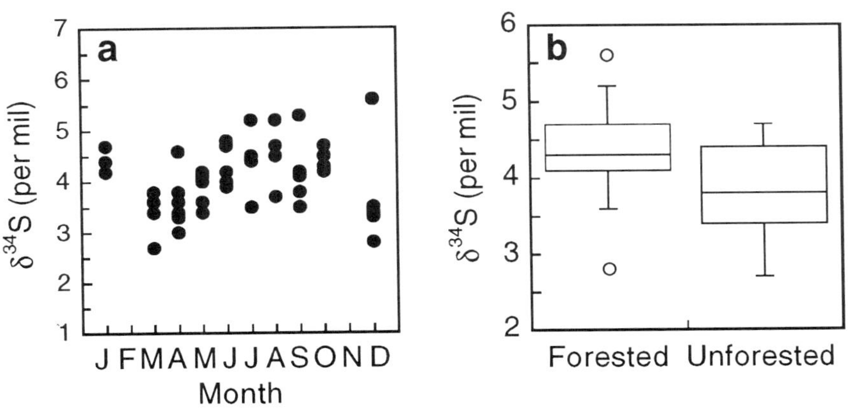

Fig. 5. Temporal and spatial trends in $\delta^{34}S$ ratios of streamwater SO_4^{2-}: a - monthly grab samples taken at various altitudes; b - box plots for sampling sites draining forested (left) and unforested (right) segments of the catchment (medians, 50 % of the variable population and ranges shown, see Figure 2 for explanation).

5.4. Altitude-Weighted $\delta^{34}S$ of Catchment Input

A single $\delta^{34}S$ ratio of atmospheric input into the whole catchment was calculated for 1995. Such a value is of interest in input/ouput isotope mass balance studies. For the calculation, the following types of data were used: altitude, type of canopy, sampling interval, precipitation total, SO_4^{2-} concentrations, the $\delta^{34}S$ ratios for each sampler nest, and area (m^2) of each type of vegetation cover at each altitude within the catchment. The slope was divided into five elevational bands: 475-650, 650-750, 750-800, 800-850 and 850-924 m a.s.l. The calculation was carried out in three steps: 1) if a band contained more than one sampler nest in the same vegetation, means were calculated for all measurements; 2) in each band for each type of vegetation, a mass-weighted $\delta^{34}S$ was calculated for 1 m^2 and the whole year; and 3) $\delta^{34}S$ was mass-weighted for the whole catchment using total S deposition into the area of each vegetation type in each band.

The resulting S isotope signature of atmospheric inputs into the Jezeří catchment was 5.9 ‰. This ratio was higher than the median $\delta^{34}S$ ratio for any of the deposition types in Figure 2. Clearly, open areas ($\delta^{34}S$ mean of 5.2 ‰) dominated the integrated input signature. The Jezeří input signature (5.9 ‰) was higher than the mean $\delta^{34}S$ of bulk deposition (5.3 ‰) for 13 different catchments located throughout the Czech Republic (Novák, 1998).

6. Conclusions

The Jezeří catchment spans an elevational gradient of 440 m. At each of the 17 sampling sites, the $\delta^{34}S$ ratios of atmospheric inputs were affected by at least three factors: type of vegetation, altitude and time of the year. The differences in mean $\delta^{34}S$ ratios for different types of vegetation (open area, beech, spruce) were robust enough to be statistically different ($p < 0.05$) on the yearly basis and across all altitudes. Sulfate concentrations increased and $\delta^{34}S$ ratios decreased in the order: bulk deposition - beech throughfall - spruce throughfall - surface runoff. The differences were caused by enhanced interception of pollution-derived, low-$\delta^{34}S$ SO_2 by spruce needles and deciduous canopies. At the base of the mountain slope, $\delta^{34}S$ of bulk deposition, beech and spruce throughfall were close to 5.2 ‰; with increasing altitude $\delta^{34}S$ of bulk deposition increased, and $\delta^{34}S$ of spruce throughfall decreased in a nearly symmetrical, fan-like fashion (Figure 3d). At 850 m a.s.l., SO_4^{2-} concentrations and $\delta^{34}S$ ratios converged toward intermediate values marking the upper limit of the common winter inversions that trap industrial emissions in the basin south of the slope. The role of distant vs. local pollution sources along the slope can be assessed only when the $\delta^{34}S$ signatures imported from the upwind area in Germany and Poland are known. In 1995, the whole-year, mass- and altitude-weighted input $\delta^{34}S$ ratio into the catchment was 5.9 ‰. This value was most affected by the presence of clearings in areas where spruce died back in the 1970's.

Acknowledgments

This work was supported by the Czech Granting Agency (Grant no. 205/93/2404 to MN) and the European Union (CANIF, ERB IC20 CT960024). We thank Dr. J. Kirchner of the University of California, Berkeley and R. Krejčí for assistance.

References

Bottrell, S.H. and Novák, M.: 1997, *J. Ecol.* **85**, 125.

Černý, J.: 1995, *Water, Air, Soil Pollut.* **85**, 589.

de Caritat, P., Krouse, H.R. and Hutcheon, I.: 1997, *Water, Air, Soil Pollut.* **94,** 191.

Gebauer, G., Giesemann, A., Schulze, E.-D. and Jäger, H.J.: 1994, *Plant Soil* **164,** 267.

Havel, M., Krejčí, R. and Černý, J.: 1996, *Decrease in Acidic Atmospheric Deposition in the Krušné Mts.*, Czech Granting Agency, Final Report (205/93/0675), Prague.

Heaton, T.H.E., Spiro, B. and Robertson, S.M.C.: 1997, *Oecologia* **109**, 600.

Krouse, H.R. and Grinenko, V.A.: 1991, *Stable Isotopes: Natural and Anthropogenic Sulphur in the Environment,* SCOPE 43, John Wiley & Sons, Chichester.

Mayer, B., Feger, K.-H., Giesemann, A.: 1995, *Biogeochemistry* **30**, 31.

Novák, M.: 1998, *Proceedings of the 9th Int. Conference on Water-Rock Interaction, Taupo, New Zealand, March 30 - April 3, 1998*. Balkema, Rotterdam.

Novák, M., Bottrell, S.H., Fottová, D., Buzek, F., Groscheová, H. and Žák, K.: 1996, *Environ. Sci. Technol.* **30**, 3473.

Novák, M., Bottrell, S.H., Groscheová, H., Buzek, F., Černý, J.: 1995, *Water, Air, Soil Pollut.* **85**, 1641.

Novák, M., Wieder, R.K. and Schell, W.R.: 1994, *Limnol. Oceanogr.* **39**, 1172.

Ohizumi, T., Fukuzaki, N. and Kusakabe, M.: 1997, *Atmos. Environ.* **31**, 1339.

Saltzman, E.S., Brass, G.W. and Price, D.A.: 1983, *Geophys. Res. Lett.* **10**, 513.

Yanagisawa, F. and Sakai, H.: 1983, *Anal. Chem.* **55**, 985.

ORGANIC SULFUR AND THE RETENTION OF NUTRIENT CATIONS IN FOREST SURFACE SOILS

HENRY G. SPRATT, JR.

Department of Biological and Environmental Sciences, University of Tennessee at Chattanooga, Chattanooga, TN 37403, USA

Abstract. Sulfur and cation transformations were studied in A-horizon soils from 9 plots in Deer Run State Forest (DRSF) in southeastern Missouri on two dates, and from 27 plots of the Missouri Ozark Forest Ecosystem Project (MOFEP) near DRSF, seasonally for three years. The DRSF plots were chosen based on history of timber harvest; 2-3 yr and 8-10 yr post clear-cut, with controls about 40 yt post cut. Total C and S, organic sulfur (OS), exchangeable K^+ and Mg^{2+}, and rates of microbial OS production varied by season and year in surface soils from MOFEP plots. Exchangeable K^+ concentrations were correlated with total C, total S, and OS. In clear-cut DRSF A-horizon soils, OS production was greatly reduced ($p < 0.05$) compared with controls. Exchangeable K^+ and Mg^{2+} also were much lower in soils from clear-cut plots than in soils from control plots ($p < 0.01$). Cation exchange for OS compounds with sulfhydryl groups was only marginally greater than for organic matter of similar structure with no sulfhydryl groups. Overall, litter input appears to be a key to maintaining long-term surface soil OS, K^+, and Mg^{2+} concentrations.

Keywords: base cations, exchangeable K, exchangeable Mg, forest soils, organic sulfur

1. Introduction

Retention of nutrient cations, such as K^+ and Mg^{2+}, by forest soils is essential to long-term sustainability of forest ecosystems. Of great concern to many researchers is the effect of acidic precipitation on forest soil nutrient cation leaching (Johnson *et al.*, 1982; Kennedy, 1986; Watwood *et al.*, 1993). The principal mechanism for nutrient cation leaching in soils subjected to high levels of acidic precipitation is the exchange of nutrient cations for H^+ associated with SO_4^{2-} from precipitation (Mitchell *et al.*, 1992). Retention of SO_4^{2-} within the soil by either biotic or abiotic means has been associated with reduced cation leaching in a number of forest ecosystems (Figure 1, Wilklander, 1980; Johnson *et al.*, 1982; Richter *et al.*, 1983; Rechcigl and Sparks, 1985; Johnson, 1992; Mitchell and Lindberg, 1992; Mitchell *et al.*, 1992; Watwood *et al.*, 1993).

Cation exchange sites in forest surface soils can be mineral (e.g., clays) or organic. Organic cation exchange sites include carboxyl or phenolic residues (Hausenbuiller, 1978), and sulfhydryl groups (Stevenson, 1986). Because the conversion of SO_4^{2-} into organic sulfur (OS) in soils subjected to acidic precipitation has been associated with retention of nutrient cations, and sulfhydryl groups of OS can serve as a form of organic cation exchange site, the role OS plays in the retention of cations may be of interest. Watwood *et al.* (1993) demonstrated that mineralization of OS in A-horizon soils from diverse forested sites was associated with ecosystem-wide cation loss. This suggests that either critical cation exchange sites are lost with the mineralization of OS, or the flush of SO_4^{2-} as a result of OS mineralization is enough to strip nutrient cations from other cation exchange sites present in the soil. Therefore, soils that exhibit large losses of OS also might be subjected to loss of nutrient cations.

Water, Air, and Soil Pollution **105:** 305–317, 1998.

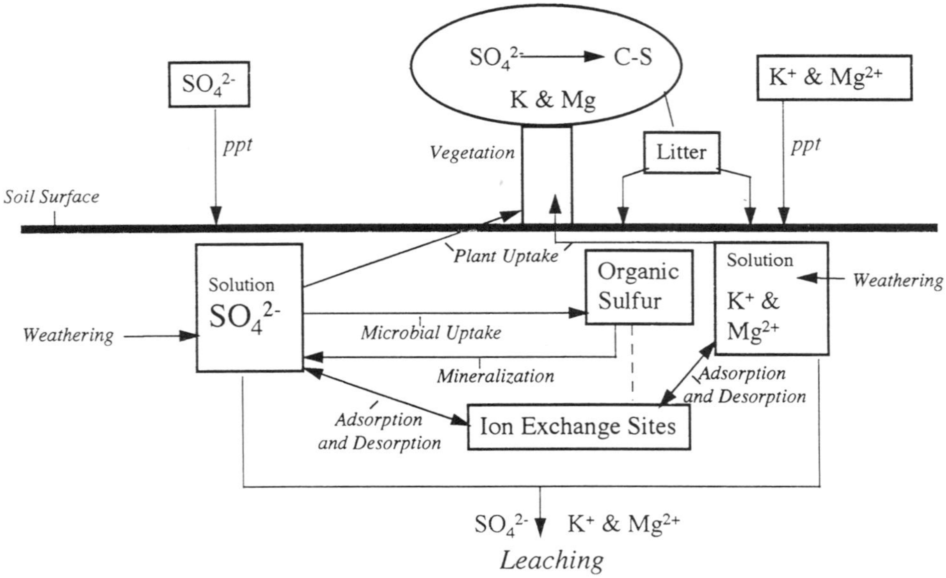

Fig. 1. Links between forest S and nutrient cation cycling (modified from Mitchell *et al.* 1992).

Ecosystem disturbance often leads to changes in the nutrient status of the system. Spratt (1997a) documented very large changes in the OS content of A-horizon soils in Missouri Ozark forests from which timber was harvested by clear-cutting. By analyzing data from both previously clear-cut Missouri sites and seasonal data from plots that are located in a nearby long-term ecosystem-level study (the Missouri Ozark Forest Ecosystem Project, MOFEP, sponsored by the Missouri Department of Conservation), this study focuses on the conversion of SO_4^{2-} to OS by microorganisms and on the role that this OS plays in retention of nutrient cations in surface soils.

2. Methods

Surface soils were collected from nine plots in Deer Run State Forest (DRSF) in southeastern Missouri on two sample dates (September and October 1992), and from 27 permanent plots, part of MOFEP near DRSF, seasonally from August 1993 until May 1996 (Figure 2). The DRSF plots had three different experimental treatments based on previous timber harvest; control (about 40 yr post-cut), 2-3 yr post-clear-cut, and 8-10 yr post-clear-cut. The MOFEP plots were about 40 yr post-cut. Further information regarding these sample sites and MOFEP may be found in Spratt (1997a,b), and in Brookshire *et al.* (1997).

Ozark Forest Soils are derived from cherty dolomites and sandstones, and tend to be well drained. The soils sampled were either ultisols or alfisols, and were characterized by thin A-horizons (ranging from 0.5 to 10 cm) overlying thick B-horizons. All sites in DRSF and MOFEP were south-southwest facing slopes (approximately mid-slope).

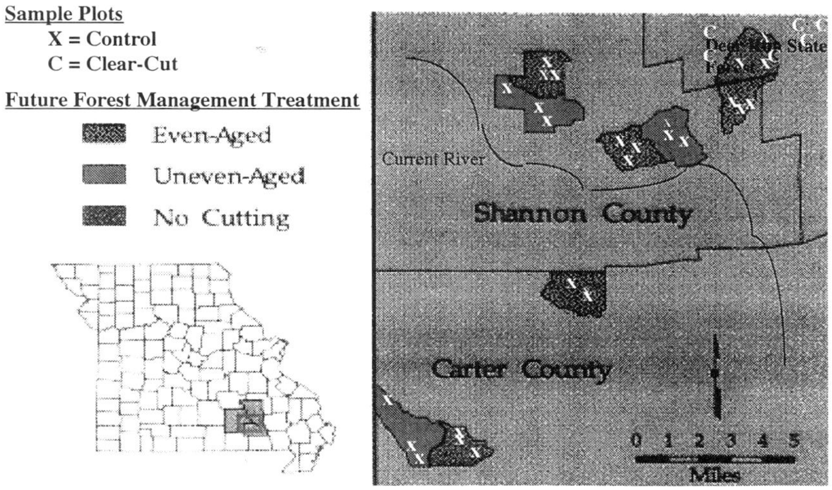

Fig. 2. Sample sites and future treatments; Missouri Ozark Forest Ecosystem Project (MOFEP) and Deer Run State Forest (DRSF).

Dominant vegetation for these sites (Kabrick *et al.*, 1997) included: *Quercus alba* (white oak), *Q. velutina* (black oak), *Q. coccinea* (scarlet oak), *Pinus echinata* (shortleaf pine), *Cornus florida* (flowering dogwood), and *Vaccinium angustifolium* (lowbush blueberry). Regeneration in Ozark forest clear-cuts occurs naturally. Although not specifically determined for the DRSF plots, regeneration on other experimental clear-cut plots in the vicinity of DRSF generally promotes the growth of oak species (Annand and Thompson, 1997). In the short term (4-5 yr post harvest), however, the number of stems ha^{-1} of shrub species tends to be much greater on clear-cut sites than in mature stands.

A-horizon soils were collected by removing the overlying litter layer and cutting into the organic A-horizon with a sharp spatula. A-horizon soil samples were placed in sterile Whirl-pak bags. All soil samples were stored on ice until arrival at the laboratory. In the laboratory, the soils were stored at 5 °C and, within 3 days of collection, processed according to the scheme in Figure 3. Sieved soils were subdivided into three fractions, one for measurement of extractable SO_4^{2-}, a second for percent moisture determination, and a third to measure $^{35}S\text{-}SO_4^{2-}$ incorporation. After determining the percent moisture, dried soils were used for determination of total C, total S, and exchangeable cation bases.

Incorporation of SO_4^{2-}-S into soil organic matter was monitored using a modified ^{35}S-tracer method (Watwood and Fitzgerald, 1988; Spratt 1997a,b). About 1 μCi of $^{35}SO_4^{2-}$ containing a total of 8 pmol of SO_4^{2-} (ICN Biomedicals, Inc.) was added to approximately 0.5 g of sieved soil in conical centrifuge tubes. The soils were incubated at temperatures determined for surface soils on the sample collection dates, under aerobic conditions, for 48 hr. The soils were then placed in a -80 °C freezer to arrest any additional transformations of SO_4^{2-} in the soils. The ^{35}S present in the soils after incubation was sequentially extracted using distilled H_2O, a salt mixture, HCl, and NaOH to quantify the radiolabel present in the water soluble sulfate (WS) and adsorbed sulfate (AdS) pools, and the OS fraction (Watwood and Fitzgerald, 1988; Spratt, 1997a,b).

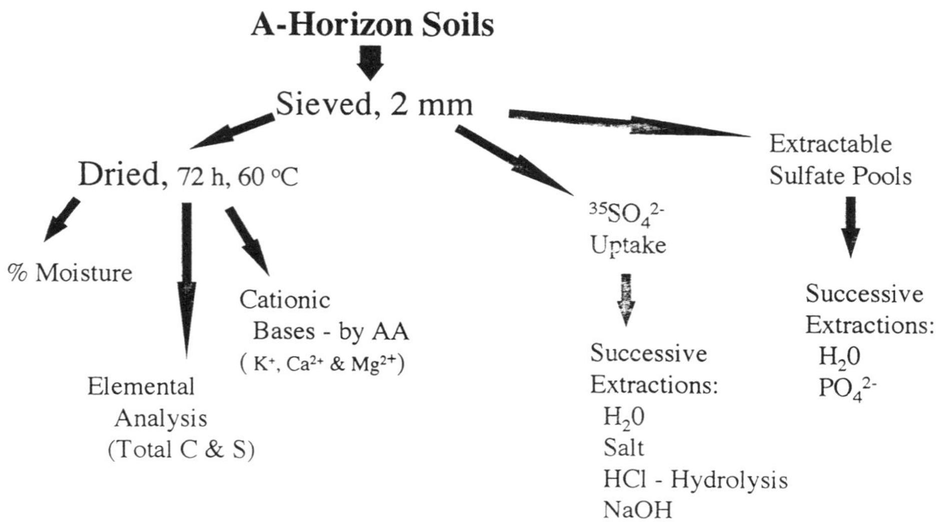

Fig. 3. Flow chart of sample processing for S transformation and exchangeable K^+ and Mg^{2+} studies in DRFS and MOFEP surface soils.

Radiolabel used in the DRFS study was quantified using a Beckman LS 5000 TA liquid scintillation counter, while a Wallac 1409 liquid scintillation counter was for all MOFEP samples. A biodegradable scintillation cocktail was used for all samples (Packard - Ultima Gold XR). Quenching of the samples was accounted for using quench curves, taking advantage of external quench monitoring techniques (Beckman's H-number; Wallac's SQPE number).

Soil total C (TC) and total S (TS) were quantified by combustion of about 200 mg of dried soils in a LECO CNS 2000 elemental analyzer. Sulfamethazine was used to standardize the instrument, with an National Institute of Standards and Technology-traceable soil standard used for drift correction. Sulfate in the WS pools of these soils was determined using a filtration extraction procedure (Spratt, 1997a). Pools of AdS were determined on the soil remaining in the syringe-filter assembly after the WS fraction had been washed from the soil. Concentrations of both WS and AdS were determined using ion chromatography (Watwood and Fitzgerald, 1988). Organic S (OS) present in the soil was calculated by difference (OS = TS - (WS + AdS)).

Exchangeable bases (K^+ and Mg^{2+}) present in A-horizon soils were determined using an ammonium acetate (1 M NH_4OAC, pH 7.0) extraction procedure (Simard, 1993) followed by atomic absorption spectrometry (Perkin-Elmer 1100B instrument for all DRSF samples; Varian Spectr AA10 for all MOFEP samples).

The relative ability of A-horizon soils, amended with various types of organic matter, to retain K^+ and Mg^{2+} was determined using an adsorption/desorption filtration procedure. Approximately 0.5 g of dried soil was added to the filter of a maxi-spin centrifugal filter device (50 cm^3, Osmonics, Livermore, CA). Organic compounds were added to the soils in a 1:40 ratio (dry mass basis) and included: 2-thiouracil, uracil, cellulose (fibrous, long, Sigma Chemical Co., St. Louis, MO), and lignin-alkali (kraft, Aldrich Chemical Co., Milwaukee, WI). These organic compounds were selected based on their degree of complexity and the presence or absence of sulfhydryl groups (2-thiouracil and lignin). As described by Stevenson (1986), compounds with sulfhydryl groups might be part of the larger fraction of soil OS that has been only poorly characterized. All of these organic compounds were found to be retained in the soils under conditions of the experiments. Following addition of the organic compounds to the soils, 5 mL of solutions containing either 2 mM Mg^{2+} or 0.4 mM K^+ were added to soil in the centrifugal filters. These concentrations of K^+ and Mg^{2+} were calculated as the approximate input of base via 30 days of precipitation in the Missouri Ozarks to an area equivalent to that occupied by the soil present in the centrifugal filter. After mixing, the soils were allowed to adsorb or desorb the bases for 16 hr. The filters were then centrifuged (5 min; 1,500 × g), collecting filtrate in the tube base. All soils were rinsed one additional time with 5 mL of distilled H_2O, and the K^+ and Mg^{2+} concentrations in the filtrates were determined by atomic absorption spectrometry.

Analysis of variance with a factorial design was used to determine significant trends in the data ($p < 0.05$). Pairwise *t*-tests were used for multiple comparisons when significant differences were found by the ANOVA. Relationships among variables were examined using a Pearson correlation analysis. Unless otherwise indicated, data are expressed on a per gram dry mass basis.

3. Results

Total S in MOFEP and DRSF control A-horizon soils was composed primarily of OS (95.2 % - 97.2 % of the TS, Table I, Figure 4). Large variations in both TS and OS occurred in MOFEP soils over the 3-yr study ($p = 0.001$). Soil TS concentrations ranged from 10 to 65 μmol g^{-1}, while OS concentrations ranged from 9 to 64 μmol g^{-1}.

Mean production of OS by soil microorganisms ranged from 2 to 20 nmol g^{-1} da^{-1} for all MOFEP A-horizon soils sampled (Figure 4). Although slight seasonal differences were evident in the data, the greatest difference in these data was observed in the interaction between year and season ($p = 0.020$).

Exchangeable K^+ and Mg^{2+} in MOFEP A-horizon soils exhibited noticeable seasonal differences ($p = 0.050$, and $p = 0.003$, respectively). Exchangeable K^+ concentrations ranged from 7 to 35 μmol g^{-1}, while exchangeable Mg^{2+} concentrations ranged from 12 to 76 μmol g^{-1} over the sample period (Figure 5). Exchangeable K^+ also had notable variation from year to year ($p = 0.001$). Seasonal trends in K^+ were found to correlate with OS and TC content of the A-horizon soils (r = 0.564 and r = 0.560, respectively). A very slight correlation was found between Mg^{2+} and TC (r = 0.154), but no correlation was observed between Mg^{2+} and OS.

Clear-cutting timber had noticeable effects on A-horizon soil TC , TS, and OS concentrations. DRSF plots that were clear-cut lost notable amounts of total C, compared with the controls ($p = 0.044$, Table I), with losses ranging from 22 % to 32 % for 2-3 yr and 8-10 yr post-harvest plots, respectively. A-horizon soil TS and OS loses from clear-cut plots were also evident ($p = 0.003$ and $p=0.029$, respectively). Loss of TS ranged from 28 % to 44 % of the control values for 2-3 yr and 8-10 yr post-harvest plots. OS losses ranged from 23 % to 41 % of the control values for 2-3 yr and 8-10 yr post-harvest plots.

TABLE I
Total C (TC), total S (TS), and organic S (OS) concentrations in A-horizon soils from experimental and control plots in Deer Run State Forest (mean values over two dates ± standard errors; n=30).

Treatment	Total C	Total S	Organic S
Control	9.9 ± 1.0	15.2 ± 1.5	13.7 ± 1.4
2-3 yr post-harvest	7.7 ± 0.9[a]	10.9 ± 1.4[b]	10.6 ± 1.4[c]
8-10 yr post-harvest	6.7 ± 0.9[a]	8.5 ± 1.2[b]	8.1 ± 1.2[c]

[a] significant treatment effect (compared with control, $p = 0.0438$)
[b] significant treatment effect (compared with control, $p = 0.0027$)
[c] significant treatment effect (compared with control, $p = 0.0292$)

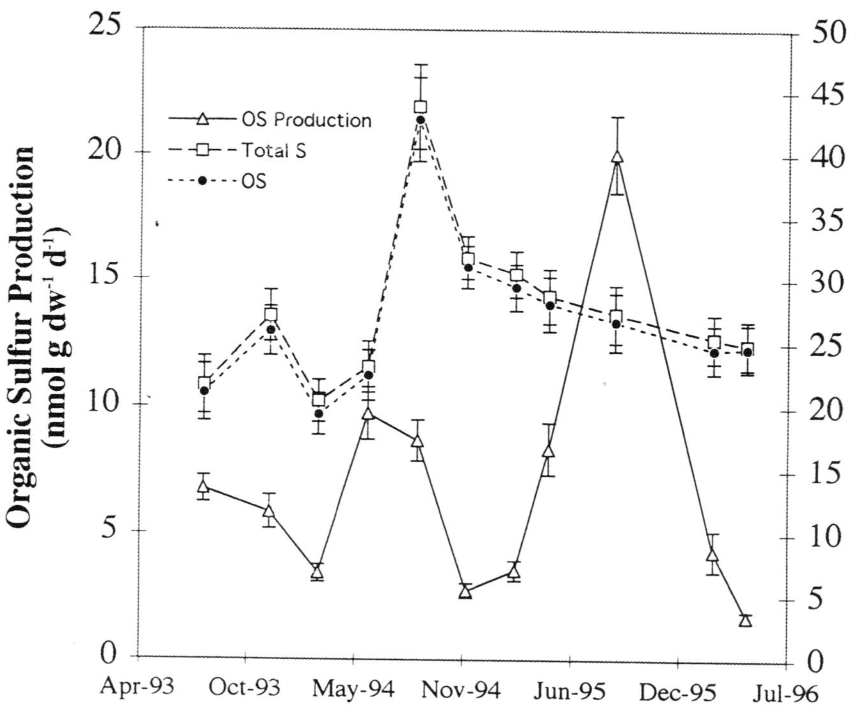

Fig. 4. Total S (TS) and organic S (OS) concentrations, and organic S production rates in MOFEP A-horizon soils, August 1993 to May 1996.

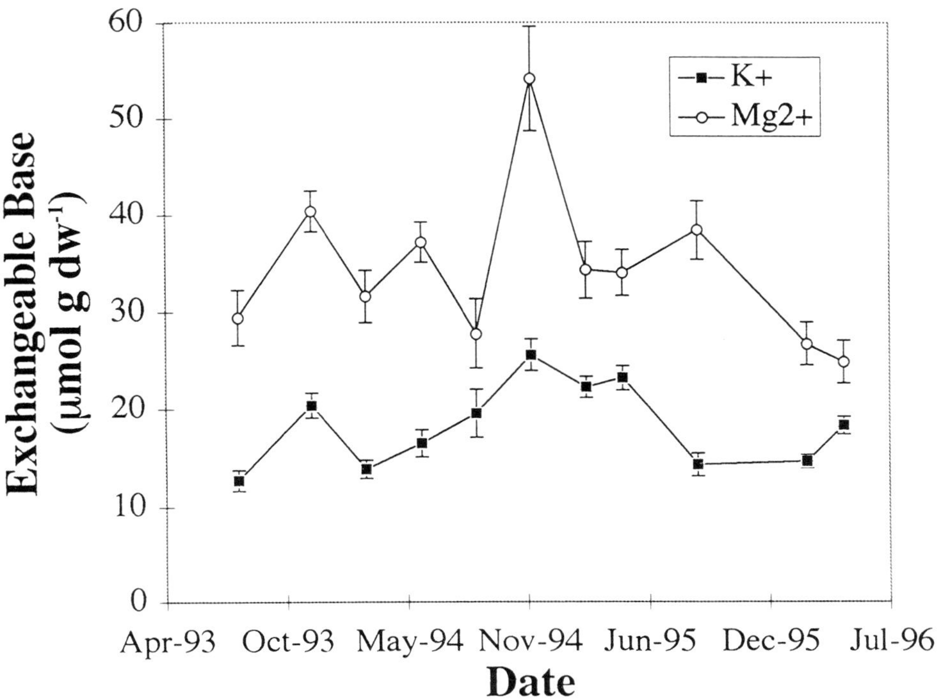

Fig. 5. Exchangeable K^+ and Mg^{2+} in MOFEP A-horizon soils, August 1993 to May 1996.

Microbial production of OS in A-horizon soils of DRSF plots was much lower in the disturbed plots than in the control plots. For the two fall sample dates, microorganisms in the DRSF A-horizon soils from the control plots produced OS at a mean rate of 9.7 nmol g^{-1} da^{-1} (Table II). These results compare favorably with rates of OS production observed in MOFEP plots during fall sampling periods (see Figure 4). Rates of OS production in DRSF A-horizon soils from the two harvested sites fell by 77-78% compared with rates for the control plots ($p < 0.050$).

TABLE II
Organic S (OS) production (nmol g^{-1} dry mass da^{-1}) in A-horizon soils from experimental and control plots in Deer Run State Forest (mean values over two dates ± standard errors; n=30).

Treatment	Organic S production (nmol g^{-1} da^{-1})
DRSF control	9.7 ± 1.8
DRSF 2-3 yr post-harvest	2.2 ± 0.4
DRSF 8-10 yr post-harvest	2.1 ± 0.7

Concentrations of exchangeable K^+ and Mg^{2+} in clear-cut DRSF A-horizon soils were also much lower than those in control plots. Losses of K^+ in A-horizon soils ranged from 22 % to 41 % for 2-3 yr and 8-10 yr post-harvest plots (Table III). Exchangeable Mg^{2+} concentrations in A-horizon soils of clear-cut plots exhibited greater loss compared with controls, ranging from 51 % to 68 % ($p < 0.001$).

TABLE III
Exchangeable K^+ and Mg^{2+} in A-horizon soils from experimental and control plots in Deer Run State Forest (mean values over two dates ± standard errors; n=30).

Treatment	Exchangeable K^+ (μmol g^{-1} dry mass)	Exchangeable Mg^{2+} (μmol g^{-1} dry mass)
DRSF control	13.7 ± 1.4	18.9 ± 2.0
DRSF 2-3 yr post-harvest	10.7 ± 1.4[a]	6.0 ± 0.8[b]
DRSF 8-10 yr post-harvest	8.1 ± 1.2[a]	9.2 ± 1.3[b]

[a] significant treatment effect (compared with control, $p < 0.001$).
[b] significant treatment effect (compared with control, $p < 0.001$).

Timber clear-cutting has resulted in a gradual increase in the C:S and C:OS ratios in A-horizon soils of DRSF 2-3 yr and 8-10 yr post-harvest plots, compared with controls (Table IV). For C:S, the trend toward greater C concentrations has been constant from control to 8-10 yr post-harvest ($p < 0.001$). C:OS also was markedly greater 8-10 yr post-harvest ($p < 0.001$), however, the greatest change in this ratio occurred between 2-3 yr and 8-10 yr post-harvest.

TABLE IV
Elemental ratios (mol mol^{-1}) in Deer Run State Forest A-horizon soils (mean values over two dates ± standard errors; n=30).

Treatment	C:TS	C:OS	C:K	C:Mg	TS:K	TS:Mg	OS:K	OS:Mg
DRSF control	651.2 ± 24.5	697.8 ± 29.1	1769.2 ± 136.5	746.9 ± 142.3	2.76 ± 0.21	1.10 ± 0.19	2.63 ± 0.21	1.02 ± 0.18
DRSF 2-3 yr post-harvest	731.6 ± 31.6	750.9 ± 32.4	2337.2 ± 195.4	1436.7 ± 147.2	3.23 ± 0.26	2.00 ± 0.22	3.16 ± 0.26	1.95 ± 0.21
DRSF 8-10 yr post-harvest	829.0 ± 34.5	871.5 ± 35.8	1979.4 ± 191.4	838.7 ± 95.2	2.42 ± 0.24	1.06 ± 0.14	2.32 ± 0.24	1.01 ± 0.13

Clear-cutting also resulted in interesting trends in the ratios of C:K, C:Mg, TS:K, TS:Mg, OS:K, and OS:Mg in DRSF A-horizon soils (Table IV). In all cases, the elemental molar ratios were greatest in the 2-3 y post-harvest plots (all $p < 0.010$), with the ratios returning to approximately those of control values 8-10 y post-harvest.

Amendment of MOFEP A-horizon soils with different types of organic matter resulted in the loss of K^+ for all of the organic compounds tested (Figure 6). Large losses of K^+ were observed for uracil- ($p = 0.047$), cellulose- ($p = 0.026$), and lignin-amended ($p =$

0.062) soils, compared with the controls. Organic matter amendment of these soils had the opposite effect on retention of Mg^{2+} (Figure 6). Slightly increased retention of Mg^{2+} was observed for cellulose amendment (p = 0.060), while lignin amendment resulted in nearly 10 % greater retention of Mg^{2+} than the control (p = 0.001). The mean soil solution pH in controls of this study was 5.63 ± 0.19 (± standard deviation; n=6). Experimental soil solution pH values ranged from 5.42 to 5.93.

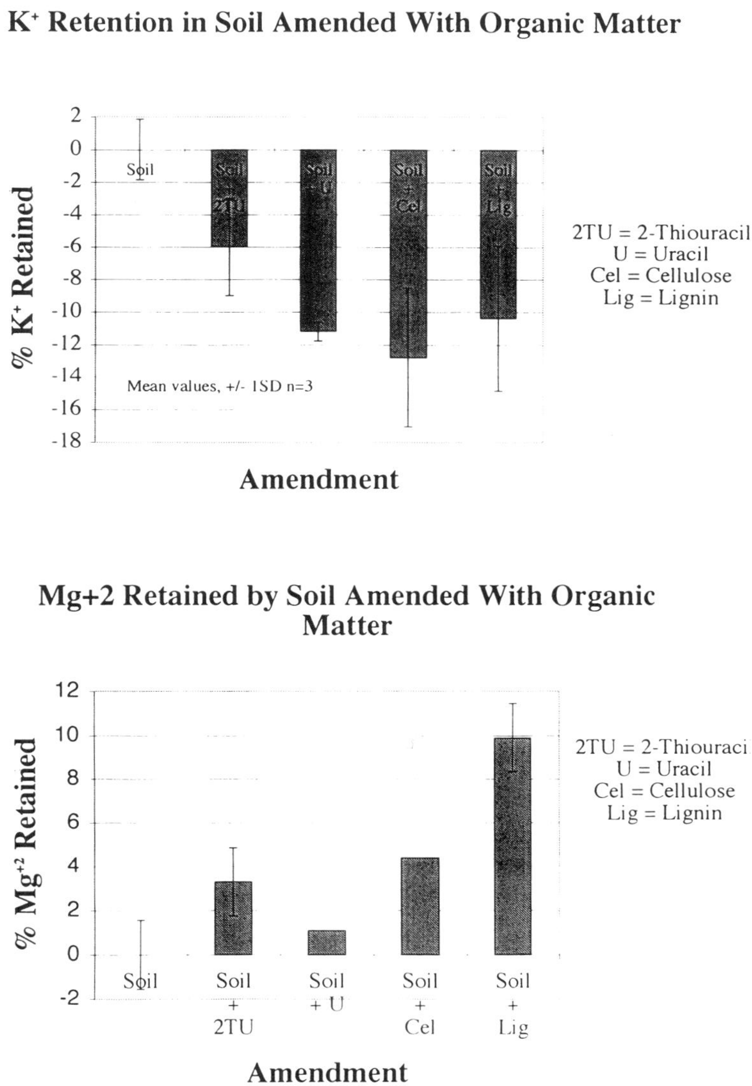

Fig. 6. K^+ and Mg^{2+} retention in soil amended with organic matter.

4. Discussion

Organic S has been found to make up the greatest portion of soil TS in most forests studied, including the surface soils of Missouri Ozark forests in this study. Seasonal and year by year trends in the OS pools of forest A-horizon soils observed here indicate that the system is dynamic, involving both microbial and vegetational inputs (see Figure 1). Microbial production of OS in DRSF soils has been demonstrated by Spratt (1997a). Many others have demonstrated that microorganisms are required for the production of OS in different forest soils (David *et al.*, 1982; Fitzgerald *et al.*, 1983; Schindler *et al.*, 1986; Strick *et al.*, 1982; Strickland and Fitzgerald, 1984; Swank *et al.*, 1984; Watwood *et al.*, 1993). The relative importance of microbial OS production to A-horizon soil OS pools appears to be much less than that of leaf litter. The largest change in A-horizon soil OS observed in this study occurred for MOFEP plots between early June and late September 1994 (leaf drop was well on its way, see Figure 4). During this period the concentration of OS in A-horizon soils nearly doubled (an increase of 20 μmol g^{-1} in 3.5 mo). During this time period, the rate of microbial OS production for MOFEP A-horizon soils decreased by nearly 15 %. This decrease in microbial OS production rate may have reflected surface soil temperature differences between June (24 °C) and September (18 °C), which would be expected of a microbial process. During the summer months, however, the average surface soil temperatures determined for nearby sites remained fairly constant, approximately 20.6 °C, from June to September (Chen *et al.*, 1997). This suggests that microbial OS production in surface soils should be fairly constant during the summer. Integrating the highest rate of production of OS by the A-horizon soil microbial community during this time period (June 1994 data) over time, it would take approximately 2,000 days to produce 20 μmol g^{-1} of OS. This calculation is at best an underestimation of potential OS production by microorganisms in these soils, but it serves to emphasize the importance of plant-derived OS to surface soil OS pools.

In undisturbed Missouri Ozark forest A-horizon soils a large percentage of the variation in exchangeable K^+ over the three-year study could be explained by variation in OS and total C. Only some of the variation in A-horizon soil exchangeable Mg^{2+} could be explained by variation in OS and total C. Determining how organic matter affects these bases in A-horizon soils is one of the major objectives of this research.

Watwood *et al.* (1993) have demonstrated that forest ecosystems that lose OS from their A-horizon soils also lose nutrient cations, such as Ca^{2+}, K^+ and Mg^{2+}. In this study, natural seasonal and yearly variation in OS, K^+ and Mg^{2+} and timber clear-cutting all resulted in the loss of OS, K^+ and Mg^{2+} from A-horizon soils. For soils from different forest types, retention of SO_4^{2-} has been found to reduce leaching of cations (Johnson *et al.*, 1982; Rechcigl and Sparks, 1985; Richter *et al.*, 1983; Wilklander, 1980; Watwood *et al.*, 1993). In the forest surface soils studied here, uptake by vegetation and microbial production of OS appear to be the principal methods of SO_4^{2-} retention. Because the major contributor of OS to A-horizon soils studied here appears to be vegetation, a mechanism for loss of K^+ and Mg^{2+} in the DRSF clear-cut sites may involve reduced litter contributions to the soil over the years following harvest. Thinner litter layers were observed in DRSF clear-cut sites several years post-harvest (Spratt, 1997a). This would result in less plant derived OS and K^+ and Mg^{2+} reaching the A-horizon soils each year. Microbial OS production may not substitute for the lost OS contributions from vegetation for several possible reasons. First, after 2 years most of the easily utilizable organic substrates that these microbes would use would be gone (Lundgren, 1982; Pietikäinen and Fritze, 1995), which would lower potential growth of these cells. Second, reduced concentrations of K^+ and Mg^{2+} also may negatively affect microbial growth rates. Finally, the soils of these south-southwest facing slopes would potentially go through much greater variation in moisture content - being wetter during wet periods due to lost evapotranspiration, and dryer during dry periods due to greater evaporation - following cutting. As a result, less of the incoming K^+ and Mg^{2+} from precipitation or plant litter would be available to the microbial community. These factors, in turn, would reduce the amount of OS that microorganisms could produce in the soil. Because the vegetation's role in OS production is greatly reduced in the clear-cut sites, the system gradually loses OS and K^+ and Mg^{2+} over time. Data suggesting that this mechanism may be at work in the DRFS sites may be found in comparisons of the C:S ratios for A-horizon soils. As OS has been lost from the clear-cut soils over time, any new organic matter added to the soil appears to be reduced in S content. Whether this loss of OS is reversible, since control sites have previously been cut, and the C:S ratio and OS content of control soils is much greater than the recently cut soils, is yet to be seen.

Another change that may modify A-horizon soil OS in DRSF clear-cut sites is the quality of litter in the years immediately following harvest. The reported density of herbaceous species in clear-cut sites 4-5 years post harvest is much greater than in mature forests, which are dominated by oak species (Annand and Thompson, 1997). This shift from oak to herbaceous OS may affect cation exchange in surface soils. Analysis of the OS present in herbaceous litter compared with oak litter is currently underway.

One other potential mechanism that may help explain the relationship between OS and nutrient cation retention in forest A-horizon soils might be that sulfhydryl groups of OS serve as better cation exchange sites than other forms of ionizable organic matter present in the soil (e.g., phenolics). Retention of K^+ and Mg^{2+} by organic matter of different complexity yielded conflicting evidence regarding the role that OS plays as a cation exchanger in these surface soils. In comparing retention of K^+ and Mg^{2+} for organic substances tested here, different results were obtained for the two cations. For K^+ it was not clear whether 2-thiouracil behaved more like the unamended soil with regard to K^+ retention than uracil due to the presence of sulfhydryl groups. Even if the sulfhydryl groups of 2-thiouracil have some effect on K^+ retention in these soils, this doesn't explain why K^+ was not retained by the soil. It is possible that conditions in these experiments may have lead to the removal of exchangeable K^+ from these soil/organic compound mixtures regardless of the amounts of organic matter added, as a characteristic of ultisols and alfisols is their limited K supplying power (Hausenbuiller, 1978). Retention of Mg^{2+} was much greater in soils amended with organic matter than in controls. Marginally greater retention of Mg^{2+} was observed in 2-thiouracil-amended soils than in uracil-amended soils, suggesting that sulfhydryl groups provided useful ion exchange sites for Mg^{2+}. Addition of lignin led to much greater Mg^{2+} retention within the soil. Lignin contains both sulfhydryl and other ionizable organic sites, making comparisons of cation retention difficult.

Evidence that sulfhydryl groups of OS may not be especially important in the retention of cations as cation exchange sites in the surface soils tested here may be found in comparisons of the relative amounts of soil TC and TS. Total C is present in these A-horizon soils at concentrations from 600 to 900 times greater than OS. This suggests that the larger organic pool containing no S, by its very size, may outweigh any additional ion exchange benefit obtained by having sulfhydryl groups present. Studies of other forms of organic matter with and without sulfhydryl groups may help determine the exact role that OS plays as a cation exchange site in forest soils. One certainty, however, for A-horizon soils studied here the rapid seasonal cycling of OS suggests that SO_4^{2-} uptake/retention by vegetation and microorganisms may be the most important mechanism leading to retention of cations.

Other forest management techniques currently being implemented on a wider scale might help reduce the loss of nutrients from forest surface soils. One such technique, selective harvest, has the potential advantage of retaining some of the forest canopy following harvest. This might help provide a litter source that would help keep the OS, K^+ and Mg^{2+} content of surface soils higher. Beginning in May 1996, one-third of the MOFEP sites studied here had timber harvested using selective cutting techniques, one-third of the sites were clear-cut, and one-third of the sites were left as controls. Follow up studies on these sites underway now should help provide useful information regarding the impact these timber harvest techniques have on the relationship between surface soil OS and the availability of K^+ and Mg^{2+}.

Acknowledgments

This work was supported by grants from the Missouri Department of Conservation and the U.S. Department of Agriculture, NRICGP. The help of Ms. Anne Spratt with statistical analyses was especially appreciated.

References

Annand, E.M. and Thompson, F.R.: 1997, *J. Wildl. Manage.* **61**, 159.

Brookshire, B.L., Jensen, R. and Dey, D.C.: 1997, 'The Missouri Ozark Forest Ecosystem Project: Past, Present and Future', in B.L. Brookshire and S.R. Shifley (eds.), *Proceedings of the Missouri Ozark Forest Ecosystem Project Symposium: An Experimental Approach to Landscape Research,* Gen. Tech. Rep. GTR NC-193, U.S. Department of Agriculture, Forest Service, North Central Forest Experiment Station St. Paul, MN (in press).

Chen, J., Xu, M. and Brosofske, K. D.: 1997, 'Microclimatic Characteristics in Southern Missouri Ozarks', in B.L. Brookshire and S.R. Shifley (eds.), *Proceedings of the Missouri Ozark Forest Ecosystem Project Symposium: An Exper-imental Approach to Landscape Research,* Gen. Tech. Rep. GTR NC-193, U.S. Department of Agriculture, Forest Service, North Central Forest Experiment Station St. Paul, MN (in press).

David, M.B., Mitchell, M.P. and Nakas, J.P.: 1982, *Soil Sci. Soc. Am. J.* **46,** 847.

Fitzgerald, J.W., Ash, J.T.,and Strickland, T.C.: 1983,*Can. J. For. Res.* **13,** 1077.

Hausenbuiller, R.L.: 1978, *Soil Science, Principles and Practices,* 2nd edition, Wm. C. Brown Co., Dubuque, IA.

Johnson, D.W., Henderson, G.S., Huff, D.D., Lindberg, S.E., Richter, D.D., Shriner, D.S., Todd, D.E. and Turner, J.: 1982, *Oecologia* **54,** 141.
Johnson, D.W.: 1992, 'Base Cation Distribution and Cycling', in D.W. Johnson and S.E. Lindberg (eds.), *Atmospheric Deposition and Forest Nutrient Cycling*, Springer-Verlag, New York.
Kabrick, J., Larsen, D. and Shifley, S.: 1997, 'Analysis of Pretreatment Woody Vegetation and Environmental Data for the Missouri Ozark Forest Ecosystem Project', in B.L. Brookshire and S.R. Shifley (eds.), *Proceedings of the Missouri Ozark Forest Ecosystem Project Symposium: An Exper-imental Approach to Landscape Research,* Gen. Tech. Rep. GTR NC-193, U.S. Department of Agriculture, Forest Service, North Central Forest Experiment Station St. Paul, MN (in press).
Kennedy, I.R.: 1986, *Acid Soil and Acid Rain: The Impact on the Environment of Nitrogen and Sulphur Cycling,* John Wiley & Sons, Inc., New York.
Lundgren, B.: 1982, *Soil Biol. Biochem.* **14**, 537.
Mitchell, M.J. and Lindberg, S.E.: 1992, 'Sulfur Chemistry, Deposition, and Cycling in Forests', in D.W. Johnson. and S.E. Lindberg (eds.), *Atmospheric Deposition and Forest Nutrient Cycling,* Springer-Verlag, New York.
Mitchell, M.J., David, M.B. and Harrison, R.B.: 1992, 'Sulphur Dynamics of Forest Ecosystems', in R.W. Howarth, J.W.B. Stewart and M.V. Ivanov (eds.), *Sulphur Cycling on the Continents*, John Wiley & Sons, New York, pp. 215-254.
Pietikäinen, J. and Fritze, H.: 1995, *Soil Biol. Biochem.* **27**, 101.
Rechcigl, J.E. and Sparks, E.L.: 1985, *Comm. Soil Sci. Plant Anal.* **16,** 653.
Richter, D.D., Johnson, D.W. and Todd, D.E.: 1983, *J. Environ. Qual.* **12**, 263.
Schindler, S.C., Mitchell, M.J., Scott, T.J., Fuller, R.D. and Driscoll, C.T.: 1986, *Soil Sci. Soc. Am. J.* **50,** 457.
Simard, R.R.: 1993, 'Ammonium Acetate-extractable Elements', in M.R. Carter (ed.), *Soil Sampling and Methods of Analysis*, Lewis Publishers, Boca Raton.
Spratt, H.G., Jr.: 1997a, *Soil Biol. Biochem.* **29**, 1639.
Spratt, H.G., Jr.: 1997b, 'Aspects of Carbon and Sulfur Transformations in MOFEP Surface Soils: Pretreatment Conditions', in B.L. Brookshire and S.R. Shifley (eds.), *Proceedings of the Missouri Ozark Forest Ecosystem Project Symposium: An Experimental Approach to Landscape Research,* Gen. Tech. Rep. GTR NC-193, U.S. Department of Agriculture, Forest Service, North Central Forest Experiment Station St. Paul, MN (in press).
Stevenson, F.J.: 1986, *Cycles of Soil - Carbon, Nitrogen, Phosphorus, Sulfur, Micronutrients,* John Wiley & Sons, Inc., New York.
Strick, J.E., Schindler, S.C., David, M.B., Mitchell, M.J. and Nakas, J.P.: 1982, *Northeast. Environ. Sci.* **1,** 161.
Strickland, T.C. and Fitzgerald, J.W.: 1984, *Biogeochemistry* **1,** 79.
Swank, W.T., Fitzgerald, J.W. and Ash, J.T.: 1984, *Science* **223,** 182.
Watwood, M.E. and Fitzgerald, J.W.: 1988, *Soil Sci. Soc. Am. J.* **52,** 1478.
Watwood, M.E., Sommer, AS. and Fitzgerald, J.W.: 1993, *Soil Sci. (Trend. Agric. Sci.)*, **1**,103.
Wilklander, L.: 1980, 'Interaction between Cations and Anions Influencing Adsorption and Leaching', in T.C. Hutchinson, and M. Havas (eds.), *Effects of Acid Precipitation on Terrestrial Ecosystems,* Plenum Press, New York.

RECOVERY OF SURFACE WATERS IN THE NORTHEASTERN U.S. FROM DECREASES IN ATMOSPHERIC DEPOSITION OF SULFUR

C.T. DRISCOLL[1], G.E. LIKENS[2] and M.R. CHURCH[3]

[1]*Department of Civil and Environmental Engineering, Syracuse University, Syracuse NY 13244, USA*
[2]*Institute of Ecosystem Studies, Box AB, Millbrook, NY 12545, USA*
[3] *U.S. EPA, 200 SW 35th Street, Corvallis, OR 97333, USA*

Abstract. A simple mass flux model was developed to simulate the response of SO_4^{2-} concentrations in surface waters to past and anticipated future changes in atmospheric deposition of SO_4^{2-}. Values of bulk (or wet) SO_4^{2-} deposition and dry deposition of S determined from measured air concentrations and a deposition velocity were insufficient to balance watershed SO_4^{2-} export at the Hubbard Brook Experimental Forest, NH and for a regional survey of watersheds in the northeastern U.S. We propose two explanations for the unmeasured S source: 1) a significant underestimation of dry S deposition, and/or 2) internal watershed S sources, such as weathering and/or mineralization of soil organic S. Model simulations based on these two mechanisms agreed closely with measured stream SO_4^{2-} concentrations at Hubbard Brook. Close agreement between measured and model predicted results precluded identification of which of the two mechanisms controlled long-term trends in stream SO_4^{2-}. Model simulations indicated that soil adsorption reactions significantly delayed the response of stream water to declines in SO_4^{2-} inputs since 1970, but could not explain the discrepancy in watershed S budgets. Extrapolation of model predictions into the future demonstrates that uncertainty in the source of the S imbalance in watersheds has important implications for assessments of the recovery of surface water acid neutralizing capacity in response to anticipated future reductions in SO_2 emissions.

Keywords: acid neutralizing capacity, Hubbard Brook Experimental Forest, mass balance model, reduced SO_2 emission, surface water sulfate, sulfur deposition

1. Introduction

Sulfate (SO_4^{2-}) is the major anion in precipitation and in low acid neutralizing capacity (ANC) surface waters that have been influenced by acidic deposition. Sulfate, in the form of sulfuric acid, is the major source of acidity contributing to the acidification of surface waters in the northeastern U.S. (Driscoll, 1991). It has been assumed generally that SO_4^{2-} concentrations in low ANC surface waters are derived from atmospheric deposition, and that inputs of SO_4^{2-} to glaciated watersheds in the northeastern U.S. are transported relatively conservatively through soil with drainage waters (Galloway *et al.*, 1983; Likens *et al.*, 1990; Rochelle *et al.*, 1987; Church *et al.*, 1989).

Several watershed studies in the northeastern U.S. have shown that stream outflow of SO_4^{2-} exceeds S inputs from wet (bulk) deposition and estimates of dry deposition based on net throughfall or inferred from measured air concentrations and application of a deposition velocity model (inferential method; Norton *et al.*, 1988; Driscoll *et al.*, 1995; Mitchell *et al.*, 1996). Lovett *et al.* (1992) compared three approaches to estimating dry deposition of S at the Hubbard Brook Experimental Forest (HBEF). These methods included: 1) dry deposition of S using the inferential method, 2) measurements of net throughfall S, and 3) the discrepancy in watershed mass balance of S. These three methods showed large differences in estimates of dry deposition of S, with the watershed mass balance approach giving substantially higher values than the other two methods.

Water, Air, and Soil Pollution **105:** 319–329, 1998.

It is critical to understand the controls on SO_4^{2-} concentrations in order to assess the rate and extent of recovery of low ANC surface waters in response to decreases in atmospheric deposition of SO_4^{2-}. There are three possible explanations for the discrepancy in watershed S mass balances reported in the above studies: 1) dry deposition of S measured by the inferential method or net throughfall highly underestimates the actual inputs of dry S deposition for watersheds in the northeastern U.S. (Norton *et al.* 1988; Church *et al.*, 1989; Likens *et al.*, 1990); 2) SO_4^{2-} that accumulated in soil during an earlier period of elevated S deposition is desorbed (or released) to drainage waters (Driscoll *et al.*, 1995; Mitchell *et al.*, 1996); and 3) SO_4^{2-} is supplied to drainage waters through mineral weathering or from mineralization of organic S in soil (Novák *et al.*, 1996). In this study we used a biogeochemical model, PnET-CN/CHESS, to investigate these three explanations for long-term trends in surface water SO_4^{2-} concentrations at the HBEF and for a regional survey of watersheds in the northeastern U.S.

2. Methods

Abiotic processes regulating SO_4^{2-} concentrations in surface waters were modeled using output from the CHESS (Santore and Driscoll, 1995) and PnET-CN (Aber *et al.*, 1997) models. A mass flux model was developed using hydrologic data and SO_4^{2-} concentrations from the linked PnET-CN/CHESS model. Measured or estimated monthly fluxes of precipitation and atmospheric deposition of S were routed to a hypothetical forest system with SO_4^{2-} adsorption properties calibrated to mineral soil samples collected at the study sites. Sulfate adsorption reactions were simulated using CHESS parameterized for soils from the study sites. Simulated surface water SO_4^{2-} concentrations were compared with monthly samples for watershed 6 (w6) at the HBEF for the period 1964-65 to 1993-94, and with synoptic data for watersheds in the northeastern U.S. from the U.S. Environmental Protection Agency Direct/Delayed Response Project (DDRP; Brakke *et al.*, 1988; Church *et al.*, 1989).

During winter months, precipitation fluxes of water and SO_4^{2-} from atmospheric deposition were added to snowpack storage. Precipitation falling as rain plus snowmelt, if available, was partitioned between infiltration, or "fast flow" after subtracting evaporative and transpiration losses using PnET-CN. Water that infiltrated the soil surface was assumed to be mixed completely with existing soil water, while fast flow water was assumed to have negligible soil contact and directly entered the stream channel. The composition of the soil water was used to calculate SO_4^{2-} adsorption reactions and dissolved SO_4^{2-} concentrations using CHESS. Water and dissolved SO_4^{2-} then was drained from the soil water pool and mixed with "fast flow" water in the stream channel.

We have observed a strong empirical relationship between estimates of SO_2 emissions for the eastern U.S. (including the Ohio Valley) and annual volume-weighted concentration of SO_4^{2-} in bulk precipitation and stream water at the HBEF (Figure 1; Butler and Likens, 1991). Based on the relationship for bulk deposition, historical (1850-1964) concentrations of SO_4^{2-} were assumed to be proportional to emissions of SO_2 in the eastern U.S. Estimates of post-Industrial Revolution SO_2 emissions were used to reconstruct historical bulk deposition inputs of SO_4^{2-} (Husar *et al.*, 1991; Nizich *et al.*, 1996). Measured bulk deposition inputs were used from the start of deposition monitoring (1964-65) at the HBEF to the present. For the DDRP watersheds, we estimated wet deposition of SO_4^{2-} using the spatial model of Ollinger *et al.* (1993). Historical wet deposition of SO_4^{2-} for the DDRP watersheds also was assumed to be proportional to historical patterns of SO_2 emissions for the eastern U.S.

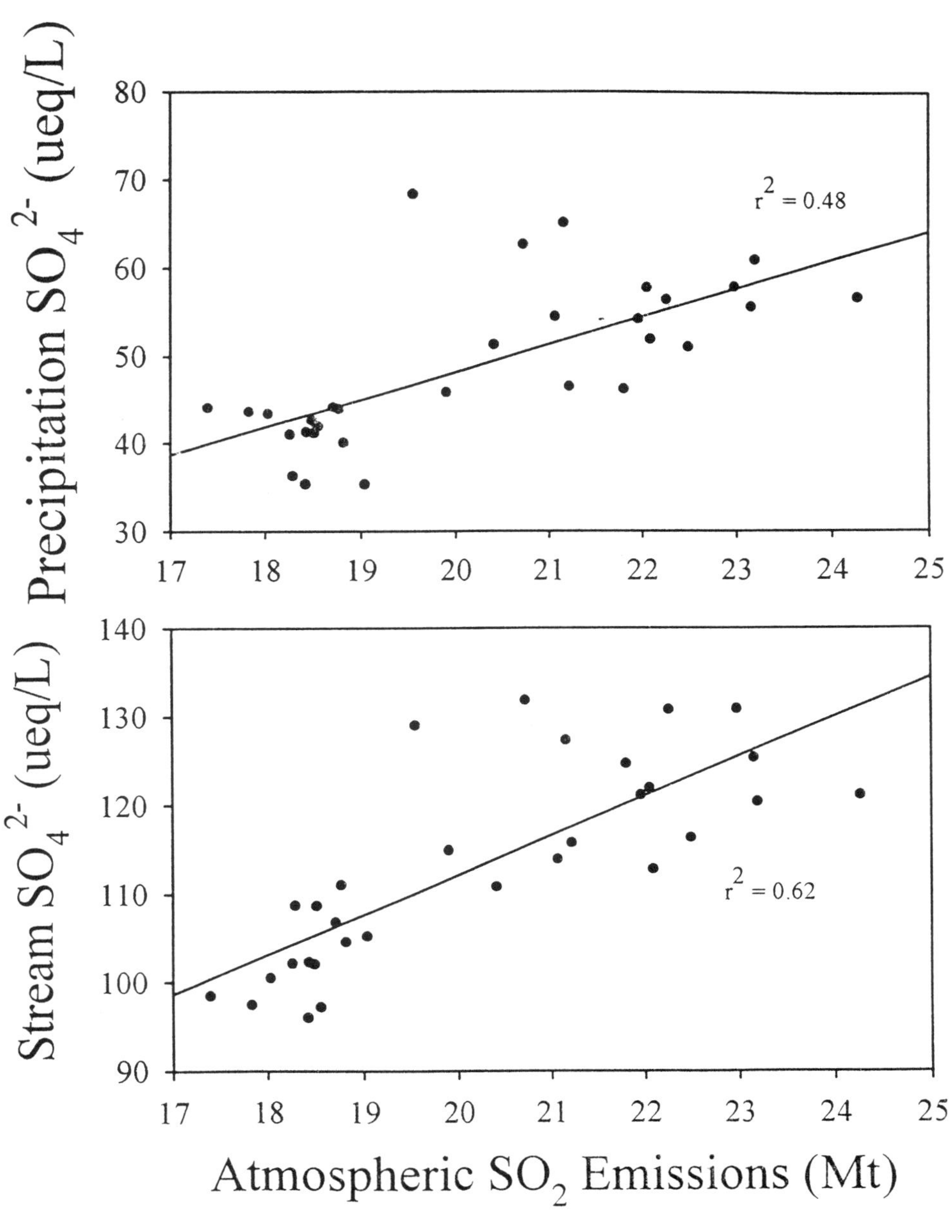

Fig. 1. Annual volume-weighted concentrations of SO_4^{2-} in bulk precipitation and stream water at the Hubbard Brook Experimental Forest, NH as a function of annual SO_2 emissions in the eastern U.S. (Nizich *et al.*, 1996).

Two other sources of SO_4^{2-} inputs were considered in addition to measured wet (or bulk) deposition. Dry deposition is a significant source of SO_4^{2-} in forest catchments of the eastern U.S. For example, at the HBEF, dry S deposition determined using the inferential method was 21 % of bulk deposition inputs (112 eq ha^{-1} yr^{-1}; Lovett *et al.*, 1992). In this study, we assumed that changes in SO_2 emissions for the eastern U.S. would result in the same relative change for dry S deposition as bulk (or wet) SO_4^{2-} deposition. Dry S deposition was calculated as a constant percentage of bulk (or wet) SO_4^{2-} deposition over the entire historical period for the HBEF and the DDRP watersheds. Dry deposition inputs to the DDRP watersheds were estimated using the Ollinger *et al.* (1993) model.

Despite the discrepancy in S budgets of forest catchments of the northeastern U.S. (Norton *et al.*, 1988; Likens *et al.*, 1990), relatively little attention has been given to either the weathering of soil S minerals (e.g. FeS) or the mineralization of soil organic S pools (Novák *et al.*, 1996) as a source of SO_4^{2-} to streamwater. Likens *et al.* (1990) considered and calculated that the release of S by mineral weathering was small at the HBEF. The soil S pool at Hubbard Brook is 46 kmol ha^{-1}, with 93 % of the S in the mineral soil occurring as an organic form and 6 % as adsorbed SO_4^{2-} (Mitchell *et al.*, 1989). In this study we lumped weathering and mineralization of organic S as a possible internal source of S contributing to the watershed SO_4^{2-} budget.

Sulfate adsorption reactions were calibrated to laboratory adsorption data measured for mineral soil at the HBEF (Fuller *et al.*, 1987) and the DDRP watersheds (Church *et al.*, 1989) using a simple Langmuir adsorption model (Sposito, 1989). For the HBEF simulations, the soil site density (adsorption maximum) for SO_4^{2-} was 30 μeq g^{-1}, the log soil adsorption constant was 3.32, the soil mass per unit area was 300 kg m^{-2}, and the soil depth was 1 m (Johnson *et al.*, 1991). Because there are no historical records of soil SO_4^{2-} pools to use as initial conditions for model simulations, we relied on the long-term accumulation of adsorbed soil SO_4^{2-} as calculated by the CHESS model.

3. Results and Discussion

The model, PnET-CN/CHESS, was used to evaluate whether measured/estimated values of atmospheric S deposition (bulk and dry) and soil SO_4^{2-} adsorption reactions could simulate the long-term trends of surface water SO_4^{2-} concentrations at the HBEF (Table I; Figure 2; Case 1). Watershed supply of SO_4^{2-} from weathering/mineralization was assumed to be zero for this simulation. Model simulations qualitatively depicted higher stream SO_4^{2-} concentrations in the 1960's and the decreases that have occurred since the early 1970's. However, model simulations substantially underestimated stream concentrations and loss of SO_4^{2-} at the HBEF. The mean difference between annual volume-weighted measured and model calculated concentrations over the simulation period was 32 μeq L^{-1}. Simulations for the DDRP watersheds (n = 126) showed a spatial pattern of decreasing concentrations of SO_4^{2-} in surface waters from west-to-east across the northeastern U.S. that qualitatively was consistent with both the spatial pattern of atmospheric S deposition (Ollinger *et al.*, 1993) and measured concentrations in surface waters (Figure 3). As shown for the HBEF, model predictions underestimated SO_4^{2-} concentrations in surface waters for the northeastern U.S. (for 1984) with a mean difference between mean and model predicted concentrations of SO_4^{2-} of 28 μeq L^{-1}.

TABLE I
Inputs of S for simulations of stream SO_4^{2-} at the HBEF.

	Bulk SO_4^{2-} Deposition	Dry S Deposition	Weathering/Mineralization of S
Case 1	Measured bulk deposition	Measured for site (21 % of bulk deposition)	None
Case 2	Measured bulk deposition	Increased to balance annual SO_4^{2-} budget for the first four years or study (53 % of bulk deposition)	None
Case 3	Measured wet or bulk deposition	Measured for site (21 % of bulk deposition)	Increased to balance annual SO_4^{2-} budget for the first 4 yr of study (264 eq ha^{-1} yr^{-1})

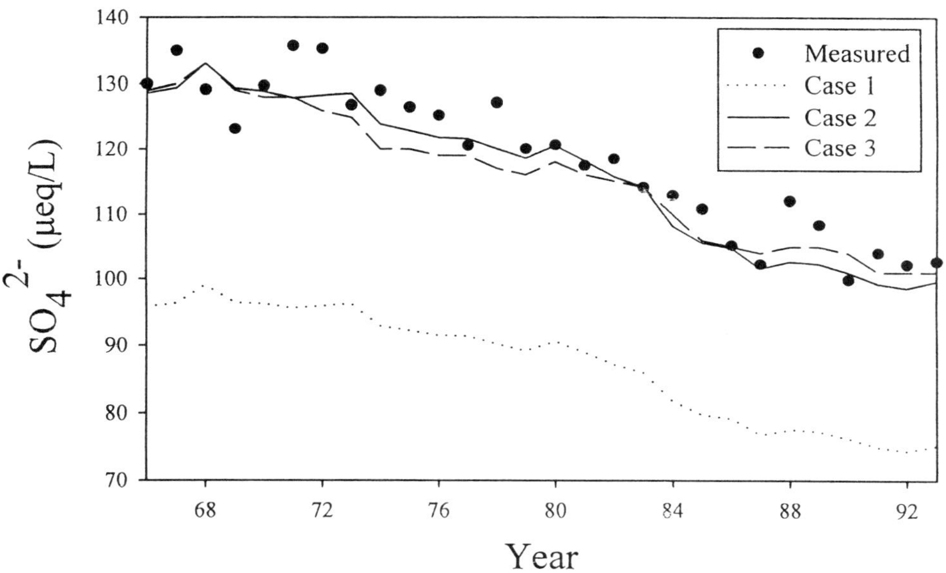

Fig. 2. A comparison of measured and model predictions of annual volume-weighted concentrations of SO_4^{2-} in streamwater of watershed 6 at the Hubbard Brook Experimental Forest, NH. Model predictions were made assuming: measured inputs of bulk and dry S deposition (21 % of bulk deposition) and no weathering/mineralization inputs of S (Case 1), the discrepancy in the watershed S budget is due to unmeasured dry S deposition (53 % of bulk deposition; Case 2) and the discrepancy in the watershed S budget is due weathering/mineralization S inputs (264 eq ha^{-1} yr^{-1}; Case 3).

Additional simulations were conducted with PnET-CN/CHESS to attempt to explain the discrepancy in watershed S balance at the HBEF (Table I). First (Case 2), the discrepancy in the annual S budget was assumed to be due to an underestimate of dry S deposition. The model was "calibrated" by increasing the contribution of dry deposition of S for the first four years of the study (1964-65 to 1968-69), so measured drainage concentrations of SO_4^{2-} matched simulated values. Under this condition, dry deposition of S was increased to 53 % of bulk deposition values. Second (Case 3), we assumed that the wet deposition and dry deposition determined from the inferential method were accurate (and therefore not adjusted), but a constant weathering/mineralization flux of S (264 eq ha^{-1} yr^{-1}) was added to balance the annual budget over the initial four years of the streamwater record at the HBEF. The calibration period for both assumptions (Cases 2 and 3) was restricted to the initial four years of the monitoring record so that the remaining period of the drainage water record could be used for model evaluation.

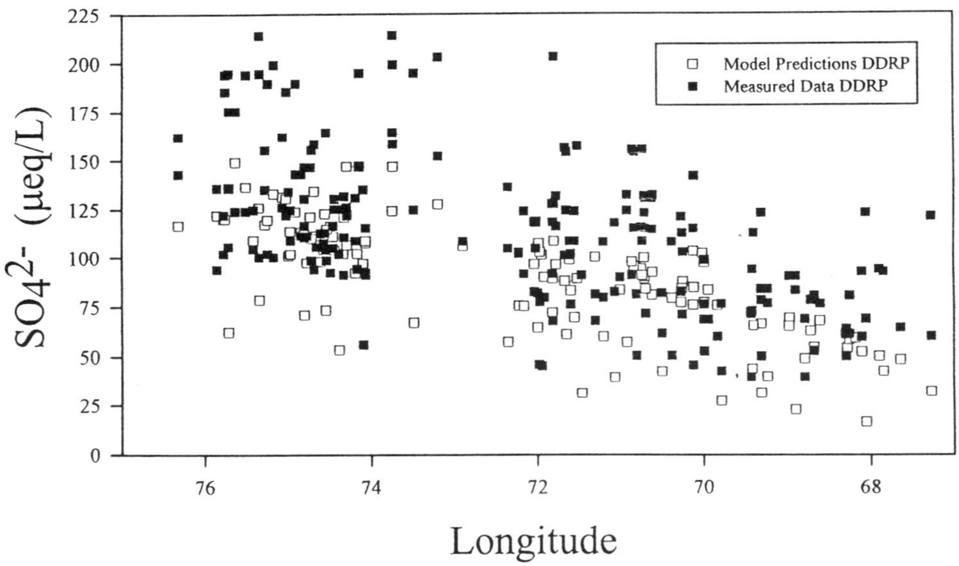

Fig. 3. Model predicted and measured concentrations of SO_4^{2-} in watersheds as a function of longitude in the northeastern U.S. for watersheds of the U.S. Environmental Protection Agency Direct/Delayed Response Program.

Under the assumptions that the discrepancy in the watershed S budget was due to: 1) additional dry deposition of S (Case 2), or 2) inputs of weathering/mineralization of S (Case 3), the simulated concentrations and trends of annual stream SO_4^{2-} concentrations compared closely with measured values (Figure 2; Case 2 and 3). Seasonal patterns in simulated SO_4^{2-} concentrations also agreed well with measured results (results not shown). Although Case 2 (additional dry S deposition) and Case 3 (additional S from weathering/ mineralization) simulations both compared favorably with measured surface water SO_4^{2-} concentrations at the HBEF, the two simulations showed different rates of decline in surface water SO_4^{2-} in response to decreases in SO_4^{2-} deposition. Because Case 2 simulations relied on dry deposition fluxes to explain the discrepancy in SO_4^{2-} budgets, these simulations showed a more pronounced decline of stream SO_4^{2-} to declines in atmospheric deposition of S over the period 1974-75 to 1993-94 (-1.74 µeq L^{-1} yr^{-1}). In Case 3, it was assumed that the "missing" source of SO_4^{2-} was derived from a constant input, these simulations show a slower response to declines in deposition (-1.37 µeq L^{-1} yr^{-1}). The measured rate of decline in stream SO_4^{2-} at the HBEF over the period 1974-75 to 1993-94 was -1.36 µeq L^{-1} yr^{-1}.

A sensitivity analysis was conducted to evaluate the role of soil adsorption parameters on the simulation of surface water SO_4^{2-} concentrations. A simple approach was used, comparing simulations with measured site density for SO_4^{2-} adsorption (30 μeq g^{-1}) with values ½ (15 μeq g^{-1}) and 2 times (60 μeq g^{-1}) the measured value. All other values of atmospheric deposition and input fluxes were unchanged from the Case 2 and 3 simulation described previously. Results of this analysis showed that changes in the value of adsorption site density affected the response of drainage water SO_4^{2-} concentrations to changes in SO_4^{2-} deposition (Figure 4). A higher value of SO_4^{2-} adsorption site density resulted in less responsive declines in surface water SO_4^{2-} due to the greater retention of SO_4^{2-} in soil which buffered the change in input fluxes. Lower values of site density resulted in less retention of SO_4^{2-} by soil and a more direct correspondence with changes in SO_4^{2-} inputs. The simulations using the measured value of site density came the closest to reproducing trends in measured surface water concentrations. This sensitivity analysis showed that the desorption of SO_4^{2-} from soil adsorption sites, by itself, could not explain the discrepancy in watershed S budgets observed at the HBEF.

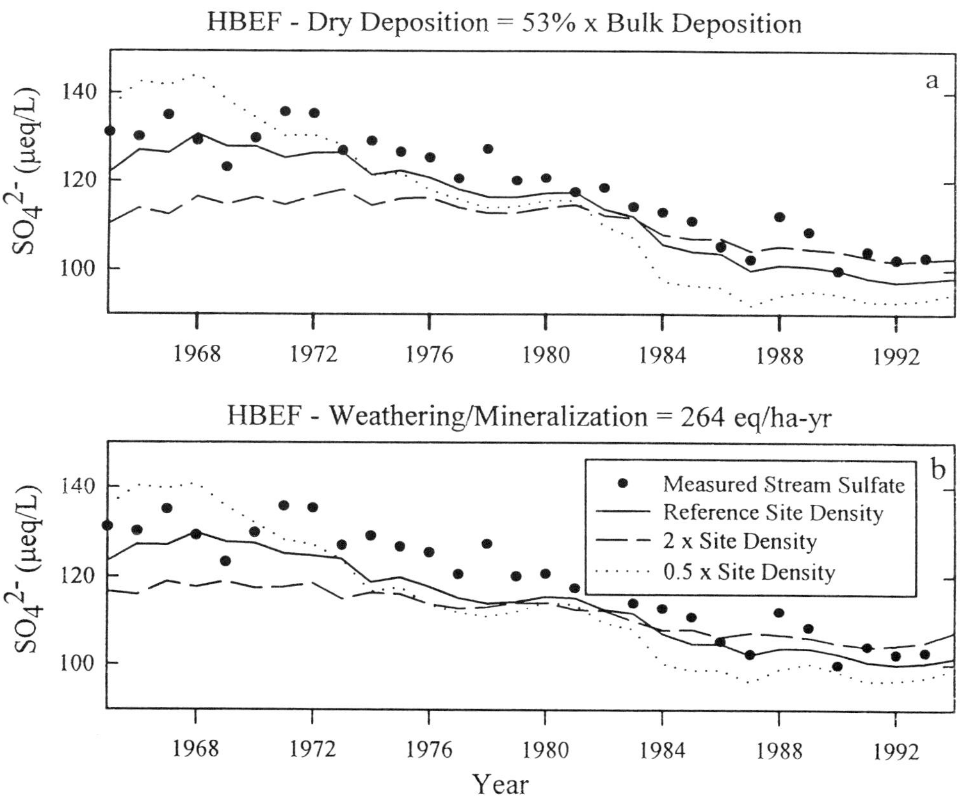

Fig. 4. Sensitivity analysis of model predictions to SO_4^{2-} adsorption site density at watershed 6 at the Hubbard Brook Experimental Forest, NH assuming the discrepancy in the watershed S budget is due to: a) unmeasured dry S deposition (53 % of bulk deposition), and b) weathering/mineralization S inputs (264 eq ha^{-1} yr^{-1}). The sensitivity analysis was conducted by doubling and halving the soil SO_4^{2-} adsorption site density. Model predictions are compared to measured values.

The results of this time-series analysis were inconclusive in determining the mechanism contributing to the discrepancy in the watershed S budgets. Although it appears that we can rule out desorption of SO_4^{2-} as the sole explanation for the discrepancy in watershed S budgets, based on our analysis it was not possible to distinguish between: 1) underestimates of dry S deposition (Case 2), and/or 2) additional inputs of S to the watershed from weathering/mineralization as the causative mechanisms. The assumption that S was derived from a constant weathering/mineralization source resulted in predictions of the rate of SO_4^{2-} decline in surface water that were very close to measured values over the period 1974-75 to 1993-94 (-1.37 vs. -1.36 μeq L^{-1} yr^{-1}). Although the difference between model simulations of stream SO_4^{2-} under the assumptions of additional S from dry deposition (Case 2) or additional S from a constant input of S depicting weathering/mineralization was fairly small over the period for which measured drainage water concentrations were available (1964-65 to 1993-94), these differences should become more pronounced as model simulations are extended into the future. Continued monitoring of atmospheric deposition and streamwater at Hubbard Brook (for approximately 5 to 10 years) and model application should reveal the mechanisms controlling stream SO_4^{2-} concentrations.

We also examined spatial patterns in the difference between measured and model-predicted SO_4^{2-} in surface waters for the DDRP watersheds to try to assess the source of missing S. Ollinger *et al.* (1993) reported a pattern of decreasing dry S deposition with increasing latitude across the northeastern U.S. We saw no significant spatial relationship (latitude or longitude) in the difference between measured and model-predicted SO_4^{2-} for the DDRP watersheds. Although the lack of a spatial pattern may be suggestive of the role of S inputs from weathering/mineralization, our analysis showed considerable variability in the difference between measured and model-predicted SO_4^{2-} for the DDRP watersheds, which may simply reflect the inherent variability among systems arrayed across a variable landscape.

Our results indicate that the ability of biogeochemical models to forecast changes in surface water chemistry in response to changing emissions of atmospheric SO_4^{2-} will depend to a large extent on the assumptions made in model development/applications. This result makes a clear statement of: 1) assumptions used, 2) model sensitivity analysis, and 3) model testing (e.g., against catchment manipulations; Cosby *et al.,* 1996) critical components of reporting and interpreting model results. To illustrate the importance of identifying the missing input of S to watersheds in the northeastern U.S. either as dry deposition (Case 2) or as a constant input attributed to weathering/mineralization (Case 3), simulations were conducted to examine the response of the HBEF to anticipated future decreases in inputs of SO_4^{2-} from atmospheric deposition. Calculations were conducted to simulate conditions under which atmospheric inputs of SO_4^{2-} were reduced at a linear rate from 1995 to 2005, such that the value at 2005 was 50 % of 1980 deposition values. Atmospheric inputs of SO_4^{2-} were assumed to be constant from 2005 until the end of the simulation period in 2030. These reductions might correspond to decreases in SO_2 emissions as mandated in the 1990 Amendments of the Clean Air Act for the U.S. For comparison, a similar scenario was conducted except a 100 % decrease in atmospheric deposition of S from 1995 to 2005 was assumed. These decreases in inputs were assumed to result in decreases in dry deposition as well as bulk deposition. For reference, the results of these scenarios were compared to a simulations in which inputs were held constant at 1995 values (Figure 5).

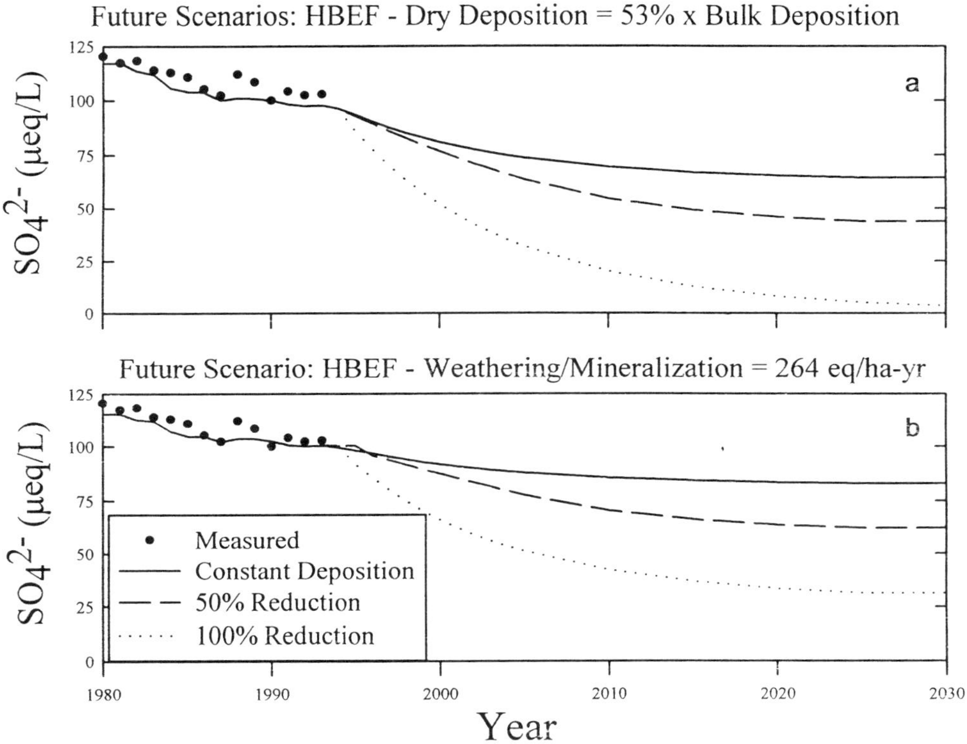

Fig. 5. Measured SO_4^{2-} concentrations and future predictions of changes in stream SO_4^{2-} in watershed 6 at the Hubbard Brook Experimental Forest, NH assuming the discrepancy in the watershed S budget is due to: a) unmeasured dry S deposition (53 % of bulk deposition) and b) weathering/mineralization S inputs (264 eq ha^{-1} yr^{-1}). Three SO_4^{2-} deposition scenarios were considered: constant deposition from 1995 values, 50 % reductions by 2005 from 1980 values, and 100 % reductions by 2005.

The simulations for the HBEF showed that when the SO_4^{2-} budget was balanced by increases in dry deposition (Case 2) or weathering/mineralization (Case 3) inputs, marked decreases in surface water SO_4^{2-} concentrations were predicted in comparison to the scenario of constant inputs of atmospheric deposition of SO_4^{2-} (Figure 5). These decreases reflect soil desorption of SO_4^{2-} as the watershed approached steady-state conditions with respect to lower atmospheric inputs. Note that for the assumed condition of elevated inputs of dry S deposition (53 % of bulk deposition; Case 2), the predicted decline of stream SO_4^{2-} concentrations was much greater than values predicted under the assumption of elevated inputs of S from weathering/mineralization (Case 3). Simulations of reduced inputs of SO_4^{2-} resulted in predictions of decreases in surface water SO_4^{2-}. Note that by 2030, predictions of surface water SO_4^{2-} for the HBEF were very different depending on whether it was assumed that additional S was supplied by dry deposition (Case 2) or by weathering /mineralization (Case 3). Model-predicted SO_4^{2-} concentration in 2030 assuming additional watershed input of S was by dry deposition (Case 2) was 64 μeq L^{-1} for constant deposition,

44 μeq L^{-1} for 50 % decrease in S deposition, and 3.5 μeq L^{-1} for 100 % decrease in S deposition; predicted SO_4^{2-} concentrations in 2030 assuming additional watershed input of S was by weathering/mineralization (Case 3) was 83 μeq L^{-1} for constant deposition, 62 μeq L^{-1} for 50 % decrease in S deposition and 31 μeq L^{-1} for 100 % decrease in S deposition.

4. Conclusions

Our calculations corroborate observations from watershed studies in the northeastern U.S. that there is a missing or poorly measured source of S to forest watersheds in the eastern U.S. (Norton *et al.*, 1988; Driscoll *et al.*, 1995; Mitchell *et al.*, 1996). The long-term trends in surface water SO_4^{2-} concentrations and the discrepancy in watershed S mass balances cannot be explained by desorption of SO_4^{2-} from soil. It appears likely that the source of missing SO_4^{2-} is from an underestimation of dry S deposition when this flux is determined by net throughfall or inferred by measured air concentrations and an assumed deposition velocity, and/or from weathering of S minerals or mineralization of S associated with soil organic mater. Unfortunately it was not possible to determine with the current long-term records which of these two sources of S supplies SO_4^{2-} to drainage waters. This information, however, is vital to understanding the rate and extent to which surface water ANC will recover in response to decreases in atmospheric deposition of S. Continued long-term monitoring and investigations of S cycling and potential sources of S for forested watersheds are critical for developing a balanced SO_4^{2-} budget and reducing the uncertainty in the predicted response of SO_4^{2-} to changes in atmospheric deposition.

Acknowledgments

This study was supported with funds from the National Science Foundation, the IBM Environmental Research Program, the A.W. Mellon Foundation and the U.S. Environmental Protection Agency (CR 820482). We thank R. Santore and N. Chopada for help with data analysis, the USDA Forest Service, W. Martin, A. Bailey, C. Eagar and D. Buso for support of the field program at Hubbard Brook, K. Driscoll, D. Mateti and D. Long for assistance in manuscript preparation, and D. Landers, J. Wigington, M. Novák and S. Norton for comments on the manuscript. This document has been reviewed in accordance with U.S. EPA policy and approved for publication. Mention of trade names or commercial products does not constitute endorsement or recommendation for use. The views and opinions expressed do not necessarily represent policy positions of the Environmental Protection Agency or any other organization. This is a contribution of the Hubbard Brook Ecosystem Study. The Hubbard Brook Experimental Forest is operated by the USDA Forest Service, Radnor, PA.

References

Aber, J.D., Ollinger, S.V. and Driscoll, C.T.: 1997, *Ecol. Modelling* **101**, 61.
Brakke, D.F., Landers, D.H., and Eilers, J.M.: 1988, *Environ. Sci. Technol.* **22**, 155.

Butler, T.J. and Likens, G.E.: 1991, *Atmos. Environ.* **25A**, 305.

Church, M.R., Thornton, K.W., Shaffer, P.W., *et al.*: 1989, *Direct/Delayed Response Project: Future Effects of Long-Term Sulfur Deposition on Surface Water Chemistry in the Northeast and Southern Blue Ridge Province.* U.S. Environmental Protection Agency, EPA/600/3-89/061, Washington, D.C.

Cosby, B.J., Norton, S.A., and Kahl, J.S.: 1996, *Sci. Total Environ.* **183**, 49.

Driscoll, C.T.: 1991, 'Northeast Overview', in D.F. Charles (ed.), *Acidic Deposition and Aquatic Ecosystems: Regional Case Studies*, Springer-Verlag, New York, pp. 129-132.

Driscoll, C.T., Postek, K.M., Kretser, W. and Raynal, D.J.: 1995, *Water, Air Soil Pollut.* **85**, 583.

Fuller, R.D., Driscoll, C.T., Lawrence, G.B. and Nodvin. S.C.: 1987, *Nature* **325**, 707.

Galloway, J.N., Norton, S.N. and Church, M.R.: 1983, *Environ. Sci. Technol.* **17**, 541A.

Husar, R., Sullivan, T. and Charles, D.: 1991, 'Historical Trends in Atmospheric Sulphur Deposition and Methods for Assessing Long-term Trends in Surface Water Chemistry', in D.F. Charles (ed.), *Acidic Deposition and Aquatic Ecosystems: Regional Case Studies*, Springer-Verlag, New York, pp. 65-82.

Johnson, C.E., Johnson, A.H., Huntington, T.G. and Siccama, T.G.: 1991, *Soil Sci. Soc. Am. J.* **55**, 497.

Likens, G.E., Bormann, F.H., Hedin, L.O., Eaton, J.S. and Driscoll, C.T.: 1990, *Tellus* **42B**, 319.

Lovett, G.M., Likens, G.E. and Nolan, S.S.: 1992, 'Dry Deposition of Sulfur to the Hubbard Brook Experimental Forest: A Preliminary Comparison of Methods', in S.E. Schwartz and W.G.N. Slinn (coords.) *Fifth International Conference on Precipitation Scavenging and Atmosphere-Surface Exchange. The Summer Volume (3): Application and Appraisals*, Hemisphere Publishing Corp., pp. 1391-1402.

Mitchell, M.J., Driscoll, C.T., Fuller, R.D., David, M.B. and Likens, G.E.: 1989. *Soil Sci. Soc. Amer. J.* **53**, 933.

Mitchell, M.J., Driscoll, C.T. and Raynal, D.J.: 1996, *Water Air Soil Pollut.* **88**, 355.

Nizich, S.V.; Misenheimer, D., Pierce, T., Pope, A. and Carlson. P.: 1996, *National Air Pollutant Emission Trends, 1990-1995*, U.S. Environmental Protection Agency, EPA-454/R-96-007, Office of Air Quality, Research Triangle Park.

Norton, S.A., Kahl, J.S., Brakke, D.F., Brewer, G.F., Haines, T.A. and Nodvin S.C.: 1988, *Sci. Total Environ.* **72**,183.

Novák, M., Bottrell, S.H., Fottova, D., Buzek, F., Groscheová, H. and Žák, K.: *Environ. Sci. Technol.* **30**, 3473.

Ollinger, S.V., Aber, J.D., Lovett, G.M., Millham, S.E., Lathrop, R.G. and Ellis, J.M.: 1993, *Ecol. Appl.* **3**, 459.

Rochelle, B.P., Church, M.R. and David, M.B.: 1987, *Water, Air Soil Pollut.* **33**, 73.

Santore, R.C. and Driscoll, C.T.: 1995, 'The CHESS Model for Calculating Chemical Equilibria in Soils and Solutions', in R. Loeppert, A.P. Schwab and S. Goldberg, (eds.), *Chemical Equilibrium and Reaction Models*, Soil Sci. Soc. Amer., Madison, pp. 357-375.

Sposito, G.: 1989, *The Chemistry of Soils*, Oxford University Press, New York.

TRENDS IN SULFUR EMISSION-INDUCED EFFECTS IN NORTHERN EUROPE

OLGA RIGINA[1,3] and ALEXANDER BAKLANOV[2,3]

[1]*Swedish University of Agricultural Sciences, S-901 83 Umeå, SWEDEN*
[2]*Defense Research Establishment (FOA), S-901 82 Umeå, SWEDEN*
[3]*Institute of Northern Ecological Problems, Kola Science Center RAS, 184200 Apatity, RUSSIA*

Abstract. Fennoscandia is subjected to a considerable anthropogenic load, where S still is a notable contributor. At the same time, the region is very sensitive to acidifying depositions. The Regional Acidification Information and Simulation model (RAINS) was applied for a regional assessment of S deposition and loads in Fennoscandia from 1960 through1995. Different variants of sources were considered: 1) all Europe, 2) north-west Russia (the Kola, Karelia and Leningrad county), and 3) the Nordic countries (Finland, Norway, Sweden). This allowed their contribution to the deposition pattern of the region to be assessed. The percentage of the ecosystems where the critical load had been exceeded was calculated for each year. For a more detailed assessment of the influence of the sources of the Kola region, estimations on the basis of local and meso-scale models were made. Comparative analysis of the calculations for the domestic and all the European sources revealed a dominating effect of S deposition from west and central Europe (Europe minus Fennoscandia) in general. The calculations showed a high, but local impact of the Kola sources on northern parts of the Fennoscandian region.

Key words: Northern Europe, Fennoscandia, Kola Peninsula, sulfur deposition, critical loads, modeling

1. Introduction

Sulfur, together with N, causes soil and water acidification and deterioration of terrestrial ecosystems. Fennoscandia (Norway, Finland, Sweden, the Kola Peninsula, Karelia, Leningrad county) falls into the most sensitive category (Downing *et al.*, 1993) and receives a considerable anthropogenic load. Although the emissions in the region are multi-component, SO_2 appears to be dominating (Kismul *et al.,*1992). Severe acidification of freshwaters and soils is occurring over large areas in southern parts of the Nordic countries (Norway, Finland, Sweden) (Brodin and Kuylenstierna, 1992), and on the Kola Peninsula (Moiseenko, 1994). Acidification is related mainly to S (Brodin and Kuylenstierna, 1992); N (NO_x + NH_x) is not yet a problem (Downing *et al.,* 1993; Løbersli and Venn, 1995), contributing 10-30 % of the acidification in southern parts (Brodin and Kuylenstierna, 1992). Regions most sensitive to S deposition include large parts of the Nordic countries and Kola Peninsula with a critical load of less than 0.3 g S m^{-2} yr^{-1} (Downing *et al.,* 1993). Regional contributions to SO_2 emission in Fennoscandia are presented in Figure 1.

Bulk SO_2 emissions from the Kola Peninsula (Figure 2), and the location of the Pechenganickel Cu-Ni smelter (7 km from the Norwegian border) and the Severonickel smelter (120 km from the Finnish border), have aroused international concern. Three expert groups under the Joint Norwegian-Russian Commission on Environmental Cooperation have considered local-scale atmospheric, aquatic, and terrestrial aspects of acidification in the near-border areas (Sivertsen *et al.*, 1992, 1994; Løbersli and Venn, 1995). In connection with a concern about the effect of transboundary emissions from the Kola to Finnish Lapland, the 5-year Lapland Forest Damage Project was launched in 1990 (Tikkanen and Niemelä, 1995).

Water, Air, and Soil Pollution **105:** 331–342, 1998.

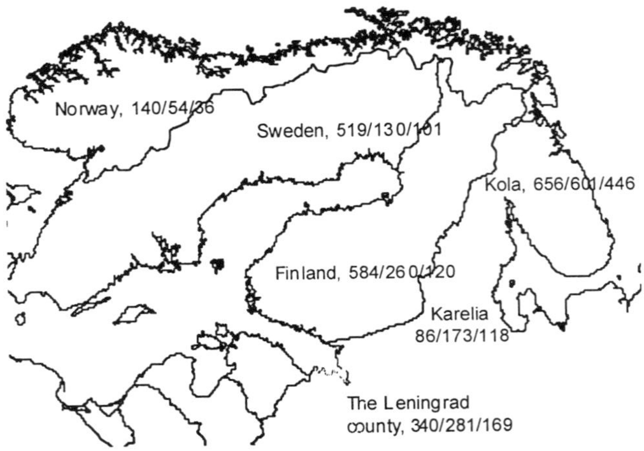

Fig. 1. The study area and the SO_2 emissions in 1980, 1990, and 1995 (kt yr^{-1}, after the Eastern Center of EMEP and RF Ministry of Environment data).

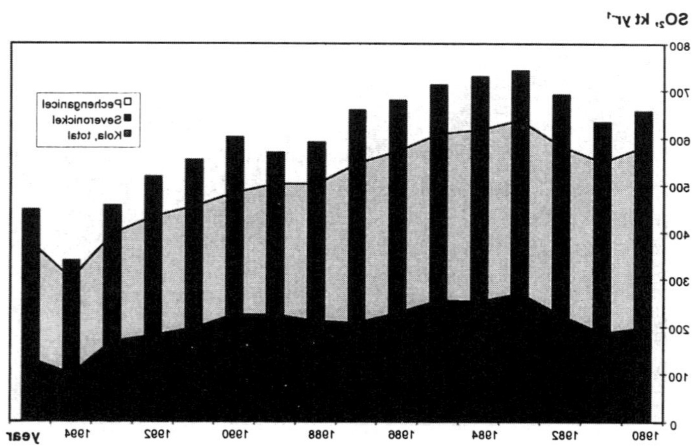

Fig. 2. SO_2 emission from the Severonickel (at Monchegorsk) and Pechenganickel (at Nickel and Zapolyarny) smelters and total for the Kola Peninsula (after the Murmansk Regional Committee of Nature Protection data).

Earlier results for the Nordic countries (Tuovinen *et al.*, 1993) showed that a critical S deposition was exceeded on an area of 150,000 km^2, including 32,000 km^2 in Finland and 19,000 km^2 in Norway. The relative contribution of the Kola smelters was estimated to range from 40 % to almost 100 %. According to Aamlid *et al.* (1995), critical levels of SO_2 for sensitive ecosystems (10-15 mg m^{-3}) were exceeded within a region including 2,376 km^2 in Russia and 869 km^2 in Norway. Earlier results (Anttila *et al.*, 1995) estimated that Kola's sources had a distinct impact over a radius of 200 km and about half of the S deposition in central Lapland originated from the Kola Peninsula. According to results of the Norwegian-Russian terrestrial expert group (Sivertsen *et al.*, 1994), areas with severe vegetation damage occupied 1,270 km^2 in Norway and 3,433 km^2 in Russia.

Sensitivity to acidification depends on the type of ecosystem. In general, lake ecosystems seem to be more sensitive to acidification than forest ecosystems (Brodin and Kuylenstierna, 1992). Bedrock lithology, soil type, land use, and rainfall are factors that determine local sensitivity (Kämäri *et al.*, 1989). For rough estimations of complex effects of acid deposition on multiple ecosystems, the critical load approach was elaborated in the late 1970s. A critical load was defined as 'the highest deposition of a pollutant that will not cause harmful effects on ecosystems according to present knowledge' (Nilsson and Grennfeld, 1988). However, recently this approach was revised and a new method based on combined assessment of the acidifying effects of S and N was introduced (Hettelingh *et al.*, 1995).

The purpose of this study was to assess the relative contribution of emission sources in northwest Russia (the Kola, Karelia and Leningrad county), especially the Kola Peninsula, to acid deposition in Fennoscandia and, thereby, to acidification of its ecosystems.

2. Materials and Methods

2.1. Regional-Scale Modeling

The 'Regional Air Pollution Information and Simulation model (RAINS) was developed at the International Institute of Applied Systems Analysis (Alcamo *et al.*, 1987) for integrated assessments of alternative strategies to reduce acid deposition in Europe. A sub-module within the acid deposition and ecosystems impact computes the S deposition caused by the emissions in each country. Contributions from each country are summed to the total deposition in any grid sell. The model uses transfer matrices based on a Lagrangian model of long-range transport of air pollutants in Europe, developed under the European Monitoring and Evaluation Program (EMEP) (Eliassen and Saltbones, 1983). The EMEP model accounts for the effects of wind, precipitation, and other meteorological and chemical variables on acid deposition and air concentrations. The use of a transfer matrix assumes that there is a linear relationship between the pollutant emission and deposition. The model contains a database on critical loads of S for 14 different European ecosystems calculated by the steady-state mass balance method compiled for each EMEP grid cell (Downing *et al.*, 1993). These critical loads are compared further with calculated pollutant deposition, to determine which areas currently receive deposition levels exceeding the critical load, and to derive for each grid the percentage of ecosystems where the critical load is exceeded.

To estimate the contribution of the emission sources of northwest Russia to the acidification of Fennoscandia, transport and deposition were computed for all European sources for 1960-1995 and separately for the sources of the Nordic countries (Finland, Norway, Sweden) and northwest Russia (the Kola, Karelia and Leningrad county) for 1980-1995. The emission data used in the calculations for the last years were updated after the Eastern Center of EMEP and the Russian Ministry of Environment. The percentage of ecosystems for which the critical loads had been exceeded was calculated for 1960-1995. The resulting maps were presented in the 150 × 150 km^2 EMEP grid.

2.2. Local and Meso-Scale Modeling

To examine local dispersion in more details, a calculation of a pollutant transport and deposition for the Kola Peninsula was performed by a Gaussian plume sector model for local scale (Baklanov and Rodyushkina, 1993) and by a numerical trajectory model for the meso-scale (Baklanov and Sivertsen, 1994). The models take into account the locations and parameters of emission sources, the vertical structure of wind (high and low sources might be transported by different winds), and the influence of topography and atmospheric stability.

A model for local-scale assessment of atmospheric pollution influences on the Kola ecosystems was realized, based on region-oriented critical levels and loads for S elaborated by the expert groups on terrestrial ecosystems within the Russian-Norwegian cooperation (Løbersli and Venn, 1995) and a regional forest impact model (Mäkelä *et al.*, 1985). The model was applied for the Severonickel impact zone as an example for 1960-1995.

3. Results and Discussion

Our calculations by RAINS for all the European sources (Figure 3) show that NO_x deposition was 0.1-0.3 g m^{-2} yr^{-1} across most of Fennoscandia for 1980-1995. Of these sources, the northern regions have the lowest deposition and the southern areas the highest. No significant contribution to NO_x deposition pattern from northern domestic sources was revealed. Nitrogen deposition in the southern areas is about three times lower of that in central Europe. Thus, the main contribution to acidification in Fennoscandia, especially in its northern part, is still mostly attributable to S. This is consistent with the conclusions ofearlier studies (Brodin and Kuylenstierna, 1992; Downing *et al.,* 1993; Løbersli and Venn, 1995).

The calculations for the period of 1960-1995 reveal a trend of increasing total S deposition in Fennoscandia until 1980. From 1980 to 1990 (Figures 4b, 5b), S deposition decreased in the southern part of the Nordic countries (owing to the west and central European as well as domestic sources), in contrast to no change in the Kola, northern Finland and Norway. During 1990-1995 (Figures 5b, 6b), the decreasing trend still took place, but it covered also northern areas of the Nordic countries and the Kola Peninsula (mainly due to a production volume decrease since 1991).

The calculations for all the European sources (Figures 4b, 5b, 6b) show three main maxima in the deposition pattern in Fennoscandia during the whole period. The two maxima: in southern Sweden and Norway (up to 1.2 g S m^{-2} yr^{-1} in 1990) and in southern Finland, Estonia and the Leningrad county (up to 1.5 g S m^{-2} yr^{-1}) both are attributed to the domestic as well as west and central European sources, whereas the third maximum in the north of Fennoscandia (up to 1.8 g S m^{-2} yr^{-1}) is influenced only by the Kola sources. These results are consistent with data of Nordlund *et al.* (1985) that the total deposition of S in Finland is highest along the coastal area of southern Finland (1.5-1.8 g S m^{-2} yr^{-1}) and decreases northwards.

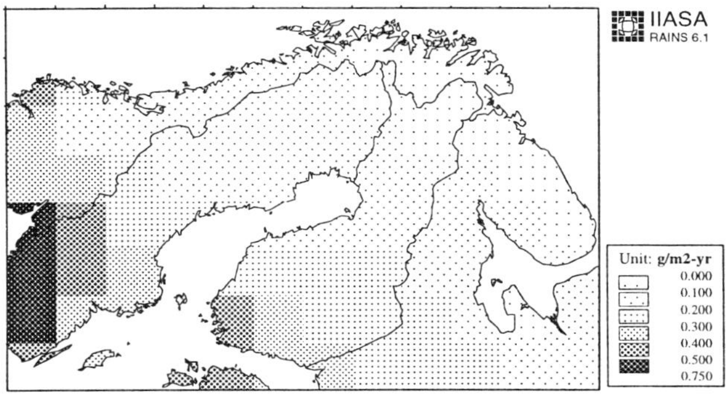

Fig. 3. NO_x deposition in Fennoscandia in 1990, calculated for all the European emission sources.

Exceedance of critical loads of S is greatest in the southern part of the Nordic countries, mostly owing to the very low critical loads there (0.04-0.14 g m^{-2} in northern Norway in contrast to 0.30-0.45 g m^{-2} in the adjacent areas of the Kola Peninsula; Downing *et al.*, 1993) and transboundary fluxes from west and central European sources. Thus, in 1980 an area with exceedance of critical loads by 87.5-100 % accounted for about 180,000 km^2 in the study area. In the 1990 pattern, there was no such class and an area with exceedance by 75-87.5 % amounted only 67,000 km^2.

The calculations for the sources of north-west Russia (Figures 4a, 5a, 6a) are consistent with a local scale of transboundary flux from the Kola Peninsula to limited areas of northern Norway and Finland (up to 70-85 % of the total S deposition). Even for areas of northern Sweden, central and north-west Finland the influence of other than Kola S sources is dominating (up to 75-85 % and 60-80 % of the total S deposition, correspondingly). The conclusions of the Lapland Forest Damage Project (Tikkanen and Niemelä, 1995) suggest that the emissions from the Kola Peninsula are not expected to cause significant damage to forests in the Finnish territory, excluding only a small area in Lapland (at Inari). They also recognized that as a whole the annual acid deposition in Lapland is moderate, around half of that in southern Finland. Our results also show a deposition pattern analogous to that of Tuovinen *et al.* (1993). However, in comparison with ours (both the regional- and meso-scale modeling), their result of the Kola relative contribution seem to be overestimated, probably due to application of a Gaussian-type model for regional scale.

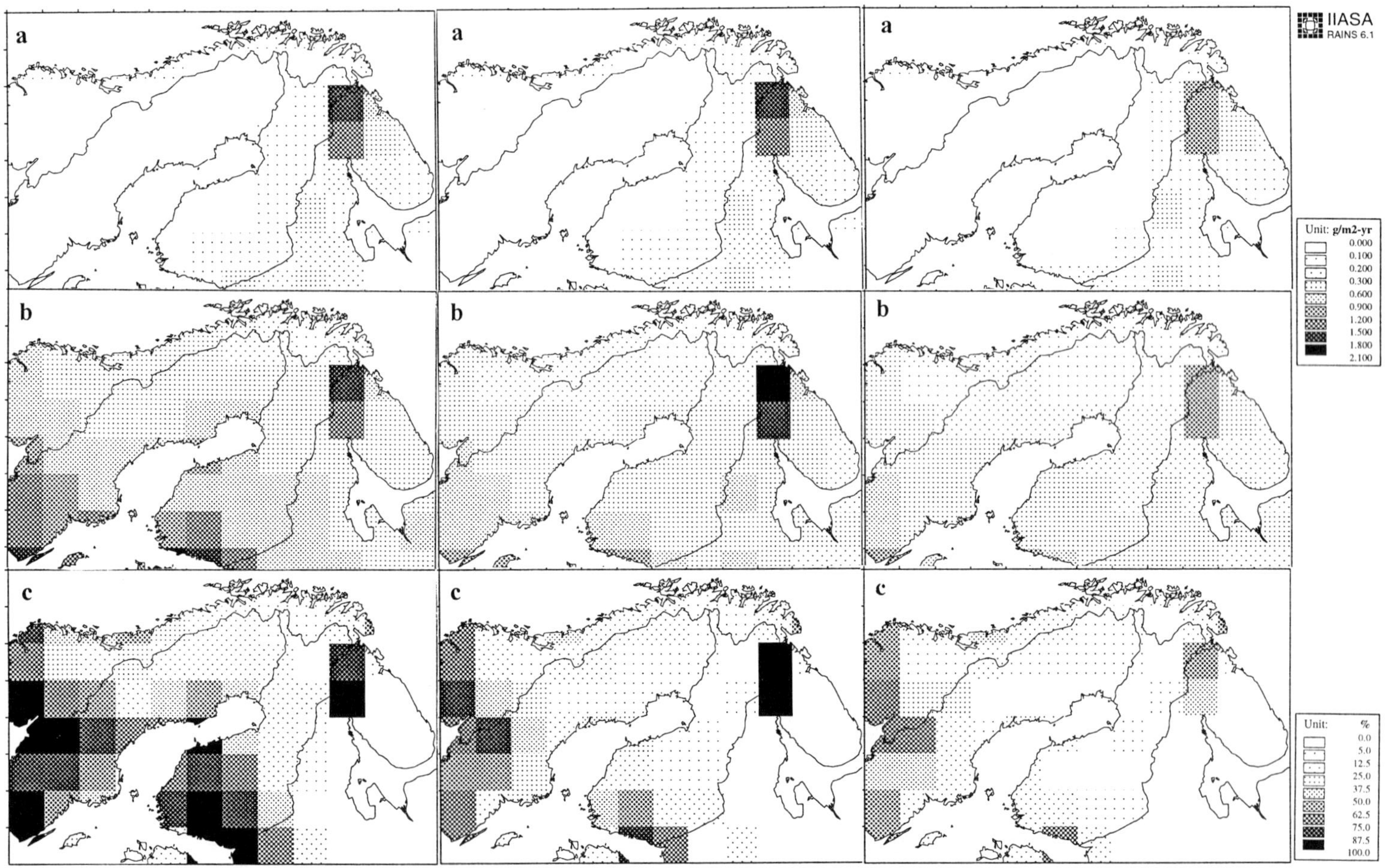

Figure 4. Sulfur deposition in Fennoscandia, 1980. Figure 5. Sulfur deposition in Fennoscandia, 1990. Figure 6. Sulfur deposition in Fennoscandia, 1995.
a) emissions from north-west Russia; b) all the European emission sources; c) persentage of ecosystems where the critical load of sulfur is exceeded.

The separate calculations for emissions from the Nordic countries (although not shown here) reveal a considerable share of domestic sources (in 1980 only in southern Finland up to 0.6-0.8 g S m^{-2} yr^{-1} and in coastal Sweden up to 0.4-0.6 g S m^{-2} yr^{-1}; in 1990, in southern Finland up to 0.3-0.6 g S m^{-2} yr^{-1} - up to 30 % of the total S deposition). The Swedish maximum disappeared after reconstruction of the Rönnskär smelter in Skellefteå in 1989/1990, and in 1995 S deposition in the Nordic countries is less than 0.2 g S m^{-2} yr^{-1}).

The contribution of the sources of Karelia and the Leningrad county on the north-west Russia was 0.4-0.8 g S m^{-2} yr^{-1} in 1980 (Figure 4a), 0.2-0.6 g S m^{-2} yr^{-1} in 1990 (both up to 40-45 % o total S deposition) (Figure 5a) and 0.2-0.5 g S m^{-2} yr^{-1} in 1995 (Figure 6a). In southern Finland, the north-west Russian sources contributed 0.2-0.3 g S m^{-2} yr^{-1} (both up to 10-20 % of the total S deposition), whereas other sources contributed from 0.6 to 1.2 g S m^{-2} yr^{-1} (Figures 4b, 5b, 6b). The domestic sources practically determine the S deposition pattern on the Kola and the neighboring areas of northern Norway and Finland; up to 1.8 g S m^{-2} yr^{-1} in 1990 and in 1995 up to 1.5 g S m^{-2} yr^{-1}. However, comparison of calculations for the Kola sources (Figure 6a) and all the European sources (Figure 6b) reveals in 1995 considerable contribution of the transborder pollution from west and central European sources (0.1-0.4 g S m^{-2} yr^{-1}) to total pollution on the Kola Peninsula.

A shortcoming of the model is the smoothing of local maxima and local-scale features of the deposition pattern, due to the coarse resolution. To assign a share of the Kola sources in transborder transport and deposition on a basis of the regional-scale calculations only is not correct enough due to a local scale of influences and the near-border locations of the large emission sources. This may alleviate the overestimated Kola contribution by Anttila *et al.* (1995). For more detailed estimation of local transborder S transport and deposition it is advisable to apply local and meso-scale models that also account for local topography, nonhomogeneity of meteorological fields, and contribution of low and high emission sources. The calculation by the meso-scale model for the Kola smelters (giving over 80 % of the Kola emissions, Figure 2) gives a more detailed pattern of S deposition on the Kola Peninsula (Figure 7).

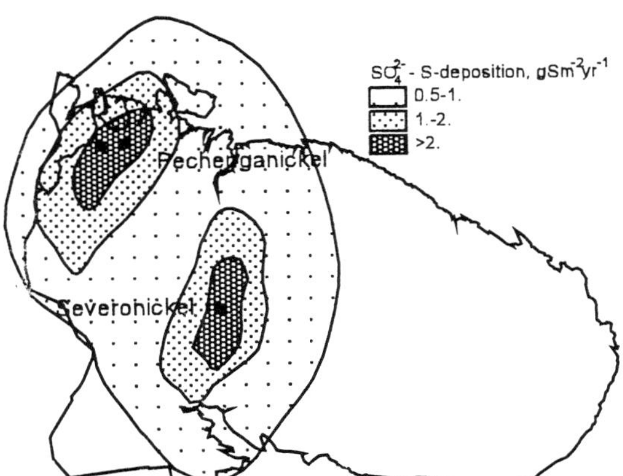

Fig. 7. Sulfur deposition (g S m^{-2} yr^{-1}) in 1990 on the Kola Peninsula based on meso-scale modeling.

Two maxima are clearly delineated: from the Pechenganickel in the north and from the Severonickel in the central Kola. Of them, the Pechenganickel exerts main influence on the neighboring areas, giving SO_4^{2-} deposition up to 1.0-2.0 g S m^{-2} yr^{-1} in northern Norway and up to 0.5-1.0 g S m^{-2} yr^{-1} in northern Finland. The Severonickel gives up to 0.5 g S m^{-2} yr^{-1} in a near-border area of northern Finland. The SO_4^{2-} deposition predicted by the model showed a good correlation with monitoring data of SO_4^{2-} in surface water (Moiseenko, 1994) across the Kola peninsula ($r = 0.813$, $n = 283$, $p=0.000$). The pattern is similar to the results of the Finnish HAKOMA model (Johansson *et al.*, 1989). Figure 8 presents total SO_4^{2-} deposition from the Pechenganickel in the impact zone in further detail. This calculation is based on the local-scale Gaussian plume model and verified by local monitoring data (Figure 9) from the both sides of the Russian-Norwegian border (Baklanov and Sivertsen, 1994; Rodyushkina, 1994).

Deposition of greater than1.0 g S m^{-2} yr^{-1} covers a very limited Norwegian territory in the near-border Finnmark (about 860 km^2) and 2,350 km^2 of the Russian territory. However, along the border areas, deposition up to 2 g S m^{-2} yr^{-1} occurs (about 6 km^2 at Svanvik). A trend of decreasing S load in the Pechenganickel area since 1980 up to now is observed. Deposition of SO_4^{2-} was decreased by 37.5 % in 1995 relative to 1980. These results were supported by monitoring data on S air pollution (Figure 10) of the Russian-Norwegian expert group on atmospheric pollution (Sivertsen *et al.,* 1994).

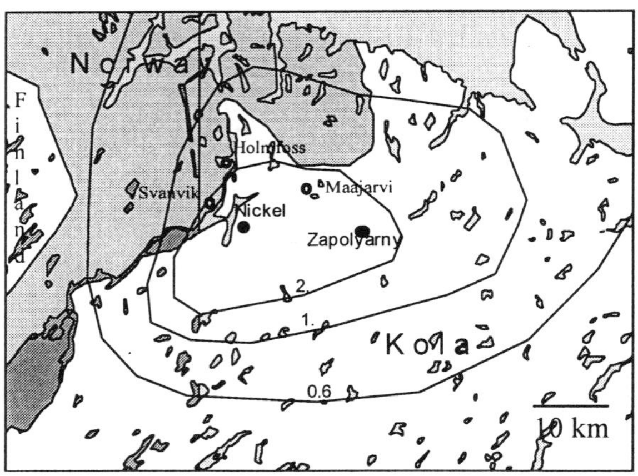

Fig. 8. Sulfur deposition from the Pechenganickel smelter in the near-border area in 1990, g S m^{-2} yr^{-1} based on local-scale model

Estimations show that SO_2 sources from the Kola give in 1995 reduction of about 32 % relative to 1980 (including 37 % at the Severonickel and 36 % at the Pechenganickel). Although as a result of the Helsinki agreement, about 30 % reduction in SO_2 emissions was implemented, the calculations of exceedance of S loads on ecosystems show that this was not enough, especially in the impact zones of the Severonickel and Pechenganickel smelters.

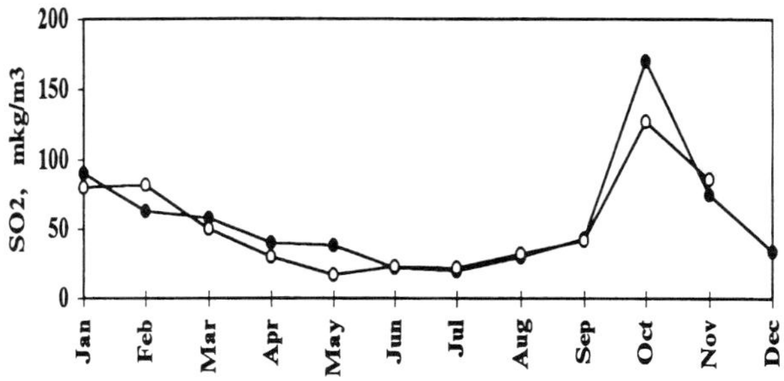

Fig. 9. Month-mean SO_2 concentration (mg m^{-3}) at the monitoring station Maajarvi in the Russian-Norwegian border area in 1992: •- modeled; ○ - measured (after Rodyushkina, 1994).

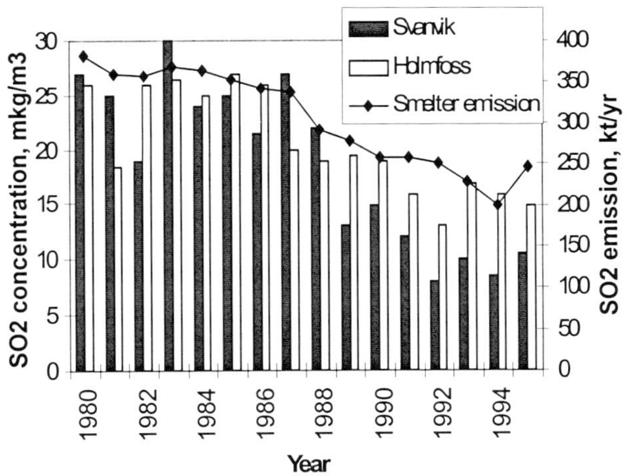

Fig. 10. Average annual SO_2 concentrations (μg m^{-3}) at the monitoring stations Svanvik and Holmfoss in the Russian-Norwegian border area and SO_2 emission from the Pechenganickel smelter (kt yr^{-1}) for 1908-1995 (updated after Sivertsen *et al.*, 1994).

The local-scale forest impact model (Figure 11) shows expansion of forest damage area around the Severonickel smelter until 1992 due to accumulation of S load. The isoline corresponding to the 1995-year damage area coincides with that of 1992, implying a stabilization of the damaged area since 1992, that is consistent with a decreasing of emission level.

Fig. 11. Spreading of forest damage area around the Severonickel smelter in 1960-1995.

4. Conclusions

The calculations indicate the highest exceedance of critical loads of S in the south of Norway, Sweden and Finland, mainly owing to the very low critical loads in the Nordic countries and transborder transports from west and central European sources. The result of the meso-scale modeling shows in more detail the deposition pattern, and a high level of S deposition in the impact zones (over 2 g S m^{-2} yr^{-1}), and a relatively short-range effect of the Kola sources. The regional-scale calculations can not take into account the local character of large separate emission sources; local-scale models can be used to enhance the analyses. The calculations of relative contributions of different regions in acidification of Fennoscandia show: a high, but local impact of the Kola sources on the northern parts of the Nordic countries; a considerable impact of other Russian sources (Karelia and the Leningrad county) in Fenno-Russian border area; and a dominating share in the S deposition in the south-west parts of Nordic countries from west and central European sources. Despite large reductions from the Kola sources in 1995 relative to 1980, calculations of exceedance of loads on ecosystems show that this is not enough, especially in the impact zones of the Severonickel and Pechenganickel smelters. Under recent decreasing emissions from the Severonickel smelter, a stabilization of spreading of the damaged forest area takes place in its impact zone since 1992 as predicted by the forest impact model.

Acknowledgments

Our gratitude to the Center for Environment Research, Umeå, Sweden for sponsorship, Drs. M. Amann, W. Schopp from the Transboundary Pollution Project at IIASA for providing with the RAINS model and kind assistance, Dr. S. Dutchak from the EMEPcenter Eastern, Moscow for data, Prof. V. Nikonov from INEP, Prof. H. Olsson and O. Hagner from SLU, Umeå, Sweden for valuable discussions. Comments from Dr. K. Wieder and the anonymous referees were greatly appreciated.

References

Aamlid, D., Tømmervik, H., Gytarsky, M., Karaban, R., Venn, K., Rindal, T., Vassilieva, N., Koptsik, G. and Løbersli, E.: 1995, 'Determination of Exceedance of Critical Levels in the Border area between Norway and Russia', in E. Løbersli and K. Venn (eds.), *Effects of Air Pollutants on Terrestrial Ecosystems in the Border Area between Russia and Norway*, Svanvik, Norway, 3-5 Oct. 1994..

Anttila, P., Barrett, K., Derome, J., Hillamo, R., Jalkanen, L., Juntto, S., Jurvelin, J., Kubin, E., Laurila, T., Lindroos, A-J., Lättilä, H., Mäkelä, T., Mäkinen, M., Pakkanen, T., Poikolainen, J., Reissell, A., Ryaboshapko, A., Saari, H., Tuovinen, J-P., Virkkula, A.: 1995, 'Air Quality and Deposition', in E. Tikkanen and I. Niemelä (eds.), *Kola Peninsula Pollutants and Forest Ecosystems in Lapland*, Gummerus Kirjapaino Oy Jyväskylä, pp. 30-33.

Alcamo, J. , Amann, M., Hettelingh, J.-P., Holmberg, M., Hordijk, L., Kämäri, J., Kauppi, L., Kornai, G., Mäkelä, A.: 1987, *Ambio* **6**, 232.

Baklanov, A. and Rodyushkina, I.: 1993, 'Pollution of Ambient Air by 'Severonickel' Smelter Complex: Observations and Modeling', in M. Kozlov, E. Haukioja, and V. Yarmishko (eds.), *Aerial Pollution in Kola Peninsula*, April 14-16, 1992, University of Turku, Finland, pp. 83-89.

Baklanov, A. and Sivertsen, B.: 1994, *Investigations of Local Transborder Air Pollution*, NILU report, Kjeller, Norway (NILU: F 19/94).

Brodin, Y.-W. and Kuylenstierna, J.: 1992, *Ambio* **21**, 332.

Downing, R., Hettelingh, J.-P. and de Smet, P. (eds.): 1993, *Convention on Long-range Transboundary Air Pollution - Calculation and Mapping of Critical Loads in Europe: Status Report 1993*, ISBN No 90-6960-047-1, Bilthoven, the Netherlands.

Eliassen, A. and Saltbones, J.: 1983, *Atmos. Environ.* **17**, 1457.

Hettelingh, J.-P., Posch, M., De Smet, P. and Downing, R.: 1995, *Water, Air Soil Pollut.* **85**, 2381.

Johansson, M., Savolainen, I. and Tähtinen, M.: 1989, 'The Finnish Integrated Acidification Model', in J. Kämäri and D. Brakke (eds.), *Regional Acidification Models: Geographic Extent and Time Development,* Springer, pp. 203-211.

Kismul, V., Jerre, J., and Løbersli, E. (eds.): 1992, *Proceedings: Effects of Air Pollutants on Terrestrial Ecosystems in the Border Area between Russia and Norway*, Svanvik, Norway, 18-20 March 1992.

Kämäri, J., Brakke, D., Jenkins, A., Norton, S. and Wright, R. (eds.): 1989, *Regional Acidification Models - Geographical Extent and Time Development*, Springer-Verlag, Berlin.

Løbersli, E. and Venn, K. (eds.): 1995, *Effects of Air Pollutants on Terrestrial Ecosystems in the Border Area between Russia and Norway*, Svanvik, Norway, 3-5 Oct. 1994.

Moiseenko, T.: 1994, *Ambio* **23**, 418.

Mäkelä, A., Materna, J., and Schöpp, W.: 1987, *Direct Effects of Sulfur on Forest in Europe - A Regional Model of Risk*, WP-87-57, IIASA, Laxenburg, Austria.

Nilsson, J. and Grennfeld, P. (eds.): 1988, 'Critical Loads for Sulfur and Nitrogen', Nord 1988:97, Nordic Council of Ministers, Copenhagen, Denmark.

Nordlund, G., Pietarinen, M. and Tuovinen, J.-P.: 1985, *Publ. of Nature Conversation Division, A:26*, The Ministry of the Environment, Helsinki, Finland.

Rodyushkina, I.: 1994, 'The Evaluation of Mean for Period Atmosphere Pollution in the Pechenga Region', in B. Sivertsen (ed.), *Air Pollution Problems in the Northern Region of Fennoscandia Including Kola*, Proceedings from the Seminar at Svanvik, Norway, 1-3 June 1993, ISBN-82-425-0612-4, pp.168-178.

Sivertsen, B., Makarova, T., Hagen, L., and Baklanov, A.: 1992, Air Pollution in the Border Areas of Norway and Russia, Summary Report 1990-1991, NILU OR:8/92.

Sivertsen, B., Baklanov, A., Hagen, L., and Makarova, T.: 1994, Air Pollution in the Border Areas of Norway and Russia, Summary report 1991-1993, NILU, ISBN: 82-425-0609-4.

Tikkanen, E. and Niemelä, I. (eds.): 1995, *Kola Peninsula Pollutants and Forest Ecosystems in Lapland*, Final report of the Lapland Forest Damage project, Finland, ISBN 951-40-1455-3.

Tuovinen, J.-P., Laurila, T., Lättilä, H., Ryaboshapko, A., Brukhanov, P. and Korolev, S.: 1993, *Atmos. Environ.* **27**, 1379.

RECENT TEMPORAL PATTERNS IN THE CHEMISTRY OF SMALL, ACID-SENSITIVE LAKES IN CENTRAL ONTARIO, CANADA

DONALD K. McNICOL[1], MARK L. MALLORY[1], CLAUDE LABERGE[2] and DANIEL A. CLUIS[3]

[1]*Canadian Wildlife Service (Ontario Region), 49 Camelot Drive, Nepean, ON, Canada, K1A 0H3*
[2]*STATEX, 3332 de la Paix, Sainte-Foy, QC, Canada, G1X 3W6*
[3]*INRS-Eau, 2700 rue Einstein, C.P. 7500, Sainte-Foy, QC, Canada, G1V 4C7*

Abstract. We monitored the chemistry of 603 small water bodies in three acid-sensitive regions of central Ontario, Canada (Algoma n=235, Muskoka n=216, Sudbury n=152) between 1988-1996 to determine whether they have responded to recent SO_2 emission reductions, and whether any chemical changes were related to lake characteristics. During the study, 27-56 % of lakes declined in SO_4^{2-} concentrations, 41-57 % declined in base cation concentrations, but only 26-28 % increased in pH or ANC (acidity status). Increases in pH were greatest in lakes with low ANC, but had weak relationships to lake color or volume. No consistent trends were observed for DOC, NO_3^--N or TP concentrations. Clearly, the long-term biological recovery of these sensitive aquatic ecosystems will depend on interactions among several environmental stressors, including acidification.

Keywords: acidification, decreasing SO_2 emissions, lake water chemistry, Ontario, sulfate deposition

1. Introduction

In Canada, acidic precipitation has been a major environmental problem for more than three decades. Sulfur dioxide (SO_2) emissions have been reduced in Canada and the U.S. by 53 and 20 %, respectively, from 1980 levels, resulting in a 46 % reduction in the area of eastern Canada receiving more than 20 kg ha^{-1} yr^{-1} wet sulfate (SO_4^{2-}) deposition. Monitoring programs currently track responses of aquatic ecosystems to reductions in SO_4^{2-} deposition (summarized in Jeffries, 1997); however, only a few examine small lakes and wetlands (e.g., Kelso *et al.*, 1992; McNicol *et al.*, 1995), despite the fact that these water bodies comprise the vast majority of aquatic resources at risk (73 % < 5 ha; Hélie *et al.*, 1993). These resources are particularly vulnerable to acidic inputs because they have low water retention times (Keller *et al.*, 1992; Yan *et al.*, 1996). More importantly, small lakes and wetlands (< 20 ha) are key habitat for wildlife, especially waterfowl (McNicol *et al.*, 1995).

Some chemical recovery (defined in this paper as increases in pH or alkalinity) does occur following SO_2 emission reductions (Newell and Skjelkvåle, 1997). For example, of 202 soft water lakes monitored in eastern Canada since 1981, lake pH has increased in 33 %, but has decreased in 11 % and remained unchanged in 56 % (Jeffries *et al.*, 1995). Sulfate concentrations generally are declining in eastern North American lakes (Jeffries *et al.*, 1995; Webster and Brezonik, 1995; Likens *et al.*, 1996; Bouchard, 1997), but other chemical responses vary dramatically by region, and include: increases in acid-neutralizing capacity (ANC) and pH (Keller *et al.*, 1992; Webster and Brezonik, 1995), declines in base cations (Couture, 1995; Likens *et al.*, 1996), in part attributable to declining airborne cations (Summers, 1995), increased nitrogen-based acidification (e.g., Driscoll and Van Dreason, 1993; Couture, 1995; Jeffries, 1995), and changes in natural organic acidity

Water, Air, and Soil Pollution **105:** 343–351, 1998.

(Bouchard, 1997). Patterns of change in lake chemistries also are influenced by climatic effects (Webster and Brezonik, 1995), including oxidation and mobilization of stored S in exposed sediments and watershed soils following drought (Jeffries *et al.*, 1995). To varying degrees, these responses have been observed in small water bodies in some regions (Keller *et al.*, 1992; Kelso *et al.*, 1992; McNicol and Mallory, 1994). Recent studies also indicate that small, rapid-flushing lakes (and the biota they support) are vulnerable to the combined effects of climate change, acidification and increased solar ultraviolet-B radiation (UV-B; Schindler *et al.*, 1996; Yan *et al.*, 1996).

Here, we examine chemical data from 603 small water bodies in three acid-sensitive regions (Algoma, Muskoka, Sudbury) of central Ontario, Canada monitored between 1988-1996. Our objectives were to describe the status of small lake chemistries in these areas, to determine whether chemical change has occurred over this period, and if so, to determine whether chemical changes were related to lake characteristics.

2. Methods

The three study areas in central Ontario are Algoma (center 47° 01'N, 83° 55'W), Muskoka (center 45° 30'N, 79° 06'W) and Sudbury (center 46° 54'N, 80° 41'W). All are underlain by Precambrian granitic bedrock covered by thin deposits of glacial ground moraine (McNicol *et al.*, 1995), and receive 20-24 kg ha^{-1} yr^{-1} wet SO_4^{2-} deposition and 900-1250 mm annual precipitation. Local nickel smelters at Sudbury affected lakes over a broad area (McNicol *et al.*, 1995), but emissions of SO_2 and metal particulates have been reduced dramatically (roughly 90 %) since the early 1970s.

For each lake, we measured open water area (ha), riparian area (ha; defined as the wetland and shrub zone between the open water and forest), and mid-lake depth (m). Lake volume was calculated as 0.464 × (area × mid-lake depth). We lacked water retention times for most lakes in Algoma or Muskoka, but in Sudbury, lakes with larger volumes had significantly longer retention times (Spearman correlation, $r_s = 0.61$, $p < 0.001$), so we used lake volume as a surrogate for water retention time for all areas. We also measured the color (Hazen units, Hu) of filtered water samples as an index of humic content.

Water samples were collected as mid-lake grabs (shallow lakes < 3.3 m) or integrated tube samples (for deeper lakes) from a helicopter in early October following turnover. Samples were collected at Algoma in five years (1988, 1992, 1994-1996), at Muskoka in five years (1990, 1991 (partial set), 1993, 1995, 1996), and at Sudbury in seven years (1990 (partial set; August shoreline grab samples), 1991, 1992 (partial set), 1993-1996). Water levels in many lakes were influenced by beaver activity. In some years, certain lakes could not be sampled because they were drained when beaver dams broke. Twenty-two chemical parameters were determined for filtered water samples (Mallory *et al.*, 1998); however, here we were only interested in: pH, alkalinity (hereafter termed acid-neutralizing capacity; ANC µeq L^{-1}), base cations (C_β, the sum of Ca^{2+}, Mg^{2+}, Na^+ and K^+ in µeq L^{-1}), SO_4^{2-} (mg L^{-1}), dissolved organic carbon (DOC, mg L^{-1}), nitrate plus nitrite (NO_3^--N + NO_2^--N, mg L^{-1}; hereafter NO_3^--N), and total phosphorus (TP, µg L^{-1}).

Trend detection analyses were conducted using a modified version of the DETECT software (Cluis *et al.*, 1989). We assumed that there was no persistence in chemical parameters because we collected only one sample per lake per year (at roughly the same time each year), and most lakes at Sudbury exchanged their water more than once yearly (Mallory *et al.*, 1998). We considered only monotonic trends (not step trends) because we had only one data point per lake per year, and because previous studies suggested a progressive change in chemistries (e.g., Keller *et al.*, 1992). For each parameter for each lake (with data in four or more years), we used Kendall's Tau to assess whether significant trends could be found ($\alpha = 0.05$). This non-parametric test is a well-suited, common approach to assessing monotonic trends with limited data, particularly because it is less sensitive to the potential effects of outliers with small sample sizes (El Shaarawi and Niculescu, 1992). We also conducted linear regressions to estimate trend slopes for each lake and parameter, and used the slopes to compare patterns of change across lake types. Lakes with data in three years also were used to calculate medians, while lakes with data in one or two years were not treated. Because our main goals were to examine patterns of change in lake chemical parameters within and among regions and lake types, we followed the approach of studies where individual site trends are aggregated to determine the proportion of sites with trends of similar direction within a region (Larsen *et al.*, 1995). Thus, we calculated the percentage of lakes demonstrating significant positive or negative trends, or no trends over the time period for each parameter for each region. While we recognize that four years of data is minimal on which to base trend analyses for individual lakes, the arbitrary choice of four or more years allowed the use of more lakes to assess regional conclusions (e.g., Kelso *et al.*, 1992).

3. Results

Although study lakes ranged from large, oligotrophic lakes to small, headwater lakes and swamps (including some bogs and fens), average physical (Table I) and chemical (Table II; median of 1995 values when all lakes were sampled in the same year) lake characteristics were similar across study areas. Lakes were small (93 % < 20 ha open water area), often shallow (60 % < 5 m deep), with a small volume (81 % < 3×10^5 m^3) and a large riparian zone (averaging more than one third the size of the open water area; Table I). This large wetland component to most water bodies was manifested by the staining (48 % with water color > 30 Hu), which indicates high humic content (median DOC near 5.0 mg L^{-1}). Most lakes were oligotrophic (TP < 10 μg L^{-1}) and acid-sensitive (median ANC ranged from 12.6 to 28.6 μeq L^{-1}; Table II). Among study areas, Sudbury lakes had higher C_β and SO_4^{2-} concentrations than the other areas, and lower pH and ANC than Algoma, whereas

TABLE I
Median (range) physiognomic characteristics of study lakes.

Parameter	Algoma (n = 235)	Muskoka (n = 216)	Sudbury (n = 152)
Open Water Area (ha)	3.8 (0.2 - 81.3)	4.2 (0.2 - 51.8)	4.0 (0.1 - 112.4)
Mid-lake Depth (m)	2.7 (0.5 - 36.0)	4.4 (0.3 - 30.5)	4.6 (0.6 - 31.4)
Volume (10^5 m^3)	2.7 (0.01 - 73.3)	2.9 (0.01 - 73.2)	3.5 (0.02 - 73.1)
Riparian Area (ha)	1.7 (0 - 42.6)	1.4 (0 - 29.4)	1.6 (0 - 18.7)
Color (Hu)	45 (10 - 165)	35 (5 - 145)	30 (5 - 175)

TABLE II
Proportion of Algoma (n=235), Muskoka (n=216) and Sudbury (n=152) study lakes demonstrating significant trends in chemical parameters.

	Parameter values		Parameter slopes		% lakes with significant trend		
Parameter	Median	Range	Median	Range	-	No Change	+
Algoma							
pH	6.04	4.01 - 7.30	0.04	-0.07 - 0.19	1	87	12
ANC (μeq L^{-1})	28.6	-108.1 - 496.8	3.40	-16.6 - 14.4	<1	77	23
C_β (μeq L^{-1})	160.6	53.5 - 688.8	-7.52	-37.53 - 9.45	57	43	<1
SO_4^{2-} (mg L^{-1})	3.86	1.19 - 12.15	-0.39	-1.39 - 0.41	56	43	<1
DOC (mg L^{-1})	6.54	1.38 - 24.30	-0.02	-0.73 - 1.27	5	92	3
NO_3^--N (μg L^{-1})	16.0	0 - 470.0	0.00	-0.03 - 0.02	<1	92	7
TP (μg L^{-1})	5.9	0.3 - 21.0	-0.05	-3.73 - 1.14	7	92	<1
Muskoka							
pH	5.85	4.12 - 6.90	0.05	-0.06 - 0.25	1	87	12
ANC (μeq L^{-1})	17.3	-75.9 - 340.2	2.75	-9.8 - 10.8	<1	79	21
C_β (μeq L^{-1})	165.5	34.1 - 1255.8	-8.28	-96.7 - 11.1	51	49	<1
SO_4^{2-} (mg L^{-1})	5.23	0.18 - 19.64	-0.39	-2.29 - 0.92	47	51	2
DOC (mg L^{-1})	5.25	1.95 - 26.02	0.01	-1.03 - 1.02	4	93	3
NO_3^--N (μg L^{-1})[2]	12.0	0 - 241.0					
TP (μg L^{-1})	7.15	1.98 - 143.0	-0.10	-23.0 - 2.5	5	92	3
Sudbury							
pH	5.68	4.01 - 7.90	0.06	-0.19 - 0.24	-	80	20
ANC (μeq L^{-1})	12.6	-90.6 - 655.1[3]	2.55	-65.2 - 175.4	<1	74	15
C_β (μeq L^{-1})	219.7	53.8 - 988.3[3]	-6.49	-126.6 - 146.0	41	59	<1
SO_4^{2-} (mg L^{-1})	8.75	1.29 - 22.19	-0.25	-1.48 - 1.41	28	71	1
DOC (mg L^{-1})	4.60	0.53 - 24.87	0.05	-1.09 - 2.02	2	92	6
NO_3^--N (μg L^{-1})	9.0	0 - 210.0	0	-0.03 - 0.01	<1	88	12
TP (+μg L^{-1})	6.0	0.8 - 386.7	-0.22	-6.4 - 23.2	5	94	1

[1] Trend determined significant using Kendall's Tau as part of a modified DETECT software (Cluis *et al.*,1989)
[2] For NO_3^--N, data were insufficient for trend analyses
[3] One lake with exceptionally high ANC and C_β was not included in the range to avoid an unrealistic depiction of Sudbury lake chemistries

Algoma lakes had the lowest SO_4^{2-} and highest DOC and NO_3^--N concentrations, and Muskoka lakes had the highest TP concentrations (Kruskal-Wallis tests, all p's < 0.03).

Trends in chemistry were similar for all three areas (Table II). Median annual changes in pH (+), ANC (+), and SO_4^{2-} concentrations (-) were in the direction expected if lakes are responding to reduced SO_4^{2-} deposition. Roughly 50 % of Algoma and Muskoka lakes showed significant declines in SO_4^{2-} concentrations since 1988 and 1990, respectively, but only 26-28 % increased in pH or ANC. At Sudbury, fewer lakes (27 %) declined in SO_4^{2-} concentration, but similar numbers (26 %) increased in pH or ANC since 1990. Rather than having larger increases in pH or ANC, lakes apparently offset SO_4^{2-} declines with declining C_β concentrations; more than 40 % of lakes at each area showed significant C_β declines. Most lakes (90 %) showed no change in DOC, NO_3^--N or TP concentrations, and had negligible median annual changes. Despite the increases in pH or ANC and decreases in SO_4^{2-}, many lakes remained damaged or acid-sensitive in 1996; 45 % of Algoma, 59 % of Sudbury, and 64 % of Muskoka lakes had pH values below 6.0 (where biological damage, such as loss of biodiversity, to some aquatic ecosystems may occur; Jeffries, 1997).

Across regions, chemical changes tended to differ among types of lakes. In Algoma and Muskoka, lakes that had ANC < 20 μeq L^{-1} (hence higher sensitivity to acidic inputs) had greater increases in pH than higher ANC lakes (ANC > 20 μeq L^{-1}; Spearman correlations, $r_s \leq -0.68$, both $p < 0.001$), although this difference was not significant at Sudbury. Declines in C_β and SO_4^{2-} concentrations tended to be higher in lakes with smaller volumes at Algoma and Muskoka ($r_s \leq -0.11$, both $p < 0.10$), but pH changes were unrelated to volume. Relationships among Sudbury lake volume and C_β, SO_4^{2-} concentrations and pH trends were in the same direction as those at Algoma and Muskoka, but were not statistically significant. In Algoma, changes in pH, C_β and SO_4^{2-} concentrations were not significantly correlated with water color. However, more highly colored lakes had greater pH increases in Muskoka ($r_s = 0.19, p = 0.005$) and smaller C_β concentration declines in Sudbury ($r_s = -0.18, p = 0.03$), although these correlations were weak.

4. Discussion

4.1. Trends in Lake Chemistries

Small lakes in acid-sensitive regions of central Ontario exhibited changing chemistries since 1988, concurrent with declines in SO_4^{2-} deposition across eastern North America (Summers, 1995). In particular, we observed widespread declines in lake water SO_4^{2-} concentrations, but these were not balanced by proportionally similar increases in pH or ANC. Instead, declines in C_β concentrations were observed, perhaps delaying the expected increases in pH and ANC (i.e., chemical recovery). Similar results have been reported near Sudbury (Keller *et al.*, 1992), Muskoka and northwestern Ontario (Jeffries *et al.*, 1995), southern Québec (Couture, 1995; Bouchard, 1997), the Adirondack region of New York (Driscoll *et al.*, 1995) and Europe (see Newell and Skjelkvåle, 1997). These studies generally confirm that, in many watersheds, concentrations of C_β exported from soils are declining, but it will take many years for pools of C_β to be replenished, and this may delay pH- and ANC-based recovery (Likens *et al.*, 1996). Moreover, declines in C_β in precipitation (Summers, 1995) means that there is less input of cations to watersheds, which may be partly responsible for the widespread observations of reduced C_β concentrations in surface waters (Bouchard, 1997), exacerbating the problem of depleted C_β pools in watershed soils.

We found that concentrations of DOC, NO_3^--N or TP did not change, consistent with the results from Clearwater Lake near Sudbury (Bodo and Dillon, 1994). In contrast, Bouchard (1997) found DOC increases in some southern Québec lakes, a possible further mechanism by which recovery of lakes was delayed. Although Jeffries (1995) did not find evidence of N-saturation or acidification in most of our study lakes, other studies have found that acidification due to N deposition is increasing in certain regions of Ontario, Québec, and the northeastern U.S. to the point that it might undermine the improvements achieved through reductions in SO_2 emissions (Driscoll and Van Dreason, 1993; Couture, 1995; Jeffries, 1995). Why our results for DOC, NO_3^--N and TP concentrations differ somewhat from those of other regions with similar acid sensitivities is unclear, but we cannot exclude the likelihood that these parameters, which tend to vary seasonally (Bodo and Dillon, 1994; Jeffries, 1995), may exhibit higher variability between years using once-annual sampling, possibly masking trends.

While we have documented some chemical recovery, roughly half of the lakes still have autumn pH values that are less than 6.0. For lakes that are not naturally acidic (generally those with little organic acidity), biological damage (e.g., loss of biodiversity) begins to occur when acidic inputs cause lake pH values to drop below 6.0 (Jeffries, 1997). During spring runoff or storm events throughout the year, many lakes above pH 6.0 in the autumn will have significantly lower pH values at other times (Bodo and Dillon, 1994). Moreover, recent experimental studies have shown that in-lake alkalinity generation, a biogeochemical process that influences the rate of chemical recovery of lakes, becomes impaired when lake pH is reduced to below 5 (Kelly *et al.* 1995). Thus, the majority of small water bodies in the Algoma, Muskoka, and Sudbury regions remain sensitive to acidic precipitation, and many remain chemically and biologically damaged (56 % < pH 6.0; 36 % < pH 5.5, 15 % < pH 5.0).

4.2. Differential Responses Among Lakes

For Algoma and Muskoka, rates of chemical change were greater for smaller lakes, supporting the contention that small lakes should respond more quickly to changes in precipitation (McNicol and Mallory, 1994; Yan *et al.*, 1996). However, the lack of significant relationships between lake volume and chemical change in Sudbury indicates that volume alone (in the absence of information on watershed area) may not be an accurate surrogate for water retention time, despite their statistical relationship. Recently, Mallory *et al.* (1998) showed that these same small, rapid-flushing lakes at Sudbury exhibited more rapid chemical improvements than larger lakes between 1983-1995.

In Algoma and Muskoka, lakes that were more sensitive to acidification (low ANC due to limited buffering capacity of underlying geology) during our initial sampling exhibited the greatest improvements in chemistry over the study. Monitoring elsewhere in Ontario has also found that increases in pH were greater for lakes with initial pH < 6.0 (Jeffries, 1997). We found no evidence that more sensitive lakes were responding faster in Sudbury (see also Mallory *et al.*, 1998). Extensive, historical damage to the most sensitive lakes at this site (usually glacial headwaters) may mean that these watersheds will require more time to recover (possibly due to hysteresis associated with higher dry S loads, or more severe climatic disruption). Bouchard (1997) showed that acidity status improved the most in southern Québec lakes with relatively high pH (median pH 6.51), but that it could be attributed to lower DOC levels in these lakes and their more rapid response to changes in atmospheric inputs based on watershed characteristics. Nonetheless, our results are encouraging, because they indicate that even very damaged lakes (ANC < 0 $\mu eq\ L^{-1}$) exhibit signs of chemical recovery when S deposition on the watershed is reduced.

The most surprising result in this study was the limited evidence of a relationship between water color and chemical recovery. We expected that differences in chemical responses might be related to lake color, because higher color (positively correlated with DOC) indicates greater wetland influences in watersheds (Jeffries *et al.*, 1995). Wetlands are typically net sinks for SO_4^{2-}, but become net sources following dry periods (LaZerte, 1993; Jeffries, 1997). In larger Muskoka lakes, pH increases were smaller in lakes with higher DOC concentrations, presumably because these lakes had a higher wetland component in their watersheds, and thus received higher export of SO_4^{2-} from watershed wetlands following dry years than low DOC lakes (Jeffries, 1997). Similarly, Bouchard (1997) found that DOC concentrations in lakes were correlated to pH and C_β concentration changes in southern Québec, with low DOC lakes showing the greatest recovery. Given these precedents and the range of lake color in our sampling (Table I), it is unclear why we found few differences in key parameters in relation to lake color in this or longer term sampling at Sudbury (Mallory *et al.*, 1998) and Algoma, and why the significant results in Muskoka were in the direction opposite to other studies (i.e., brownwater lakes had significantly greater pH increases over the time period). One possible explanation is that many of the Muskoka lakes are headwater beaver ponds. In some beaver ponds, lake productivity (particularly nutrient transformations in anoxic zones) and organic matter decomposition can be important contributors to ANC production (Cirmo and Driscoll, 1993) and may be related to the patterns of pH change we observed. Nonetheless, these conditions are also common at Algoma and Sudbury where we did not see similar patterns.

4.3. Interactions with Other Stressors

While there has been progress, albeit slow, in the response of Canada's lakes to the reductions in SO_2 emissions, acidity remains high, in part because reductions in SO_2 are too recent in many regions for lakes to respond noticeably. Even after full implementation of control programs in Canada and the U.S., critical loads of wet SO_4^{2-} deposition required to maintain lake pHs above 6.0 will be exceeded by up to 10 kg $ha^{-1}\ yr^{-1}$ in parts of central Ontario and southern Québec (Jeffries, 1997), leaving approximately 10 % of lakes in these areas acidic (down from 20 % and 12 %, respectively, before emission controls). Other environmental stressors cast further uncertainty on these predictions. There is now strong concern that climate change and increased variability will further retard chemical recovery (Wilby, 1995). Recent experiments in Ontario (Schindler *et al.*, 1996; Yan *et al.*, 1996) clearly link decreases in precipitation to pH declines of lakes. Jeffries *et al.* (1995) suggested that the drought of the mid-1980s set chemical recovery back six years at Turkey Lakes (part of our Algoma data set). If climate variability interferes with recovery, then efforts to achieve regional critical loads for the protection of aquatic biota (i.e., restoring lakes to pH 6.0) in eastern Canada may require even greater reductions in SO_2 emissions to compensate for climatic effects and stored S in watersheds. More importantly, chemical and biological recovery will be delayed even further (Wilby, 1995). There are also disturbing signs that because NO_x emissions have not changed much recently, NO_3^- deposition may undermine some of the benefits derived from the control of SO_2 emissions (Jeffries, 1995).

In addition to the concern over climate change, increased UV-B and mercury (Hg) also pose significant problems for aquatic ecosystems. UV-B penetration into lakes is a growing threat because it appears to be increasing (Kerr and McElroy, 1993) and because many aquatic biota are deleteriously affected (e.g., Schindler *et al.*, 1996), particularly at low pH (e.g., Long *et al.*, 1995). In Ontario, climate warming and associated pH declines caused a decrease in lake DOC, and this led to increased penetration of UV-B into lakes (Schindler *et al.*, 1996; Yan *et al.*, 1996). Similarly, Hg uptake has detrimental impacts on biota if sufficient quantities are absorbed from the diet (Scheuhammer, 1991). While direct Hg deposition has led to elevated environmental concentrations (e.g., Mierle, 1990), levels also appear to be elevated in lakes and biota remote from point sources but affected by acidic precipitation (Gilmour and Henry, 1991).

5. Conclusions

Most lakes in this study were small (< 20 ha) with confined watersheds, and responded quickly to changes in atmospheric inputs and precipitation. Therefore, the improvements observed probably reflect recent declines in acidifying inputs, and not hysteresis seen in larger lakes. In addition to acidic precipitation, recent studies demonstrate that recovery of aquatic ecosystems is threatened by interactions from other environmental stressors, and that ecosystem responses to reductions in one stressor (e.g. reduced acidic deposition) will be dictated by the concurrent levels of these other stressors. Clearly, while recent improvements in acidity status of small lakes in central Ontario are encouraging, these rather short temporal records are still insufficient to predict future trends. Even with full implementation of SO_2 abatement programs in Canada (achieved in 1994) and the U.S. (year 2010), critical loads of SO_4^{2-} deposition in some areas will still exceed levels required to protect many lakes. Thus, maintaining monitoring programs that address varied environmental stressors, including acidification, is critical to better understand aquatic ecosystem structure and function and to develop sound predictive models.

Acknowledgments

This study was part of an ongoing Canadian Wildlife Service investigation of the impact of anthropogenic acidification on aquatic food chains on lakes, and the ensuing effects on habitat suitability for water-dependent birds. Thanks to Don Kurylo and the staff at the GLFC LRTAP Laboratory for chemical analyses, and many field assistants for help with data collection. Funding for this project was provided by the Long Range Transport of Air Pollutants (LRTAP) program of Environment Canada, and the Canadian Wildlife Service (Ontario Region).

References

Bodo, B.A. and Dillon, P.J.: 1994, 'De-acidification Trends in Clearwater Lake near Sudbury, Ontario 1973-1992', in Hippel, K.W. et al. (eds.), *Stochastic and Statistical Methods in Hydrology and Environmental Engineering, Vol. 3*, Kluwer, Dordrecht.
Bouchard, A.: 1997, *Water, Air, Soil Pollut.* **94**, 225.
Cirmo, C.P. and Driscoll, C.T.: 1993, *Wetlands* **13**, 277.
Cluis, D.A., Langlois, C., Van Coille, R. and Laberge, C.: 1989, *Environ. Monitor. Assess.* **12**, 429.
Couture, S.: 1995, *Can. J. Fish. Aquat. Sci.* **52**, 1936.
Driscoll, C.T. and Van Dreason, R.: 1993, *Water, Air, Soil Pollut.* **67**, 319.
Driscoll, C.T., Postek, K.M., Kretser, W. and Raynal, D.J.: 1995, *Water, Air, Soil Pollut.* **85**, 583.
El Shaarawi, A.H. and Niculescu, S.P.: 1992, *Environmetrics*, **3**, 385.
Gilmour, C.C. and Henry, E.A.: 1991, *Environ. Pollut.* **71**, 131.
Hélie, R.G., Wickware, G.M. and Sioh, M.: 1993, *Quantitative Assessment of Surface Water at Risk due to Acidification in Eastern Canada*, Environment Canada, Ottawa.
Jeffries, D.S.: 1995, *Water, Air, Soil Pollut.* **85**, 433.
Jeffries, D.S., Clair, T.A., Dillon, P.J., Papineau, M. and Stainton, M.P.: 1995, *Water, Air, Soil Pollut.* **85**, 577.
Jeffries, D.S. (ed.): 1997, *Long Range Transport and Acid Deposition Assessment Report - Aquatic Effects Part, 1997*, Environment Canada.
Keller, W., Pitblado, J.R. and Carbone, J.: 1992, *Can. J. Fish. Aquat. Sci.* **49** (Supp. 1), 25.
Kelly, C.A., Amaral, J., Turner, M.A., Rudd, J.W.M., Schindler, D.W. and Stainton, M.P.: 1995, *Biogeochemistry* **28,** 115.
Kelso, J.R. M., Shaw, M.A. and Jeffries, D.S.: 1992, *Environ. Pollut.* **78**, 65.
Kerr, J.B. and McElroy, C.T.: 1993, *Science* **262**, 1032.
Larsen, D.P., Urquhart, N.S. and Kugler, D.L.: 1995, *Water Res. Bull.* **31**, 117.
Lazerte, B.D.: 1993, *Biogeochemistry* **18**, 153.
Likens, G.E., Driscoll, C.T. and Buso, D.C.: 1996, *Science* **272**, 244.
Long, L.E., Saylor, L.S. and Soule, M.E.: 1995, *Conserv. Biol.* **9**, 1301.
Mallory, M.L., McNicol, D.K., Cluis, D.A. and Laberge, C.: 1998, *Can. J. Fish. Aquat. Sci.*, in press.
McNicol, D.K. and Mallory, M.L.: 1994, *Water, Air, Soil Pollut.* **73**, 105.
McNicol, D.K., Kerekes, J.J., Mallory, M.L., Ross R.K. and Scheuhammer A.M.: 1995, *The Canadian Wildlife Service LRTAP Biomonitoring Program. Part 1. A Strategy to Monitor the Biological Recovery of Aquatic Ecosystems in Eastern Canada from the Effects of Acid Rain*, Can. Wildl. Serv. Tech. Rep. Ser. No. 245.
Mierle, G.: 1990, *Environ. Toxicol. Chem.* **9**, 843.
Newell, A.D. and Skjelkvåle, B.L.: 1997, *Water, Air, Soil Pollut.* **93**, 27.
Scheuhammer, A W.: 1991, *Environ. Pollut.* **71**, 329.
Schindler, D.W., Curtis, P.J., Parker, B.R. and Stainton, M.P.: 1996, *Nature* **379**, 705.
Summers, P.W.: 1995, *Water, Air, Soil Pollut.* **85**, 653.
Webster, K.E. and Brezonik, P.L.: 1995, *Water, Air, Soil Pollut.* **85**, 1575.
Wilby, R.L.: 1995, *Environ. Conserv.* **22**, 363.
Yan, N.D., Keller, W., Scully, N.M., Lean, D.R.S. and Dillon, P.J.: 1996, *Nature* **381**, 141.

TRENDS IN SULFATE, BASE CATIONS AND H^+ CONCENTRATIONS IN BULK PRECIPITATION AND THROUGHFALL AT INTEGRATED MONITORING SITES IN FINLAND 1989-1995

LIISA UKONMAANAHO[1], MICHAEL STARR[1] and TUIJA RUOHO-AIROLA[2]

[1]*Finnish Forest Research Institute, P.O. Box 18, FIN-01301 Vantaa, Finland*
[2]*Finnish Meteorological Institute, Sahaajankatu 22 E, FIN-00810 Helsinki, Finland*

Abstract. Temporal trends in sulfate, base cation (Ca^{2+} + Mg^{2+} + K^+), and H^+ ion concentrations in bulk precipitation and throughfall samples collected over a seven year period (1989-95) in four forested catchments in Finland are presented. The catchments are in remote locations and span the boreal zone (61-69 °N). The stands represent old, undisturbed forests, and are composed of varying proportions of Scots pine, Norway spruce and deciduous species (mainly *Betula* spp.). Monthly SO_4^{2-}-S and H^+ ion concentrations in bulk precipitation averaged over the study period and catchments were: 18.7 μmol L^{-1} and 32.3 μmol L^{-1}. The corresponding values for throughfall were: 37.4 μmol L^{-1} and 32.4 μmol L^{-1}. Sulfate and H^+ ion concentrations in bulk precipitation and throughfall both showed negative linear trends, which were significant ($p < 0.05$) for the three southernmost catchments. Concentrations and trend slope decreased northwards (e.g., bulk precipitation SO_4^{2-}-S slope estimates: -1.6 to -1.0 μmol L^{-1} yr^{-1}). The decline was greater for throughfall than for bulk precipitation, indicating a proportionally greater reduction in dry deposition than wet. The sum of base cation concentrations averaged 12.1 $\mu mol_{(+)}$ L^{-1} in bulk precipitation and 83.1 $\mu mol_{(+)}$ L^{-1} in throughfall. There were no significant trends in the sum of base cations ($p > 0.05$). It is concluded that the reported reduction in S emissions over the study period has resulted in a significant reduction in the acidity and SO_4^{2-} concentration of bulk precipitation, and this reduction has has been reflected in throughfall concentrations. The greatest reduction has taken place in the southern part of the country.

Keywords: base cations, bulk precipitation chemistry, catchments, protons, sulfate, throughfall chemistry

1. Introduction

Emissions of SO_2 from Finland have declined progressively from their peak of about 600,000 t yr^{-1} in the 1970s to about 100,000 t yr^{-1} by the mid 1990s. This is well in line with the UN-ECE 1985 protocol to reduce S emissions by at least 30 % by 1993 compared with 1980 levels and Finland's goal of 121,000 t yr^{-1} by 2000. However, only about 22 % of the S deposition in Finland is derived from domestic sources (Tuovinen *et al.* 1990; Bernes, 1993).

Precipitation chemistry mainly consists of compounds of S and N, which have an acidifying effect, and base cations (Ca^{2+}, Mg^{2+}, K^+), which have a neutralizing effect. Differences in the composition of precipitation collected in the open and that collected under boreal forest canopies have been reported in many studies (e.g., Parker, 1983; Lindberg *et al.*, 1986, Helmisaari and Mälkönen, 1989; Hyvärinen, 1990; Edmonds *et al.*, 1991; Johnson and Lindberg, 1992; Lövblad *et al.*, 1994). These differences are due to interaction with the canopy (foliar uptake *vs.* leaching) and to additional dry deposition filtered out of the atmosphere by the canopy (interception) and subsequent wash-off. The difference between open and forest precipitation varies with chemical constituent as well as with the species composition and structure of the stand (Mahendrappa, 1983; Starr and Ukonmaanaho, 1995). Spruce stands are especially effective in trapping particles and certain gases (Bredemeier, 1988).

Water, Air, and Soil Pollution **105:** 353–363, 1998.

The monitoring of throughfall and bulk precipitation has proved a useful means of assessing total deposition to forest ecosystems (Bredemeier, 1988; Forsius *et al.*, 1995; Hultberg and Grennfelt, 1992; Ivens *et al.*, 1990; Kallio and Kauppi, 1990). However, most throughfall and bulk precipitation studies have been carried out only for a few years and/or at single sites.

In this paper, we present some Finnish results from the throughfall and bulk precipitation subprograms of the UN-ECE International Cooperative Programme on Integrated Monitoring (ICP-IM) (EDC, 1993; Bergström *et al.*, 1995). The data cover a seven year period and are for eight stands, two in each of four catchments located throughout Finland. The ICP-IM program is an international ecosystem monitoring program, the overall objectives of which are to describe the present state of ecosystems and to monitor changes due to environmental change, particularly the influence of anthropogenic pollutants transported over long distances (EDC, 1993). To assess the levels of air pollution, we have decided to present concentration rather than deposition data because deposition fluxes in Finland are largely determined by the amount of precipitation, which shows a strong south-north gradient.

2. Materials and Methods

2.1. Catchments and Plots

The study was carried out at the four Finnish ICP-IM catchments: Valkea-Kotinen, Hietajärvi, Pesosjärvi and Vuoskojärvi (Table I). Bulk precipitation was collected in an open area and throughfall collected from two permanent plots located in each of the four catchments. The small forested catchments span the boreal coniferous phytogeographic zone and are situated in remote, protected, conservation areas (Table I). The forest in the catchments represent natural or semi-natural communities.

TABLE I
General features of the IM study catchments.

	Valkea-Kotinen	Hietajärvi	Pesosjärvi	Vuoskojärvi
Location	61° 14′ N	63° 09′ N	66° 17′ N	69° 44′ N
	25° 03′ E	30° 40′ E	29° 26′ E	26° 56′ E
Area, km^2	0.3	0.7	6.1	2.0
Phytogeographic zone*	Southern boreal	Middle boreal	Northern boreal	Northern boreal
Climatic zone**	Humid continental	Continental subarctic	Continental subarctic	Continental subarctic
Annual mean temperature, °C	3.1	2.0	-0.5	-1.9
Annual mean precipitation, mm	618	592	571	395
Vegetation period (>5 °C), da	112	102	82	57

* Ahti *et al.* 1968; ** Köppen-Geiger classification

Permanent plots varying in size from 25 x 25 m to 40 x 40 m were established in each catchment. The plots are situated on flat to nearly flat surfaces in areas representative of the dominant forest site types in each catchment. The plots are as homogenous as possible with regard to soil morphology, ground vegetation and tree stand characteristics (Table II). The stands are composed of varying proportions of Scots pine (*Pinus sylvestris* L.), Norway spruce (*Picea abies* Karst.) and deciduous species (mainly *Betula* spp.), representing submesic heathland forests to lichen woodlands. For further details about the catchments and plots, see Bergström *et al.* (1995).

TABLE II
Characteristics of the throughfall study stands.

Catchment	Plot	Plot size, ha	Tree density ha^{-1}	Basal area $m^2\ ha^{-1}$	Mean height m	Stem volume				Canopy coverage %	dbh[‡] cm	Age of dominant trees, years
						$m^3\ ha^{-1}$	pine* %	spruce** %	deciduous[†] %			
Valkea-	2	0.16	1231	40	16	407	7	86	7	81	26.2	190
Kotinen	3	0.12	1433	48	17	531	4	54	41	88	26	155
Hietajärvi	1	0.16	1238	22	10	199	86	0	14	61	23.4	100
	4	0.16	519	21	15	228	91	0	9	51	30.3	230
Pesosjärvi	1	0.09	922	20	9	147	2	69	29	55	29.9	320
	2	0.12	1825	25	8	192	46	39	15	52	24	240
Vuoskojärvi	2	0.09	1789	5	4	16	3	0	97	53	8.4	150
	3	0.16	431	9	6	50	98	0	2	34	27.4	180

* Scots pine *(Pinus sylvestris)*
** Norway spruce *(Picea abies)*
† deciduous species (mainly Birch, *Betula spp.* and Aspen, *Populus tremula*; but at Vuoskojärvi, Mountain birch, *B.pubescens ssp. tortuosa)*
‡ breast height diameter, 1.3 m

2.2. Bulk Precipitation and Throughfall Collectors and Sampling

Three bulk precipitation collectors (NILU type) were placed in an open area at a height of 120 cm and such that the nearest trees did not shadow the collectors; the angle to the top of the trees was <45°. Samples were collected once a week throughout the year, but combined into monthly samples for each collector for analysis.

Throughfall collectors were placed systematically at 10 m intervals around the plots. Depending on plot size, there were 10 to 16 collectors per plot. Each collector used to collect rainfall consisted of an acid-washed polyethylene funnel (collecting surface = 308 cm^2) with a 2 cm high rim connected to an acid-washed polyethylene bottle (from 1994, a polyethylene bag) (Hyvärinen, 1990). Snow samples were collected only since winter 1994/95. The snow collectors, which replaced the rainfall collectors during the winter months (October-November to April-May), each consisted of a polyethylene bucket with a collector area is 491 cm^2. The volume of throughfall collected by each collector was recorded before being poured (after melting in the case of snow samples) into a common acid-washed polyethylene canister and a subsample was taken for analysis. The rainfall throughfall samples were collected weekly and snow throughfall samples monthly.

2.3. Sample Analysis

The samples were stored in dark and cool (+4 °C) conditions until analysis began. Concentrations of Ca^{2+}, Mg^{2+} and K^+ in bulk precipitation were determined using ion chromatography and in throughfall samples by using ICP-AES. Sulfate concentrations were determined by ion chromatography and H^+ concentrations were calculated from pH measurements made potentiometrically using a combined electrode. For further details, see Bergström *et al.* (1995).

2.4. Numerical Handling and Statistical Analysis

Throughfall samples with PO_4^{3-} ion concentrations >10.5 μmol L^{-1} (>1 mg L^{-1}) were rejected on the grounds of probable contamination of the whole sample chemistry from bird droppings (Novo *et al.*, 1992). Monthly volume weighted bulk precipitation values were calculated from the monthly values of the three parallel collectors. Monthly throughfall (rainfall) volume weighted concentrations were calculated from the weekly plot composite values.

Because the sampling of throughfall throughout the year did not start until 1994, the number of monthly observations for each catchment are less than the potential 84 (7 years x 12 months). Accordingly, the number of monthly throughfall observations ranged from 40 at Vuoskojärvi to 52 at Valkea-Kotinen.

The trend over the seven year period is shown as a LOWESS (locally weighted) curve fitted to the monthly data (GraphPad PRISM™) using an algorithm adapted from Chambers *et al.* (1993). Linear trends were analyzed using all the monthly values available and the non-parametric Seasonal-Kendall trend analysis procedure to calculate an estimate of the true slope of the linear trend (Hirsch *et al.*, 1982; Gilbert, 1987). This test may be used even though there are seasonal cycles and missing values.

3. Results

The monthly amount of throughfall was, as expected, strongly correlated to the amount of bulk precipitation (Valkea-Kotinen, r = 0.93; Hietajärvi, r = 0.93; Pesosjärvi, r = 0.94; Vuoskojärvi, r = 0.95). The amount of throughfall as a percentage of the amount of bulk precipitation (throughfall yield) for each catchment ranged from 86 to 90 %. In an earlier paper (Starr and Ukonmaanaho, 1994), we showed that the throughfall yield for the same plots was correlated to stand basal area (r = -0.80) and crown cover (r = -0.81).

The monthly SO_4^{2-}-S, sum of the base cations, and H^+ ion concentrations in bulk precipitation averaged over the study period and catchments were: 18.7 μmol L^{-1}, 12.1 $\mu mol_{(+)}$ L^{-1} and 32.3 μmol L^{-1}, respectively. The corresponding values for throughfall were: 37.4 μmol L^{-1}, 83.1 $\mu mol_{(+)}$ L^{-1} and 32.4 μmol L^{-1}.

3.1. Differences among Catchments and Plots

Mean monthly concentrations calculated for only those months common to all catchments are presented in Table III. By selecting values for the same common months only, the effects of seasonal (monthly) variability are reduced and geographic comparisons among the catchments can be made.

TABLE III

Monthly volume weighted mean concentrations and standard deviation of acidity (H^+), SO_4^{2-}-S and base cations (Ca^{2+} + Mg^{2+} + K^+) in bulk precipitation and throughfall. Values are only for those months common to all catchments during 1989-95 (n=36). Means within columns followed by the same letter (upper case for bulk precipitation,BP, and lower case for throughfall, TF) are significantly different by Bonferroni tests ($p < 0.05$).

Catchment	Plot	H^+, µmol L^{-1}		SO_4^{2-}-S, µmol L^{-1}		Base cations, $\mu mol_{(+)}$ L^{-1}	
		Mean	sd	Mean	sd	Mean	sd
Valkea-Kotinen	BP	30.3 A	11.89	18.7 AB	10.3	12.5 A	7.52
	TF 2	37.9 a	20.24	56.1 abcdef	25.0	141.8 abcdef	56.24
	TF 3	30.4 b	17.68	40.5 ghij	18.4	108.1 ghijk	48.30
Hietajärvi	BP	29.8	11.62	18.7 C	8.4	9.9	6.24
	TF 1	36.3 c	25.96	28.1 ak	20.3	48.7 ag	37.21
	TF 4	38.7 d	20.05	28.1 bg	13.4	39.9 bh	29.69
Pesosjärvi	BP	26.4	11.45	12.5 A	6.9	5.3 A	2.64
	TF 1	26.5	16.48	31.2 cl	17.5	63.5 c	39.44
	TF 2	24.6	15.60	21.8 dh	25.0	27.6 di	15.68
Vuoskojärvi	BP	22.0 A	16.12	12.5 BC	7.5	10.3	15.93
	TF 2	15.5 abcde	10.30	15.6 eikl	10.3	58.7 ej	139.53
	TF 3	31.9 e	22.40	28.1 fj	21.2	59.0 fk	70.08

The mean H^+ concentrations in bulk precipitation decreased northwards, but the only significant ($p < 0.05$) difference among catchment means was that between the northernmost and southernmost catchments (Vuoskojärvi and Valkea-Kotinen). Among the throughfall plots, only plot 2 at Vuoskojärvi was significantly different from any of the other throughfall plots.

Sulfate concentrations in bulk precipitation and throughfall showed the same latitudinal decrease as H^+ concentrations. The mean for Valkea-Kotinen was significantly different from the means for Pesosjärvi and Vuoskojärvi, and that for Hietajärvi was significantly different from that for Vuoskojärvi. Throughfall SO_4^{2-} means at Valkea-Kotinen were significantly greater than the means for the other, more northerly, throughfall plots.

The mean of the sum of base cation concentrations in bulk precipitation decreased northwards as far Pesosjärvi and then increased again at Vuoskojärvi. The means for Valkea-Kotinen and Pesosjärvi were significantly different. The same pattern was shown by the throughfall means, although not as clearly. The increase in the sum of base cation concentrations at Vuoskojärvi is due to additional inputs of marine Mg^{2+} derived from the nearby Arctic Ocean (Starr and Ukonmaanaho, 1995). The significantly greater bulk precipitation and throughfall sum of base cation means at Valkea-Kotinen are further evidence of greater anthropogenic base cation pollution (dry deposition) in the southwest of Finland (Tuovinen *et al.,* 1990).

3.2. TIME TRENDS

Monthly H^+, SO_4^{2-}, and sum of base cation concentrations for the seven year study period, together with the LOWESS smoother, are presented in Figures 1, 2 and 3, respectively. The corresponding Seasonal-Kendall linear slope estimates are presented in Table IV.

H^+ and SO_4^{2-} concentrations in both bulk precipitation and throughfall have significant ($p < 0.05$) negative linear trends at all catchments except Vuoskojärvi (where the slope estimates were not significantly different from zero). Thus, there has been a significant progressive reduction in the acidity and SO_4^{2-} concentration in precipitation throughout most of Finland over the period 1989-1995. Furthermore, the size of the slope estimate decreases northwards; showing that the reductions in acidity and sulphate concentrations have been the greatest in southern Finland where the concentrations have been the highest. Because bulk throughfall includes a fraction of dry deposition (filtered) that is not included in the bulk precipitation, the steeper decline in acidity and SO_4^{2-} concentrations registered for throughfall is presumably due to a proportionally greater reduction in the dry deposition of S than wet.

The sum of base cation concentrations did not show significant linear temporal trends ($p > 0.05$). The decrease in the acidity of bulk precipitation and throughfall therefore has been due to the decline in SO_4^{2-} concentrations rather than an increase in the neutralizing effect of base cations.

TABLE IV

Seasonal-Kendall estimates of true slope coefficients of linear trends for SO_4^{2-}-S (μmol L^{-1} yr^{-1}), H^+ (μmol $L^{-1}yr^{-1}$) and base cations ($Ca^{2+} + Mg^{2+} + K^+$, $\mu mol_{(+)}$ L^{-1} yr^{-1}) (ns= non significant, $p > 0.05$).

Catchment	Plot	SO_4^{2-}-S	H^+	Base cations
Valkea-Kotinen	BP	-1.6	-1.4	ns
	TF 2	-5.5	-6.3	ns
	TF 3	-5.4	-4.2	ns
Hietajärvi	BP	-1.7	-2.0	ns
	TF 1	-3.1	-3.0	ns
	TF 4	-3.1	-3.1	ns
Pesosjärvi	BP	-1.0	-1.4	ns
	TF 1	-3.4	-4.7	ns
	TF 2	-2.9	-2.1	ns
Vuoskojärvi	BP	ns	ns	ns
	TF 2	ns	-2.4	ns
	TF 3	ns	-4.5	ns

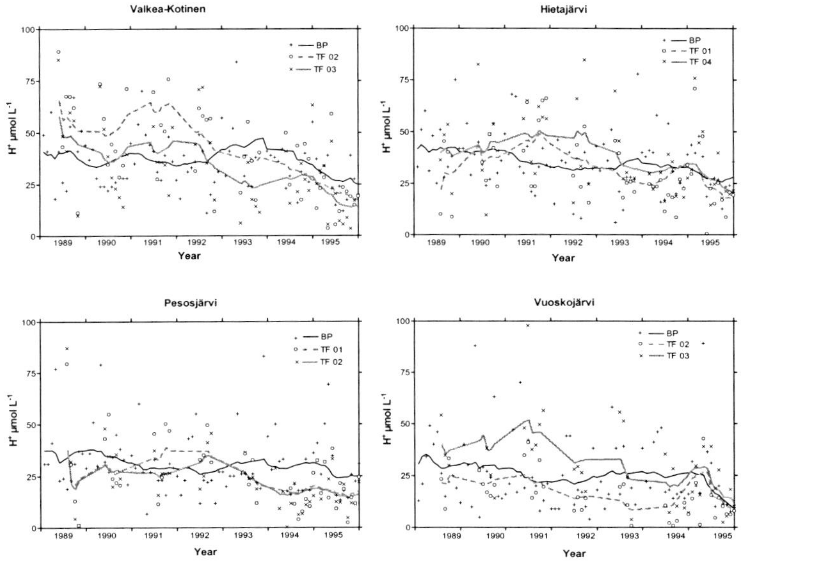

Fig. 1. Monthly H^+ ion concentrations in bulk precipitation (BP) and throughfall (TF) over ther period 1989-95 at each study catchment. For description of TF plots, see Table II. The lines are the LOWESS smoother.

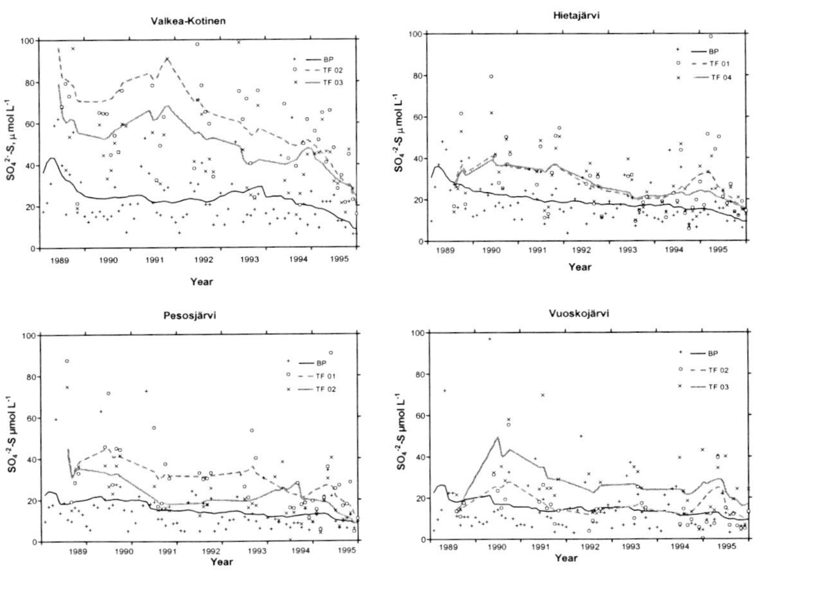

Fig. 2. Monthly SO_4^{2-} concentrations in bulk precipitation (BP) and throughfall (TF) over ther period 1989-95 at each study catchment. For description of TF plots, see Table II. The lines are the LOWESS smoother.

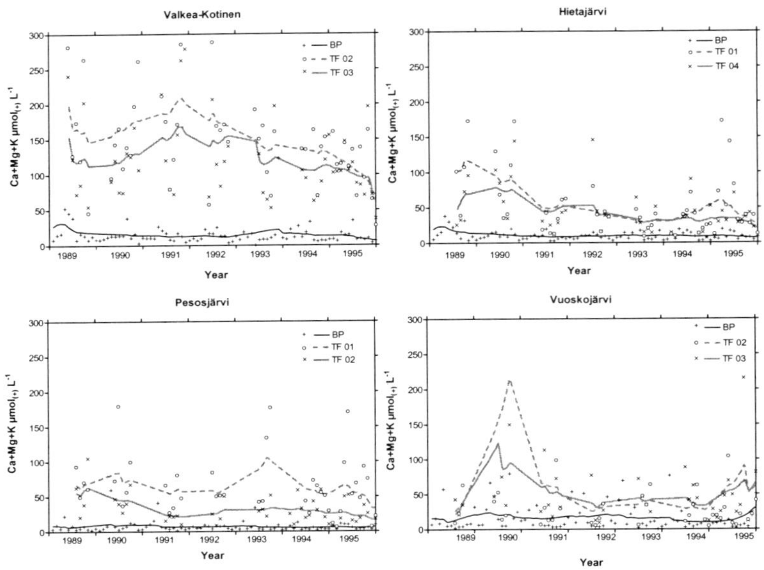

Fig 3. Monthly sum of base cations ($Ca^{2+} + Mg^{2+} + K^{+}$) concentrations in bulk precipitation (BP) and throughfall(TF) over ther period 1989-95 at each study catchment. For description of TF plots, see Table II. The lines are the LOWESS smoother.

4. Discussion

The open collectors used for bulk precipitation and throughfall collect wet and part of the dry deposition. The fraction of dry deposition collected in throughfall also includes that filtered by the forest canopy. Tuovinen *et al.* (1990) report that 10-30 % of the S collected by wet-only collectors is derived from dry deposition. However, in remote forested areas, such as our sites, this fraction can be expected to be less (Lindberg *et al.,* 1986).

Compared to bulk precipitation, H^{+} concentrations in the throughfall collected at those plots with a predominantly coniferous stand were slightly higher than for those stands with a significant deciduous tree species component (i.e., Valkea-Kotinen, TF3; Pesosjärvi, TF1 and TF2; Vuoskojärvi, TF2; see Table II). Cronan and Reiners (1983) also found that coniferous canopies tend to show a tendency toward net acidification of precipitation, indicating that acid interception rates are higher than the buffering capacity of canopy surfaces or dry deposited bases in coniferous stands. Further evidence of the effect of stand composition on throughfall chemistry is indicated by the difference between throughfall plots within catchment. The greatest difference between plots within catchment tended to be for those plots with the greatest difference in stand composition. Thus, the Vuoskojärvi plots showed the greatest difference in throughfall while the throughfall for the Hietajärvi plots were similar to each other.

Sulfate concentrations in throughfall were about double those in bulk precipitation (Table III). Because SO_4^{2-} shows little interaction with canopies (Bredemeier, 1988; Lindberg *et al.*, 1986), the SO_4^{2-} enrichment of thrughfall is primarily the result of wash-off of accumulated dry deposition (Lindberg and Lovett, 1992). The wash-off of deposited sea-salts from the canopy seems particularly important at the Vuoskojärvi catchment. The high concentrations of base cations in the bulk precipitation at the two southernmost catchments is probably related to shale burning in power stations in Estonia and the St. Petersburg region and the emission of base cation-rich aerosols (Anttila, 1990).

Even though the Seasonal-Kendall test can be used when there are missing values (Hirsch *et al.*, 1982; Gilbert, 1987), the number of throughfall observations was only about half of the 84 months making up the study period. This was primarily because snow sampling did not start until the 1994/95 winter. To test whether having a more complete data set changed our results we developed simple linear regression models of the form: TF = constant + BP for each plot to estimate snow throughfall SO_4^{2-} concentrations for those months with missing data. The models were based on the data for the two winters for which we had data. These models were all significant ($p < 0.05$) and had R^2 values ranging from 0.50 to 0.94. Using these data to make a complete data set, we then re-ran the Seasonal-Kendall test. The slope estimate was similar to that of the original run with missing data, although the negative trend for Vuoskojärvi became significant ($p < 0.05$). We therefore consider the trends shown by the original data with missing values (Table IV) to be valid.

5. Conclusions

Using bulk precipitation and throughfall collected over seven years (1989-1995) at four [illegible]ated throughout Finland we have shown statistically significant reductions in the [illegible]ncentrations of SO_4^{2-} and acidity (H^+ concentrations) in precipitation in southern [illegible]entral Finland. These reductions have been the greatest in southern Finland where pollution concentrations have been the highest. Although there has been no significant reduction in bulk precipitation and throughfall SO_4^{2-} concentrations at the northernmost site, H^+ concentrations have been reduced significantly, as they have at all sites. The reduction in the acidity of precipitation has been due to the concurrent reduction in SO_4^{2-} concentrations, as base cations concentrations have not changed significantly over the study period.

Our results compliment the observed reduction in SO_2 emissions that has taken place over Europe over the last decade. They also validate the use of a long-term integrated monitoring programs to assess the success (or otherwise) of international agreements on air pollution abatement.

Acknowledgments

We gratefully acknowledge the assistance of field technicians: Seppo Aikio, Kari Lyytikäinen, Markku Rontti and Pekka Vuori in sample collection. The study was jointly financed by the Forest Research Institute, Meteorological Institute and Ministry of the Environment.

References

Ahti, T., Hämet-Ahti, L. and Jalas, J.: 1968, *Ann. Bot. Fenn.* **5**, 378.

Anttila, P.: 1990, 'Characteristics of Alkaline Emissions, Atmospheric Aerosols and Deposition', in P. Kauppi, P. Anttila, and K. Kenttämies (eds.), *Acidification in Finland*, Springer, Berlin, pp. 111-134.

Bergström, I., Mäkelä, K. and Starr, M. (eds.): 1995, *Integrated Monitoring Programme in Finland, First National Report 1,* Ministry of the Environment, Environmental Policy Department, Helsinki.

Bernes, C.: 1993, *The Nordic Environment - Present State, Trends and Threats*, Prepared by the Nordic Council of Minister's Nordic Environment Report Group, *Nord* 1993:12, pp. 212.

Bredemeier, M.: 1988, *Water, Air, Soil Pollut.* **40**, 121.

Chambers, J.M., Cleveland, W.S, Kleiner, B. and Tukey, P.: 1983, *Graphical Methods for Data Analysis*, Wadsworth International Group, Belmont, California.

Cronan, C.S. and Reiners, W.A.: 1983, *Oecologia* **59**, 216.

EDC.: 1993, *Manual for Integrated Monitoring*, Programme Phase 1993-1996, UN-ECE Convention on Long-Range Transboundary Air Pollution, International Cooperative Programme on Integrated Monitoring of Air Pollution Effects on Ecosystems, Environmental Data Centre, National Board of Waters and the Environment, Environmental Report 5, pp. 114.

Edmonds, R.L., Thomas, T.B. and Rhodes, J.J.: 1991, *Soil Sci. Soc. Am. J.* **55**, 1685.

Forsius, M., Kleemola, S., Starr, M. and Ruoho-Airola, T.: 1995, *Water, Air, Soil Pollut.* **79**, 19.

Gilbert, R.O.: 1987, *Statistical Methods for Environmental Pollution Monitoring*, Van Nostrand Reinhold, New York.

Helmisaari, H-S. and Mälkönen, E.: 1989, *Scand. J. For. Res.* **4**, 13.

Hirsch, R.M., Slack, J.R. and Smith, R.A.: 1982, *Water Resourc. Res.* **18**, 107.

Hultberg, H. and Grennfelt, P.: 1992, *Environ. Pollut.* **75**, 215.

Hyvärinen, A.: 1990, 'Deposition on Forest soils - Effect of Tree Canopy on Throughfall,' in P. Kauppi, P. Anttila, and K. Kenttämies (eds.), *Acidification in Finland*, Springer, Berlin, pp. 199-213.

Ivens, W., Lövblad, G. and Kauppi, P. (eds.): 1990, Throughfall Monitoring as a Means of Monitoring Deposition to Forest Ecosystems- Evaluation of European Data, *Nord* 1990 120, *Miljørapport* 1990:16.

Johnson, D.W. and Lindberg, S.E. (eds.): 1992, *Atmospheric Deposition and Forest Nutrient Cycling,* Ecological Studies 91, Springer, New York.

Kallio, K. and Kauppi, L.: 1990, 'Ion Budgets of Small Forested Basins,' in P. Kauppi, P. Anttila and K. Kenttämies (eds.), *Acidification in Finland*, Springer, Berlin, pp. 811-823.

Lindberg, S.E., Lovett, G.M., Richter, D.D. and Johnson, D.W.: 1986, *Science* **231**, 141.

Lindberg, S.E., Lovett, G.M.: 1992, *Atmos. Environ.* **26A**, 1477.

Lövblad, G., Hovmand, M., Reissel, A., Westling, O., Aamlid, D., Hyvärinen, A. and Schaug, J.: 1994, *Throughfall Monitoring in the Nordic Countries*, Institute för vatten och luftvårdsforskning, Swedish Environmental Research Institute, IVL Publ. B 1132.

Mahendrappa, M.K.: 1983, *Can. J. For. Res.*, **13**, 948.

Novo, A., Buffoni, A. and Tita, M.: 1992, *Environ. Pollut.* **75**, 199.
Parker, G.G.: 1983, *Adv. Ecol. Res.* **13**, 58.
Starr, M. and Ukonmaanaho, L.: 1994, 'Stand Precipitation and Soil Water Quality at Integrated Monitoring Catchments', in E. Mälkönen and H. Sivula, (eds.), *State of Forest Health, Research Programme Interim Report*, The Finnish Forest Research Institute, Research Papers **527**, 271-284. (In Finnish).
Starr, M. and Ukonmaanaho, L.: 1995, 'Throughfall Chemistry', in I. Bergström, K. Mäkelä and M.Starr (eds.), *Integrated Monitoring Programme in Finland, First National Report 1*, Ministry of the Environment, Environmental Policy Department, Helsinki.
Tuovinen, J-P., Kangas, L. and Nordlund, G.:1990, 'Model Calculations of Sulphur and Nitrogen Deposition in Finland', in P. Kauppi, P. Anttila and K. Kenttämies (eds.), *Acidification in Finland*, Springer, Berlin, pp. 167-197.

CHANGES IN MASS ELEMENT FLUXES AND THEIR IMPORTANCE FOR CRITICAL LOADS: GEOMON NETWORK, CZECH REPUBLIC

D. FOTTOVÁ[1] and IRENA SKOŘEPOVÁ[2]

[1]*Czech Geological Survey, Klárov 3, 118 21 Prague 1, Czech Republic*
[2]*Czech Environmental Institute, Vršovická 65, 100 10 Prague 10, Czech Republic*

Abstract. Three years (1994-1996) of runoff, bulk deposition, and throughfall S and N data obtained from GEOMON, a network of 14 small forest catchments in the Czech Republic, are presented. In this paper, we assess regional differences and temporal trends in S and N fluxes under changing regional burdens. Decreases in S deposition have occurred in the emission-burdened northern and western parts of the Czech Republic, where remedial measures have been applied. For example, at the Uhlířská catchment (Jizerské hory Mtns., northern Czech Republic), throughfall S fluxes decreased from 72.7 to 48.5 kg ha^{-1} yr^{-1}. In contrast, the relatively clean central and southern parts of the Czech Republic showed an increase in S deposition; for example at the Na lizu catchment (Šumava Mtns., southern Czech Republic) throughfall S fluxes increased from 9.0 to 18.8 kg ha^{-1} yr^{-1}. In burdened catchments with damaged forests, high runoff of N was observed. Using fluxes of S and N measured at the GEOMON sites, we calculated critical loads and exceedances for both S and N in the catchments. The critical loads of S ranged from 505 to 3,631 mol L^{-1} yr^{-1} during 1994-1996; in about 80 % of the catchments the critical loads were exceeded by atmospheric deposition; half of the catchments showed an exceedance for N.

Keywords: catchment, critical loads, Czech Republic, element fluxes, GEOMON network, nitrogen, sulfur

1. Introduction

In November of 1993, an integrated monitoring of mass element fluxes in the GEOMON network of small catchments in the Czech Republic (CR) was initiated (Fottová,1995). The integration involved sampling methods and analysis. In this paper, we focus on SO_4^{2-}-S and NO_3^--N to: 1) assess regional differences and temporal trends of S and N fluxes under conditions of changing regional burden, and 2) evaluate critical loads for S and N, based on the element flux data obtained at the GEOMON network sites. Critical load is defined as the highest deposition of acidifying compounds that will not cause chemical changes leading to long-term harmful effects on ecosystem structure and function (Nilsson and Grennfelt, 1988). In addition to acidification, inputs of N (and S) may influence the eutrophication and nutrient balances of ecosystems. Critical load for N nutrition effects is defined as a quantitative estimate of an exposure to deposition of N below which harmful effects in ecosystem structure and function do not occur.

Critical loads in the small catchments were derived with the Simple Mass Balance (SMB) model. This method also was used in the mapping of critical loads for the Czecg Republic with in the framework of the European program (Posch *et al.*,1997). The calculation and mapping of critical loads of S and N have formed a basis for assessing the effects of changes in the emissions and deposition of these compounds from a long-term point of view. However, time scales of ecosystem changes due to inputs of acidity and/or N may range from short (minutes to days for responses such as episodic surface water acidification), to very long (decades to centuries, for example for soil acidification or humus accumulation). Critical loads calculated from element fluxes in small catchments permit assessment within short-term and local perspectives.

Water, Air, and Soil Pollution **105:** 365–376, 1998.

2. Methods

2.1. Catchment Locations

The 14 forest catchments of the GEOMON network located in the Czech Republic (Figure 1) were chosen such that they cover various landscape types and areas with different levels of regional emission burdens and various degrees of forest damage. Their selection was steered by requirements of hydrological definition, relative geological homogeneity, vegetation stability, and the absence of local pollution. Five of the catchments are located in the border mountains of northern and western Czech Republic, with a high regional burden and with forests showing a marked degree of damage: Uhlířská (UHL; Jizerské hory Mountains), Modrý potok (MOD; Krkonoše Mountains), U dvou louček (UDL; Orlické hory Mountains), Jezeří (JEZ; Krušné hory Mountains), and Červík (CER; Beskydy Mountains). Two adjacent catchments in the Slavkovský les Mountains of western Czech Republic, Lysina (LYS) and Pluhův bor (PLB), situated on substantially different types of bedrock, are near the above mentioned polluted area, but are still without marked damage. Six relatively clean catchments with no significant regional burden are Anenský potok (GEM), Salačova Lhota (SAL), Polomka (POM), and Loukov (LKV) (all in the Českomoravská vrchovina highlands of the central Czech Repubic), and Na lizu (LIZ) and Spálenec (SPA; Šumava Mountains, southern Czech Republic). The last catchment, Lesní potok (LES), is rather exceptional because of its vicinity to Prague.

The main types of the bedrock underlying the GEOMON network sites are granite, orthogneiss and paragneiss. The Pluhův bor catchment is exceptional with underlying serpentinite. The catchments range in size from tens of hectares to a few hundred hectares. Vegetation cover is constituted mainly by spruce; only in Jezeří and Lesní potok did an important portion of beech occurred. All of the catchments except Anenský potok and Salačova Lhota are in protected landscape areas. Polomka and Loukov were observed for two years only, as was runoff from Lesní potok.

2.2. Sampling and Analytical Procedure

In all catchments, we determined fluxes of environmentally significant elements via the input by precipitation on the open unforested area (bulk deposition), the input by throughfall, and output via surface runoff. On the basis of monthly data, the mass element fluxes for the hydrological year were evaluated. The resulting input-output budgets provided information on accumulation, in the case of positive values, and leaching, in the case of negative values. This paper is based on data collected over three hydrological years 1994 - 1996. The basic program of regular monitoring in individual catchments is the following: 1) sampling of bulk precipitation - monthly cumulative samples; 2) sampling of throughfall precipitation - a mix of samples from the regular network of nine sampling stations in each catchment (for estimating the variability in vegetation density) - monthly cumulative samples; 3) sampling of the runoff from the catchment - on a monthly basis; 4) acquiring data on the amount of monthly precipitation in the unforested segment of the catchment; 5) acquiring data on the monthly throughfall amount; and 6) a continual record of water level in the concluding profile of the catchment and determination of a daily average outflow (from the consumption curve).

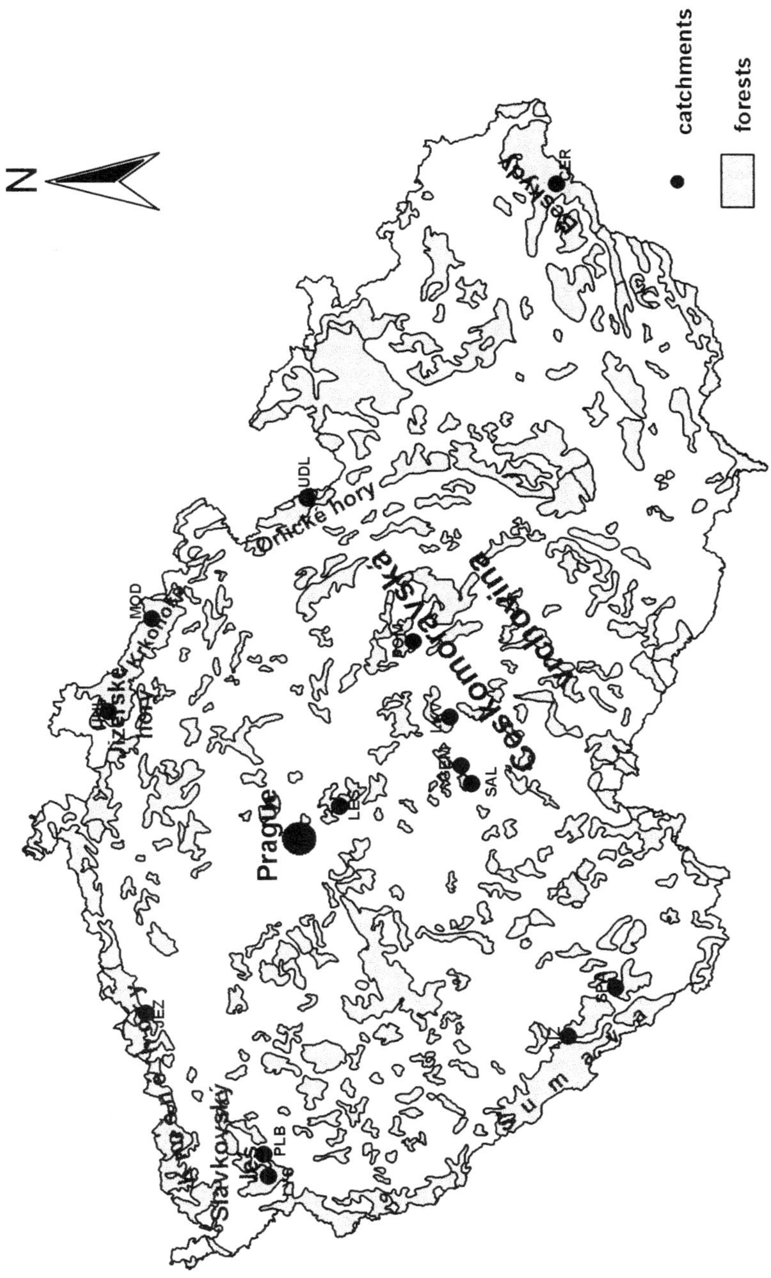

Fig. 1. Location of the 14 small catchments of the GEOMON network.

The sampling and analytical methods are uniform in the network, which markedly raises the reliability of data and its mutual comparability. In accredited laboratories of the Czech Geological Survey, the concentrations of S and N in the samples are determined by HPLC (High Pressure Liquid Chromatography) methods.

2.2. CALCULATION OF ELEMENT FLUXES

Calculations of the element fluxes - input to the catchment of a defined area by bulk deposition and throughfall, and output from the catchment by surface runoff - are carried out from the data obtained in one hydrological year (from Nov. 1 to Oct. 31 of the following year). The deposition of a particular element is counted as a sum of the twelve monthly deposition in units of mg m^{-3}, calculated by multiplying the concentration in mg L^{-1} by the total monthly precipitation (both for bulk and throughfall data) in mm:

$$D_i = 10^{-2} \times \sum_j c_i \times u_j$$

where D_i = deposition of element i on a unit area of the catchment (kg ha^{-1} yr^{-1}); c_i = concentration of the element in the cumulative monthly sample (mg L^{-1}); u_j = monthly precipitation amount (mm = L m^{-2}); j = number of samples.

To evaluate the annual runoff of an individual component, it is essential to calculate the annual discharge (in m^3) of water from daily average outflows. A weighted average is calculated from the instantaneous concentrations and outflow at the time of sampling.

It is multiplied by the annual discharge and than divided by the catchment area:

$$O_i = (V / P) \times 10^3 \times \frac{\sum_j c_i \times q_j}{\sum_j q_j}$$

where O_i = runoff of component i from the unit area of the catchment (kg ha^{-1} yr^{-1}); V = yearly discharge from catchment (m^3); P = catchment area (ha); c_i = concentration of component i in the sample (mg L^{-1}); q_j = actual outflow during sample taking (L s^{-1}); and j = number of samples.

The results for every element are presented in units of kg ha^{-1} yr^{-1}. Inputs of S to the whole catchment are estimated as throughfall deposition flux in the forested (spruce) plus bulk deposition flux in the unforested segment of the catchment. Only in the Jezeří and Lesní potok catchments, with an important portion of beech, was it necessary to take into account beech throughfall data for calculation of input to the whole catchment. For N, bulk deposition only is considered as the input. The annual budget is calculated as the difference between input and output (runoff) and gives the total accumulation (in the case of positive values) or loss (in the case of negative values) of an individual component.

2.3. Calculation of Critical Loads

The simple mass balance method (SMB) is used to calculate the critical loads (CL) of S and N in the small catchments. In the SMB model, the soil is treated as a single homogeneous compartment and its depth is equal to the root zone. The calculations of critical load include the following equations:

$$CL(A) = ANC_w + [H^+]_{crit} \times Q + [Al^{3+}]_{crit} \times Q$$
$$CLmax(S) = CL(A) + BC_d - BC_{u,crit}$$
$$CLmax(N) = N_{u,crit} + N_{i,crit} + CLmax(S) / (1-f_{de})$$
$$CLmin(N) = N_{u,crit} + N_{i,crit}$$
$$CLnut(N) = N_{u,crit} + N_{i,crit} + N_{l,crit} / (1-f_{de})$$
$$N_{l,crit} = Q \times [N]_{crit}$$

where

$CL(A)$ = critical load of acidity ($mol_c\ ha^{-1}\ yr^{-1}$)
ANC_w = alkalinity produced by weathering at a critical load ($mol_c\ ha^{-1}\ yr^{-1}$)
Q = discharge under root zone ($m^3\ yr^{-1}$)
$[H^+]_{crit}$ = soil solution H^+ concentration at a critical load ($mol_c\ m^{-3}$)
$[Al^{3+}]_{crit}$ = soil solution Al^{3+} concentration at a critical load ($mol_c\ m^{-3}$)
$Cl_{max}(S)$ = maximum critical load for S ($mol_c\ ha^{-1}\ yr^{-1}$)
BC_d = atmospheric deposition of basic cations ($mol_c\ ha^{-1}\ yr^{-1}$)
$B_{cu,crit}$ = uptake of basic cations at a critical load ($mol_c\ ha^{-1}\ yr^{-1}$)
$N_{u,crit}$ = uptake of N at a critical load
$N_{i,crit}$ = immobilization rate of N at a critical load
$N_{l,crit}$ = leaching of N at a critical load
$CL_{nut}(N)$ = critical load of nutrient N
$CL_{min}(N)$ = minimum critical load for N
$CL_{max}(N)$ = maximum critical load for N
f_{de} = denitrification fraction

The methods for calculating critical loads of N and S with the Simple Mass Balance equation model have been summarized in the CCE Status Report (Posch *et al.*,1995) and are documented more extensively in the new Mapping Manual (UBA,1996). Critical loads are calculated to express the sensitivity of ecosystems to the deposition of S and N in quantitative terms. Computing the so-called exceedance, i.e. the difference between deposition and the critical load, is done to provide a comparison between actual deposition and the critical load. We calculated the values of S exceedances for every small catchment in the GEOMON network. Critical loads of S (maximum) and critical loads of nutrient N were used for the evaluation of exceedances. All units of critical loads and exceedances are in $mol_c\ ha^{-1}\ yr^{-1}$ according to the methodology.

3. Results and Discussion

3.1. Mass Element Fluxes and Temporal Trends

Between 1994 and 1996 most of the catchments in the border mountains of the northern and western part of Czech Republic, where regional remedial measures were applied, showed a decrease in deposition of S via both input to the whole area of the catchment (including bulk for unforested and throughfall for unforested area) and throughfall separately, see Table I. Input at the Uhlířská catchment, with only 15 % forested area, fell from 36.9 to 18.7 kg S ha^{-1} yr^{-1} between 1995 and 1996, and throughfall from 72.7 to 48.5 kg S ha^{-1} yr^{-1}. At the fully forested U dvou louček catchment, deposition decreased from 82.5 to 69.9 kg ha^{-1} yr^{-1} between 1994 and 1995. Throughfall deposition below spruce canopies at the Jezeří catchment in the Krušné hory Mts. was rather exceptional in that it increased from 62.7 to 86.7 kg ha^{-1} yr^{-1} between 1995 and 1996. During that time a new wave of forest dieback has started in that area. In contrast to the prevailing decrease in the emission area of the northern and western part of the Czech Republic, deposition of S in relatively clean areas (with a lower portion of dry deposition) in the central part of the Czech Republic (Českomoravská vrchovina highlands) and in the southern Šumava Mountains. markedly increased between 1994 and 1996. At the Na lizu catchment, S deposition increased from 8.2 to 17.2 kg ha^{-1} yr^{-1}, and at the Salačova Lhota catchment S deposition increased from 21.3 to 40.1 kg S ha^{-1} yr^{-1}. The highest yearly amount of precipitation in all three years was obtained in 1995.

The output of S from the emission area is high and the resultant budget is generally negative because of the removal of adsorbed S. The highest output 86.2 kg ha^{-1} yr^{-1} occurred at the Jezeří catchment in the wettest year, 1995. In the relatively clean part of Czech Republic, the budget was mostly positive. Only at the Loukov catchment, due to the occurrence of quartz veins with sulfides, and at the Na lizu catchment because of low deposition, was a negative budget noticed. Decreases in runoff were observed at the Lysina catchment (from 30.2 to 20.9 kg S ha^{-1} yr^{-1}) and at thePlůhov bor catchment (from 39.6 to 25.2 kg S ha^{-1} yr^{-1}).

Deposition of N via bulk deposition fell between 1994 and 1996 only at the Uhlířská catchment (see Table II); temporal trends were not observed at any of the other catchments. The maximum input was at the U dvou louček catchment (13.0 kg N ha^{-1} yr^{-1}) in 1994. In the cleaner part of of the Czech Republic, N deposition ranged from about 3 - 5 kg ha^{-1} yr^{-1}; at the Na lizu catchment, it increased from 2.8 to 8.8 kg N ha^{-1} yr^{-1}. Nitrogen in runoff from the catchments of the western and the northern part of of the Czech Republic with damaged forests was very high, probably a consequence of the decrease of loss of the ability of declining or dead forests to uptake N as nutrient. At the Uhlířská and Modrý potok catchments, even negative budgets occurred, despite relatively high deposition. The highest N output in the GEOMON network was from the U dvou louček catchment (16.1 kg ha^{-1} yr^{-1} in 1995). On the contrary, in the catchments of the relatively clean central and southern parts of Czech Republic (Na lizu, Spálenec, Anenský potok, Salačova Lhota, Lesní potok, and Loukov), where forests did not lose the ability to uptake N, the runoff was below 3 kg ha^{-1} yr^{-1}, and often was below 1 kg N ha^{-1} yr^{-1}. Exceptional is Polomka with an output above 4 kg ha^{-1} yr^{-1}. The reason is contamination from underground water from nearby agricultural localities. At two catchments (Plůhov bor and Lysina) with no marked damage, a moderate gradual increase in runoff of N was observed, at the Plůhov bor catchment from 1.2 to 2.5 kg N ha^{-1} yr^{-1}. The lowest output of N (0.2 kg ha^{-1} yr^{-1}) occurred at the Loukov catchment.

TABLE I

Fluxes of SO_4^{2-}-S (kg ha^{-1} yr^{-1}; precipitation in mm) at the GEOMON catchments.

UHL	prec.	input	output	budg.	thrg.	LIZ	prec.	input	output	budg.	thrg.
1994	1171	33.6	46.7	-13.1	70.7	1994	911	8.2	14.1	-5.9	9.0
1995	1370	36.9	62.0	-25.1	72.6	1995	1142	11.2	18.3	-7.1	12.1
1996	1231	18.7	46.7	-28.0	48.5	1996	986	17.2	20.2	-3.0	18.8
MOD	prec.	input	output	budg.	thrg.	SPA	prec.	input	output	budg.	thrg.
1994	1620	28.2	33.8	-5.6	35.7	1994	705	12.9	12.2	0.7	13.2
1995	1837	38.5	38.8	-0.3	49.5	1995	1041	20.8	15.8	5.0	21.4
1996	1492	21.2	23.5	-2.3	26.7	1996	924	25.4	21.4	4.0	26.4
UDL	prec.	input	output	budg.	thrg.	GEM	prec.	input	output	budg.	thrg.
1994	1333	82.5	32.6	49.8	82.5	1994	607	17.7	6.7	11.0	18.4
1995	1565	69.9	32.9	37.0	69.9	1995	798	20.0	15.2	4.8	20.8
1996	1188	62.0	26.2	35.8	62.0	1996	607	22.8	10.6	12.2	23.7
CER	prec.	input	output	budg.	thrg.	SAL	prec.	input	output	budg.	thrg.
1994	1137	30.0	39.0	-9.0	30.0	1994	571	21.3	4.0	17.3	21.3
1995	1037	22.6	44.3	-21.8	22.6	1995	688	25.8	11.0	14.8	25.8
1996	1244	34.2	48.0	-13.8	34.2	1996	589	40.1	9.7	30.4	40.1
JEZ	prec.	input	output	budg.	thrg.	LES	prec.	input	output	budg.	thrg.
1994	982	33.7	63.7	-30.0	66.7	1994	637	27.3	-	-	27.3
1995	1059	26.3	86.2	-59.9	62.7	1995	771	27.9	40.7	-12.8	27.9
1996	951	28.7	67.3	-38.7	86.7	1996	593	24.6	49.2	-24.6	25.5
LYS	prec.	input	output	budg.	thrg.	POM	prec.	input	output	budg.	thrg.
1994	992	24.6	30.2	-5.6	31.6	1994	-	-	-	-	-
1995	1154	24.2	27.2	-3.0	30.4	1995	749	29.4	36.4	-7.1	30.9
1996	827	24.2	20.9	3.4	31.1	1996	658	46.1	42.5	3.6	48.7
PLB	prec.	input	output	budg.	thrg.	LKV	prec.	input	output	budg.	thrg.
1994	992	25.7	39.6	-13.9	25.7	1994	-	-	-	-	-
1995	1154	23.1	33.3	-10.2	23.1	1995	954	25.8	64.3	-38.5	25.8
1996	827	18.4	25.2	-6.9	18.4	1996	636	27.2	65.9	-38.6	27.2

prec. - precipitation; budg. - budget (input minus output); thrg. - throughfall

TABLE II

Fluxes of NO_3^--N (kg ha^{-1} yr^{-1}; precipitation in mm) at the GEOMON catchments.

UHL	prec.	input	output	budg.	LIZ	prec.	input	output	budg.
1994	1171	8.2	8.8	-0.6	1994	911	2.8	2.6	0.2
1995	1370	6.9	7.2	-0.3	1995	1142	5.1	2.1	3.0
1996	1231	6.1	8.9	-2.8	1996	986	8.8	3.0	5.8
MOD	prec.	input	output	budg.	SPA	prec.	input	output	budg.
1994	1620	7.3	13.3	-6.0	1994	705	3.1	0.5	2.6
1995	1837	8.3	14.5	-6.2	1995	1041	4.6	0.7	3.9
1996	1492	5.0	8.6	-3.6	1996	924	4.2	0.6	3.6
UDL	prec.	input	output	budg.	GEM	prec.	input	output	budg.
1994	1333	13.0	10.9	2.1	1994	607	4.2	0.6	3.6
1995	1565	10.4	16.1	-5.7	1995	798	4.9	1.4	3.5
1996	1188	10.2	9.4	0.8	1996	607	5.0	0.8	4.2
CER	prec.	input	output	budg.	SAL	prec.	input	output	budg.
1994	1137	4.6	3.0	1.6	1994	571	3.2	0.2	3.0
1995	1037	4.6	2.9	1.7	1995	688	4.2	0.3	3.9
1996	1244	6.1	4.6	1.5	1996	589	5.2	0.3	4.9
JEZ	prec.	input	output	budg.	LES	prec.	input	output	budg.
1994	982	6.4	4.4	2.0	1994	637	3.3	-	-
1995	1059	6.9	5.3	1.6	1995	771	3.4	1.5	1.9
1996	951	7.3	4.6	2.7	1996	593	3.3	1.0	2.3
LYS	prec.	input	output	budg.	POM	prec.	input	output	budg.
1994	992	4.7	1.0	3.7	1994	-	-	-	-
1995	1154	5.6	1.1	4.5	1995	749	3.7	4.2	-0.5
1996	827	5.2	1.4	3.8	1996	658	3.7	4.9	-1.2
PLB	prec.	input	output	budg.	LKV	prec.	input	output	budg.
1994	992	4.7	1.2	3.5	1994	-	-	-	-
1995	1154	5.6	1.6	4.0	1995	954	7.6	0.2	7.4
1996	827	5.2	2.5	2.7	1996	636	7.3	0.4	6.9

prec. - precipitation; budg. - budget (input minus output)

3.2. Critical Loads and Exceedances

The critical loads of S acidity for an ecosystem are considered to be the highest deposition of acidifying compounds that will not cause chemical changes leading to long-term harmful effects on ecosystem structure and function. Catchments of higher critical load values are assumed to be more resistant than catchments with lower critical load values to negative effects caused by S atmospheric deposition. The atmospheric deposition of S represents the main source of ecosystem acidification in the Czech Republic. Critical load values higher than 2,000 mol_c S ha^{-1} yr^{-1} (32 kg S ha^{-1} yr^{-1}) were obtained for the Uhlířská, Modrý potok, and U dvou louček catchments in 1994 and 1995 and for the Uhlířská, U dvou louček, and Červik catchments in 1996 (Table III). All of these catchments are located in the mountain areas where the largest discharge of water occurs, and therefore they have the highest critical load values. The differences in critical load values between 1994 and 1996 were caused by different runoff and deposition levels of basic cations. These catchments belong to localities with a high regional burden, and are characterized by forests showing a marked degree of damage. These catchments with higher critical loads would seem to be more resistant to acidic deposition than those with lower critical loads, but the forest damage is a result of exceedance due to atmospheric deposition. For high deposition levels in the catchments of the burdened area, the values of exceedances were high as well. The maxima of S exceedances ranged from 3,186 $mol_c ha^{-1} yr^{-1}$ to 4,495 $mol_c ha^{-1} yr^{-1}$, and were obtained for Jezeří catchment in the Krušné hory Mountains, which has suffered from chronic heavy atmospheric pollution for many decades. In this area, coniferous forests dominated by *Picea abies* have nearly disappeared. Some of the catchments in the areas of no significant burden also showed high exceedances caused by the relatively low critical load values and increased atmospheric deposition of S, e.g. Salačhova Lhota, Anenský potok, and Polomka (see Table III). The Salačhova Lhota catchment exhibited an extreme increase of S exceedance between 1994 and 1996, increasing from 710 mol_c S $ha^{-1} yr^{-1}$ in 1994 to 1,551 mol_c S $ha^{-1} yr^{-1}$ in 1996.

TABLE III

Critical loads of S and their exceedances (mol_c S $ha^{-1} yr^{-1}$) for at the GEOMON catchments.

	1994			1995			1996		
	CL_{max}	Dep	Ex	CL_{max}	Dep	Ex	CL_{max}	Dep	Ex
UHL	2,570	4,410	1,840	3,631	4,540	909	2,581	3,019	438
MO	2,259	2,228	0 (-31)	2,709	3,093	384	1,885	1,662	0 (-223)
UDL	3,449	5,143	1,694	2,885	4,367	1,482	2,587	3,864	1,277
CER	1,542	1,870	328	1,777	1,410	0 (-367)	2,066	2,132	66
JEZ	976	4,162	3,186	1,230	3,920	2,690	925	5,420	4,495
LYS	1,078	1,971	893	1,270	1,897	627	991	1,936	945
PLB	1,557	1,603	46	1,597	1,445	0 (-152)	1,488	1,143	0 (-345)
LIZ	998	559	0 (-439)	1,226	755	0 (-471)	1,292	1,169	0 (-123)
SPA	1,630	825	0 (-805)	1,034	1,337	303	1,503	1,644	141
GE	468	1,147	679	540	1,298	758	535	1,475	940
SAL	620	1,330	710	700	1,613	913	946	2,498	1,552
LES	505	1,701	1,196	711	1,744	1,033	802	1,589	787
PO	-	-	-	736	1,931	1,195	1,037	3,034	1,997
LKV	-	-	-	1,303	1,609	306	1,430	1,697	267

CL_{max} - critical load of S; Dep - deposition of S; Ex - exceedance of critical load by deposition

Both S and N compounds contribute to the total deposition of acidity. In addition to acidification, inputs of N may influence the eutrophication and nutrient balances of ecosystems. The critical load for N nutrition effects is defined as a quantitative estimate of an exposure to deposition of N below which harmful effects in ecosystem structure and function do not occur according to present knowledge. To protect forest ecosystems against eutrophication, critical loads of nutrient N and their exceedances had to been calculated (Table IV). The critical loads of nutrient N in the catchments of the GEOMON network varied from 435 mol_c N ha^{-1} yr^{-1} (Anenský potok in 1994) to 982 mol_c N ha^{-1} yr^{-1} (Modrý potok in 1994). The highest exceedances of critical loads of nutrient N (901 - 516 mol_c N ha^{-1} yr^{-1}) in the period of 1994 - 1996 were observed in at the U dvou louček catchment. The Salačova Lhota catchment showed the maximum of N exceedance in 1996 for increased atmospheric deposition of N. In contrast to exceedances of critical loads of S, the exceedances of critical loads of nutrient N reached values lower than 1,000 mol_c N ha^{-1} yr^{-1} (14 kg N ha^{-1} yr^{-1}). Very low or no exceedances of N were found for the Uhlířská, Červik, Plůhov bor, Spálenec, Lesní potok, and Polomka catchments. The other catchments, except U dvou louček and Salačova Lhota discussed above, showed exceedances in the range from 112 (at Na lizu) to 791 mol_c N ha^{-1} yr^{-1} (at Modrý potok).

TABLE IV

Critical loads of N and their exceedances (mol_c N ha^{-1} yr^{-1}) for at the GEOMON catchments.

	1994			1995			1996		
	CL_{nut}	Dep	Ex	CL_{nut}	Dep	Ex	CL_{nut}	Dep	Ex
UHL	859	1,096	237	833	766	0 (-67)	810	876	67
MO	982	1,142	160	793	1,584	791	752	711	0 (-41)
UDL	944	1,845	901	893	1,615	722	874	1,391	517
CER	843	613	0 (-230)	826	764	0 (-62)	834	894	60
JEZ	838	1,126	288	824	1,364	540	815	1,199	384
LYS	744	402	0 (-42)	731	920	189	714	814	100
PLB	869	702	0 (-167)	852	920	68	844	814	0 (-30)
LIZ	792	340	0 (-452)	789	901	112	789	1,194	405
SPA	698	418	0 (-280)	688	774	86	814	574	0 (-240)
GE	435	684	249	436	811	375	434	794	360
SAL	609	627	18	610	775	165	612	1,221	609
LES	717	384	0 (-333)	734	547	0 (-187)	740	461	0 (-279)
PO	-	-	-	733	576	0 (-157)	751	780	29
LKV	-	-	-	795	1,378	583	795	1,302	507

CLnut - critical load of N; Dep - deposition of N; Ex - exceedance of critical load by deposition

4. Conclusion

In the border mountains of northern and western part of Czech Republic, remedial measures caused a decrease in the regional S burden. Between 1994 and 1996, decreases occurred both in throughfall deposition and in total input to the catchments, including bulk and throughfall deposition of S. At the Uhlířská catchment (Jizerské hory Mountains), input fell from 36.9 to 18.7 kg ha^{-1} yr^{-1} from 1995 to 1996. It caused a new occurrence of forest dieback. In contrast to the prevailing decrease in emission area in the northern and western part of Czech Republic, in the relatively clean central and southern parts of the Czech Republic (Českomoravská vrchovina highlands and Šumava Mountains), a markedly increased S deposition both via throughfall and input was observed. At the Na lizu catchment, input increased from 8.2 to 17.2 kg S ha^{-1} yr^{-1} while throughfall fluxes increased from 9.0 to 18.8 kg S ha^{-1} yr^{-1}; at the Salačova Lhota catchment (100 % forested) S fluxes increased from 21.3 to 40.1 kg S ha^{-1} yr^{-1} both in input and in throughfall. N input decreased moderately at the Uhlířská catchment only, and increased mostly at the Na lizu catchment, from 2.8 to 8.8 kg N ha^{-1} yr^{-1}.

Output of S was high in burdened areas because of desorption from soils. Only from the Lysina and Plůhov bor catchments did S output deacrease (by around 10 kg S ha^{-1} yr^{-1}). In burdened areas with damaged forests, high runoff of N was observed (the highest 16.1 in U dvou louček in 1995) because the trees have lost their ability to uptake N as nutrient. On the contrary in the relatively clean central and southern parts of the Czech Republic, the runoff of N was mostly below 1 kg ha^{-1} yr^{-1}.

The critical loads of S and N for forested areas of the catchments were calculated using the data on mass element fluxes and complementary data on weathering, soil properties and the consumption of base cations by the vegetation. The primary aim of the critical load approach has been to give a scientific background for policy analysis of environmental effects of S and N emission reduction alternatives. In addition to the indirect effects of geochemical changes in soils, consideration of the direct effects of air concentrations on vegetation and human health may become important for the scientific support of protocol negotiations that aim to reduce different pollutants. Calculated values of critical loads were compared with the actual deposition figures and exceedances were calculated. The critical loads of S with values higher than 2,000 mol_c S ha^{-1} yr^{-1} (32 kg S ha^{-1} yr^{-1}) were obtained for the Uhlířská, Modrý potok, and U dvou louček catchments in 1994 and 1995 and for the Uhlířská, U dvou louček, and Červik catchments in 1996. The critical loads of nutrient N in the catchments of the GEOMON network varied from 435 to 982 mol_c N ha^{-1} yr^{-1} in years 1994 - 1996. The maxima of S exceedances (ranging from 3,186 mol_c ha^{-1} yr^{-1} to 4,495 mol_c ha^{-1} yr^{-1}) were obtained at the Jezeří catchment in the Krušné hory Mountains for extreme atmospheric deposition of S. The highest exceedance of nutrient N critical loads (517-901 mol_c N ha^{-1} yr^{-1} in the period of 1994 - 1996) was observed at the U dvou louček catchment. The Salačova Lhota catchment, located in the relatively clean area, exhibited the extreme increase of the S exceedance between 1994 and 1996, increasing from 710 mol_c ha^{-1} yr^{-1} of S in 1994 to 1551 mol_c ha^{-1} yr^{-1} of S in 1996, attendant with an increase of atmospheric deposition of S; an increase in N exceedances also was observed at the Salačova Lhota catchment during this period.

Acknowledgments

The results were obtained with support of the Grant Agency of the Czech Ministry of the Environment and the Grant Agency of the Czech Republic. Beside the Czech Geological Survey, other Czech institutes cooperate on the program - Czech Hydrometeorological Institute, Forestry and Game Institute, Naturmanagement Volary, Nature Protection Services of the Železné hory Mountains, and Geological Institute of the Academy of Sciences of the Czech Republic. The text was corrected and partially translated by R. Withers from the Czech Environmental Institute.

References

Fottová, D.: 1995, *Environ. Monit. Assess.* **34,** 215.

Nilsson, J. and Grennfelt, P.(eds.):1988, *Critical Loads for Sulphur and Nitrogen, NORD 1988: 97*, Nordic Council of Ministers, Copenhagen.

Posch, M., de Smet, P.A.M., Hettelingh, J. P. and Downing, R.J. (eds.): 1995, *Calculation and Mapping of Critical Tresholds in Europe: Status Report 1995*, RIVM Report no. 259101004, Bilthoven.

UBA: 1996, *Manual on Methodologies and Criteria for Mapping Critical Levels/Loads and Geographical Areas where they are Exceeded*, UN/ECE Convention on Long-range Transboundary Air Pollution. Federal Environmental Agency (Umweltbundesamt), Texte 71/96, Berlin.

THE RELATIONSHIP BETWEEN STREAM CHEMISTRY AND WATERSHED LAND COVER DATA IN THE MID-ATLANTIC REGION, U.S.

ALAN T. HERLIHY[1], JOHN L. STODDARD[2], and COLLEEN BURCH JOHNSON[3]

[1]*Department of Fisheries and Wildlife, Oregon State University, c/o U.S. EPA, Corvallis, OR 97333, USA*
[2]*U.S. EPA Environmental Research Laboratory, Corvallis, OR 97333, USA*
[3]*OAO Corporation, c/o U.S. EPA, Corvallis, OR 97333*

Abstract. In order to investigate the relationship between stream chemistry and watershed land cover at the regional scale, we analyzed data from 368 wadeable streams sampled in the mid-Atlantic region of the U.S. during spring 1993-1994. Study sites were selected using a probability sample and the digitized version of the 1:100,000 scale USGS map stream network as the sample population. Both classified Thematic Mapper (TM) and USGS Land Use/Land Cover (LULC) data were used to quantify land cover in the study watersheds. On average, the most common land cover was forest (77 %) followed by agriculture (20 %), and urban (1 %). Multiple regression analysis showed that concentrations of Cl^-, nutrients, acid neutralization capacity, and base cations were the analytes most strongly related to watershed land cover. Despite large differences in resolution and age of the TM and LULC data sources, similar results were obtained with the two sources. Using a greater number of land cover subclasses did not greatly improve the land cover-chemistry relationships. Ecoregions with predominantly forested land cover had weaker relationships than ecoregions with more agricultural and/or urban land cover. In studies or databases without land cover information, Cl^- concentration is a good surrogate indicator for general human disturbance in the watershed.

Keywords: land cover, land use, mid-Atlantic region, stream water chemistry, watershed

1. Introduction

Many studies have illustrated the effects of different land uses and land covers on stream water quality on a local or basin scale (e.g., Liegel *et al.*, 1991; Mueller *et al.*, 1995; Jordan *et al.*, 1997). The expectation that such relationships exist has formed the basis for the design of many water quality monitoring efforts (e.g., Puckett, 1995). The strength and nature of these relationships rarely have been examined over large regions however, due to the paucity of large, regional data sets, and the effort involved in quantifying land cover for large numbers of sites.

In 1993, the U.S. Environmental Protection Agency (EPA) began a stream pilot survey in the Mid-Atlantic region of the U.S., as part of the Environmental Monitoring and Assessment Program (EMAP). Over the first two years of the survey, 368 wadeable stream sites were sampled for stream water chemistry, and the land cover composition in the contributing watersheds was quantified using remote imagery and geographic information system (GIS) software. The objectives of the present study were to examine the relationship between land cover and stream chemistry on a regional scale to determine: 1) the strength of the relationship across the region and at smaller spatial scales; 2) what kind of land cover data is the most useful; 3) which chemical variables most related to land cover; and 4) the effects of different spatial aggregation schemes (drainage basins versus ecoregions).

The Mid-Atlantic region of the U.S. is a roughly 200,000 km^2 area comprising the states of Delaware, Maryland, Pennsylvania, Virginia, and West Virginia (Figure 1) and

Water, Air, and Soil Pollution **105:** 377–386, 1998.

consists of five major, somewhat linear, southwest-northeast trending physiographic provinces. From east to west, they are the Coastal Plain, Piedmont, Blue Ridge Mountains, Valley and Ridge, and Appalachian Plateau. Streams are the dominant aquatic resource in this region. The Coastal Plain is a relatively flat, sandy plain bordering the Atlantic Ocean. The Piedmont Province is a peneplain composed primarily of metamorphic and plutonic rocks. Both the Piedmont and Coastal Plain are influenced strongly by agriculture and urbanization. The Blue Ridge Mountains make up a narrow ridge lying west of the Piedmont and are composed of metamorphosed sedimentary and complex gneissic and plutonic bedrock; forests are the dominant land cover. The Valley and Ridge Province is composed of folded sedimentary rock layers, resulting in structural valleys separated by narrow, parallel, linear ridges. The ridges typically are forested and are composed of resistant sandstones and shales, whereas agriculture and urban lands occur on the limestone and shale formations of the valley bottoms. The Appalachian Plateaus Province is separated from the Valley and Ridge Province to the east by a steep escarpment (viz., Allegheny Front). The rocks in the Appalachian Plateaus Province are younger than those in the other provinces, and clastic conglomerates, sandstones, and shales predominate. The mountains of the Appalachian Plateaus are the result of severe plateau dissection. The more rugged northern and central portions of the Appalachian Plateaus are forested predominantly, while the lower gradient western Appalachian Plateaus is more influenced by agriculture and urbanization.

2. Methods

In this paper, we present data compiled from EMAP pilot surveys conducted in the mid-Atlantic region in 1993 and 1994 (Figure 1). Each year, sample sites were selected using a randomized sampling design with a systematic spatial component. The sample frame was based on the digital 1:100,000 scale U.S. Geological Survey (USGS) hydrologic maps that were incorporated into EPA's River Reach File (Version 3). The survey was restricted to wadeable streams defined as 1st, 2nd, or 3rd order (Strahler) streams on the 1:100,000 scale maps. Sample probabilities were set so that roughly equal numbers of 1st, 2nd, and 3rd order stream sites would appear in the sample. Among the study sites, watershed areas ranged from 0.05 to 594 km^2. Streams were sampled once in the spring (April 15-July 1), avoiding storm episodes, for a suite of biological, chemical, and physical habitat indicators (U.S. EPA, 1995). In all, we present data collected at 368 different streams representative of 182,000 km of wadeable streams in the mid-Atlantic region.

A 4 L cubitainer and four 60 mL syringes of stream water were collected in flowing water near the middle of each selected stream site. The syringes were sealed with a Luer-lock valve to prevent gas exchange. All samples were placed on ice and sent by overnight courier to the analytical laboratory. The syringe samples were analyzed for pH, dissolved inorganic carbon (DIC), and monomeric aluminum (Al), and the cubitainer samples were split into aliquots and preserved (filtration and/or acidification) within 48-72 hr of collection. Detailed information on the analytical and sample processing procedures can be found in U.S. EPA (1987). In brief, base cations, Fe, and Mn concentrations were determined by atomic absorption, SO_4^{2-}, NO_3^-, and Cl^- concentrations by ion chromatography, dissolved organic carbon (DOC) concentrations by persulfate oxidation and a carbon analyzer, turbidity by nephelometer, dissolved Si and NH_4^+ concentrations by colorimetry, and total N and P concentrations by persulfate oxidation and colorimetry.

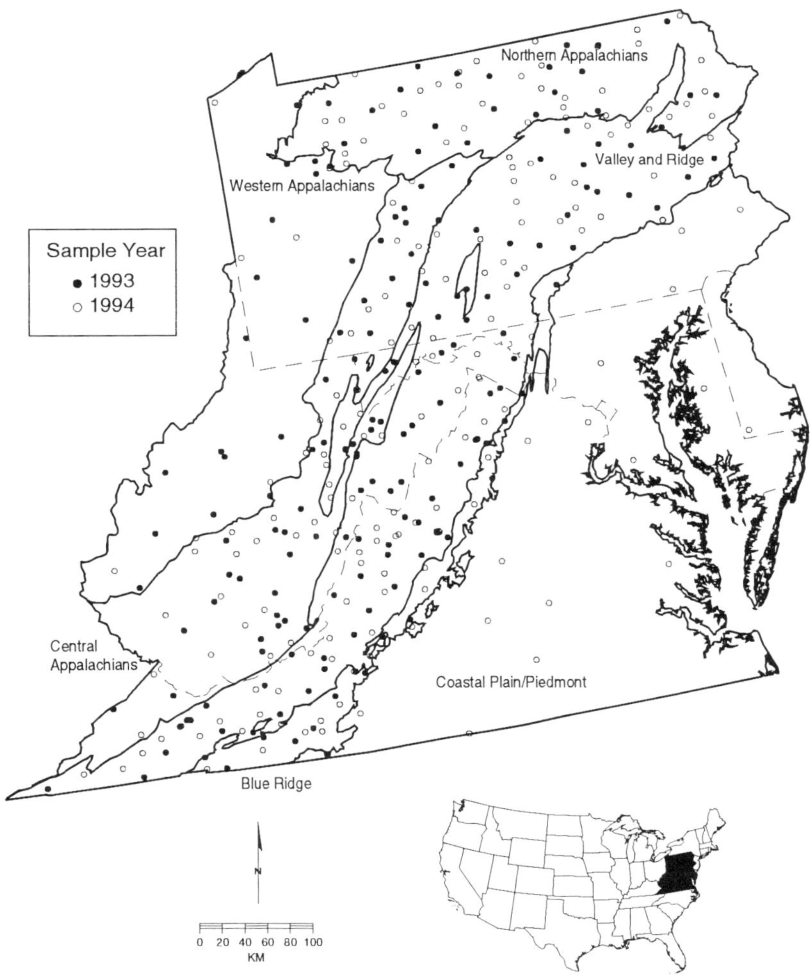

Fig. 1. Location of sample sites in 1993 and 1994 EMAP stream surveys.

For each site, the corresponding watershed was delineated by hand using topographic contours on 1:24,000 to 1:250,000 scale USGS maps then digitized into a GIS. GIS overlay and computational techniques were used to calculate land cover percentages from two data sources: USGS Land Use and Land Cover (LULC) digital maps and classified satellite Thematic Mapper (TM) imagery. LULC data were mapped using high-altitude photography with a minimum mapping resolution of 4-16 ha (depending on land cover type; U.S. Geological Survey, 1990). In the mid-Atlantic, LULC maps were compiled in the 1970s and early 1980s, thus the information is 15-25 years old. TM imagery was taken from 1991-1993 and has a 30 m resolution. Spectral clusters were classified into land-cover classes similar to those in the LULC data (Vogelmann *et al.*, in press). Details of the methods and data sources for the EMAP watershed land cover information can be found in Johnson *et al.* (in review). For this study, we aggregated the LULC and TM information into five major classes that cover nearly all of the region: 1) forests (deciduous, evergreen, mixed); 2) agriculture (row crop, pasture, probable row crop); 3) urban (low/high intensity); 4) wetland (emergent, woody); and 5) barren (quarry, beach, coal mines, transitional). These follow Anderson level I categories (Anderson, 1976). The 14 TM subclasses listed after each major class above, also were used.

In this study, we examined two different spatial aggregation schemes: ecoregions and drainage basins. The smallest spatial aggregations were controlled by available sample sizes in the survey data; at least 15-20 samples in each unit were needed to assure valid statistical analyses. We used the ecoregion map of Woods *et al.* (1996) to form seven aggregate ecoregions: the Piedmont/Coastal Plain (combined due to sample size limitations), Blue Ridge, separate Valley and Ridge ecoregions, and Western, Northern, and Central Appalachian Plateau ecoregions (shown in Figure 1). For another level of spatial analysis, we aggregated the ecoregions into two classes: Lowlands (Piedmont/Coastal Plain, Valley, and Western Appalachians), and Highlands (Ridge, Blue Ridge, Northern and Central Appalachians). The drainage basin aggregations were based on USGS hydrologic unit boundaries. The first level of distinction was Atlantic versus Ohio River drainages. Sub-basins (six-digit hydrologic units) were also used where sample size allowed.

Stream chemistry-land cover relationships were examined by multiple regression analyses (SAS, 1985) with stream chemistry as the dependent variable, and the land cover classes as the independent variables. The general form of the statistical model is thus; chemistry variable = $a_1(LULC_1) + ... + a_n(LULC_n)$ + error. Examination of the chemistry variables has shown that they generally are not normally distributed so they were $\log_{10}$-transformed before statistical analysis to reduce skewness. As a check, we ran the regressions on ranks for both chemistry and land cover. As the rank regressions showed similar patterns and magnitude of correlations as the regressions done on the log transformed data, we are comfortable with the results of the regression analyses.

3. Results and Discussion

For the multiple regression results in Table I, we used all 5 of the major land cover classes. Generally, however, a model using any two of the land cover classes (as long as one of them was either % forest or % agriculture) had an R^2 almost as high as the complete five land cover variable model. Thus, a large number of different two or three land cover variable models could be put together that had virtually the same R^2. This is likely due to the fact that the land cover has to sum to 100 % and that forest and agricultural land cover account for most of the land cover in most sites. To facilitate comparisons, we report the regression results using all 5 classes in the model. All of the tested water quality variables (Table I) were significantly related to land cover, but R^2 values ranged from 0.05 to 0.50. Surprisingly, given the different ages and resolutions of the land cover databases, there were only small differences between the models based on TM and LULC data. Although individual watersheds often have different land cover percentages based on TM vs. LULC data, both data sets generated similar stream chemistry-land cover relationships across the region. The variables that were the most strongly related to land cover are concentrations of nutrients, sum of base cations, acid neutralizing capacity (ANC), and (strongest of all) Cl^- concentration (Table I).

TABLE I

Multiple regression results relating stream chemistry to watershed land cover categories. Values are the proportion of variance explained (R^2) for a model using all five major land cover categories derived from either TM or LULC data as the independent variables. All correlations were significant at $p < 0.05$.

Dependent variable	TM data	LULC data
Chloride	0.48	0.47
Base Cations (Ca, Mg, Na, K)	0.40	0.36
Nitrate	0.39	0.34
Acid Neutralizing Capacity	0.30	0.30
Total Phosphorus	0.31	0.31
Ammonium	0.25	0.34
Dissolved Organic Carbon	0.20	0.13
Turbidity	0.18	0.16
Hydrogen Ion (pH)	0.17	0.15
Manganese	0.12	0.17
Sulfate	0.11	0.16
Silicon	0.07	0.06
Aluminum	0.06	0.08

Strong relationships between NO_3^- concentrations and both agricultural (Osborne and Wiley, 1988; Jordan *et al.*, 1997) and urban (Field *et al.*, 1996) land covers also have been described by other researchers and would be expected given the nutrient enrichment associated with these land covers. Langland *et al.* (1995) compiled annual loading data from 127 sites in the Chesapeake Bay drainage. They found that basins with a greater percentages of agricultural land produced greater nutrient and sediment yields. Also, nutrient yields from predominantly urban land were less than those from predominantly agricultural land. Jordan *et al.* (1997) showed a significant relationship between N species and agriculture, but not between P and agriculture in 17 watersheds in the mid-Atlantic Coastal Plain. In our data, both total N and total P concentrations were related to land cover. Liegel *et al.* (1991) reported relationships between ANC and base cation concentrations with developed (agriculture + urban) lands in a probability survey of low ANC lakes in the Northeastern U.S. The strength of the ANC-land cover relationship in our data is reflective of the high ANC values observed in agricultural/urban systems versus lower ANC values in more forested systems. As noted by Herlihy *et al.* (1993), systems with ANC less than 50 $\mu eq\,L^{-1}$ were observed only in greater than 85 % forested watersheds in the mid-Atlantic Highlands.

Two of the variables showed particularly strong relationships with land cover, NO_3^- and Cl^-. Attempts to use the 14 TM land cover subclasses to improve the regression models for these variables showed only limited gains in variance explained for Cl^- concentration (R^2 increased from 0.48 for the five major class model to 0.53 for the 14 subclass model). There was no gain in variance explained for NO_3^- concentration using all 14 subclasses as the independent variables.

Nitrate concentrations were inversely related to the percentage of forest cover in each watershed (Figure 2). On average, forest plus agriculture cover comprise 97 % of the land cover in the region (77 % and 20 %, respectively). Urban lands added about 1 % while wetlands and barren were each less than 1 % (means). As a result, watersheds with less forest cover have a strong tendency to be dominated by agricultural land cover (and to a small extent, urban land cover).

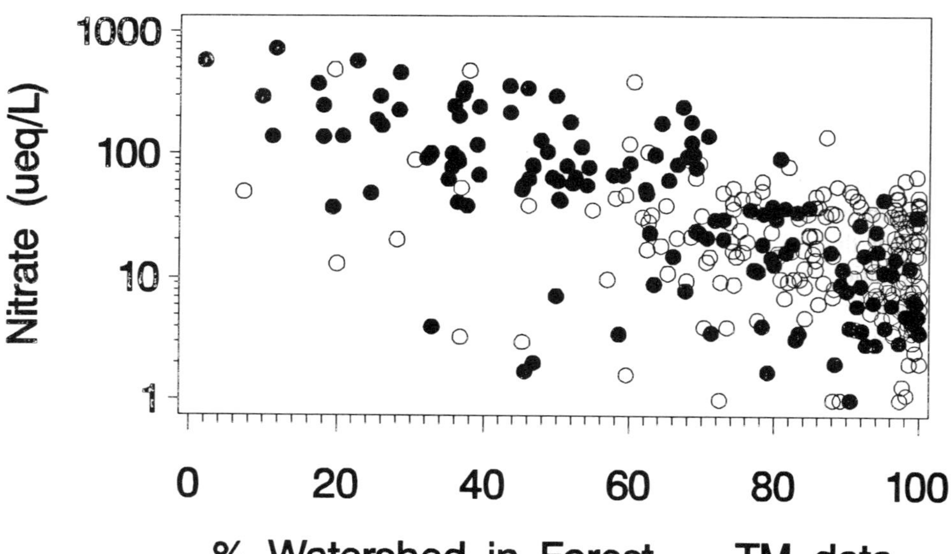

Fig. 2. Relationship between stream NO_3^- concentration (log axis) and watershed forest land cover (TM data) for Lowland (solid circles, R^2=0.51) and Highland (open circles, R^2=0.11) ecoregions.

In highland ecoregion streams, where forest cover was high (e.g., greater than 80 % in Figure 2), variability in NO_3^- concentrations was also high (0 to 100 $\mu eq\,L^{-1}$). Because much of the Mid-Atlantic region receives very high atmospheric N deposition, there is a concern that watersheds may become saturated and begin to "leak" N (Stoddard, 1994). Although our data show that many of the forest-dominated watersheds release N as NO_3^- into streams, the high NO_3^- concentration variability and alternative N cycling mechanisms (e.g., insect outbreaks, Webb *et al.*, 1995) make the atmospheric deposition linkage inconclusive.

One advantage of a data set with a large sample size is the ability to quantify relationships at multiple spatial scales. We evaluated the strength of the stream chemistry-land cover relationships at two different ecoregion and river basin scales (Table II). While the relationship between NO_3^- concentration and land cover is clear at the scale of the entire Mid-Atlantic (Table I), many of the smaller areas (either ecoregions or river basins) show much stronger relationships. Agricultural and urban lands consistently increased NO_3^- concentrations, but the magnitude of the effect varied among regions.

Models predicting NO_3^- concentration from land covers have the lowest R^2 in ecoregions that are dominated by forest land cover (e.g., the Northern Appalachians and the Ridges; Table II). Most of the watersheds in these regions are nearly 100% forested (Figure 2) and there is very little "signal" in the land cover percentages that can be used to predict NO_3^- concentrations. For the lowland ecoregions, agriculture is more common and the land cover composition is more diverse, therefore, relationship models have a higher R^2. In many cases, aggregating sites by river basins rather than ecoregions produces more precise models, presumably because the broad elevation gradients in these basins contain wide ranges in land cover. This is consistent with McDaniel *et al.* (1987) who reported no relationship between water quality and terrestrial ecosystem patterns in Kansas, but distinct relationships in Georgia and hypothesized that this was due to more uniform environmental conditions in Kansas.

Strong correlations between Cl^- concentrations and land cover occurred at virtually all scales of spatial aggregation (Table II). As in the case of NO_3^-, there is a tendency for the stream chemistry-land cover relationships in the predominantly forested ecoregions to have the lowest R^2 (e.g., Ridges and Northern Appalachians). However, the variability in Cl^- concentrations is much lower than that for NO_3^- concentrations in watersheds with greater than 80% forest cover. As a result, multiple regression models show a high correlation within all spatial aggregations and land cover is a good predictor of Cl^- concentration (Figure 3).

Most previous researchers have attributed elevated surface water Cl^- concentrations to the use of road deicing salts (e.g., Mattson and Godfrey, 1994; Siver *et al.*, 1996), but inputs from numerous other anthropogenic sources, including industry (Prowse, 1987), fertilizer use, animal wastes (Schnabel *et al.*, 1993) and sewage (Prowse, 1987) are also possible. Because of its variety of sources (virtually all of which are traceable to human activities within a watershed), Cl^- concentration seems to be a general indicator of any non-forested land whereas NO_3^- concentration is more strongly associated with agriculture. Chloride by itself is not often of major environmental concern. In studies or databases without land cover information, however, Cl^- concentration could be used as a good surrogate indicator for general human disturbance in the watershed.

TABLE II

Results of multiple regression analysis for Cl^- and NO_3^- (both log transformed) with all five major TM land cover categories as independent variables in various mid-Atlantic spatial aggregations. All models are significant ($p < 0.05$) unless noted otherwise (n.s.).

		R^2	
Geographic Area	n	$Log_{10}(Cl^-)$	$Log_{10}(NO_3^-)$
All Highlands	217	0.30	0.16
Blue Ridge	25	0.58	0.43
Central Appalachians	76	0.50	0.27
Northern Appalachians	59	0.46	0.08 (n.s.)
Ridges (from Valley and Ridge)	57	0.40	0.16 (n.s.)
All Lowlands	151	0.61	0.52
Piedmont/Coastal Plain	18	0.89	0.82
Valleys (from Valley and Ridge)	108	0.71	0.53
Western Appalachians	25	0.53	0.60
Atlantic Basin	212	0.56	0.42
Delaware River	16	0.83	0.74
James River	38	0.80	0.64
Potomac River	55	0.63	0.44
Susquehanna River	92	0.58	0.38
Ohio Basin	155	0.41	0.36
Allegheny River	31	0.43	0.41
Kanawha River	57	0.51	0.26
Monongahela River	32	0.59	0.59
Main Stem Drainages	17	0.88	0.49
Tennessee River	18	0.92	0.68

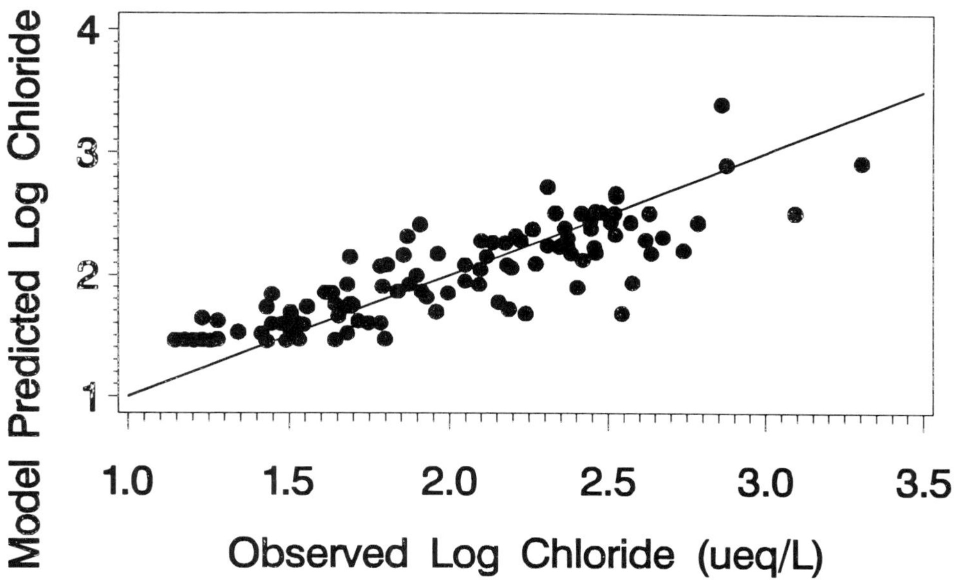

Fig. 3. Relationship of Cl^- concentration predicted from land cover multiple regression vs. observed Cl^- concentration for the Valley ecoregion, line is 1:1 line ($R^2 = 0.71$, slope = 0.71, intercept = 0.59, n = 108).

4. Conclusions

Multiple regression analysis of EMAP stream data in the mid-Atlantic showed that Cl^-, nutrients, ANC, and base cations are the chemical variables most strongly related to watershed land cover. Despite large differences in resolution and age, virtually identical results were obtained with TM and LULC data sources. Using 14 land cover subclasses rather than 5 broad classes did not greatly improve relationships with stream chemistry. Ecoregions with predominantly forested land cover had weaker relationships than those with more agricultural and/or urban land. Similar results were obtained using either ecoregions or drainage basins as spatial aggregation units. In studies or databases without land cover information, Cl^- concentration is a good surrogate indicator for general human disturbance in the watershed.

Acknowledgments

This research was funded by the U.S. Environmental Protection Agency through a cooperative agreement with Oregon State University (#CR821738) and contracts with Dynamac International Corp. (68-C8-005) and OAO Corporation (68-W5-0065). It has been subjected to the Agency's peer and administrative review and approved for publication. Mention of trade names or commercial products does not constitute endorsement or recommendation for use. We thank Patti Haggerty for generating the site map, T. Moser and the journal peer reviewers for their helpful comments on the preparation of this manuscript, and all EMAP personnel involved with generating and assembling the data.

References

Anderson, J.R., Hardy, E.E., Roach, J.T. and Witmer, R.E.: 1976, *A Land Use and Land Cover Classification system for Use with Remote Sensor Data*, USGS Prof. Paper 964.

Field, C.K., Siver, P.A., and Lott, A.: 1996, *J. Environ. Qual.*, **25**, 325.

Herlihy, A.T., Kaufmann, P.R., Church, M.R., Wigington, P.J., Webb, J.R. and Sale, M.J.: 1993, *Water Resourc. Res.*, **29**, 2687.

Johnson, C.B., Weaver, G.W. and Larsen, D.P.: *Environ. Manag.*, in review.

Jordan, T.E., Correll, D.L. and Weller, D.E.: 1997, *J. Environ. Qual.*, **26**, 836.

Langland, M.J., Lietman, P.L. and Hoffman, S.: 1995, *Synthesis of Nutrient and Sediment Data for Watersheds within the Chesapeake Bay Drainage Basin*, U.S. Geological Survey Water-Resources Investigations Report 95-4233.

Liegel, L., Cassell, D., Stevens, D., Shaffer, P. and Church, R.: 1991, *Environ. Manag.*, **15**, 269.

Mattson, M.D. and Godfrey, P.J.: 1994, *Environ. Manag.*, **18**, 767.

McDaniel, T.W., Hunsaker, C.T. and Beauchamp, J.J.: 1987, *Environ. Manag.*, **11**, 507.

Mueller, D.K., Hamilton, P.A., Helsel, D.R., Hitt, K.J. and Ruddy, B.C.: 1995, *Nutrients in Ground Water and Surface Water of the United States - An Analysis of Data through 1992*, U.S. Geological Survey Water-Resources Investigations Report 95-4031.

Osborne, L.L. and Wiley, M.J.: 1988, *J. Environ. Manag.*, **26**, 9.

Prowse, C.W.: 1987, *Water, Air, Soil Pollut.*, **32**, 277.

Puckett, L.J.: 1995, *Environ. Sci. Technol.*, **29**, 408A.

SAS.: 1985, *SAS/STAT Guide for Personal Computers, Ver. 6*, SAS Inst. Inc., Cary, NC.

Schnabel, R.R., Urban, J.B. and Gburek, W.J.: 1993, *J. Environ. Qual.*, **22**, 589.

Siver, P.A., Canavan, R.W., Field, C.K., Marsicano, L.J. and Lott, A.: 1996, *J. Environ. Qual.*, **25**, 334.

Stoddard, J.L.: 1994, 'Long-term Changes in Watershed Retention of Nitrogen: Its Causes and Aquatic Consequences', in L.A. Baker (ed.), *Environmental Chemistry of Lakes and Reservoirs*, American Chemical Society, Washington, DC, pp. 223-284.

U.S. Environmental Protection Agency: 1987, *Handbook of Methods for Acid Deposition Studies: Laboratory Analysis for Surface Water Chemistry*, EPA 600/4-87/026, U.S. Environmental Protection Agency, Washington, D.C.

U.S. Environmental Protection Agency: 1995, *Environmental Monitoring and Assessment Program - Surface Waters: Field Operations and Methods for Measuring the Ecological Conditions of Wadeable Streams*, U.S. Environmental Protection Agency, Cincinnati, OH.

U.S. Geological Survey: 1990, *Land Use and Land Cover Digital Data from 1:250,000- and 1:100,000-scale maps. Data Users' Guide 4*, U.S. Geological Survey, Reston, Virginia.

Vogelmann, J.E., Soho, T. and Howard, S.M.: *Photogram. Eng. Remote Sens,* in press.

Webb, R., Cosby, J., Eshleman, K. and Galloway, J.: 1995, *Water, Air, Soil Pollut.*, **85**, 535.

Woods, A.J., Omernik, J.M., Brown, D.D. and Kiilsgaard, C.W.: 1996, *Level III and IV Ecoregions of Pennsylvania and the Blue Ridge Mountains, the Ridge and Valley, and the Central Appalachians of Virginia, West Virginia, and Maryland*, EPA/600R-96/077, U.S. Environmental Protection Agency, Corvallis, Oregon.

HYDROLOGY AND CHEMISTRY OF THE CHOPTANK RIVER BASIN

T.R. FISHER, K.-Y. LEE, H. BERNDT, J.A. BENITEZ, and M.M. NORTON

Horn Point Environmental Laboratories, University of Maryland - CEES, Cambridge, MD 21613, USA

Abstract. The Choptank River basin is a coastal plain catchment dominated by agriculture (52 % of land use). We summarize an 11 year data set of discharge and chemistry from a gauged subbasin. Discharge exhibited seasonal variations driven by seasonal evapotranspiration. There were double seasonal maxima of pH, NH_4^+, NO_3^-, total N, Fe, and total P concentrations in late spring and fall as the saturated zone rose and fell within the soil. Significant interannual variability in discharge was the result of rainfall variation. There were positive nterannual trends in NO_3^- concentrations and negative interannual trends in NH_4^+ and PO_4^{3-} concentrations. These data were combined to estimate N and P export coefficients of 3-11 kg N ha^{-1} yr^{-1} and 0.14-0.66 kg P ha^{-1} yr^{-1}, driven primarily by interannual variations in discharge. These export coefficients are low compared to other coastal plain watersheds dominated by agriculture and may be responsible for the small anthropogenic effects in the Choptank estuary compared to other Chesapeake drainages.

Keywords: Chesapeake Bay, Choptank River, iron, nitrogen, phosphorus, river water chemistry

1. Introduction

The eutrophication of North American estuaries has been caused by nutrient enrichment from their surrounding watersheds. Enhanced supplies of nitrogen (N) and phosphorus (P) have resulted from the past 300 years of watershed disturbance, beginning in the 17th century and continuing to the present as the primary land use changed from forest to intensively managed agriculture and urban. As the human population increased, greater amounts of N and P entered aquatic systems of all types via atmospheric inputs of N (e.g., Fisher and Oppenheimer, 1991), agriculture and animal husbandry (Beaulac and Reckhow, 1982), and sewage discharges (Peierls *et al.*, 1991).

The Chesapeake Bay has been particularly impacted by anthropogenic activities in its watershed. Long and overdeepened during the lower sea level of the last glacial epoch (Meade, 1972; Schubel and Hirschberg, 1978), the Chesapeake Bay has a water residence time of 6-9 mo, which encourages consumption of N and P entering the estuary (Harding *et al.*, 1986; Fisher *et al.*, 1988). As a result, Chesapeake Bay is an efficient system for trapping the anthropogenically enhanced quantities of nutrients entering from the airshed, watershed, and coastal ocean (Fisher *et al.*, 1988, 1992).

Not all regions of the Chesapeake Bay have been influenced uniformly by anthropogenic activity. The main axis has been enriched considerably by development of its watershed in Pennsylvania and New York, resulting in higher nutrient inputs, higher phytoplankton biomass, and loss of submerged macrophytes (Orth and Moore, 1983; Harding, 1994). While some tributaries in Maryland and Virginia have been affected severely (e.g., the Patuxent River), other tributaries, particularly those on the coastal plain, have been less impacted (e.g., the Choptank River; Boynton *et al.*, 1995). In this paper, we summarize the hydrology and chemistry of a gauged Choptank River subbasin to examine temporal variation, to compare export of N and P to export from other coastal plain basins, and to evaluate whether anthropogenic effects are less in this basin than elsewhere.

Water, Air, and Soil Pollution **105:** 387–397, 1998.

2. Study Site

Located on the eastern shore of the Chesapeake Bay, the Choptank basin lies in the coastal plain on the Delmarva peninsula (Figure 1). Good soils and low relief have resulted in a history of agricultural land use, beginning with the earliest European colonization. Land use is dominated by agriculture (52 %) and forests (26 %, Mayers, 1998), and soils have high P sorption (Richardson, 1985). The region is underlain by the surficial Columbia

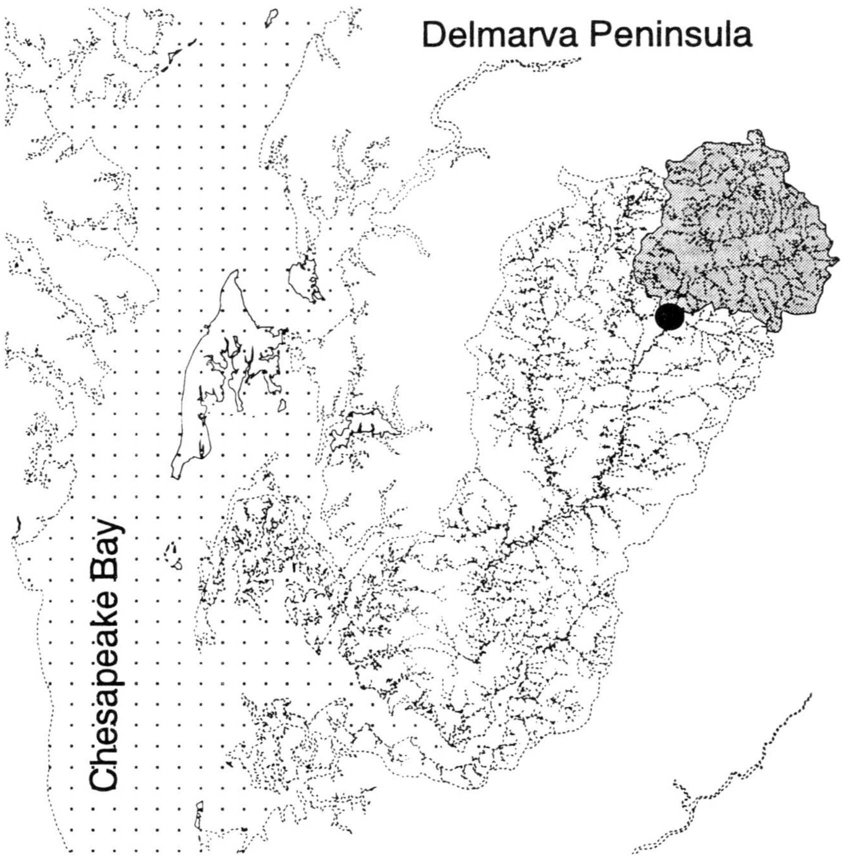

Fig. 1. Position of the Choptank River basin along the shore of the Chesapeake Bay. The gauged catchment is shaded gray; the location of the gauging station is shown by a black dot.

aquifer, with a nearly continuous aquiclude. Local groundwater flow paths are less than 2 km through soils with high hydraulic conductivity (Bachman and Wilson, 1984; Carpenter, 1983; Hamilton *et al.*, 1993). Within the Choptank basin, the terrestrial catchment (1,777 km^2, Figure 1) drains into an estuarine tributary (280 km^2) of the Chesapeake Bay. The gauged portion of the basin (gray area in Figure 1) is located in the upper northeastern section with an area of 293 km^2, representing 16 % of the terrestrial catchment. The gauge is maintained by the U.S. Geological Survey (USGS gauge # 01491000) and is located well inland because of the flat topography and penetration of tides.

3. Methods

Rain data were obtained from the U.S. Weather Service, and discharge data were obtained from USGS Water-Data Reports. Daily water samples were collected manually at the gauging station, and samples were stored without filtration or preservatives in acid-washed bottles at 4 °C in the dark for up to a month. Conductivity and pH were measured with probes calibrated with standard solutions. Conductivity values were corrected to 25 °C, and pH data were transformed to H^+ concentrations prior to statistical analyses.

From each set of monthly samples, a subset was chosen for chemical analyses. We selected samples representing the largest discharge event (usually ~ 1 week in duration) as well as a minimum of two samples per week. This subset, typically 8-12 samples per month, was filtered using Gelman A/E or Whatman GFF glass fiber filters. Filtrate was analyzed manually for ferrozine-reactive iron (Fe), using the method of Stookey (1970). Colorimetric methods were used for ammonium (NH_4^+), nitrite + nitrate (NO_3^-), and phosphate (PO_4^{3-}, Strickland and Parsons, 1972; Whitledge *et al.*, 1981). Total N and P (TN, TP) were measured beginning in August 1981 by manual persulfate combustion of unfiltered samples followed by colorimetric NO_3^- and PO_4^{3-} analyses (Valderrama, 1981).

Artifacts associated with the storage procedure have been evaluated previously (Fisher, 1987). Briefly, at the median storage time of 3 wk, biases were +1 μmol L^{-1} for dissolved inorganic N, -1.5 μmol L^{-1} for Fe, and +0.1 μmol L^{-1} for PO_4^{3-}. Three QA/QC procedures indicated that errors in chemistry were always within ±10 % and usually ±5 %.

Each observed concentration value on day i (C_i) was weighted by the flow at the gauge on day i (Q_i) relative to the average flow on all days with samples, Q_{av}. Volume-weighted means, $<C_{vw}>$, and variances, $V(C_{vw})$, for a given time period were computed as:

$$<C_{vw}> = (\Sigma(C_i \cdot Q_i))/\Sigma Q_i \quad (1)$$

$$V(C_{vw}) = \Sigma(C_i - <C_{vw}>)^2/(n-1) \quad (2)$$

where summation is over n observations in the time period, and C_i and Q_i are the ith paired observation of concentration and discharge. Data were compiled and analyzed by month and by water year (WY, Oct. 1 of the previous calendar year to Sept. 30).

Linear regressions, correlations, and *t* statistics were used for testing the significance of relationships or differences. Significance of tests were indicated as not significant (NS, $p > 0.05$), significant (*, $0.05 > p > 0.01$), or highly significant (**, $0.01 > p$).

4. Results

Responses to precipitation occurred primarily within 1 wk (Figure 2). For example, a storm on 16Dec1982 (WY day 77) deposited 1.7 cm of rain, and maximum flow occurred the following day. Over the next week, discharge declined exponentially toward previous baseflow levels, prolonged slightly by a smaller event on days 80-81. Nitrate concentrations were inversely correlated with discharge ($r^2 = 0.50^*$), whereas P concentrations were positively correlated with discharge ($r^2 = 0.85$-0.95^{**}). This was a common pattern on time scales of less than 1 wk. Below we have eliminated this short-term variability by volume weighting over longer periods to illustrate variations on monthly and interannual time scales.

Discharge exhibited a common regional pattern. Although monthly precipitation is uniform, average monthly discharge was greatest in March and April and least in August and September (Table I). This is typically the result of a seasonal maximum of evapotranspiration in summer reducing groundwater to its lowest level, followed by seasonal groundwater recharge and saturation (Novotny and Olem, 1994). On an interannual basis, stream discharge varied primarily as a function of rainfall (Table II). For all 11 water years, annual discharge was significantly correlated with annual precipitation ($r^2 = 0.69$).

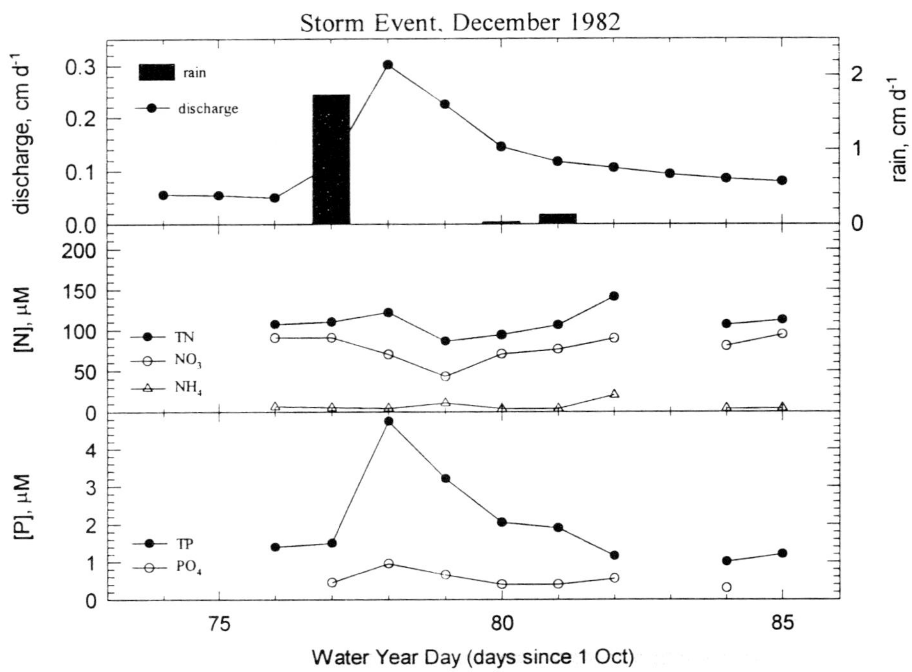

Fig. 2. Rain event in the Choptank basin at Greensboro, Maryland. Phosphorus is mobilized during events by erosion; total N is reduced due to dilution of groundwater NO_3^-.

TABLE I

Monthly volume-weighted means (± standard errors) in precipitation, discharge, conductivity, pH and concentrations of NH_4^+, NO_3^-, total N, dissolved Fe, PO_4^{3-}, and total P.

Month	Precipitation ($cm\ mo^{-1}$)	Discharge ($cm\ mo^{-1}$)	Conduc. ($\mu mho\ cm^{-1}$)	pH	Concentration ($\mu mol\ L^{-1}$) NH_4^+	NO_3^-	Total N	Fe	PO_4^{3-}	Total P
Oct	7.58 ± 0.49	1.63 ± 0.49	138 ± 42	6.35 ± 0.31	1.7 ± 0.3	51 ± 4	104 ± 12	8.4 ± 1.2	0.87 ± 0.18	1.8 ± 0.3
Nov	9.64 ± 0.74	1.91 ± 0.44	175 ± 24	6.84 ± 0.09	3.4 ± 0.1	60 ± 5	117 ± 22	13.0 ± 1.6	1.20 ± 0.20	1.9 ± 0.4
Dec	7.57 ± 1.11	3.13 ± 0.78	152 ± 22	6.07 ± 0.07	5.2 ± 0.9	70 ± 7	115 ± 19	7.3 ± 1.1	0.86 ± 0.24	2.5 ± 0.7
Jan	8.27 ± 0.64	3.36 ± 0.74	121 ± 19	5.95 ± 0.16	2.5 ± 0.4	90 ± 9	131 ± 18	7.6 ± 1.1	0.98 ± 0.22	1.7 ± 0.4
Feb	7.72 ± 0.50	4.48 ± 0.53	117 ± 26	6.13 ± 0.34	4.4 ± 0.9	80 ± 5	126 ± 10	6.7 ± 0.6	0.75 ± 0.08	2.0 ± 0.4
Mar	9.94 ± 0.90	6.16 ± 1.08	123 ± 11	6.01 ± 0.11	3.8 ± 0.1	65 ± 6	120 ± 16	9.0 ± 1.0	0.95 ± 0.12	2.1 ± 0.4
Apr	10.29 ± 0.90	6.03 ± 0.29	127 ± 4	6.55 ± 0.04	4.8 ± 1.5	55 ± 5	107 ± 13	13.0 ± 2.0	1.20 ± 0.20	2.3 ± 0.5
May	12.60 ± 1.14	4.67 ± 0.92	121 ± 12	6.55 ± 0.14	7.5 ± 1.4	73 ± 11	140 ± 37	19.0 ± 2.0	1.30 ± 0.20	3.2 ± 1.0
Jun	8.27 ± 0.99	2.72 ± 0.74	145 ± 11	6.78 ± 0.05	6.6 ± 1.7	90 ± 16	140 ± 33	19.0 ± 2.0	1.20 ± 0.30	3.6 ± 1.1
Jul	11.08 ± 0.84	1.45 ± 0.41	144 ± 9	6.65 ± 0.05	4.3 ± 0.7	68 ± 6	116 ± 14	14.0 ± 3.0	1.40 ± 0.40	2.4 ± 0.4
Aug	9.71 ± 0.86	1.01 ± 0.27	168 ± 16	6.52 ± 0.06	4.8 ± 0.9	54 ± 5	96 ± 12	12.0 ± 2.0	1.30 ± 0.30	2.1 ± 0.4
Sep	9.26 ± 0.91	1.04 ± 0.32	104 ± 48	6.28 ± 0.34	4.3 ± 2.2	43 ± 6	112 ± 35	8.4 ± 3.2	1.10 ± 0.30	2.0 ± 0.6

TABLE II

Annual volume-weighted means (± standard errors) in precipitation, discharge, conductivity, pH and concentrations of NH_4^+, NO_3^-, total N, dissolved Fe, PO_4^{3-}, and total P.

Month	Precipitation ($cm\ yr^{-1}$)	Discharge ($cm\ yr^{-1}$)	Conduc. ($\mu mho\ cm^{-1}$)	pH	Concentration ($\mu mol\ L^{-1}$) NH_4^+	NO_3^-	Total N	Fe	PO_4^{3-}	Total P
1980	114 ± 1	45.9 ± 0.4		5.7 ± 0.7	6.7 ± 1.0	66 ± 3		17.1 ± 1.0	1.80 ± 0.20	
1981	102 ± 5	23.9 ± 0.2		6.1 ± 0.6	8.7 ± 1.3	59 ± 5	96 ± 38	16.0 ± 2.0	0.94 ± 0.10	3.06 ± 1.50
1982	102 ± 5	32.8 ± 0.3	118 ± 34	6.3 ± 1.6	6.1 ± 0.6	77 ± 6	110 ± 8	10.9 ± 1.0	0.89 ± 0.09	1.89 ± 0.20
1983	129 ± 8	53.7 ± 0.4	99 ± 58	6.0 ± 0.5	6.0 ± 0.5	59 ± 7	109 ± 13	11.2 ± 2.0	1.03 ± 0.20	2.51 ± 0.50
1984	141 ± 9	61.8 ± 0.4	132 ± 34	6.8 ± 0.8	2.4 ± 0.4	73 ± 7	133 ± 14	7.8 ± 0.8	0.79 ± 0.12	2.58 ± 0.40
1985	109 ± 5	16.4 ± 0.2	98 ± 44	6.4 ± 0.8	3.8 ± 1.1	61 ± 6	126 ± 18	5.8 ± 1.4	0.64 ± 0.14	2.79 ± 0.70
1986	77 ± 5	28.3 ± 0.4	134 ± 15	6.2 ± 0.3	1.7 ± 0.3	77 ± 9	128 ± 15	5.1 ± 0.5	0.70 ± 0.16	1.76 ± 0.20
1987	100 ± 5	37.0 ± 0.4	128 ± 12	6.2 ± 0.1	3.7 ± 0.5	82 ± 5	112 ± 13	9.0 ± 0.9	0.52 ± 0.13	1.71 ± 0.20
1988	96 ± 2	20.2 ± 0.2	170 ± 31	5.8 ± 0.6	5.8 ± 0.7	79 ± 7	117 ± 12	9.5 ± 1.0	0.45 ± 0.13	5.62 ± 2.20
1989	149 ± 9	60.5 ± 0.4	154 ± 36	5.7 ± 0.6	4.0 ± 0.6	75 ± 6	111 ± 15	15.4 ± 1.2	0.83 ± 0.19	3.45 ± 1.60
1990	116 ± 8	43.2 ± 0.3	119 ± 7	5.3 ± 0.4	2.2 ± 0.6	89 ± 6	123 ± 10	12.4 ± 1.1	0.30 ± 0.09	1.75 ± 0.20

Conductivity and pH exhibited seasonal patterns but no systematic interannual trends (Tables I and II). There were double seasonal peaks in November (conductivity ≈ 170 μmho cm^{-1}, pH ≈ 6.8), as discharge was increasing, and again in June (pH ≈ 6.8) or August (conductivity ≈ 170), as discharge approached its seasonal minimum. pH and conductivity were depressed slightly during winter at low temperatures and in early fall because of occasional hurricanes. Therefore, the Choptank River at Greensboro may be described as slightly acidic (pH 5.5 - 6.5) and moderately conductive (100-180 μmho cm^{-1}).

There were similar patterns for N concentrations (Figure 3). Seasonally, NH_4^+, NO_3^-, and total N concentrations exhibited double peaks, particularly NO_3^-, but these were shifted compared to conductivity and pH (Table I). There was a winter peak in N (December-January), depressed concentrations during high spring flow, and a seasonal maximum in late spring (May-June). The lowest concentrations were observed in late summer or fall (August-October) at low flow. There were several significant interannual trends in N concentrations. Ammonium concentrations declined at ~0.4 μmol L^{-1} yr^{-1}, and NO_3^- concentrations increased at ~2 μmol L^{-1} yr^{-1}. However, there was no significant trend for total N concentrations. Nitrate comprised the largest portion of total N, averaging 57 %; NH_4^+ represented 4 %. Volume-weighted monthly and interannual statistics are given in Tables I and II.

Dissolved Fe and P concentrations exhibited seasonal patterns similar to those of N concentrations (Figure 4, upper panels). There were double peaks in late fall or winter (November - December), during the period of increasing discharge, and again in late spring (May-June), following the peak in discharge. The lowest concentrations occurred in winter. Although PO_4^{3-} concentrations only generally followed this seasonal pattern, it was particularly evident for Fe concentrations, which increased three-fold from February to May. Interannually, Fe and TP exhibited no significant trends over WY 1980 to 1990, although both showed higher concentrations at the beginning and end of the decade, with lower values in 1985-1987. In contrast, PO_4^{3-} concentrations declined significantly and consistently during the entire period, particularly between WY 1980 and 1981. Phosphate comprised an average of 42 % of the total P, generally declining as a fraction of the total P over the decade. At the monthly time scale, Fe was positively correlated with both PO_4^{3-} ($r^2 = 0.61$) and total P concentrations ($r^2 = 0.65$); however, at the interannual time scale, only Fe and PO_4^{3-} concentrations were weakly correlated ($r^2 = 0.35$).

5. Discussion

We observed three significant time scales for variation in discharge and chemistry. Storm events dominated at time scales of less than 1 wk (Figure 2). At monthly time scales, discharge exhibited spring and fall maxima (Table I), accompanied by variation in water chemistry (Figures 3-4). Although we have no direct evidence, we hypothesize that the concentration maxima associated with rising and falling discharge is due to flushing of the vadose zone of soils. At the interannual time scale, variability was due primarily to interannual variation in precipitation (Table II). There were no significant trends among the hydrological parameters or among the chemical parameters. Causes of the trends in NH_4^+, NO_3^-, and PO_4^{3-} concentrations at this time scale are not readily apparent, although increasing NO_3^- could be due to time lags in groundwater NO_3^- leaching resulting from historic increases in fertilizer use (Nixon and Pilson, 1983).

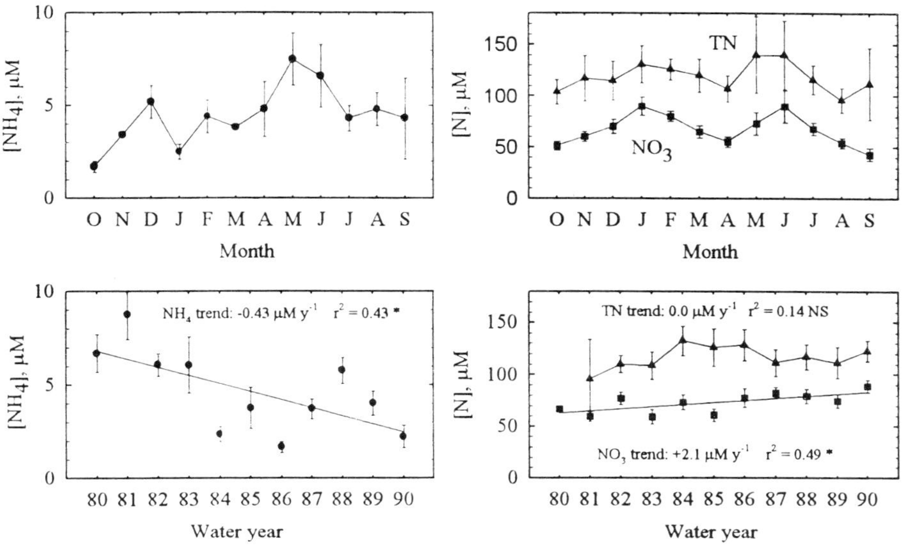

Fig. 3. Volume-weighted means and standard errors for NH_4^+, NO_3^-, and TN (right panels). Statistics were computed by month (top panels) and by water year (bottom panels) with interannual trends.

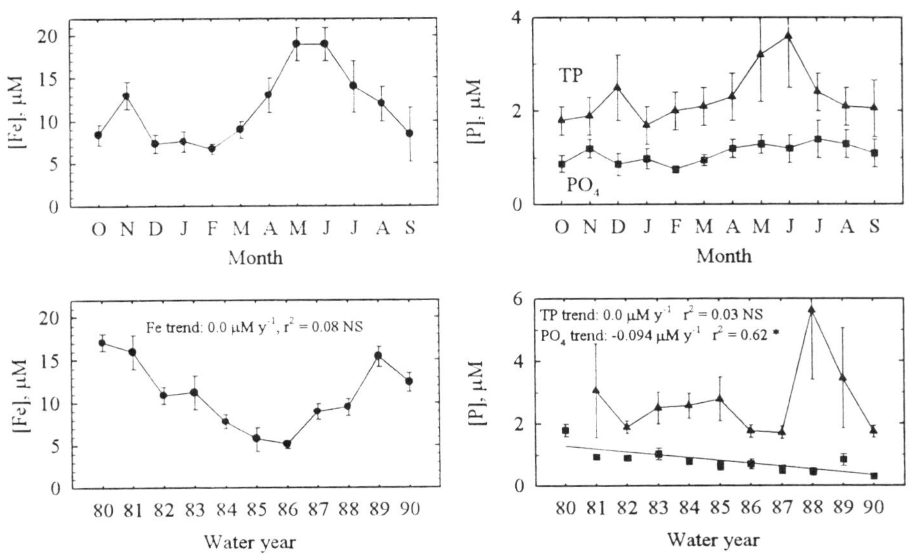

Fig. 4. Volume-weighted means and standard errors tor dissolved iron (left panels), phosphate, and total P (right panels). Statistics were computed by month (top panels) and by water year (bottom panels) with interannual trends.

The data presented above have been used to compute annual losses of N and P. Export from the gauged basin (E, kg N or P $time^{-1}$) was expressed as:

$$E = <C_{vw}> \bullet \Sigma Q_j \tag{3}$$

where $<C_{vw}>$ is the annual volume-weighted mean concentration (eq. 1), and ΣQ_j is the total water flow (j = total days in the year). Export varied from 85 - 340 Mg N and 4.1 - 19 Mg P yr^{-1} (Table III). The data also were normalized by basin area and expressed as export yield coefficients (Y, kg N or P ha^{-1} $time^{-1}$):

$$Y = E/A \tag{4}$$

where A is the drainage area in hectares. This form is particularly useful for comparisons with data from other drainages. Export coefficients of total N and total P from the gauged area of the Choptank River varied from 2.9 - 12 kg N and 0.14 - 0.65 kg P ha^{-1} yr^{-1} (Table III).

Export of N and P were primarily a function of the strong seasonal and interannual variations in water discharge. The smaller variations in concentrations (Figures 3-4) were dominated by those of discharge. As a result, export of N and P peaked in late winter to spring (February - April) and was minimal in summer and early fall (July - October). Interannually, years with high discharge (Table II) were years with the highest export.

TABLE III
Export and export coefficients for the gauged watershed of the Choptank River at Greensboro, MD.

Water Year	Total N export (Mg yr^{-1})	N export coefficient (kg ha^{-1} yr^{-1})	Total N export (Mg yr^{-1})	N export coefficient (kg ha^{-1} yr^{-1})
1981	93.9	3.21	6.64	0.23
1982	148.0	5.06	5.62	0.19
1983	239.0	8.17	12.2	0.42
1984	336.0	11.50	14.50	0.49
1985	84.7	2.89	4.14	0.14
1986	149.0	5.09	4.51	0.15
1987	169.0	5.77	5.75	0.20
1988	96.7	3.30	10.30	0.35
1989	276.0	9.44	18.90	0.65
1990	217.0	7.42	6.87	0.23

Yield coefficients from a literature survey of North American coastal plain watersheds are summarized in Figure 5. The data reflect differences in land use, particularly agriculture and urban areas, which have much higher N and P yields (5-80 kg N and 0.5 - 20 kg P ha^{-1} yr^{-1}) than forests (1-6 kg N and 0.1 - 0.9 kg P ha^{-1} yr^{-1}; Beaulac and Reckhow, 1982). As a result, export

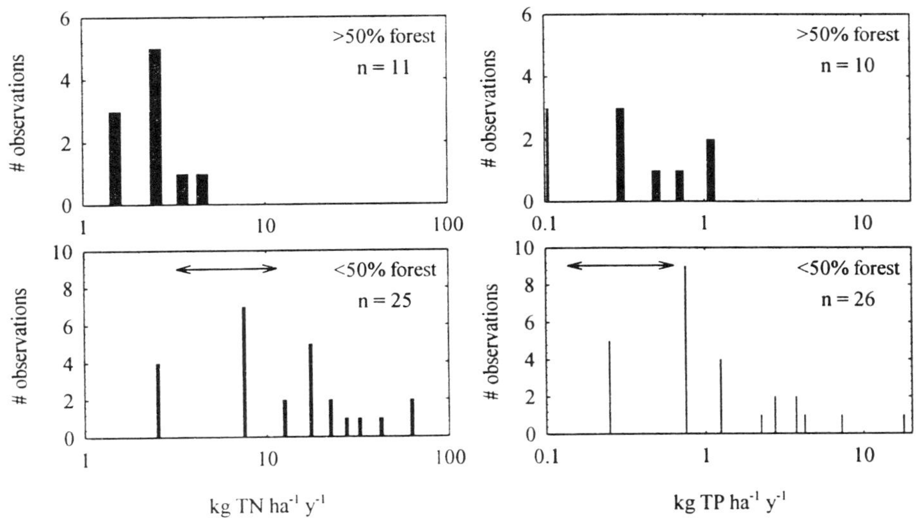

Fig. 5. Export yield coefficients from North American coastal plain watersheds. Data were obtained from search of the literature published during 1977-1996 and were segregated into two classes with >50 % and <50 % forest. Arrows in lower panels show the range of export coefficients calculated for the Choptank River (Table III). A list of publication sources is available from the authors.

coefficients usually increase as the percentage of forest cover decreases (e.g., Vitousek, 1977; Fisher, 1987; Mayers, 1998). Accordingly, we have separated the data into two groups: those with >50 % forest cover (primarily forested basins) and those with <50 % forest cover (primarily developed basins). Primarily developed basins have larger ranges and significantly greater mean yields, and distributions are skewed toward higher values.

The range of N and P yields from the Choptank River at Greensboro are shown as arrows in the lower panels of Figure 5. These fall within the lower range of values observed in other North American coastal plain basins with <50 % forest land cover, particularly for P. The low yield coefficients support the conclusion of Boynton *et al.* (1985) that export of N and P from the Choptank River basin is small and explains the smaller anthropogenic impacts of N and P within this tributary. However, the gauged area of the Choptank River occupies only 16 % of the basin and is the most distant portion from the estuary (Figure 1). It will be important to test whether the gauged basin is representative of the remainder of the basin, which tends to have more intensive land uses closer to estuarine waters.

Acknowledgments

We wish to thank E. Handley, R. Crump, P. Carlson, L. Yarbro, D. Mowbray, S. Pearce, E. Doyle, T. Coley, A. Gustafson, L. May, M. L. Malone, and M.-L. Chang. USGS supplied data and programs, and funding was provided by the Maryland Department of Natural Resources, the U.S. Fish and Wildlife Service, the National Science Foundation Ecosystem Studies Program, and NASA MTPE.

References

Bachman, L.J. and Wilson, J.M.: 1984, *Hydrogeology and Selected Water-well Records, Chemical Analyses, Water Level Measurements, Lithologic Logs, and Geophysical Logs*, Report of Investigations No. 40, Maryland Geological Survey, Department of Natural Resources, Baltimore.

Beaulac, M.N. and Reckhow, K.H.: 1982, *Water Resourc. Bull.* **18**, 1013.

Boynton, W.R., Garber, J.H., Summers, R. and Kemp, W.M.: 1995, *Estuaries* **18**, 285.

Carpenter, D.H.: 1983, *Characteristics of Streamflow in Maryland*, Report of Investigations No. 35, Maryland Geological Survey, Department of Natural Resources, Baltimore.

Fisher, D.C. and Oppenheimer, M.: 1991, *Ambio* **30**, 102.

Fisher, T.R.: 1987, *Nutrient Inputs to the Choptank River: Monthly and Interannual Variations in Water Discharge, and N, P, and Fe Concentrations at the USGS Gauging Station in Greensboro, MD for Water Years 1980-1986*, Report 3A to the U.S. Fish and Wildlife Service under cooperation agreement 14-16-005-86-9035 with the University of Maryland, Center for Estuarine and Environmental Studies.

Fisher, T.R., Harding Jr., L.W., Stanley, D.W. and Ward, L.G.: 1988, *Estuar. Coast. Shelf Sci.* **27**, 61.

Fisher, T.R., Peele, E.R., Ammerman, J.W. and Harding Jr., L.W.: 1992, *Mar. Ecol. Prog. Ser.* **82**, 51.

Hamilton, P.A., Denver, J.M., Phillips, P.J. and Shedlock, R.J.: 1993, *Water-quality Assessment of the Delmarva Peninsula, Delaware, Maryland, and Virginia - Effects of Agricultural Activities on, and Distribution of Nitrate and other Inorganic Constituents in the Surficial Aquifer*, U.S. Geological Survey Open File Report 93-40, Towson, Maryland.

Harding Jr., L.W.: 1994, *Mar. Ecol. Prog. Ser.* **104**, 267.

Harding Jr., L.W., Meeson, B.W. and Fisher, T.R.: 1986, *Est. Coast. Shelf Sci.* **23**, 773.

Mayers, M.: 1998, *The Impact of Forest on Stream Water Quality Draining two Agricultural Watersheds on the Delmarva Peninsula*, Ph.D. dissertation, University of Maryland.

Meade, R.H.: 1972, *Transport and Deposition of Sediments in Estuaries*, Geological Society of America Memoir 133.

Nixon, S.W. and Pilson, M.E.Q.: 1983, 'Nitrogen in Estuarine and Coastal Marine Ecosystems', in E.J. Carpenter and D.G. Capone (eds.), *Nitrogen in the Marine Environment*, Academic Press, New York, pp. 565-648.

Novotny, V. and Olem, H.: 1994, *Water Quality: Prevention, Identification, and Management of Diffuse Pollution*, Van Nostrand Reinhold, New York.

Orth, R.J. and Moore, K.A.: 1983, *Science* **222**, 51.
Peierls, B.L., Caraco, N.F., Pace, M.L. and Cole, J.J.: 1991, *Nature* **350**, 386.
Richardson, C.J.: 1985, *Science* **228**, 1427.
Schubel, J.R. and Hirschberg, D.J.: 1978, 'Estuarine Graveyards, Climatic Change, and the Importance of the Estuarine Environment', in M.L. Wiley (ed.), *Estuarine Interactions*, Academic Press, New York, pp. 285-303.
Stookey, L.L.: 1970, *Anal. Chem.* **42**, 779.
Strickland, J.D.H. and Parsons, T.R.: 1972, *A Practical Handbook of Seawater Analysis*, Fisheries Research Board of Canada, Ottawa.
Valderrama, J.C.: 1981, *Mar. Chem.* **10**, 109.
Vitousek, P.M.: 1977, *Ecol. Monogr.* **47**, 65.
Whitledge, T.E., Malloy, C., Patton, C.J. and Wirick, C.D.: 1981, *Automated Nutrient Analyses in Seawater*, Brookhaven National Laboratory Report No. 51398.

FLUVIAL HYDROGEOCHEMISTRY IN A SUB-TROPICAL COASTAL WATERSHED IN BRAZIL

R.O. FIGUEIREDO and A.R.C. OVALLE

Laboratório de Ciências Ambientais - CBB - UENF - Campos dos Goytacazes, RJ, 28015-620, Brazil

Abstract. Hydrogeochemical fluxes and their relationship to environmental factors were evaluated in three catchments of the Maricá-Guarapina fluvio-lagoonar system located on the coast of Rio de Janeiro State, Brazil. Sampling was performed biweekly from March to December 1993 at the outlets of three sub-basins with different land use distributions. Water discharge, physico-chemical parameters, major cations and anions, and dissolved inorganic N and P were analyzed. Despite an increase in contributions of weathering processes westward, the major constituent concentrations covaried temporally among catchments, indicating similarity of the involved biogeochemistry processes. Moreover, there is a spatial dissolved nutrient pattern, with concentrations increasing westward in association with urbanization and farming. Land use classes were arranged according to their potential impacts as follows: forest < small properties < farming < urbanization.

Keywords: biogeochemistry, Brazil, land use, Maricá-Guarapina lagoonar system, stream chemistry, watershed

1. Introduction

Rio de Janeiro, a Brazilian state, is characterized by numerous coastal lagoons, ranging from freshwater to hypersaline systems. Their associated watersheds are responsible for a great part of freshwater and nutrient inputs, which are essential for the biogeochemical dynamic of the lagoons. These systems are important for tourism, fishing, industry and water supply, but the lack of environmental protection in the development planning for these regions has compromised ecosystem function, with drastic consequences to the quality of life of the local communities (Nunes *et al.*, 1989; Guimarães, 1989).

Drainage basins commonly are used as basic units for environmental research and planning, because drainage basins represent areas with easily definable topographic boundaries where all landscape components interact: atmosphere and vegetation, plants and soils, bedrock and groundwater, stream or lake, and surrounding land (Moldan and Černý, 1994). Fluvial chemical composition is a result of water, rock, soil and vegetation interactions, reflecting most of the biogeochemical processes occurring within the environmental mosaic within a particular drainage basin.

An effective tool used for hydrogeochemical data interpretation is the calculation of mass flux, which is defined as the total amount per time of an element or compound entering or leaving the watershed. The most important inputs into the catchment are: atmospheric deposition (composed of wet deposition, dry deposition, and cloud deposition), weathering input (release from bedrock) and direct anthropogenic input (from liming, farming, urbanization, industrialization, etc.). Also important for ecosystem functioning, but difficult to quantify from watershed studies, are internal fluxes redistributing the mass within the system (between soil pools and above-ground vegetation, for instance). The main output is runoff by streams. However, other important outputs occur: gaseous emissions from the soil and vegetation, harvest or timber removal, and groundwater seepage through fractured bedrock (Černý *et al.* 1994).

Water, Air, and Soil Pollution **105:** 399–407, 1998.

In this context, land use can promote quantitative and qualitative changes in geochemical processes, often increasing the fluvial water constituents (cations, anions and nutrients), especially where increasing human population density has a direct effect on the effluent volume in streams (Coombes, 1987). Studying the environmental geochemistry of the Maricá-Guarapina lagoonar system, Barroso (1991) presented results indicating the necessity of specific studies related to the land use contribution to the fluvial hydrogeochemistry.

We investigated the effects of land use change on the fluvial hydrogeochemistry in this sub-tropical coastal watershed, in which the original vegetation characterizes one of the most endangered ecosystems in Brazil - the Atlantic Coastal Forest. The main objectives of this study are to evaluate dissolved inorganic material fluxes and total suspended solid transport from watersheds to the lagoons of the Maricá-Guarapina system, and to identify the environmental factors that control the fluvial hydrogeochemistry.

2. Materials and Methods

The Maricá-Guarapina lagoonar system is located in Rio de Janeiro State, Brazil, between the latitudes 22°53' and 22°58'S, and the longitudes 42°40' and 43°00'W. The system covers 35,000 ha and is composed of a series of interconnected lagoons and an associated drainage basin with three major streams: Vigário (VIG), Ubatiba (UBA) and Caranguejo (CAR) (Figure 1). The climate is characterized by rainy summers and dry winters, with a mean annual temperature of 24 °C and rainfall about 1,200 mm yr^{-1}. An orographic effect increases this total eastwards (Barbiere, 1986). The precipitation measured in 1993 by a station of the Brazilian National Institute of Meteorology (INMET) in Maricá (latitude - 22° 55' S; longitude - 42° 49' W) was 1,050 mm. This value was less than the historical annual mean characterizing a dry year. During the study period (March to December) the precipitation was 889 mm.

The drainage basin is defined by two distinct physiographic features: 1) a Quaternary alluvial plain sector dominated by red and yellow ferruginous soils, with kaolinite and illite-vermiculite as their most abundant clay minerals; and 2) a slope sector associated with pre-Cambrian lenticular banded gneiss and migmatite outcrops, replaced by homogeneous fine textured granite at the eastern portion of the basin, also prevailing the same ferruginous soils, with the main clay minerals identified as kaolinite and gibbsite (Perrin, 1986; Santos, 1987; Ovalle *et al.*, 1990).

Land use was characterized by LANDSAT-TM image interpretation and field checking. The most important land use classes were named as: forests (Atlantic Coastal Forest fragments mostly on slope areas); farming (properties larger than 1 ha, including pastures and sugar cane crops); urbanization (housing with deficient public service infrastructure); and small properties (defined as small farms not larger than 1 ha, mainly used for recreational purposes and subsistence crops and breeding) (Table I). There is a relevant difference between the type of urbanization in the Vigário and Ubatiba sub-basins. In the Vigário sub-basin, housing density is higher and domestic sewage disposal represents a point source into the stream channel, while in the Ubatiba sub-basin, the urbanization is dispersed in several small settlements that use individual septic tanks as sewage treatment, thus representing a non-point source via groundwater input to the streams.

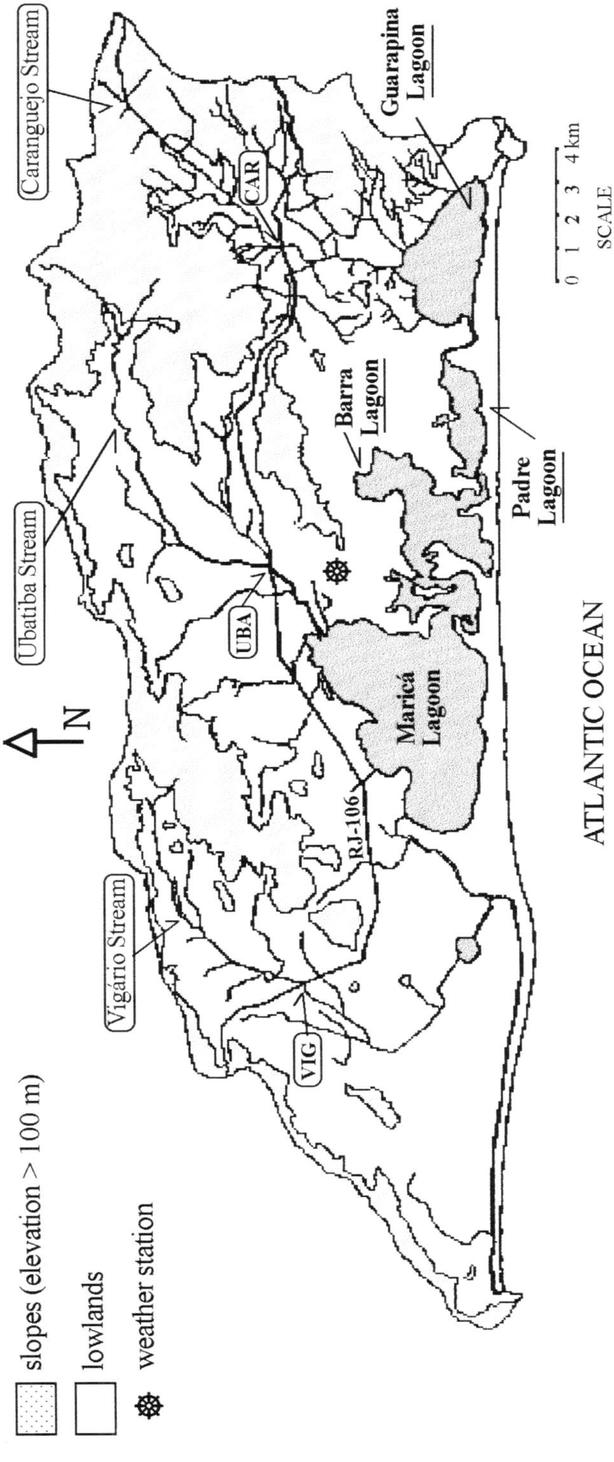

Fig. 1. The Maricá-Guarapina system; sampling stations are indicated as VIG, UBA and CAR.

TABLE I
Land use classes percentages.

	Sub-basin		
Land Use Class	Vigário	Ubatiba	Caranguejo
Forests	32%	23%	48%
Farming	65%	75%	27%
Urbanization	3%	2%	<1%
Small Properties	<1%	<1%	25%

Three sampling stations were located below bridges of RJ-106 road (Figure 1), defining three sub-basins: the Vigário catchment (VIG), measuring 2,220 ha; the Ubatiba catchment (UBA), measuring 6,850 ha; and the Caranguejo catchment (CAR), with 2,030 ha. Stream water samples were collected biweekly from March to December during 1993 in polyethylene bottles (1 L) and packed in a cooler (4 ºC) for transport. Conductivity, pH and Eh were measured *in situ,* using portable equipment. All water samples were filtered through 0.45 μm membrane filters and stored frozen before chemical analyses. Total suspended solids (TSS) were determined by gravimetry. Cations (Na^+, K^+, Ca^{2+} and Mg^{2+}) were analyzed using an atomic absorption spectrophotometer, whereas inorganic dissolved nutrients (PO_4^{3-}-P, NO_3^--N, NO_2^--N and NH_4^+-N) and SiO_2 were analyzed according Golterman *et al.* (1978). Total alkalinity (HCO_3^-) was measured by Gran plot titration, SO_4^{2-} by indirect titration with EDTA (Carmouze, 1994), and Cl^- by flow injection analysis (Zagatto *et al*., 1981). The instantaneous discharges of the streams were determined for each sampling date by measuring current speeds (G.O. current meter - Model 2030) and the cross-sectional areas of the fluvial channels. Instantaneous mass fluxes were obtained multiplying concentration values by instantaneous discharges, and then normalized to area.

3. Results and Discussion

Instantaneous discharge varied from 22 to 778 L s^{-1}, and the Caranguejo stream accounted for 57 % of the total fluvial water amount entering the lagoons. This value is four times higher than for the Vigário stream (14 %) and two times higher than for the Ubatiba stream (29 %).

Physico-chemical parameters exhibited no seasonal pattern. Their means and ranges for the studied period are in Table II. All values of these parameters decrease eastwards (from the Vigário stream to the Caranguejo), a trend that may be associated with lithological and pedological differences among the sub-basins.

None of the major dissolved constituents exhibited significant inverse relationships with discharge due to dilution by rain water. In general, major constituents covaried temporally among watersheds indicating a similarity of the involved processes. Increasing concentrations westwards (from Caranguejo to Vigário), except for SO_4^{2-}, suggest a higher migration rate between the soil-rock complex and the hydrologic system in the western portion of Maricá basin (Table III). In this context, concentration ratios of major cations to SiO_2 were used as a tracer for the lithologic control in the weathering processes. Because Cl^- is a minor trace element in the rock types of the watersheds, we assumed that Cl^- in streamwater originated primarily from the atmosphere, and used Cl^- concentrations to estimate the non-marine fraction of the major cations. The ratio of the non-marine major cations to SiO_2 presented a higher value for the Vigário sub-basin (6.3) compared to the other sub-basins (Ubatiba, 4.0; Caranguejo, 3.6), suggesting a different geochemical characteristic of the material source. In fact, the Caranguejo sub-basin, and to a less extent the Ubatiba sub-basin, have a large part of the lithology composed of homogeneous granitic rocks, compared with the dominance of banded gneisses and migmatites in the Vigário sub-basin (Perrin, 1986; Santos, 1987). Spatial variability of the major cations and SiO_2 in fluvial waters could be associated, in part, with variation in lithology. Similarly, alkalinity variation among sub-basins could be related partly to weathering of different materials.

In contrast, SO_4^{2-} exhibited a pattern of increasing concentration eastward (Table III), which cannot be linked to weathering processes because the dominant regional lithology does not contain S as a major constituent. The main source of SO_4^{2-} may be the atmosphere, enhanced by orographic rains at the Caranguejo sub-basin. Another possible source is linked to crop and pasture fertilization (Comans *et al.*, 1987), but farming is not the dominant land use within the Caranguejo catchment (Table I).

TABLE II
Physico-chemical mean values (n = 20) and their ranges.

		Stream		
		Vigário	Ubatiba	Caranguejo
Conductivity ($\mu S\ cm^{-1}$)	Mean	396	174	138
	Range	270 - 750	130 - 240	101 - 177
Eh (mV)	Mean	194	156	86
	Range	148 - 227	16 - 205	55 - 161
pH	Mean	6.75	6.62	6.48
	Range	6.30 - 7.11	6.08 - 6.93	5.87 - 7.02

Although concentrations of PO_4^{3-}-P, NH_4^+-N, NO_3^--N, NO_2^--N and TSS may indicate potential land use effects, none of these constituents exhibited any temporal pattern among catchments (Table IV). Anthropogenic interventions in the catchments probably are affecting natural processes causing variations in streamwater nutrient concentrations, especially for non-point sources (Richards and Baker, 1993). There was, however, a spatial pattern among streams, with concentrations of most constituents (except TSS) increasing from Caranguejo to Vigário (Table IV). This pattern can be related to land use, because the importance of farming and urbanization increases at sub-basin landscapes from Caranguejo to Vigário (Table I). Domestic sewage disposal directly into the stream channel seemed to be a substantial factor for Vigário concentrations, despite the apparent minor areal importance of urban land use compared with farming. The small settlements with septic tank treatment systems in the Ubatiba basin apparently minimized the effect of sewage on streamwater chemistry. Even with a farming area larger than Vigário, Ubatiba had lower concentrations of the dissolved inorganic nutrients, indicating that urbanization with point sewage input is the most critical land use in the studied area. Although the concentrations of nutrients in Ubatiba were smaller than in Vigário, they were much higher than in Caranguejo. Because Caranguejo has a smaller area of farming and a different kind of housing (small properties), the role of urbanization and farming inputs on the concentrations of the nutrients is corroborated.

TABLE III

Major constituents mean concentrations (n = 20) and their ranges; values in meq L^{-1} (except SiO_2 in mmol L^{-1}).

		Stream		
		Vigário	Ubatiba	Caranguejo
HCO_3^-	Mean	2.15	0.86	0.40
	Range	0.83 - 4.36	0.30 - 1.52	0.04 - 0.64
Na^+	Mean	2.54	1.07	0.63
	Range	1.49 - 3.63	0.81 - 1.40	0.53 - 0.71
K^+	Mean	0.20	0.11	0.08
	Range	0.11 - 0.27	0.06 - 0.19	0.02 - 0.14
Ca^{2+}	Mean	0.46	0.29	0.22
	Range	0.31 - 0.71	0.22 - 0.49	0.16 - 0.24
Mg^{2+}	Mean	0.51	0.24	0.16
	Range	0.29 - 0.73	0.13 - 0.32	0.13 - 0.19
SiO_2	Mean	0.29	0.22	0.16
	Range	0.19 - 0.41	0.16 - 0.31	0.14 - 0.21
Cl^-	Mean	1.70	0.76	0.45
	Range	1.08 - 1.98	0.61 - 0.94	0.40 - 0.51
SO_4^{2-}	Mean	0.51	0.56	0.62
	Range	0.31 - 0.99	0.47 - 0.74	0.32 - 0.81

TABLE IV

Mean concentrations of nutrients and total suspended sediment (n = 20) and their ranges; values in µmol L^{-1} (except TSS in mg L^{-1}).

		Stream		
		Vigário	Ubatiba	Caranguejo
PO_4^{3-}-P	Mean	5.9	3.2	1.4
	Range	3.3 - 9.9	1.7 - 6.2	0.5 - 3.2
NH_4^+-N	Mean	135.0	24.2	5.1
	Range	47.3 - 347.9	3.4 - 86.1	3.4 - 9.1
NO_3^--N	Mean	48.6	15.1	4.5
	Range	5.7 - 124.1	4.5 - 30.4	2.3 - 9.7
NO_2^--N	Mean	11.4	3.4	0.7
	Range	2.1 - 24.8	0.8 - 7.9	0.3 - 1.1
TSS	Mean	54	29	38
	Range	11 - 488	4 - 254	9 - 121

Except for HCO_3^-, the instantaneous mass flux estimates normalized by area for the three sub-basins usually were highest at Caranguejo, followed by Vigário, being always lowest at Ubatiba (Table V). Despite its lower discharge, Vigário represented a significant contribution to the associated lagoons, probably due to an intense release of elements by weathering, as discussed previously. The HCO_3^- flux at Vigário seems to be not only related to a natural source, but also to land use. Describing biogeochemical processes in natural waters, Bourg and Bertin (1993) and Lyons and Bird (1995) suggest that, instead of mineral weathering, some of the HCO_3^- produced might be the result of the mineralization of organic carbon. Alkalinity can be modified by any process involving H^+, OH^- and weak acid production or consumption. Organic matter (OM) mineralization accompanying N metabolism promotes H^+ ion consumption, thus increasing alkalinity, as is shown in the following reaction: $(OM)\text{-}(NH_3)_n + nH^+ \Rightarrow nNH_4^+$. In fact Ubatiba, and mainly Vigário, presented high HCO_3^- and NH_4^+-N concentrations (Tables III and IV), which may indicate the occurrence of this process, probably related with urbanization and farming (Table I).

TABLE V
Instantaneous flux means (g m^{-2} s^{-1}) in the studied sub-basins; instantaneous water discharge means in L s^{-1}.

	Stream		
	Vigário	Ubatiba	Caranguejo
Discharge	73	174	312
TSS	39.6	19.0	60.2
Na^+	17.4	5.7	21.9
K^+	2.5	1.3	4.4
Ca^{2+}	3.2	1.6	6.3
Mg^{2+}	1.9	0.6	2.9
SiO_2	8.6	5.1	24.1
Cl^-	18.1	6.7	24.7
SO_4^{2-}	7.6	7.0	45.0
HCO_3^-	40.6	11.1	39.9
PO_4^{3-}-P	0.06	0.03	0.06
NH_4^+-N	0.54	0.06	0.10
NO_3^--N	0.22	0.03	0.10
NO_2^--N	0.06	< 0.01	< 0.01

Elevated SO_4^{2-} fluxes at Caranguejo can be associated with its higher elevation (Figure 1) and forest cover (Table I), which increase atmospheric deposition due to the direct cloud contribution (Moldan and Černý, 1994). Adsorption/desorption processes in podzols soils also may affect stream SO_4^{2-} concentrations (Courchesne and Hendershot, 1989). The higher values of TSS measured in Vigário might be related to land use, whereas in Caranguejo topography and higher discharge can enhance physical denudation.

Inorganic N fluxes reflected clearly the land use impact on fluvial hydrogeochemistry (Table V), as Vigário fluxes were up to four times higher than Caranguejo fluxes, despite their differences in discharges. This finding also demonstrates the critical aspect of the urbanization that takes place at Vigário, where NH_4^+-N concentrations of up to 340 mM are dangerous for human drinking and for fish populations (Smith *et al.*, 1992). At the Ubatiba sub-basin, where individual septic tank treatment prevails and most of the effluents are introduced into groundwater, consumption by biota and denitrification processes in riparian zones most likely had an important contribution to the fate of N compounds (Antoine and Bensons-Evans, 1988). Moreover, high fluxes of PO_4^{3-}-P caused by domestic sewage also were expected for Vigário. However, the observed low flux suggests that an immobilization of P by fluvial sediments and macrophytes is occurring at this catchment. The elevated PO_4^{3-}-P flux at Caranguejo may be associated with its lithology, as apatite is a common mineral in granitic rocks of the study area (Santos, 1987).

The change in lithology among the sub-basins affected weathering processes regulating migration rates between soil-rock complex and the hydrologic system, being remarkable for major cations, Si and P. Land use classes were arranged according to their potential impacts as follows: forest < small properties < farming < urbanization. Particularly urbanization at Vigário had a major influence on fluxes of N compounds, with values that were five times higher than the world average export rates in major rivers (Meybeck, 1993). An orographic effect, increasing precipitation amount eastwards, associated with forest cover, probably enhanced the atmospheric deposition of SO_4^{2-} in the Caranguejo sub-basin, and consequently the fluvial flux.

The results of the present study suggest that a serious environmental deterioration has been occurring in Vigário and Ubatiba catchments. If such a deterioration process is repeated at Caranguejo, which is the main contributor to the associated lagoons, the environmental impact to the lagoonar system will be even higher. Therefore, the conservation of this system requires an efficient program of environmental planning and management for its catchments that must recommend the following procedures: implement an effective domestic sewage treatment system; avoid high housing density; stimulate small properties instead of large farms; and preserve the forest fragments.

Acknowledgments

This study was supported by Fundação Estadual do Norte Fluminense (FENORTE), Coordenação de Aperfeiçoamento de Pessoal de Ensino Superior (CAPES) and Conselho Nacional de Desenvolvimento Científico e Tecnológico (CNPq).

References

Antoine, S.E. and Bensons-Evans, K.: 1988, *Acta Hydrochim. Hydrobiol.* **16**.

Barbiere, E.B.: 1986, *Distribuição da Pluviosidade ao Longo do Trecho Niterói-Cabo Frio (RJ)*, in Proceedings of the 38th SBPC Annual Meeting held in Curitiba, Brazil, p. 4.

Barroso, L.V.: 1991, *Hidrogeoquímica das Bacias Hidrográficas do Sistema Lagunar de Maricá-Guarapina (RJ)*, M.Sc. Dissertation, Universidade Federal Fluminense.

Bourg, A.C. and Bertin, C.: 1993. *Environ. Sci. Technol.* **27,** 661.

Carmouze, J.P.: 1994, *O Metabolismo dos Ecossistemas Aquáticos*, Edgard Blücher, São Paulo.

Černý, J., Billet, M.F. and Cresser, M.S.: 1994, 'Element Budgets', in B. Moldan and J. Černý (eds.), *Biogeochemistry of Small Catchments: A Tool for Environmental Research,* John Wiley & Sons, Chichester, pp. 189-205.

Comans, R.N.J., Van Der Weidjen, C.H. and Vriend, S.P.: 1987, *Environ. Geol. Water Sci.* **9,** 119.

Coombes, C.: 1987, 'Current Practices in Controlling the Effects on Fresh Waters of Urbanization in Milton Keynes Area', in J.L.F.G. Solbé (ed.), *Effects of Land Use on Fresh Waters*, Ellis Horwood Publishers, Chichester, pp. 26-40.

Courchesne, F. and Hendershot, W.H.: 1989, *Can. J. Soil Sci.* **69,** 337.

Golterman H.L., Clymo, R.S. and Ohnstad, M.A.M.: 1978, *Methods for Physical and Chemical Analysis of Fresh Waters*, 2nd ed., Blackwell Scientific Publications, Oxford.

Guimarães, E.M.M.: 1989, 'Impact on Primitive Fishing Community in Maricá County caused by Urbanization', in C. Neves (ed.), *Coastlines of Brazil,* American Society of Civil Enginners, Charleston, pp. 240-248.

Lyons, W.B. and Bird, D.A.: 1995, *J. South Amer. Earth Sci.* **8**, 97.

Meybeck, M.: 1993, 'C, N, P and S in Rivers: From Sources to Global Input', in R.Wollast, F.T. Mackenzie and L.Chou (eds.), *Interactions of C, N, P and S in Biogeochemical Cycles and Global Change*, Springer-Verlag, pp. 163-193.

Moldan, B. and Černý, J.: 1994, *Biogeochemistry of Small Catchments: A Tool for Environmental Research,* John Wiley & Sons, Chichester.

Nunes, M.F.S.Q.C., Guimarães, E.M.M. and Ramos M.C.L.: 1989, 'Environmental Evaluation of the Coastal Area of the Maricá District: A Methodological Essay', in C. Neves (ed.), *Coastlines of Brazil,* American Society of Civil Enginners, Charleston, pp. 259-266.

Ovalle, A.R.C., Barroso, L.V., DePaula, F.C.F., Perrin, P. and Bidone, E.D.: 1990, *Acta Limnol. Brasil.* **3**, 887.

Perrin, P.: 1986, 'Geomorphic Evolution of the Catchment Basin of Maricá (Rio de Janeiro): An Attempted Correlation with Sea-level Variations', in Rabassa, J. (ed.), *Quaternary of South America and Antarctic Peninsula*, São Paulo, pp. 251-268.

Richards, R.P. and Baker, D.B.: 1993, *J. Great Lakes Res.* **19**, 200.

Santos, M.E.P.: 1987, *Características Litogeoquímicas do Granito Cajú: Elementos Maiores, Menores e Traços (Cu, Pb, Zn e Ni) da Região de Maricá, Rio de Janeiro,* M.Sc. Dissertation. Universidade Federal Fluminense.

Smith, S.J., Sharpley, A.N., Berg, W.A., Naney, J.W. and Coleman, G.A.: 1992, *J. Environ. Qual.*, **21,** 595.

Zagatto, E.A.G., Jacintho, A.O., Reis, B.F., Krug, F.J., Bergamin Filho, H., Pessenda, L.C.R., Mortatti, J. and Giné, M.F.: 1981, *Manual de Análise de Plantas e Águas Empregando Sistemas de Injeção em Fluxo.* USP/CENA, São Paulo.

BERYLLIUM CHEMISTRY IN THE LYSINA CATCHMENT, CZECH REPUBLIC

P. KRÁM[1,2], J. HRUŠKA[2] and C.T. DRISCOLL[1]

[1]*Department of Civil and Environmental Engineering, Syracuse University, Syracuse, NY 13244, USA,*
[2]*Department of Environmental Geochemistry, Czech Geological Survey,*
Klárov 3, 11821 Prague 1, Czech Republic

Abstract. The environmental chemistry of beryllium (Be) was investigated at the Lysina catchment in western Bohemia, Czech Republic, a forest ecosystem with high loadings of acidic atmospheric deposition. The catchment supports Norway spruce plantations; it is underlain by leucogranite and the soils are Spodosols. Average concentrations of Be were high in groundwater (3.3 μg L^{-1}) and in stream water (1.5 μg L^{-1}), in comparison to the drinking water standard of the Czech Republic (0.2 μg L^{-1}). Chemical equilibrium calculations suggest that aquo-beryllium Be^{2+} was the prevailing inorganic species in drainage waters at the site. Atmospheric deposition of Be (45 μg m^{-2} yr^{-1}) was small in comparison to drainage outflow (586 μg m^{-2} yr^{-1}) at Lysina. Elevated Be concentrations in drainage water appear to be the result of the mobilization of Be from soils and weathered bedrock due to acidic atmospheric deposition. Increased mobility of Be due to acidification may have serious ecological consequences in acid-sensitive areas with terrestrial pools of available Be.

Keywords: acid deposition, beryllium, catchment, groundwater, soil solution, stream water

1. Introduction

Beryllium (Be) is characterized by a low atomic number (4), low atomic weight (9.01 amu), and a small ionic radius (0.031 nm). Its corresponding charge-to-ionic radius ratio (64.5 nm^{-1}) is the highest of any element, but is similar to that of Al (60.0 nm^{-1}). From these values, the chemical similarity of Be and Al is evident (Panzer, 1985). Beryllium has a valence of +2 in aqueous solution (Baes and Mesmer, 1976). The average concentration of Be in the Earth's crust is 2.8 μg g^{-1}. The major Be mineral is beryl ($Be_3Al_2(Si_6O_{18})$), a hexagonal cyclosilicate; rare minerals include chrysoberyl ($BeAl_2O_4$), bertrandite ($Be_4(OH)_2Si_2O_7$), phenacite ($Be_2SiO)_4$, euclase ($BeAl(SiO_4)(OH)$), and gadolinite ($YFeBe_2(SiO_4)_2O_2$); Klein and Hurlbut, 1985). Beryllium may be the most toxic of the non-radioactive elements (Martin, 1986; Wong, 1994). Consequently, drinking water standards for Be are very strict, e.g. 0.2 μg L^{-1} in the Czech Republic (Anonymous, 1992), or 1.0 μg L^{-1} (EPA Maximum Contaminant Level) in the USA (Pontius, 1993).

The amount of Be in the environment may be enriched by anthropogenic activities, including coal combustion (Moore, 1991) and atmospheric deposition of fly ash (Kubizňáková, 1987a; Moldan and Schnoor, 1992) or leaching from slag and ash deposits (Kubizňáková, 1987b). Beryllium is liberated from naturally occurring soil minerals with decreases in pH (Moldan and Schnoor, 1992) and may be mobilized in acidic solution together with Al during mineral dissolution or by desorption from oxyhydroxides of iron (Fe), previously formed at higher pH values (Veselý *et al.*, 1989).

We are aware of no detailed catchment studies of Be. The objective of this study was to evaluate concentrations, aqueous speciation, and fluxes of Be in a forest catchment, exhibiting acidification of drainage waters by atmospheric deposition.

Water, Air, and Soil Pollution **105:** 409–415, 1998.

2. Methods and Site Descriptions

Beryllium fluxes in bulk precipitation, throughfall, and stream water were examined from November 1, 1991 to October 31, 1992. Fluxes of major elements from this period were evaluated in Krám and Hruška (1994). Soil solution and groundwater chemistry from the same period, and stream water chemistry from 1991-1994 also were examined. Details of field sampling were described in Krám *et al.* (1995). Flameless (graphite furnace) atomic absorption spectrophotometry (AAS) was used to determine Be. The aqueous speciation of Be and solubility of Be minerals were calculated with the MINEQL$^+$ chemical equilibrium model (Schecher and McAvoy, 1991).

The study was conducted in the Lysina catchment (area 0.273 km^2), located 12 km south from the Sokolov Coal Basin in a mountainous region of the Slavkov Forest (Slavkovský les) in the Czech Republic. The site is forested by Norway spruce *(Picea abies)* plantations. Prevailing soils are Spodosols, which developed on the coarse-grained leucogranite of albite-lithium mica type. Glacial sediments are not present in the region. Characteristics of the Lysina catchment (including maps) were summarized in Buzek *et al.* (1995) and Krám *et al.* (1997). The catchment lies 10 km south of the industrial area of Sokolov.

The arithmetic mean of Be concentrations in soft brown coal of the Sokolov Basin was 85.1 $\mu g\ g^{-1}$ (Bouška and Pešek, 1997). The coal is mined from open pits and is burned in nearby power plants. Atmospheric concentrations of Be were 0.03-0.11 ng m^{-3} in the Sokolov Basin during 1982-1986 (Kubizňáková, 1987a). Concentrations of Be in leucogranite at Lysina ranged from 3 to 5 $\mu g\ g^{-1}$ (Breiter, 1988).

3. Results and Discussion

The volume-weighted annual concentration of Be in bulk precipitation (0.040 $\mu g\ L^{-1}$) was somewhat less than the volume-weighted annual concentration of Be in canopy throughfall (0.060 $\mu g\ L^{-1}$) for the 1992 water year (Figure 1). Soil solutions at Lysina were extremely acidic. Volume-weighted mean pH was 3.4 in O horizon solutions and 3.2 in E horizon solutions. Arithmetic mean pH values were 4.4 in C horizon solutions and 5.2 in shallow groundwater. Mean annual volume-weighted concentrations of Be in soil solutions were 0.21 $\mu g\ L^{-1}$ in O horizon leachate, and 0.41 $\mu g\ L^{-1}$ in E horizon waters (Figure 1). Soil solution concentrations of Be markedly increased in the lower mineral soil (arithmetic mean of C horizon leachate, 1.1 $\mu g\ L^{-1}$; arithmetic mean of shallow ground water, 3.3 $\mu g\ L^{-1}$) during the same period. Mean annual discharge-weighted pH of stream water at Lysina was 3.9 with a corresponding mean annual discharge-weighted Be concentration of 1.45 $\mu g\ L^{-1}$. The corresponding arithmetic mean concentration of Be in stream water was 1.51 $\mu g\ L^{-1}$ in the 1992 water year.

The bulk precipitation flux of Be to the catchment was 45 $\mu g\ m^{-2}\ yr^{-1}$ for the 1992 water year. In comparison, the throughfall flux of Be was 42 $\mu g\ m^{-2}\ yr^{-1}$. Thus, it appears that dry deposition and vegetational recycling were minor at Lysina. Stream water output of Be (586 $\mu g\ m^{-2}\ yr^{-1}$) was more than one order of magnitude higher than the atmospheric deposition input during the 1992 water year. These mass balance calculations suggest that drainage losses of Be from Lysina were largely mobilized from pools within the catchment.

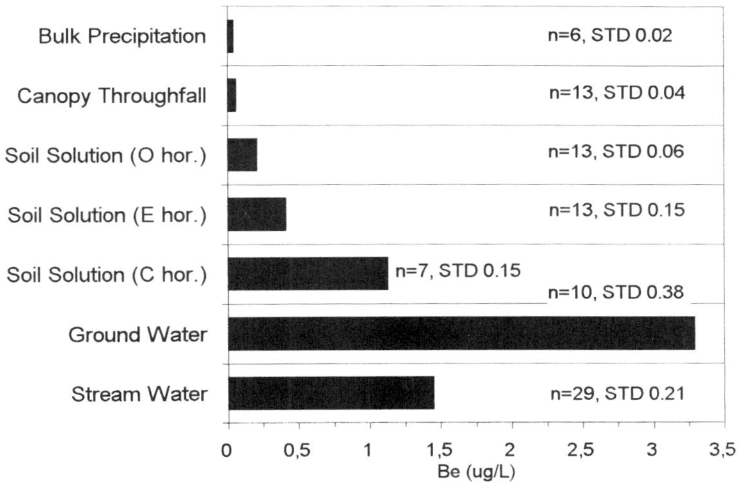

Fig. 1. Mean concentrations of dissoved Be in waters at Lysina in 1992 water year (November 1, 1991 - October 31, 1992), n is number of samples, STD is standard deviation.

In this study there was no strong relationship between Be concentrations in stream water and discharge, although concentrations of Be were low (1.02-1.26 μg L^{-1}) at the highest discharge observed (> 20 L sec^{-1}). Moreover, no strong relationship was evident between Be concentrations and either pH (Figure 2a) or Al concentrations (Figure 2f). In contrast, stream concentrations of Al increased with decreasing pH (for September 1989-November 1994, Al_{TOT} (mg L^{-1}) = -2.033×pH + 9.61, r^2 = 0.83, n=244; Figure 2b). Concentrations of Be in stream water were positively correlated with total fluoride (F) concentrations (for April 1991-November 1994, Be (μg L^{-1}) = 2.137×F_{TOT} (mg L^{-1}) + 0.685, r^2 = 0.35, n = 145; Figure 2c). Beryllium was positively correlated with concentrations of SO_4^{2-} (for April 1991-November 1994, Be (μg L^{-1}) = 0.0531×SO_4^{2-} (mg L^{-1}) + 0.172, r^2 = 0.61, n = 146; Figure 2d).

Over the study period (water years 1992-1994) there were decreases in the concentration of Be, SO_4^{2-} and F^- in stream water. During the study period, concentrations of H^+, Al and dissolved organic carbon (DOC) increased with higher flow (Figure 3).

Stream waters at Lysina were highly undersaturated with respect to $Be(OH)_2$ (mean saturation index (log Q_p/K_p) = -8.98). Chemical equilibrium calculations indicated that total dissolved Be in stream water largely occurred as Be^{2+} (93.0 %), with BeF^+, $BeSO_4$, $Be(OH)^+$, and $BeCl^+$ contributing 4.5, 2.0, 0.4, and 0.1 % of total dissolved Be, respectively. Other inorganic complexes of Be (i.e., $Be(SO_4)_2^{2-}$, BeF_2, BeF_3^-, $Be(SO_4)_2^{4-}$) were negligible. At Lysina, elevated concentrations of Al (annual discharge-weighted stream water concentrations 1.79 mg L^{-1} in 1992 water year) and associated complexation of F, decreased F available for complexation with Be. Strong complexation of F by Al decreased concentrations of BeF^+ (Veselý *et al.*, 1989).

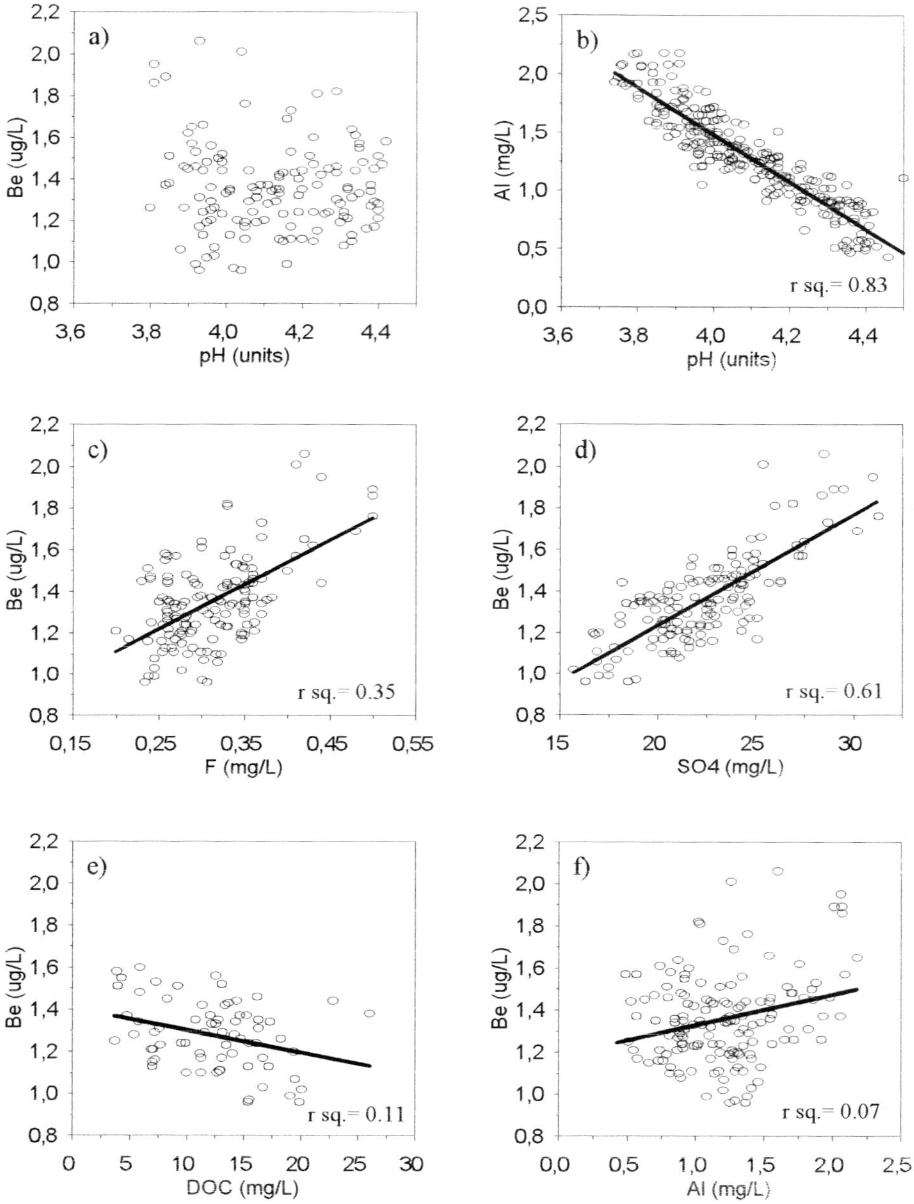

Fig. 2. Relationships of Be concentrations with pH (a), total F (c), SO_4^{2-} (d) and Al (f) in the stream water at Lysina in April 1991 - November 1994. Relationship of Al concentrations with pH (b) in September 1989 - November 1994. Relationship of Be with dissolved organic carbon (e) in April 1993 – November 1994.

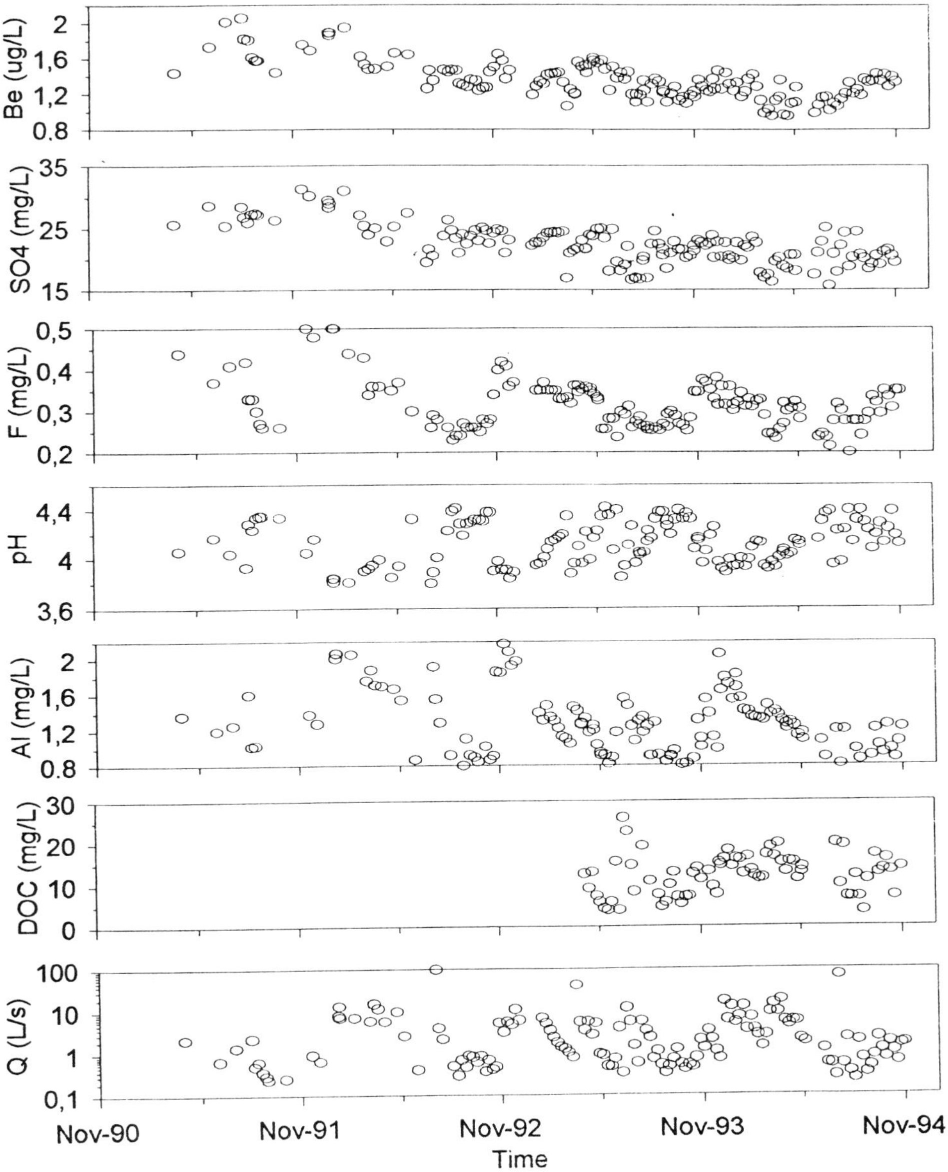

Fig. 3. Temporal patterns of stream water pH and concentrations of Be, SO_4^{2-}, F^-, Al and dissolved organic carbon. Temporal patterns of instantaneous runoff at the time of sampling.

Low Be concentrations in atmospheric deposition suggest that elevated concentrations of Be in drainage waters at Lysina largely resulted from mobilization by acid leaching within the lower mineral soil and weathered leucogranite. Although regional surveys suggest that concentrations of Be increase with decreasing pH, coincident with Al concentrations (Veselý, 1994), this study indicates that the environmental chemistry of Be at Lysina is very different from that of Al. The source of stream water Be seems to be a mixture of acidic soil water with relatively low Be, and less acidic ground water high in Be. Unlike Al, Be is not a strongly hydrolyzing metal and Lysina drainage waters were highly undersaturated with respect to the solubility of $Be(OH)_2$. As a result it is not surprising that stream concentrations of Be were not pH-dependent. It is unclear why there was a relatively strong relationship between Be and both F^- and SO_4^{2-}. The limited complexation of Be by F^- and SO_4^{2-} in the drainage water suggests that the relationship is not causal. It seems likely that soil Be is largely associated with soil cation exchange sites. Inflow of acidic water and associated mobile anions (i.e., SO_4^{2-}), to the mineral soil may enhance the mobility of soil Be, resulting in elevated Be concentrations in stream water. A gradual decline of atmospheric deposition of SO_4^{2-} and acidity during the study period may have decreased the mobility of Be from the soil exchange complex.

The median Be concentration (0.6 $\mu g\ L^{-1}$ for pH ~3.9) from a regional survey of streams in the Czech Republic was much lower than concentrations at Lysina. However, Be concentrations at Lysina were well within the regional range (0.3-3.5 $\mu g\ L^{-1}$) of values obtained for low pH streams (~3.9) in this survey (Veselý, 1994). Increased mobility of Be due to anthropogenic acidification may have serious toxicological consequences in aquatic and terrestrial ecosystems in acid-sensitive areas.

Acknowledgments

Support for this research was provided by the Czech Geological Survey and the U.S. National Science Foundation. Chemical analyses of Be were done by K. Martínková. Most of the other water analyses were done by L. Dempírová, M. Mikšovský, V. Valný, and P. Pokorný. We appreciate the field monitoring of Z. Dedek and the technical help of J. Hošek. We thank two anonymous reviewers for helpful comments.

References

Anonymous: 1992, *ČSN 73 6053*, Vydavatelství Norem, Prague.

Baes Jr., C.F. and Mesmer, R.E.: 1976, *The Hydrolysis of Cations*, John Wiley and Sons, New York.

Bouška, V. and Pešek, J.: 1997, *Distribution of Sulfur and Arsenic in Coal of the Sokolov Basin*, Research Report, Charles University, Prague (in Czech).

Breiter, K.: 1988, *Silicate Analyses - Database*, Czech Geological Survey, Prague.

Breiter, K.: 1996, "Granites of the Slavkov Forest," in *Conference Proceedings: Mining Past and Future of the Slavkov Forest*, Okresní Muzeum Sokolov, Sokolov, Czech Republic, pp. 25-44 (in Czech).

Buzek, F., Hruška, J. and Krám, P.: 1995, *Water, Air, Soil Pollut.* **79**, 391.

Klein, C., and Hurlbut, Jr, C.S.: 1985, *Manual of Mineralogy after J.D. Dana, 20th Edition*, John Wiley and Sons, New York.
Krám, P. and Hruška, J.: 1994, *Appl. Hydrogeol.* **2/94**, 50.
Krám, P., Hruška, J., Driscoll, C.T. and Johnson, C.E.: 1995, *Water, Air, Soil Pollut.* **85**, 1831.
Krám, P., Hruška, J., Wenner, B.S., Driscoll, C.T. and Johnson, C.E.: 1997, *Biogeochemistry* **37**, 173.
Kubizňáková, J.: 1987a, "Behavior of Beryllium in the Sokolov District", in B. Moldan and T. Pačes (eds.), *GEOMON Extended Abstracts*, Czech Geol. Survey, Prague, pp. 188-190.
Kubizňáková, J.: 1987b, *Water, Air, Soil Pollut.* **34**, 363.
Martin, R.B.: 1986, "Bioinorganic Chemistry of Metal Ion Toxicity", in H. Sigel (ed.), *Metal Ions in Biological Systems, Vol. 20, Concept on Metal Ion Toxicity,* Marcel Dekker, New York.
Moldan, B. and Schnoor, J.L.: 1992, *Environ. Sci. Technol.* **26**, 14.
Moore, J.W.: 1991, "Beryllium", in *Inorganic Contaminants of Surface Water - Research and Monitoring Priorities*, Springer Series on Environmental Management, Springer-Verlag, New York, pp.50-56.
Panzer, R.E.: 1985, "Beryllium, Magnesium, Calcium, Strontium, Barium, and Radium", in A.J. Bard, R. Parsons and J. Jordan (eds.), *Standard Potentials in Aqueous Solutions,* Int. Union Pure Appl. Chem., Marcel Dekker, New York, pp. 675-686.
Pontius, F.W.: 1993, *Am. Water Works Assoc. J.* **85**, 42.
Schecher, W.D. and McAvoy, D.C.: 1991, *MINEQL$^+$: A Chemical Equilibrium Program for Personal Computers - User's Manual, Version 2.1.*, Environ. Res. Software, Edgewater, Maryland.
Veselý, J.: 1994, "Effects of Acidification on Trace Metal Transport in Fresh Waters", in C.E.W.. Steinberg and R.F. Wright (eds.), *Acidification of Freshwater Ecosystems,* John Wiley & Sons, Chichester, pp. 141-151.
Veselý, J., P. Beneš, and K. Ševčík: 1989, *Water Res.* **23**, 711.
Wong, C.Y., and J.D. Woolins: 1994, *Coord. Chem. Rev.*, **130**, 243.

APPLICATION OF THE REGIONAL MERCURY CYCLING MODEL (RMCM) TO PREDICT THE FATE AND REMEDIATION OF MERCURY IN ONONDAGA LAKE, NEW YORK

S. GBONDO-TUGBAWA and C.T. DRISCOLL

Department of Civil and Environmental Engineering, Syracuse University, Syracuse, NY 13244, USA

Abstract. Onondaga Lake exhibits elevated concentrations of total mercury (Hg_T) and methyl-mercury (MeHg) in the water column, sediments and fish tissue due to industrial inputs, wastewater discharge and urban runoff. The steady-state Regional Mercury Cycling Model (RMCM) was calibrated to Onondaga Lake and applied to evaluate various remediation scenarios. Because of detailed data available for Onondaga Lake, the RMCM was effectively calibrated. Model predictions of water column and fish concentrations of Hg generally agreed with measured values. The model underestimated concentrations of Hg in sediments. Mass balance calculations show that inputs of Hg_T largely originate from tributary and wastewater inflows to the lake. In contrast, MeHg is largely derived from internal production. Model calculations suggest that elimination of Hg inputs from wastewater effluent and of drainage from a former chlor-alkali facility could greatly decrease Hg concentrations in fish tissue.

Keywords: budget, mass balance, mercury, regional model, Onondaga Lake

1. Introduction

Contamination of soil, sediment or water with mercury (Hg) is a serious environmental problem. Mercury bioaccumulates in aquatic food chains to the extent that consumption of contaminated fish may be hazardous to birds, mammals, and humans. Monomethyl Hg (MeHg) is the form of Hg that is absorbed readily from water by phytoplankton and subsequently bioconcentrates via the aquatic food chain (Mason *et al.,* 1995).

A crucial determinant of potential effects of Hg in aquatic ecosystems is the cycling of various species of Hg. Cycling consists of inputs to and outputs from the ecosystem, and the transport and transformations of Hg within the ecosystem. These processes are affected by site-specific factors, such as water chemistry, sediment transport and hydrodynamics. Inputs of Hg to aquatic ecosystems could originate from surface and ground water, industrial discharge or wastewater facilities, sediments and atmospheric deposition. Inorganic Hg ($Hg^{2+}{}_T$) is the predominant form entering these ecosystems and may form complexes with inorganic ligands (e.g., Cl^-, OH^-), bind with dissolved organic carbon, or sorb to particulate matter (Driscoll *et al.,* 1994). Inorganic Hg can be reduced to elemental Hg (Hg^o), which can volatilize to the atmosphere or be methylated to MeHg abiotically (Lee *et al.,* 1985) or by sulfate reducing bacteria in anoxic zones (Rudd, 1995). Monomethyl Hg can be further methylated to dimethyl Hg or can be demethylated by microbial processes.

Onondaga Lake received Hg discharges from a chlor-alkali manufacturer from 1946 to 1970 (USEPA 1973) and was closed to fishing in 1970 due to Hg contamination of fish tissue. Although the degree of contamination has decreased, a large fraction of the legal-size fish continues to exceed the U.S. Food and Drug Administration (FDA) limit of 1 $\mu g\,g^{-1}$ (NYSDEC, 1987). The lake has been reopened to recreational angling (since 1986), but an advisory for the consumption of fish taken has remained in effect through the present. Recent studies have found high concentrations of total Hg (Hg_T) and MeHg in the lake

Water, Air, and Soil Pollution **105:** 417–426, 1998.

water, with Hg_T concentrations (2-35 ng L^{-1}; Bloom and Effler, 1990) exceeding values reported for uncontaminated surface waters (0.9-1.9 ng L^{-1} for Wisconsin seepage lakes; Fitzgerald and Watras, 1989) by about an order of magnitude. Lake sediments are highly contaminated with Hg_T (Klein and Jacobs, 1995). Concern has been expressed that atmospheric deposition may be an important source of Hg to Onondaga Lake.

The primary objective of this study was to evaluate the cycling of Hg in Onondaga Lake under ambient conditions and various remediation scenarios of external Hg loading. The specific objectives were to: 1) determine external inputs of Hg_T and MeHg to the lake and compare these values to the internal rates of accumulation of Hg_T and MeHg in the hypolimnion of the lake, 2) use the Regional Mercury Cycling Model (RMCM; Tetra Tech Inc., 1996) to simulate mass fluxes of Hg_T and MeHg into, within and out of the lake, and 3) simulate the effects of various remediation strategies of reduced external loads on the concentrations of Hg species in water, sediments and fish.

2. Materials and Methods

2.1. Site Description

Onondaga Lake (lat. 43°06'54"N; long. 76°14'34"W) is located adjacent to the city of Syracuse, in Onondaga County, New York (Figure 1). The lake (1.4×10^8 m^3 volume, 12 m mean depth, 20.5 m maximum depth) is a hypereutrophic, dimictic lake and it flushes rapidly, 2.5 to 5.0 times annually (Devan and Effler, 1984). The lake water is well buffered with a pH of between 7 to 8.5 (Driscoll *et al.,* 1993). The major hydrologic inputs are Ninemile Creek and Onondaga Creek, which together account for about 62 % of the annual inflow to the lake, the Onondaga County Metropolitan Wastewater Treatment Plant (METRO), which is the third largest source of water to the lake making up 19 % of the annual inflow, Ley Creek, Harbor Brook, and Bloody Brook.

2.2. Sampling and Analytical Procedures

Clean sampling techniques (Fitzgerald and Watras, 1989; USEPA, 1995a) were used to collect monthly water samples from October 1995 to September 1996 in selected tributaries including Ninemile Creek downstream at Lakeland and upstream at Amboy, Onondaga Creek, Harbor Brook, Ley Creek, and the effluent discharged from METRO. Analytical protocols described in USEPA (1995b) and Gill and Fitzgerald (1987) were used to analyze the tributary water samples for Hg_T. Methyl Hg analysis was done in accordance with Hovart *et al.* (1993) and Liang *et al.* (1994). Mean daily flow measurements for the major tributaries were obtained from the U.S. Geological Survey.

Because the minor tributaries, lake water and sediments, and groundwater inputs to the lake were not sampled during this period, Hg values obtained in recent studies (Bloom and Effler, 1990; Jacobs *et al.,* 1995; Klein and Jacobs, 1995; Wang and Driscoll, 1995) were used. The minor tributaries are not gauged, so flow rates for these tributaries were estimated using the empirical relationship in Effler (1996). Concentrations of Hg in fish for the sampling period were obtained from the New York State Department of Environmental Conservation (R. Sloan, 1997, NYSDEC, unpublished data).

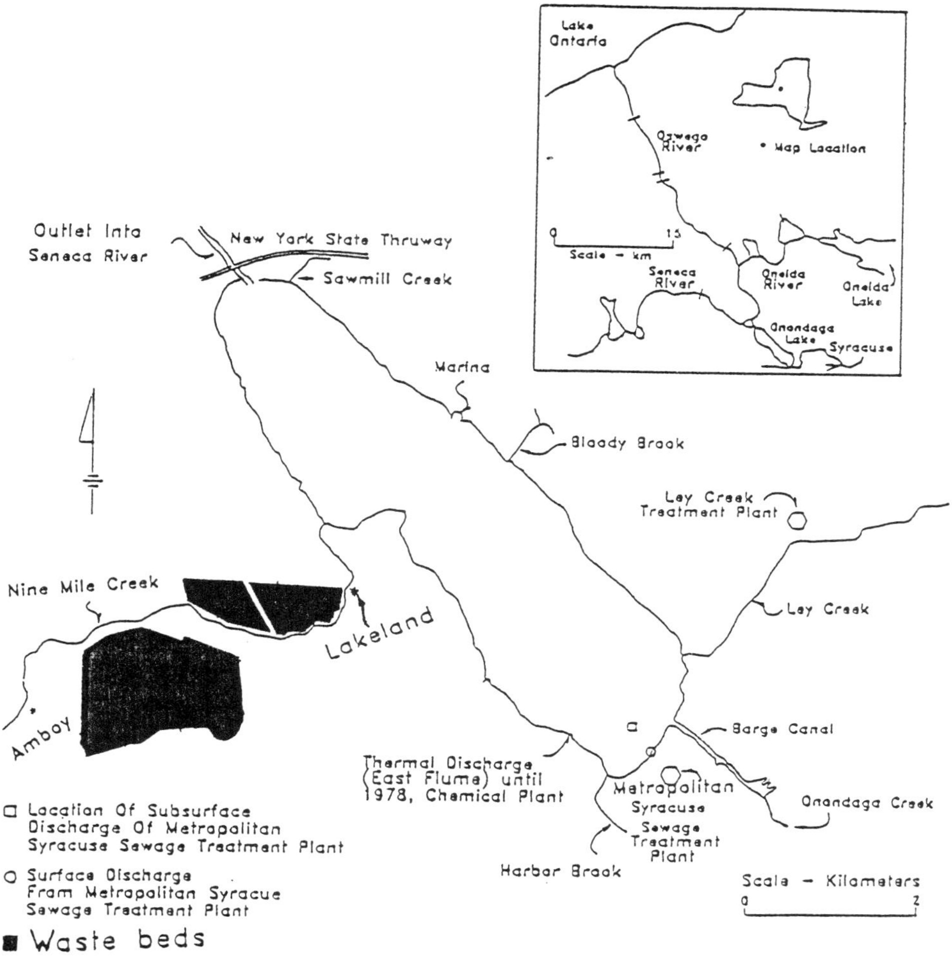

Fig. 1. Map of Onondaga Lake showing input waters (after Effler, 1996).

2.3. MODELING FLUX AND MODELING APPROACHES

Mass loadings from the input waters were estimated using the program FLUX (Walker, 1987). FLUX evaluates the flow/concentration relationships developed from the sample record onto the entire flow record to calculate total mass loading and associated error statistics. The areal mass fluxes of Hg_T and MeHg measured for Ninemile at Amboy were considered as background values for the Onondaga Lake watershed. The difference in mass fluxes of Hg_T and MeHg for Ninemile Creek at Lakeland and Amboy were assumed to be due to inputs from the former chlor-alkali facility, which was located along the stream reach. The areal mass fluxes of Hg_T and MeHg at Onondaga Creek, Ley Creek and Harbor Brook, which drain through the city of Syracuse, less the background values obtained for Ninemile Creek at Amboy were assumed to be due to Hg inputs from urban runoff.

The accumulation of Hg_T and MeHg in the hypolimnion of Onondaga Lake during the ice free period was estimated from the slopes of the regression analysis of Hg_T and MeHg pools for 1992 as a function of time on a lake surface area basis, using data obtained from Jacobs *et al.* (1995). These values were compared to the corresponding external fluxes estimated from the loadings from the tributaries.

The Regional Mercury Cycling Model (RMCM) was used to simulate the mass balance of Hg_T and MeHg in the lake. This is a steady-state mechanistic model, designed to simulate the biogeochemical cycling of Hg in lakes. Input parameters include lake/watershed physical characteristics, water chemistry, atmospheric deposition and aquatic biomass. Most of these parameters were obtained from Effler (1996). Atmospheric deposition was estimated based on values of 20 ng Hg_T L^{-1} (Glass *et al.*, 1991) and 0.016 ng MeHg L^{-1} (Fitzgerald *et al.*, 1991) in wet deposition, applied to the lake surface area. A dry deposition flux of Hg_T of 3.5 μg^{-2} yr^{-1} (Fitzgerald *et al.*, 1991) was assumed in the model. Groundwater flux was estimated from concentrations of Hg in groundwater (Jacobs *et al.*, 1995) applied to the expected groundwater flow into the lake (Effler, 1996). Mercury concentrations at Ley Creek were used to estimate mass fluxes from minor tributaries. A detailed description of the RMCM is provided by Tetra Tech Inc. (1996).

The model was calibrated using Hg^o concentrations in the water column observed by Bloom and Effler (1990). Because the model tends to predict high Hg^o concentrations, rates of demethylation (1×10^{-4} to 1×10^{-5} ng MeHg $^{-3}$ da^{-1}) and Hg^{2+}_T reduction (3×10^{-2} to 3×10^{-5} da^{-1}) were adjusted from default values so that model simulated values matched measured water column Hg^o concentrations. Default values for the model were used for all other parameters (Tetra Tech, 1996). The model was used to predict Hg concentrations in the water column, sediments and smallmouth bass in the lake. These values were compared to measured values observed in recent studies. Juvenile yellow perch were assumed to be prey species for smallmouth bass. To assess the possible Hg remediation scenarios, we considered hypothetical simulations with the calibrated RMCM. These scenarios included: 1) business as usual (current Hg loadings), 2) elimination of the inputs of Hg from METRO, 3) elimination of inputs of Hg from the former chlor-alkali facility, 4) elimination of inputs of Hg from both the former chlor-alkali facility and METRO, 5) elimination of inputs of Hg from the former chlor-alkali facility, METRO and urban runoff, and 6) inputs of Hg from only atmospheric deposition to the watershed.

3. Results and Discussion

3.1. Mass Loads from Tributaries to Onondaga Lake

Mass loadings of Hg_T and MeHg calculated using FLUX showed that the total loads to the lake from the major tributaries were 9,498 g yr^{-1} and 310 g yr^{-1}, respectively. METRO effluent and drainage from the former chlor-alkali facility were major contributors of Hg_T to the lake. Localized inputs from groundwater sources also were responsible for Hg_T in Onondaga Creek. The inputs from leachates and groundwater sources to both Ninemile Creek and Onondaga Creek were diluted during the spring snow melt period when flow rates were high. METRO effluent accounted for about half the external MeHg load (49 %; Table I) to the lake. In Harbor Brook and Ley Creek, watershed characteristics and sulfate reduction played important roles in the dynamics of MeHg.

TABLE I

Mass loadings and percent contribution of Hg_T and MeHg to Onondaga Lake (g yr^{-1}).

Input	Hg_T	MeHg
Background	2039 (21 %)	59.0 (19 %)
Urban Runoff	867 (9 %)	59.9 (19 %)
METRO	3473 (37 %)	151.2 (49 %)
Former Chlor-alkali Facility	3119 (33 %)	39.8 (13 %)
Total	9498	310.0

3.2. Rates of Hypolimnetic Accumulation of Mercury

The internal rates of Hg_T and MeHg accumulation in the hypolimnion of Onondaga Lake (Table II) were estimated for the summer stratification period (May - September 1992). These values were obtained from the slopes of the regression of the hypolimnetic Hg pool expressed in g m^{-2} on a lake surface area basis as a function of time. The internal rate of Hg_T release in the hypolimnion was 0.45 μg m^{-2} da^{-1}. This value is higher than the rate estimated by Driscoll and Wang (1996; 0.28 μg m^{-2} da^{-1}). The rate of $Hg^{2+}{}_T$ release from hypolimnetic sediments was 0.22 μg m^{-2} da^{-1}. The rate of MeHg release (0.23 μg m^{-2} da^{-1}) was about half the estimated rate of Hg_T release. The source of $Hg^{2+}{}_T$ and MeHg supplied to the hypolimnetic waters is unclear. The release of $Hg^{2+}{}_T$ (Hg_T-MeHg) to the hypolimnion may be due to desorption of Hg from recently deposited particles, mineralization of Hg in recently deposited organic matter, or from sediments. The accumulation of sulfide in the hypolimnion may facilitate release of $Hg^{2+}{}_T$ from particles through the formation of aqueous complexes (Wang and Driscoll, 1995). The supply of MeHg to the hypolimnion also may be due to release from recently deposited particulate matter or internal production of MeHg.

The external input of $Hg^{2+}{}_T$ from the tributaries (2.08 μg m^{-2} da^{-1}) was about an order of magnitude higher than release of $Hg^{2+}{}_T$ to the hypolimnion (0.22 μg m^{-2} da^{-1}). In contrast the external inputs of MeHg (0.07 μg m^{-2} da^{-1}), was considerably less than the rate of

TABLE II

Comparison of rates of external Hg input to internal Hg release/production from hypolimnetic sediments on a lake surface area basis for Onondaga Lake (in μg m^{-2} da^{-1}).

	Hg_T	Hg^{2+}_T	MeHg
External	2.15	2.08	0.07
Internal	0.45	0.22	0.23

Note: $[Hg^{2+}_T] = [Hg_T] - [MeHg]$

hypolimnetic accumulation (0.23 μg m^{-2} da^{-1}). The magnitude of these values suggests that external inputs of Hg^{2+}_T are sorbed to particles, deposited to sediments, and subsequently, released to the water column in the hypolimnion. Moreover, a portion of the Hg^{2+}_T supplied by external inputs appears to be methylated in the anoxic zone during the summer stratification period. The rate of MeHg accumulation in the hypolimnion of Onondaga Lake compares to the rate of MeHg production reported for the ice-free conditions in the Wabigoon/English River Lake System (0.037-0.55 μg m^{-2} da^{-1}; Parks *et al.,* 1989).

3.3. Model Evaluation and Mass Balances of Hg_T and MeHg for Onondaga Lake

Results from the calibration of the RMCM to current inputs of Hg_T and MeHg were compared to measurements made in previous studies of Hg in Onondaga Lake (Table III). This was done in order to appraise the performance of the model in predicting concentrations of Hg_T and MeHg. Concentrations of Hg_T and MeHg simulated by the model in the epilimnion, hypolimnion, and in fish lie within the range of observed values from previous studies. Concentrations of Hg_T in the sediments were significantly lower than values observed by Klein and Jacobs (1995). The RMCM simulates only labile (reactive) Hg (R. Harris, personal communication, June 1997). Therefore, if sediments contain a high fraction of non-labile Hg (Hg that does not readily react with the aqueous phase), the discrepancy between measured and model-calculated values could be explained.

TABLE III

Comparison of model-predicted concentrations of Hg in the water column, sediments and fish with measurements made in previous studies of Onondaga Lake.

Media	Species	Model Value	Jacobs *et al.* (1995)	Wang and Driscoll (1995)	Bloom and Effler (1990)	Klein and Jacobs (1995)	NYS DEC (1987)
Epilimnion (ng L^{-1})	Hg^o	0.014			0.01 - 0.24		
	MeHg	0.5	0.2 - 1.0		0.4 - 2.0		
	Hg_T	8.5	3.0 - 9.4	4.0 - 15.7	7.1 - 18.8		
Hypolimnion (ng L^{-1})	Hg^o	0.5			0.02 - 0.07		
	MeHg	3.1	0.4 - 12		1.61 - 6.68		
	Hg_T	29.5	5.3 - 21.7	1.4 - 34.9	10.1 - 25.7		
Sediments (μg g^{-1})	MeHg	0.14					
	Hg_T	0.29				1 - 1.8	
Smallmouth bass (μg g^{-1})	Hg	1.10					0.95 - 1.4

The RMCM was used to simulate mass fluxes of Hg_T and MeHg in the lake (Table IV). Fluxes of Hg_T and MeHg obtained from previous studies are also included in Table IV for comparison. The total input of Hg_T to the lake during the study period was 815 µg m^{-2} yr^{-1} with the major tributaries, including METRO effluent, contributing about 96 %. Atmospheric deposition directly to the lake surface, which includes both dry and wet deposition, accounted for about 2.6 % of the input fluxes. Other inputs from minor tributaries not sampled and groundwater were estimated to contribute less than 1 % of the input fluxes of Hg_T to the lake. The sinks of Hg_T include the lake outflow, volatilization and sediment burial. Outflow of Hg_T from the lake was estimated to be 524.5 µg m^{-2} yr^{-1} (77 % of the Hg_T sinks), while 154.9 µg m^{-2} yr^{-1} (22 % of the Hg_T sinks) was retained through net sedimentation or burial.

TABLE IV
Comparison of mass balances of Hg_T and MeHg (µg m^{-2} yr^{-1}) for Onondaga Lake for October 1995-September 1996, with results reported in previous studies.

	This study		Driscoll and Wang (1996)	Henry et al. (1995)	
Sources/Sinks	Hg_T	MeHg	Hg_T	Hg_T	MeHg
Major Tributaries	789.3	25.8	450.8	1,133.3[1]	21.7[1]
Atmospheric Input	21.0	0.2	10.8	36.7	0.5
Other Inflows	4.9[2]	0.2[2]	3.3[2]	1.7[3]	0.1[3]
Outflow	514.5	32.3	466.7	233.3	20.0
Volatilization	0.2	0	12.5	1.3	0
Net Methylation	-	146.7	-	-	52.5
Burial[4]	152.9	132.6	9.2	936.7	50.3

[1] Includes minor tributaries
[2] Input other minor tributaries and groundwater
[3] Groundwater inputs only
[4] Net of settling, resuspension and diffusion from sediments

The major tributaries including METRO supplied 25.8 µg m^{-2} yr^{-1} MeHg to Onondaga Lake, about 14.9 % of the total inputs (172.8 µg m^{-2} yr^{-1}). The major source of MeHg in the lake was net internal production (146.7 µg m^{-2} yr^{-1}; 84.8 % of the total MeHg inputs). Atmospheric deposition and inputs from minor tributaries and groundwater were minor sources contributing less than 1 % to the MeHg inputs of the lake. Comparing the rate of hypolimnetic accumulation of MeHg estimated above (0.23 µg m^{-2} da^{-1}) with the net MeHg production estimated using the RMCM (172.8 µg m^{-2} yr^{-1}; i.e. 0.47 µg m^{-2} da^{-1}), suggests that the model may provide a reasonable estimate of in-lake MeHg production. Our value of the rate of MeHg accumulation in the hypolimnion undoubtedly underestimates the true value as methylation also occurs in the epilimnetic sediments and/or in the hypolimnetic sediments during the nonsummer months (October to April). The major sinks for MeHg in the lake were net sedimentation or burial (132.6 µg m^{-2} yr^{-1}; 80 % of the total MeHg sink) and outflow (32.3 $\mu g\ m^{-2}\ yr^{-1}$; 20 % of the total MeHg sink).

These fluxes were compared with previous studies conducted in Onondaga Lake. The Hg_T fluxes were intermediate between the values reported in the 1992 study by Driscoll and Wang (1996) and Henry *et al.* (1995), while MeHg fluxes were higher than those reported by Henry *et al.* (1995). Differences in fluxes could be due to the method of flux estimation, the model used and the water quality parameters used in calibration. In all mass balance studies, inputs from the tributaries, including METRO, were the major source of Hg_T to the lake. While this study and that of Driscoll and Wang (1996) indicate that outflow was the major sink of Hg_T, Henry *et al.* (1995) reported that net sedimentation was the major sink. For MeHg, net methylation was the major source and sediment burial the major sink. These observations are consistent with Henry *et al.* (1995).

3.4. Impact of Reduced External Inputs of Mercury to Onondaga Lake

With significant inputs of Hg originating from the former chlor-alkali facility, METRO and urban runoff, the effects of reducing these inputs on Hg concentrations in the epilimnion, hypolimnion, sediments and fish in the lake were investigated using the RMCM. The fish species used in the simulations was smallmouth bass and Hg concentrations for the five-year age class were reported. The various scenarios investigated were: 1) current inputs, 2) removal of the METRO inputs, 3) removal of the inputs from the former chlor-alkali facility, 4) combined removal of METRO inputs and inputs from the former chlor-alkali facility, 5) combined removal of METRO, inputs from the former chlor-alkali facility and urban runoff, and 6) only atmospheric deposition of Hg to the lake watershed. Results of the concentration and fluxes for these scenarios are given in Tables V and VI, respectively.

Model simulations showed that reductions in tributary loads for the various scenarios resulted in decreases in Hg concentrations in the lake water, sediments and fish. Because the model calculates steady-state values, and because water quality parameters are assumed to remain constant, percentage reduction in concentrations and fluxes predicted closely followed the percentage load reduction. When inputs from both METRO and the former chlor-alkali facility were eliminated (Scenario 4), decreasing input fluxes of Hg_T and MeHg by 69 % and 61 %, respectively, the predicted concentrations of Hg_T and MeHg in the water column, sediments and fish were reduced by about 66 % and there was a 67 % reduction in net methylation for the lake. The model calculations suggest that atmospheric deposition contributes little Hg to Onondaga Lake. Results of this analysis suggest that reduction of major inputs to Onondaga Lake could substantially reduce fish Hg concentrations. Note that different sources of Hg_T to the lake undoubtedly exhibit different bioavailability (e.g., wastewater effluent vs. leachate from the former chlor-alkali facility). While these differences are undoubtedly important, they are not reflected in the model calculations.

Acknowledgments

Support for this study was provided by the Onondaga County Department of Drainage and Sanitation. We appreciate the assistance of R. Harris, E. Henry, S. Effler, S. Doerr, R. Goel, R. Sloan, N. Ringler, J. Mastriano, J. Powers and J.Suryadevara. Manuscript is contribution No. 175 of the Upstate Freshwater Institute.

TABLE V

Simulated concentrations of Hg in water (ng L^{-1}), sediments (μg g^{-1}) and fish (μg g^{-1}) in Onondaga Lake for the various scenarios considered.

	Epilimnion			Hypolimnium			Sediments		Fish
Scenario	Hg^{o}	MeHg	Hg_T	Hg^{o}	MeHg	Hg_T	MeHg	Hg_T	MeHg
1	0.014	0.51	8.54	0.45	3.09	29.45	0.140	0.288	1.101
2	0.014	0.31	5.54	0.42	1.94	19.11	0.088	0.184	0.686
3	0.014	0.37	5.88	0.43	2.19	20.31	0.099	0.200	0.789
4	0.014	0.17	2.83	0.39	1.04	9.94	0.047	0.096	0.375
5	0.013	0.12	2.06	0.38	0.72	7.32	0.033	0.068	0.257
6	0.01	0.03	0.44	0.37	0.14	1.89	0.001	0.014	0.050

TABLE VI

Major fluxes of Hg_T and MeHg (μg m^{-2} yr^{-1}) in Onondaga Lake for the various scenarios considered.

	Inflow streams		Outflows		Burial		Net Methylation
Scenario	Hg_T	MeHg	Hg_T	MeHg	Hg_T	MeHg	MeHg
1	789.3	25.8	514.5	32.3	152.9	132.6	146.7
2	500.7	13.2	332.6	20.0	98.9	88.5	94.9
3	530.1	22.5	351.2	23.3	104.4	99.7	100.1
4	241.5	9.9	168.9	11.0	50.2	47.5	48.2
5	169.5	4.9	123.4	7.5	36.7	33.0	35.2
6	25.7	1.4	25.7	1.4	7.1	6.4	7.4

References

Bloom N.S. and Effler S.W.: 1990, *Water, Air, Soil Pollut.* **53**, 251.

Devan, S.P. and Effler, S.W.: 1984, *J. Env. Eng. Div. ASCE* **110**, 93.

Driscoll, C.T., Effler, S.W., Auer, M.T. , Doerr, S.M. and Penn, M.R.: 1993, *Hydrobiologia* **253**, 61.

Driscoll, C.T., Yan, C., Schofield, C.L., Munson, R. and Holsapple, J.: 1994, *Environ. Sci. Technol.* **28**, 136A.

Driscoll, C.T. and Wang, W.: 1996, 'Mercury Budget for Onondaga Lake', in Effler, S.W. (eds.) *Limnology and Engineering Analysis of a Polluted Urban Lake: Prelude to Environmental Management of Onondaga Lake, New York*, Springer-Verlag, New York, pp. 352-359.

Effler, S.W. (ed.): 1996, *Limnological and Engineering Analysis of a Polluted Urban Lake: Prelude to Environmental Management of Onondaga Lake, New York*, Springer-Verlag, New York.

Fitzgerald, W.F., and Watras, C.J.: 1989, *Sci. Tot. Environ.* **87/88**, 223.

Fitzgerald, W.F., Mason, R.P. and Vandal, G.M.: 1991, *Water, Air, Soil Pollut.* **56**, 745.

Gill, G.A. and Fitzgerald, W.F.: 1987, *Mar. Chem.* **20**, 227.
Glass, G.E., Sorenson, J.A., Schmidt, K.W.; Rapp, G.R.; Yap, D. and Fraser, D.: 1991, *Water, Air, Soil Pollut.* **56**, 235.
Henry, E.A., Dodge-Murphy, L.J., Bigham, G.N., Klein, S.M. and Gilmour, C.C.: 1995, *Water, Air, Soil Pollut.* **80**, 509.
Hovart, M., Liang, L. and Bloom, N.S.: 1993, *Anal. Chem.* **47**, 592.
Jacobs L.A., Klein, S.M. and Henry, E.A.: 1995, *Water, Air, Soil Pollut.* **80**, 553.
Klein, S.M. and Jacobs, L.A.: 1995, *Water, Air, Soil Pollut.* **80**, 1035.
Lee, Y.H., Hultberg, H. and Andersson, I.: 1985, *Water, Air, Soil Pollut.* **25**, 391.
Liang, L., Hovart, M., Bloom, N.S. 1994, *Talanta* **41**, 371.
Mason, R.P., Reinfelder, J.R. and Morel, F.M.M.: 1995, *Water, Air, Soil Pollut.* **80**, 915.
New York State Department of Environmental Conservation (NYSDEC).: 1987, *An Overview of Mercury Contamination in the Fish of Onondaga Lake; Technical Report 87-1 (BEP)*, New York State Department of Environmental Conservation, Division of Fish and Wildlife, Albany, NY.
Parks, J.W., Lutz, A., and Sutton, J.A.: 1989, *Can. J. Fish. Aquat. Sci.* **46**, 2184.
Rudd, J.W.M.: 1995, *Water, Air, Soil Pollut.* **80**, 697.
Tetra Tech Inc.: 1996, *Regional Mercury Cycling Model: A Model for Mercury Cycling in Lakes (R-MCM Version 1.0b Beta)*, Tetra Tech Inc., Lafayette, CA.
United States Environmental Protection Agency (USEPA): 1973, *Report of Mercury Source Investigation Onondaga Lake, New York and Allied Chemical Corporation, Solvay, New York*, National Field Investigations Center, Cincinnati, OH, and Region II, New York.
United States Environmental Protection Agency (USEPA).: 1995a, *Method 1631, Mercury in Water by Oxidation, Purge and Trap, and Cold Vapor Atomic Fluorescence Spectrometer*, U.S. Environmental Protection Agency, Washington, DC.
United States Environmental Protection Agency (USEPA).: 1995b, *Method 1669, Method for Sampling Ambient Water for Determination of Metals at EPA Ambient Criteria Levels*. U.S. Environmental Protection Agency, Washington, DC.
Walker, W.W.: 1987, *Empirical Methods for Predicting Eutrophication in Impoundments. Report 4: Phase III: Applications Manual. Technical Report E-81-9*, U.S. Army Engineer Waterways Experimental Station, Vicksburg, MS.
Wang, W. and Driscoll, C.T.: 1995, *Environ. Sci. Technol.* **29**, 2261.

FACTORS CONTROLLING MERCURY TRANSPORT IN AN UPLAND FORESTED CATCHMENT

TIMOTHY SCHERBATSKOY[1], JAMES B. SHANLEY[2] and GERALD J. KEELER[3]

[1]*School of Natural Resources, University of Vermont, Burlington, VT 05405, USA*
[2]*U.S. Geological Survey, P.O.Box 628, Montpelier, VT 05601, USA*
[3]*Air Quality Laboratory, University of Michigan, Ann Arbor, MI, USA*

Abstract. Total mercury (Hg) deposition and input/output relationships were investigated in an 11-ha deciduous forested catchment in northern Vermont as part of ongoing evaluations of Hg cycling and transport in the Lake Champlain basin. Atmospheric Hg deposition (precipitation + modeled vapor phase downward flux) was 425 mg ha^{-1} during the one-year period March 1994 through February 1995 and 463 mg ha^{-1} from March 1995 through February 1996. In the same periods, stream export of total Hg was 32 mg ha^{-1} and 22 mg ha^{-1}, respectively. Thus, there was a net retention of Hg by the catchment of 92 % the first year and 95 % the second year. In the first year, 16.9 mg ha^{-1} or about half of the annual stream export, occurred on the single day of peak spring snowmelt in April. In contrast, the maximum daily export in the second year, when peak stream flow was somewhat lower, was 3.5 mg ha^{-1} during a January thaw. The fate of the Hg retained by this forested catchment is not known. Dissolved (< 0.22 μm) Hg concentrations in stream water ranged from 0.5-2.6 ng L^{-1}, even when total (unfiltered) concentrations were greater than 10 ng L^{-1} during high flow events. Total Hg concentrations in stream water were correlated with the total organic fraction of suspended sediment, suggesting the importance of organic material in Hg transport within the catchment. High flow events and transport with organic material may be especially important mechanisms for the movement of Hg through forested ecosystems.

Keywords: biogeochemistry, catchment, input/output budget, Lake Champlain watershed, mercury, forest, stream

1. Introduction

Mercury contamination of fresh waters and their biota has become a widespread and serious problem in many parts of the world, including northeastern North America. There is widespread concern that atmospheric sources of Hg may be responsible for increasing Hg burdens in Lake Champlain (Scherbatskoy *et al.,* 1997; Watzin 1992; Vasu and McCullough, 1994), but mechanisms of transport and accumulation are not well understood. The Lake Champlain basin (Figure 1) is characterized by a large (19:1) ratio of land to lake surface area, of which 62 % is forested and 27 % is agricultural (Budd and Meals, 1994). This relatively large forested land area has the potential to capture atmospheric pollutants and transfer them to soils, surface waters and ground water by several mechanisms. Terrestrial ecosystems accumulate atmospheric pollutants and transfer them to the soil and surface waters by wet and dry deposition, throughfall (precipitation passing through the canopy), litterfall (deposition of senescent foliage), and interception of cloud water at higher elevations. Forest canopies in this region have foliage surface areas up to ten times greater than the land area they cover, providing Hg concentration in throughfall that averages twice that of precipitation and annual Hg flux in litterfall that is greater than annual wet deposition (Rea *et al.,* 1996).

Annual Hg deposition (wet + dry) in the Lake Champlain basin is over 125 mg ha^{-1} (Scherbatskoy *et al.,* 1997). This Hg is primarily of anthropogenic origin, and if these emissions were to cease now, it would require 15-20 years to return to background Hg

Water, Air, and Soil Pollution **105:** 427–438, 1998.

levels (Mason *et al.,* 1994). Examining the behavior of Hg in forested catchments is important to understanding the present and long-term role of atmospheric Hg in the Lake Champlain basin.

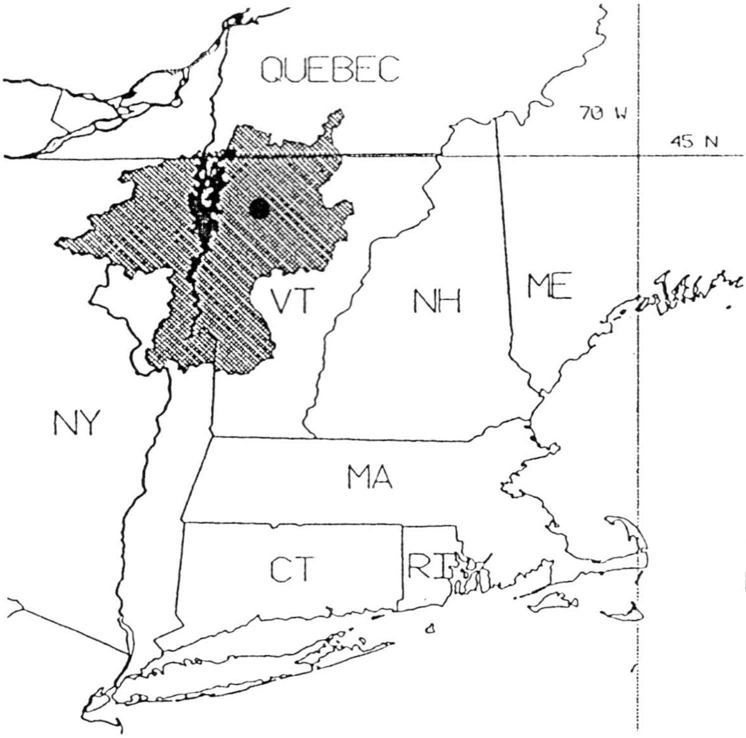

Fig. 1. Regional map of Lake Champlain and its drainage basin, with the study site (dot) shown in the eastern portion of the basin.

Information on levels or behavior of Hg in forested ecosystems in the northeastern United States is limited (Scherbatskoy *et al.,* 1997; Rea *et al.,* 1996). There is growing evidence that forested ecosystems have an important and complex role in the transport of atmospheric Hg to aquatic systems (Mierle and Ingram, 1991; Lee *et al.,* 1994; Bishop *et al.,* 1995; Johnson and Lindberg, 1995; Hurley *et al.,* 1995; Krabbenhft *et al.,* 1995). Recent studies of the behavior of Hg in terrestrial ecosystems have identified several processes that affect Hg cycling, including organic complexing in soils and water, foliar exchange and out-gassing, and accumulation in plant tissues (Schuster, 1991; Lee *et al.,* 1994; Hanson *et al.,* 1995; Zhang *et al.,* 1995; Rea *et al.,* 1996). Swain *et al.* (1992) showed the importance of watershed:lake area ratio in Hg loading to lake sediments. From these indications, it appears that anthropogenic Hg deposition, organic acidic soils, forest wetlands, hydrologically important snowmelt dynamics, and the large catchment basin feeding Lake Champlain contribute to Hg transport in the terrestrial and stream components in this basin.

The research presented here is part of a larger program addressing concerns about Hg and other toxic metals in the aquatic, terrestrial and atmospheric environments of the Lake Champlain basin. Our goals in this paper were to characterize atmospheric inputs and stream export of Hg in a small upland forested catchment, and to identify mechanisms controlling Hg transport.

2. Site Description and Methods

This research was carried out at Nettle Brook, an 11 ha mixed hardwoods catchment on the western slope of Mt. Mansfield in the town of Underhill, Vermont, 34 km east of Lake Champlain (Figure 1). Nettle Brook is operated by the Vermont Forest Ecosystem Monitoring program (VForEM) as part of their long-term integrated studies on responses of forested ecosystems to environmental change (Wilmot and Scherbatskoy, 1994). Elevation in the catchment ranges from 445 to 664 m, aspect is southwest, and slope averages 35 % (range 5-75 %). Soils are moderately well-drained Peru extremely stony loams underlain by mica-albite-quartz schist bedrock. Average annual rainfall is 1,100 mm. January daily minimum temperature is -16 °C and July daily maximum temperature is 24 °C. Annual SO_4^{2-} and NO_3^- deposition averages 21 and 16 kg ha^{-1}, respectively.

Since October 1993, stream water samples have been collected at Nettle Brook at a continuously gauged V-notch weir for analysis of Hg, DOC, and suspended sediment. During spring snowmelt periods, daily or twice-daily water samples were collected at the weir; during the rest of the year, samples were collected approximately monthly.

Precipitation was collected daily in a MIC-B wet-only precipitation collector (MIC Co., Richmond Hill, Ontario) modified for Hg sampling (Landis and Keeler, 1997), at the VForEM air quality monitoring station at the Proctor Maple Research Center (400 m elevation), 3.1 km northwest of Nettle Brook. Ambient vapor and particulate samples were collected every sixth day for 24 hr (8 AM cycle) on gold-coated bead vapor traps and glass-fiber particulate filters, respectively. Dry deposition flux of Hg vapor to this forest was estimated according to Scherbatskoy *et al.* (1997) as amended by Lindberg (1993), providing an approximation of the downward flux within a factor of 2 (Lindberg *et al.*, 1992). Upward flux (Hg evasion from soils and vegetation) is unknown for this site (indeed is poorly quantified in general), but could be significant (Hanson *et al.* 1995). Snowmelt was collected from a 1 m^2 Teflon-coated lysimeter installed on the ground beneath the snowpack. Meltwater was collected at mid-day from 24 March-14 April, 1994.

Samples were shipped by overnight courier to the University of Michigan Air Quality Laboratory in Ann Arbor, MI, where analysis for total Hg was performed by cold vapor atomic fluorescence spectrometry. Stream samples were filtered through 0.22 μm nitrocellulose filters (Millipore MF) to separate operationally defined dissolved and particulate fractions. Ultra-clean sampling, handling and storage techniques were used throughout all procedures. Field and laboratory analytical blanks were analyzed (20 % of all samples) for quality assurance purposes. Analytical accuracy was within 5 % of known standards and precision error was less than 8 %. The Hg loading error was ±5 % for wet deposition and the uncertainty of the modeled dry deposition estimates was about a factor of two (Lindberg *et al.* 1992). Analytical and quality assurance details are given in Burke *et al.* (1995). No Hg speciation was performed.

DOC was determined by ultraviolet persulfate oxidation with infrared detection at the USGS laboratory in Albany, NY. Suspended sediment concentration was determined gravimetrically from the filtrant on a sample replicate passed through a 0.7 μm glass fiber membrane. The detection limit was approximately 2 mg L^{-1}. The organic fraction of sediment was determined by loss-on-ignition at 550 °C.

Stream Hg flux was calculated from hourly average flows and modeled Hg concentrations. Dissolved Hg was weakly correlated with flow ($r^2 = 0.13$, $p = 0.0022$) and this relationship was used to derive hourly dissolved Hg concentrations throughout the period. Total Hg flux was calculated by flow class. For flow less than 0.5 mm hr^{-1}, the concentration:discharge relation was significant ($r^2 = 0.21$, $p = 0.0002$), and was used to derive total Hg concentrations. For flows greater than 0.5 mm hr^{-1}, geometric means of sample total Hg concentrations for each of three flow classes were used in flux calculations (Figure 2). Upper and lower error limits for the fluxes were determined from the 95 % confidence limits of the regression for the <0.5 mm hr^{-1} flow class, and from the high and low sample Hg concentrations for the higher flow classes. Sampling was frequent enough during the 1994 snowmelt period (March 14-April 30) to determine fluxes directly from sample total Hg concentrations, with linear interpolation of total Hg concentrations between sampling times. Error limits for this period were calculated as above, but adjusted as necessary to keep the lower error limit between 20 % and 50 % less than the interpolated flux, and the upper limit within 20 % of the interpolated flux.

3. Results

3.1. HYDROLOGY

Precipitation inputs were similar in the two study years; in each year slightly more than half of the precipitation input ran off as streamflow (Table I). The distribution of precipitation and runoff was quite different in the two years, however. The first year (March 1994 - February 1995) was dominated by a major snowmelt event in April (Figure 3), as an unusually large snowpack melted quickly. Moderate spring rains sustained relatively high flows into the early summer, but low rainfall thereafter touched off a moderate drought, which extended into the second year of the study. The 1995 snowpack was unusually small and led to a small and early snowmelt in March. Below-average precipitation produced June streamflow that was more typical of late summer. A series of heavy convective storms in July and August of 1995 generated several high flow events. Autumn rainfall was well above average and produced high sustained flow and some notable peak flows. A deep snowpack developed early in the winter of 1995-96, but most of it left the catchment in a major January thaw (Figure 3).

3.2 HG IN ATMOSPHERIC DEPOSITION

Atmospheric concentrations and deposition of Hg are summarized in Table II for 1993-95. Wet and dry deposition fluxes of Hg are summarized on a seasonal basis for 1994-95 in Figure 4. Vapor phase Hg (which dominates dry deposition) tends to be relatively constant throughout the year, but greater deposition velocities in the summer months lead to greater dry deposition. Both precipitation amount and Hg concentration in precipitation tend to be greatest in the summer months, resulting in the highest wet deposition of Hg during this period. Additional information on Hg deposition at this site is in Scherbatskoy *et al.* (1997). These data are comparable in magnitude and trend to values for rural northern Michigan (Burke *et al.,* 1995; Hoyer *et al.,* 1995; Keeler *et al.,* 1995).

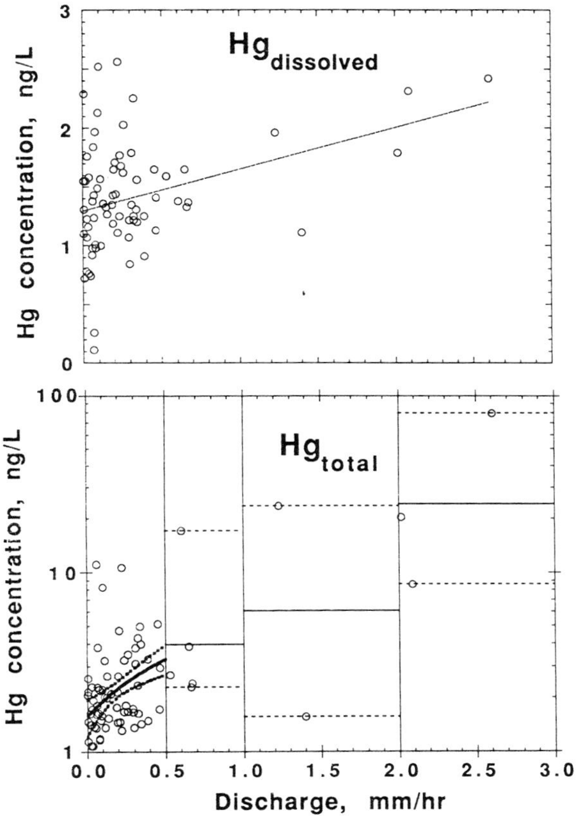

Fig. 2. Concentration:discharge relations for dissolved Hg (upper panel) and total Hg (lower panel) for two years of study. Regression line in upper panel is for all samples. Regression line in lower panel is for 1995 snowmelt period only.

TABLE I
Inputs and outputs of water and mercury at Nettle Brook catchment for the two study years.

Date	Precipitation (mm)	Streamflow (mm)	Hg input (mg ha^{-1} yr^{-1})			Hg export (mg ha^{-1} yr^{-1})			Retention
			Wet	Dry	Total	Particulate	Dissolved	Total	
Mar 94 - Feb 95	1139	587	74.5	350	425	23.6	8.4	31.9	92 %
Mar 95 - Feb 96	1109	568	92.7	370	463	13.4	8.2	21.7	95 %

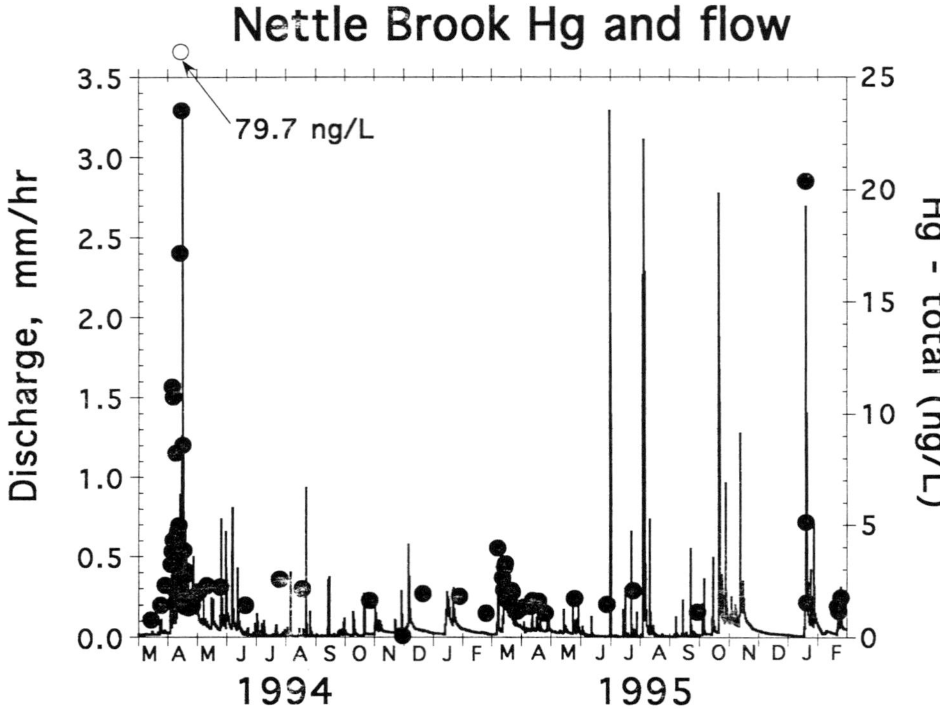

Fig. 3. Flow (mm hr^{-1}) and stream water total Hg concentration (ng L^{-1}) at the Nettle Brook weir from March 1994 through February 1996. Water was sampled for Hg analysis on a daily basis during spring snowmelt and periodically during the remainder of the year.

TABLE II

Descriptive statistics for concentrations of Hg in the atmosphere and precipitation at the air quality monitoring site in Underhill, VT for the period January 1993 through December 1995.

	Mean	Std. Dev.	n	Minimum	Maximum	Median
Vapor (ng m^{-3})	1.72	0.55	204	0.51	6.86	0.05
Aerosol (pg m^{-3})	10.21	6.75	210	0.89	43.20	7.70
Precipitation (ng L^{-1})	6.49 [1]	6.08	410	0.09	59.09	6.51

[1]volume-weighted

3.3. Hg in Snow Meltwater

Snow meltwater collected during the 1994 spring melt period had an average total Hg concentration of 4.8 ng L^{-1}, with a range of 2 to 9.2 ng L^{-1}, similar to the Hg concentration of incoming snowfall (Scherbatskoy *et al.*, 1997). These concentrations were in the same range as stream water samples collected during this period (Figure 5), except for peak snowmelt, suggesting that stream water Hg chemistry could be explained by meltwater inputs.

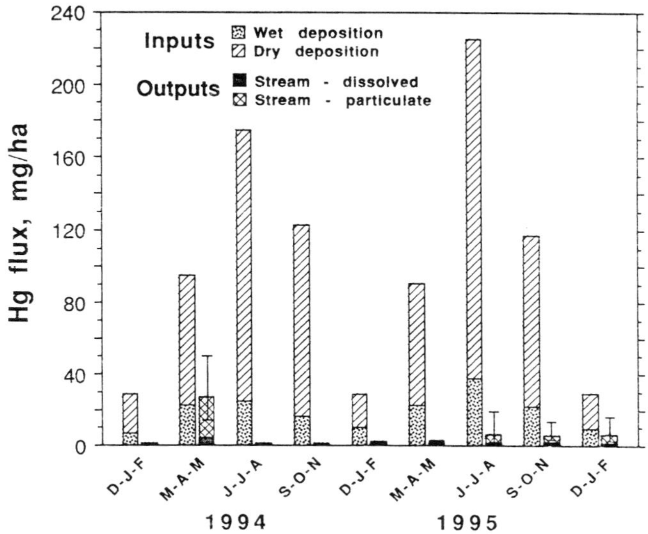

Fig. 4. Seasonal input (wet and dry deposition) and stream export (dissolved and particulate) of total Hg in Nettle Brook catchment from December 1993 through February 1996, expressed as mg ha^{-1}. Error bars on export fluxes represent the upper and lower error limits for the Hg fluxes (see Methods); where no bar is shown, error was < ±0.7 mg ha^{-1}. The wet deposition error was ±5 %, while the uncertainty of the modeled dry deposition estimates was about a factor of two (Lindberg *et al.* 1992).

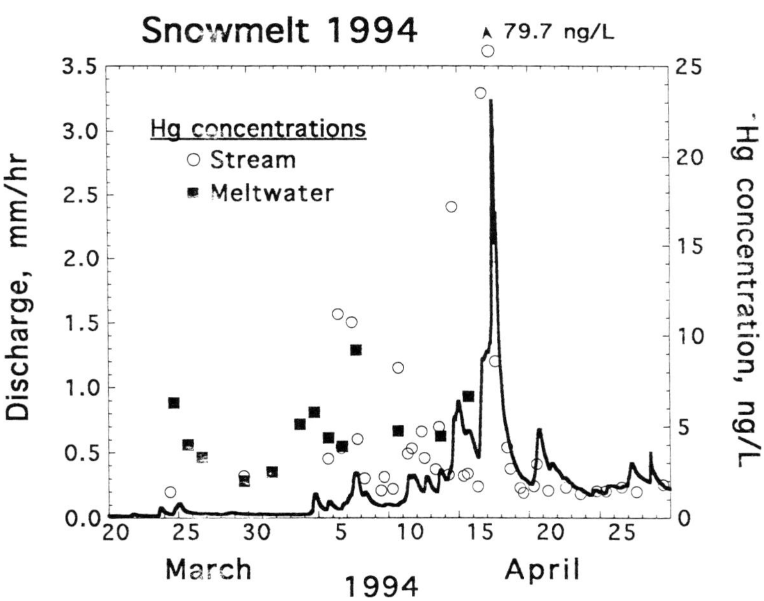

Fig. 5. Flow (mm hr^{-1}) and stream water total Hg concentration (ng L^{-1}) for the 1994 snowmelt period, with meltwater Hg concentrations from snowmelt lysimeter samples.

3.4. Hg in Stream Water

Dissolved Hg concentrations in stream water were consistently low, ranging only from 0.5 to 2.6 ng L^{-1}, and there was no apparent seasonal pattern. Total Hg concentrations ranged from 1.1 to 79.7 ng L^{-1} (Figure 3). The highest Hg concentrations occurred in the 1994 spring snowmelt, reaching 10 ng L^{-1} during the initial minor daily flow increases, and reaching a maximum at peak snowmelt. The only other period sampled where Hg concentrations exceeded 5 ng L^{-1} was during high flow in the 1996 January thaw, when total Hg concentration increased to greater than 20 ng L^{-1}. During the smaller than average 1995 snowmelt event, total Hg concentration approached 5 ng L^{-1}. At all other sampling times, total Hg generally remained less than 3 ng L^{-1}. Note, however, that we sampled only two of the five high-flow events and only once each year in autumn (Figure 3).

3.5. Hg and Suspended Sediment

Total Hg concentration increased with increasing suspended sediment concentration (r^2 = 0.41, n = 28, p = 0.0002). The r^2 improved to 0.50 when one outlier was removed. The particulate fraction (total – dissolved) of total Hg in stream water was highly correlated with the organic fraction of suspended sediment ($r^2 = 0.85, p < 0.0001$; Figure 6). A single high Hg value exerts leverage, but the regression remains strong if that point is removed ($r^2 = 0.53$, $p < 0.0001$). All but three of the samples that had non-detectable suspended sediment (<2 mg L^{-1}) had less than 1 ng L^{-1} particulate Hg.

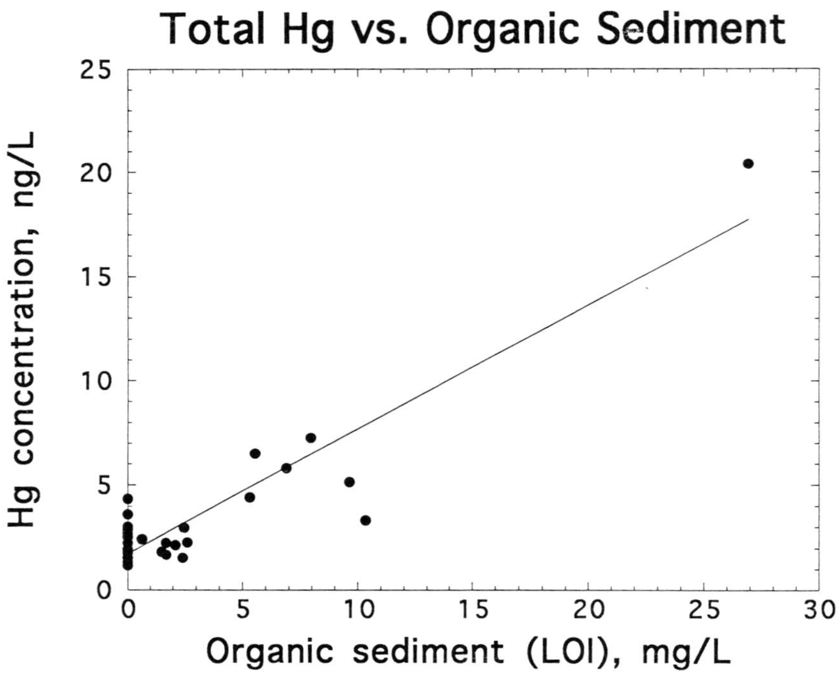

Fig. 6. Total Hg concentration as a function of organic matter fraction of suspended sediment for all samples analyzed for suspended sediment.

3.6. Hg Export

The flux of total Hg in stream water was strongly linked to high flow events. In the first year of study, about half the annual Hg flux occurred on the single day of peak snowmelt. About 84 % of the annual flux of total Hg occurred with only 56 % of the annual streamflow during the 3-month period of March-May (Figure 4). Mercury flux was minimal through most of the 1995 snowmelt because of drought conditions and associated low snowpack. High-flow events in July and August and the January thaw of 1996 were responsible for a significant share of the Hg flux in the second year.

Dissolved Hg flux in stream water varied less than the highly episodic total Hg flux. However, because dissolved Hg concentration was positively correlated with flow, dissolved Hg flux also was dominated by high flow periods (e.g., March-May 1994). During low flow periods, dissolved Hg flux accounted for most of the Hg flux from the basin. On an annual basis, dissolved Hg accounted for 26 % of the total Hg flux in the first year of study and 38 % in the second year (calculated from Table I).

3.7. Hg Budget

Total Hg input from wet and dry deposition substantially exceeded Hg export in stream water (Figure 4). During spring 1994, export was 30 % of deposition due to the large snowmelt event in April. By contrast, Hg export was only 2 % of Hg input during the summer-fall period in 1994. On an annual basis, the forested ecosystem retained 92 % of atmospheric Hg during the first year of the study and 95 % the second year (Table I).

4. Discussion

The forested Nettle Brook catchment is a net sink for atmospherically deposited Hg. Within the large uncertainties associated with our flux determinations because of limited event sampling, we found that between 60 and 90 % of deposited Hg is retained. This retention rate is similar to the 80 % to 90 % retention found in Sweden (Aastrup *et al.*, 1991; Johanssen *et al.*, 1991), but less than the 92 % retention determined by Krabbenhoft *et al.* (1995) at a Wisconsin site. Of the Hg exported in streamflow, the low but consistent dissolved Hg concentration accounts for about one-third of the annual flux. A large portion of the annual Hg export in stream water occurs as particulate Hg during a short time interval in extreme hydrologic events.

Hydrologic events at Nettle Brook have generated some extremely high total Hg concentrations. The two largest events sampled had greater than 20 ng L^{-1} Hg, and one had nearly 80 ng L^{-1} Hg. Among several catchments in Sweden, by contrast, maximum Hg concentrations did not exceed 15 ng L^{-1} (Johanssen *et al.*, 1991; Lee and Iverfeldt, 1991; Hultberg *et al.*, 1995), except for 21 ng L^{-1} in runoff from a peat bog (Westling, 1991).

The tendency of Hg concentration to increase with flow also was found by Bishop *et al.* (1995) during snowmelt in Sweden, although their stream water Hg concentration increased to only 6 ng L^{-1}. Bishop *et al.* (1995) found that 37 % of total annual Hg flux occurred in a three week spring snowmelt period. Other studies have given little attention to event sampling, so Hg behavior during high-flow events cannot be compared. Event-related Hg transport clearly lowers overall Hg retention at Nettle Brook. Some studies suggest that wetlands are responsible for Hg or Me-Hg (methyl Hg) mobilization (Driscoll *et al.,* 1994; Hurley *et al.,* 1995; Branfireun *et al.,* 1996). Nettle Brook has relatively little wetland area, but this may be compensated for by the high particulate transport of Hg resulting from the steep, mountainous slopes.

What becomes of Hg retained by the forested catchment? Some Hg may return to the atmosphere by reduction to Hg°, transport of Hg° in the plant transpiration stream, and evasion from soils (Siegel and Siegel, 1988; Hanson *et al.,* 1995; Johnson and Lindberg, 1995). Some Hg must accumulate in forest soils and biota, where it is actively cycled. Litterfall adds an estimated 130 mg ha^{-1} annually to the forest floor at this site (Rea *et al.,* 1996). Leaf litter decomposes over a few to many years, releasing the Hg for potential aqueous transport, most likely as organically bound complexes (e.g., humic acids).

Just how is Hg mobilized during large events? If Hg is sequestered primarily in the forest floor in association with organic matter (Mierle and Ingram, 1991; Aastrup *et al.* (1991), one would expect that flushing of the soil would deliver Hg to the stream as a dissolved organically-bound complex. However, we observed little or no increase in dissolved Hg during events. Moreover, Hg and DOC were poorly correlated, in contrast to the positive Hg-DOC relation found elsewhere (Driscoll *et al.,* 1994; Bishop *et al.,* 1995). The episodic Hg increase in stream water was dominated by particulate Hg and thus appears to be linked to sediment transport. The process supplying sediment may be soil erosion by overland flow in riparian areas, streambank erosion, or mobilization of in-stream sediments.

The association between total Hg concentration and the organic fraction of suspended sediment can be explained by the high affinity of Hg(II) for organic substances in forest soils, as reported by several authors (Lindqvist, 1991; Mierele and Ingram, 1991; Schuster, 1991). Krabbenhoft *et al.* (1995) found that total Hg in stream water increased sharply with releases of DOC, while Me-Hg did not, in a small forested stream system in Wisconsin. Bishop *et al.* (1995) found that the increase in Hg concentration during the spring snowmelt was associated with total organic carbon.

Episodic fluxes of Hg during large events are not limited to upland catchments. Results of synoptic sampling of some larger streams in the Lake Champlain basin (Scherbatskoy *et al.*, 1997) also suggest a strong relation of increasing Hg with increasing flow. As at Nettle Brook, there appears to be a low but constant supply of dissolved Hg in the larger streams, but the dominant Hg loading to Lake Champlain from the rivers probably occurs in large events, such as the 1994 snowmelt or the August 1995 flood. It is not known whether these elevated particulate Hg concentrations in the larger rivers reflect transport from steep upland headwater catchments such as Nettle Brook or more local sources such as runoff from agricultural or disturbed lands.

5. Conclusions

The Nettle Brook catchment exports less than one-tenth (8 % and 5 % during two study years) of the Hg that enters in atmospheric (wet + dry) deposition. Of the Hg that is exported in streamflow, between one-quarter and two-fifths (26 % and 38 %) is in the dissolved form, representing a slow but continual export. The remaining Hg is in particulate form and is exported mostly on an episodic basis; fully one-half of the total annual export in the first year of study occurred on the single day of peak snowmelt. Particulate Hg in stream water is strongly associated with the organic fraction of suspended sediment. Episodic pulses of particulate Hg are evident at downstream reaches of larger rivers as well, suggesting that mobilization of Hg during large events may be the dominant source of Hg loading to Lake Champlain.

Acknowledgments

Support was provided from EPA and NOAA through the Cooperative Institute of Limnology and Ecosystems Research agreement no. NA37RJ0204. We wish to acknowledge the University of Vermont Proctor Maple Research Center and the assistance of Joanne Cummings, Andrea Donlon, and Jon Denner. This study was undertaken in cooperation with the Vermont Forest Ecosystem Monitoring Program.

References

Aastrup, M., Johnson, J., Bringmark, E., Bringmark, I. and Iverfeldt, A.: 1991, *Water, Air, Soil Pollut.* **56**, 155.
Bishop, K., Lee, Y.-H., Pettersson, C. and Allard, B.: 1995, *Water, Air, Soil Pollut.* **80**, 445.
Branfireun, B.A., Heyes, A. and Roulet, N.T.: 1996, *Water Resourc. Res.* **32**, 1785.
Budd, L. and Meals, D.: 1994, *Lake Champlain Nonpoint Source Pollution Assessment*, Lake Champlain Basin Program, Grand Isle, Vermont.
Burke, J., Hoyer, M., Keeler, G. and Scherbatskoy, T.: 1995, *Water, Air, Soil Pollut.* **80**, 353.
Driscoll, C.T., Yan, C., Schofield, C.L.,, Munson, R.K. and Holsapple, J.G.: 1994, *Environ. Sci. Technol.* **28**, 136A.
Hanson, P.J., Lindberg, S.E., Tabberer, T.A., Owens, J.G. and Kim, K.-H,: 1995, *Water, Air, Soil Pollut.* **80**, 373.
Hoyer, H., Burke, J. and Keeler, G.: 1995, *Water, Air, Soil Pollut.* **80**, 199.
Hultberg, H., Munthe, J. and Iverfeldt, Å.: 1995, *Water, Air, Soil Pollut.* **80**, 415.
Hurley, J. P., Benoit, J. M., Babiarz, C. L., Shafer, M. M., Andren, A. W., Sullivan, J. R., Hammond, R. and Webb, D.A.: 1995, *Environ. Sci. Technol.* **29**, 1867.
Johanssen, K., Aastrup, M., Andersson, A., Bringmark, I. and Iverfeldt, Å.: 1991, *Water, Air, Soil Pollut.* **56**, 267.
Johnson, D.W. and Lindberg, S.E.: 1995, *Water, Air, Soil Pollut.* **80**, 1069.
Keeler, G., Glinsorn, G. and Pirrone, N.: 1995, *Water, Air, Soil Pollut.* **80**, 159.
Krabbenhoft, D.P., Benoit, J.M., Babiarz, C.L., Hurley, J.P. and Andren, A.W.: 1995, *Water, Air, Soil Pollut.* **80**, 425.
Landis, M. S. and Keeler, G.J.: 1997, *Environ. Sci. Technol.*. **31**, 2610.
Lee, Y.-H., Borg, G., Iverfeldt, Å. and Hultberg, H.: 1994, in C.J Watras and J.W. Huckabee (eds.), *Mercury Pollution: Integration and Synthesis*, Lewis Publishers, Chelsea, MI, pp. 329-341.
Lee, Y.-H. and Iverfeldt, Å.: 1991, *Water, Air, Soil Pollut.* **56**, 309.

Lindberg, S.E., Meyers, T.P., Taylor, G.E., Turner, R.R., and Schroeder, W.H.: 1992, *J. Geophys. Res.* **97**, 2519.
Lindberg, S.E.:1993, *J. Geophys. Res.* **97**, 14677.
Lindqvist, O.: 1991, *Water, Air, Soil Pollut.* **55**, 73.
Mason, R. P., Fitzgerald, W. F. and Morel, F.M.M.: 1994, *Geochim. Cosmochim. Acta* **58**, 3191.
Mierle, G. and Ingram, R.: 1991, *Water, Air, Soil Pollut.* **56**, 349.
Rea, A. W., Keeler, G. J. and Scherbatskoy, T.: 1996, *Atmos. Environ.* **30**, 3257.
Scherbatskoy, T., Burke, J.M., Rea, A.W. and Keeler, G.J.: 1997, in J.E. Baker (ed.), *Atmospheric Deposition of Contaminants to the Great Lakes and Coastal Waters*, Society of Environmental Toxicology and Chemistry Special Publication Series, SETAC Press, Pensacola, FL, pp. 245-258.
Schuster, E.: 1991, *Water, Air, Soil Pollut.* **56**, 667.
Siegel, S.M. and Siegel, B.Z.: 1988, *Water, Air, Soil Pollut.* **40**, 443.
Swain, E.B., Engstrom, D.R., Brigham, M.E., Henning, T.A. and Brezonik, P.L: 1992, *Science* **257**, 784.
Vasu, A.B. and McCullough, M.L.: 1994, *First Report to Congress on Deposition of Air Pollutants to the Great Lakes Waters, EPA-453/R-93-055*, U.S. EPA Office of Air Quality and Planning Standards, Research Triangle Park, North Carolina.
Watzin, M.C.: 1992, *A Research and Monitoring Agenda for Lake Champlain: Proceedings of a Workshop, Lake Champlain Basin Program Technical Report No. 1*, U.S. EPA, Boston, Massachusetts.
Westling, O.: 1991, *Water, Air, Soil Pollut.* **56**, 419.
Wilmot, S.H. and Scherbatskoy, T.: 1994, *Vermont Monitoring Cooperative Annual Report for 1993, VMC Annual Report Number 3,* Vermont Department of Forests, Parks, and Recreation, Waterbury, Vermont.
Zhang, L., Planas, D. and Qian, J.-L.: 1995, *Water, Air, Soil Pollut.* **81**, 153.

HISTORICAL VARIATIONS IN LEAD FLUXES IN THE PYRENEES (NORTHEAST SPAIN) FROM A DATED LAKE SEDIMENT CORE

L. CAMARERO[1], P. MASQUÉ[2], W. DEVOS[3], I. ANI-RAGOLTA[2], J. CATALAN[1], H.C. MOOR[3], S. PLA[1] and J.A. SANCHEZ-CABEZA[2]

[1]*Department of Ecology, University of Barcelona, E-08028 Barcelona, Spain*
[2]*Department of Physics, Autonomous University of Barcelona, E-08193 Bellaterra, Spain*
[3]*Swiss Federal Laboratories for Material Testing and Research (EMPA), CH-8600 Dübendorf, Switzerland*

Abstract. Variations in Pb concentration in lake sediments reflect changes in the fluxes of this element in the past. A sediment core from a lake in the Pyrenees (Lake Redó at 2,240 m a.s.l., NE Spain) was studied, with the aim of reconstructing past environmental and climatic conditions in the lake and its catchment area. The core was dated using both ^{210}Pb and ^{14}C. A surface peak of Pb concentration, which was about 10 times higher than the background level, was found. This peak is attributed to mining activities since the beginning of this century. Although Pb pollution due to the combustion of gasoline is expected to be present, no evidence can be deduced from Pb isotope ratios of sediment due to the masking effect of Pb from mines. A second peak appeared in a deeper layer, with a maximum Pb concentration of about 17 times higher than the background level. The origin of this peak is unclear. All evidence suggests that the ancient peak is due to mining operations in the central Pyrenees area, which were not documented historically.

Keywords: Lake Redó, lake sediments, lead, metal pollution, mining, ^{210}Pb-dating, Pyrenees, stable Pb isotopes

1. Introduction

Peak concentrations of Pb attributable to contemporary sources commonly are found in surface or near-surface sediments of lakes all over the world (Petit *et al.*, 1984; Davis and Galloway, 1993; Renberg *et al.*,1994; Ritson *et al.*, 1994; Gobeil *et al.*, 1995). This Pb mainlycomes from industrial pollution and from the combustion of leaded gasoline. In countries where unleaded gasoline was introduced in the late 1960s, a decline in Pb content in the sedimentary record has been reported (Moor *et al.*, 1996). In most of these studies, it is also common to find a Pb concentration rise in sediments starting two millennia ago, owing to mining and smelting activities during ancient Greek and Roman times first, and during Medieval and Renaissance times later (Nriagu, 1996). Evidence of northern hemisphere pollution spread from these ancient industries has been found in ice cores from Greenland (Hong *et al.*, 1994).

What had not been found up to now (to our knowledge) is a profile showing a pre-industrial peak of Pb concentration in sediments higher than the modern peak. This finding is even more surprising if we consider that Lake Redó (where we recorded such a profile) is a relatively remote, pristine, high mountain lake. The aim of this paper is to discuss the most probable origin of the Pb fluxes in the lake, in particular that causing the ancient concentration peak.

Water, Air, and Soil Pollution **105:** 439–449, 1998.

2. Methods

2.1. Study Site

Lake Redó is located in the central Pyrenees (42° 38' 33" N, 0° 46' 13" E) at 2,240 m a.s.l. The lake has a surface area of 24 ha, a maximum depth of 73 m, and a mean depth of 32 m. The average water residence time is more than 4 yr. Its catchment area is a small granodiorite basin (155 ha). More than 50 % of the catchment area is composed of bare rock. Soils are generally less than 30 cm deep, and vegetation consists of high mountain grasslands. More details on the physical features and chemistry of Lake Redó may be found in Catalan (1989, 1992).

2.2. Sampling

A sediment core was taken from the deepest part of the lake by means of a gravity corer in August 1994. The core was extruded immediately and sliced in 3 mm sections. A subsample was stored in polyethylene bags for ^{210}Pb dating, and another subsample for diverse chemical analyses was kept deep frozen in a polypropylene vial. A soil core was sampled in Lake Redó's-catchment area in July 1995. The soil core was 30 cm deep and was sliced every 5 cm for the determination of total Pb content, stable Pb isotope ratio, and ^{210}Pb content. Galena samples were collected at the Cierco mines, situated further down the valley where Lake Redó is situated, and were analyzed for their stable Pb isotope ratio. Preparation of sediment and soil samples for chemical and isotopic ratio analyses consisted of drying in the oven at 90 °C, followed by grinding in an agate mortar. Samples for ^{210}Pb dating were dried at 45 °C.

2.3. Chemical analyses

A sequential extraction procedure was applied to sediments to partition their components into six operationally defined fractions. Although these fractions are not clearly associated with specific geochemical phases (Kheboian and Bauer, 1987; Gobeil *et al.*, 1995), the procedure leaches progressively less labile chemical species. The phases indicated below are assumed to be the main contributors to each fraction, but this assignments must be accepted with caution. About 200 mg of dry sediment were distributed in polypropylene assay tubes and extracted sequentially with: 1) 2 mL of 1 M ammonium acetate (pH 7) - this step leaches mainly soluble and exchangeable forms; 2) 2 mL of 1 M ammonium acetate acidified to pH 5 with acetic acid - carbonates are dissolved in this step; 3) 5 mL of 0.04 M hydroxylamine hydrochloride in 2 5% acetic acid at 96 °C - mostly Fe and Mn oxides are leached; 4) 2 mL of 1.4 M ammonium acetate in 9 % HNO_3, after digestion with 1.6 mL of H_2O_2 at 85 °C - organic matter is digested in this step; 5) 8 mL 0.2 M NaOH at 95 °C - biogenic Si, mainly opal from diatom frustules, is dissolved. This procedure was adapted from Tessier *et al.*(1979), with an additional step to extract biogenic Si taken from Engstrom and Wright (1984).

All samples were analyzed by inductively coupled plasma spectrography for major elements and trace metals in a PolyScan 61E Thermo Jarrell Ash instrument. Nine procedure blanks were performed (Table I).

2.4. DATING

^{210}Pb was determined in each section of the top 8 cm of the profile through deterination of its daughter ^{210}Po in equilibrium. The samples were digested in a microwave oven in a HNO_3:HCl:HF:H_2O mixture (5:2:3:10 mL). Then, 25 mL of boric acid was added to convert the excess fluorides into tetrafluoroborates by heating. The resulting solutions were evaporated until they were almost dry and the residual was dissolved in 1 M Hcl. Polonium isotopes spontaneously deposited onto silver disks at 60-70 °C for 8 hr while being stirred (Sanchez-Cabeza *et al.*, in press). Polonium isotopes were counted on alpha spectrometers equipped with low background SSB detectors (EG&G Ortec Model SSB 450 R). Excess ^{210}Pb for each section was calculated by subtracting the supported fraction, in equilibrium with ^{226}Ra, from the total ^{210}Pb. Supported ^{210}Pb, namely 58.0 ± 1.3 Bq kg^{-1}, was estimated from the mean value of seven deep sections where excess ^{210}Pb was not present. ^{210}Pb dating was achieved by application of the Constant Rate of Supply (CRS) model (Appleby and Oldfield 1978) in a modified version (Minimum Variance, CRS/MV, Tobin and Schell, 1988; Sanchez-Cabeza *et al.*, 1994).

^{14}C dating was performed by Beta Analytic Inc. on 12 bulk sediment samples, using their standard method. ^{14}C was measured in an accelerator-mass-spectrometer. The Pretoria calibration procedure was used for date calibrations (Vogel *et al.*, 1993).

TABLE I

Pb concentration in the five extractable fractions and the acid extraction and isotopic ratios in the acid extraction and the galena minerals. The extractions used for total Pb (sequential extraction: sum of fractions 1 to 5 which accounts for biogenic silica, and sum of fractions 1 to 4 without biogenic silica) and for Pb isotope ratio analyses (H_2O_2-HNO_3 extraction) are compared (differences are expressed as a percentage).

Depth (mm)	Concentration in fraction (mg kg^{-1})									Isotopic ratios	
	1	2	3	4	5	S 1-4	S 1-5	acid extr.	% diff.	$^{206}Pb/^{207}Pb$	$^{208}Pb/^{206}Pb$
6	11	86	141	51	26	290	316	298	5.7	1.172 ± 0.0023	2.095 ± 0.0029
45	2	11	21	16	10	50	60	45	25.0	1.198 ± 0.0019	2.064 ± 0.0033
69	3	15	38	26	12	82	94	64	31.9	1.185 ± 0.0021	2.078 ± 0.0037
98	0	10	22	22	11	54	64	60	6.3	1.188 ± 0.0014	2.084 ± 0.0041
142	15	138	223	111	28	487	514	495	3.7	1.173 ± 0.0023	2.099 ± 0.0031
201	0	3	15	9	5	27	32	18	43.7	1.217 ± 0.0024	2.048 ± 0.0042
299	0	0	16	13	7	29	36	17	59.4	1.224 ± 0.0019	2.038 ± 0.0036
Blanks	1 ± 1	1 ± 1	0 ± 0	1 ± 1	1 ± 1			1 ± 1			
Bossost-Liat galena *										1.152	2.117
Cierco galena										1.179 ± 0.0015	2.093 ± 0.0040

* From Cardellach *et al.* in press (analytical errors not reported)

2.5. Lead Isotope Ratio Analysis

About 50 mg each from the sediment and soil samples were subjected to medium-pressure microwave digestion (5 min at 250 W, 8 min at 600 W, 7 min at 350 W) with 4 mL 65 % HNO_3 and 1 mL 30 % H_2O_2. The same procedure was followed for 4 blanks. After digestion, the samples were filtered to remove the insoluble fraction and diluted. From the galena sample, 30 mg were dissolved in 10 mL of hot 65 % HNO_3 and diluted. Lead isotope ratios as well as Pb concentrations were measured in these solutions with a Perkin Elmer Sciex Elan 6000 inductively coupled plasma mass spectrometer (ICP-MS). The accuracy of concentration determinations was checked by digesting and analyzing NIST standard reference material 2704 (Buffalo River sediment). After blank subtraction and correction of the Pb signals for dead time losses, the measured Pb isotope ratios were corrected for instrumental mass discrimination by measuring a standard solution of NIST standard reference material SRM981 (common Pb isotopic standard). The accuracy of the Pb isotope ratio determinations was checked by measuring Pb isotope ratios in SRM981. The experimentally determined concentration and isotope ratios in the standard reference material corresponded to the certified ratio within the experimental uncertainty.

3. Results

The time scale relative to depth, established by combining ^{210}Pb and ^{14}C data , is shown in Figure 1. Radiocarbon dates resulted in a time scale shifted by 940 ± 40 yr BP, the apparent age of the surface sediment sample. Consequently, all radiocarbon dates were corrected by subtracting the apparent surface age. This agreed closely with the ^{210}Pb dating: the core sample from 18 mm depth was estimated to be from 1904 (± 2 yr) by the ^{210}Pb method, and from 1905 (± 150 yr) by the ^{14}C method. Estimated sedimentation rates changed with depth: from surface to 30 mm it was on average 0.23 mm yr^{-1} (spanning 1875 AD - 1994 AD), with two periods of higher sedimentation rates identified during the first half of the twentieth century (Ani-Ragolta, 1996); from 30 to 171 mm sedimentation was 0.08 mm yr^{-1} (394 AD - 1875 AD), and from 171 to 458 mm it was 0.03 mm yr^{-1} (8171 BC - 394 AD). These rates were estimated by linear fitting of depth against the age of ten, four and five dated samples, respectively. They were fairly constant within each period, with the exception of the two episodes during the twentieth century.

The Pb concentration profile in Lake Redó is displayed in Figure 1. Concentrations shown in the graph are the sum of fractions 1-5, hereafter referred to as total extractable Pb. Atmospheric Pb of anthropogenic origin is assumed to be within these fractions. The total extractable Pb values obtained with the sequential extraction procedure correlate with those obtained with the acid extraction method used for isotope ratio determination (Table I). The higher concentrations in sequentially extracted samples are attributable to the fact that this procedure includes a final step that releases the biogenic Si. This biogenic Si fraction is seldom considered when analyzing extractable Pb and other trace metals, but it may contain these elements. Therefore, the Pb concentrations from the acid extraction used for isotope analysis is closer to the sum of fractions 1-4, which were also extracted in acid media. We assume that the isotope ratios measured correspond to the Pb in fractions 1-4. We do not expect the isotope ratios of the total extractable Pb to be significantly

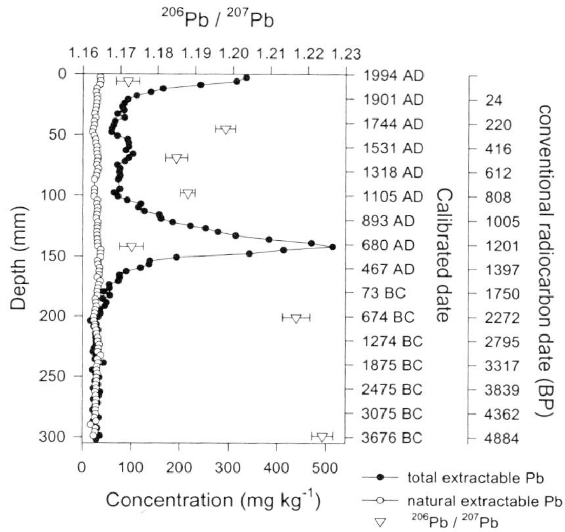

Fig. 1. Lead concentration profiles in sediments: total extractable Pb and natural Pb (as estimated from Al concentrations) profiles, and the $^{206}Pb/^{207}Pb$ ratio at certain levels are plotted against the depth and time scales obtained by ^{210}Pb and ^{14}C dating.

different from those we have measured because the Pb not leached in the acid extraction is a small amount. On the other hand, although some authors have found differences in the Pb isotope composition between the organic and mineral fractions in soils (Erel and Patterson, 1994), such differences between different lake sediment fractions other than the mineral residue (which was not considered in this study), have not been found in other studies (Gobeil *et al.,* 1995).

Background concentrations of total extractable Pb in the deeper part (below 200 mm) of the core were 20 - 30 mg kg^{-1}. In the upper part of the core, Pb concentration became higher, with two outstanding peaks at the surface (316 mg kg^{-1}) and at 142 mm depth (514 mg kg^{-1}). A third peak of lower concentration (94 mg kg^{-1}) appeared between 50 and 70 mm. The dating shows that the deepest peak started to rise rapidly at about 467 AD, reached its maximum at 658 AD and declined again to background levels at 1105 AD. This coincides with the period in which worldwide Pb production was lower after the eclipse of the Roman Empire and before the discovery of the Pb and Ag mines in central Europe, according to the widely accepted reconstruction of Settle and Patterson (1980).

Anomalously high Pb contents in the granite rocks, as caused for instance by inclusions of Pb and other metal deposits of hydrothermal origin (Johnson *et al.,* 1996), have been described in this area of the Pyrenees. To rule out the possibility that increased weathering of these minerals within the catchment area during this period might cause natural accumulation of Pb in sediments, we compared the Pb/Al and Pb/Ba ratios along the profile. Lead, Al and Ba are closely associated in the mineral composition of Pyrenean granites (J. Tritlla, pers. comm.), and one would expect their ratios to remain constant despite changes in weathering rates. Both ratios (as well as the Pb concentration) were on average much lower and more constant in the section of the core between 200 and 300 mm (Pb/Al = $10.5 \pm 2.4 \times 10^{-4}$; Pb/Ba = $20.9 \pm 5.6 \times 10^{-2}$) than in the upper section (Pb/Al = $46.6 \pm 34.3 \times 10^{-4}$; Pb/Ba = $87.3 \pm 66.3 \times 10^{-2}$). We thus considered that the values for the ratios in sediments deeper than 200 mm were those of the weathered mineral, and the natural contribution of Pb from the weathering of rocks in the catchment area for the entire core was reconstructed by multiplying the Al and Ba concentrations in sediments by the respective ratios. The reconstructed profiles were practically the same using either Al or Ba. Although Al and Ba could be also pollutants, the variations in concentration with depth did not follow that of Pb, and the possible contamination had a negligible effect on the calculations: the estimated natural Pb concentration was all through the core practically constant and identical to the background levels measured.

Figure 1 shows the natural Pb concentration as estimated from Al, compared to the total extractable Pb concentration. The difference between the two profiles indicated Pb enrichment, which can only be attributed to the contribution from anthropogenic sources through atmospheric deposition. In our core, the Pb excess started to rise in about 674 BC, which corresponded with the rise of Pb concentration in Greenland ice (Hong *et al.*, 1994) and the introduction of coinage (Settle and Patterson, 1980). The modern Pb peak started in about 1900 AD. Although Pb pollution caused by the combustion of leaded gasoline also must contribute to this peak, it is basically a record of the mining activities (Table II) in the nearby northern areas of Bossost and Liat and the southern area of Cierco. Opening of the mines at the beginning of this century (Johnson *et al.*, 1996; Cardellach *et al.*, in press) caused the yearly increase percentage of Pb rise. Between about 1943 and 1963, the period of Spanish post-war industrial development in which both northern and southern

TABLE II

Lead increase rates related to mining activities. The total increase rate was obtained by calculating the increase between two consecutive sediment samples (3 mm slices). The yearly rate is an average for the period between samples, and resulted from dividing the total rate by the number of years in each period.

Period		Pb increase rate (%)		
start	end	total	yearly	Mining activities
1984	1994	6.1	0.6	
1963	1984	30.7	1.5	In 1981, Cierco mines closed
1943	1963	46.6	2.3	During the 1950's, Liat and Bossost mines closed
1932	1943	17.9	1.6	
1916	1932	26.7	1.7	In 1930, Cierco mines opened
1905	1916	19.9	1.8	
1900	1905	8.3	1.7	At beginning of century, mines in Bossost and Liat opened
1889	1900	4.9	0.4	

mines were exploited, the Pb increase rate reached its maximum. Between about 1963 and 1984 only the Cierco mines remained open, causing a lower Pb increase rate, which reached again a minimum after the closure of the last mines in 1981.

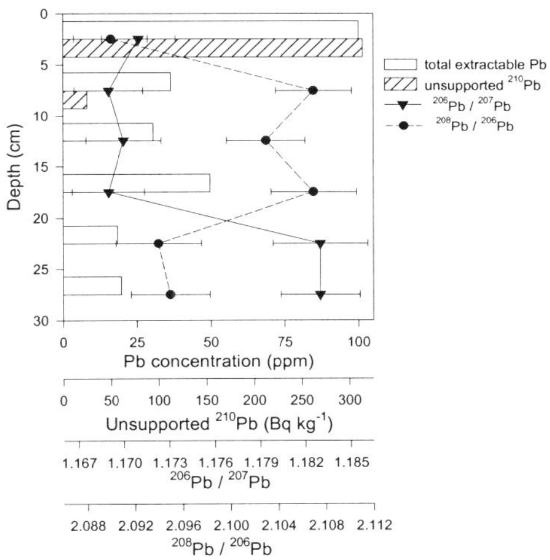

Fig. 2. Total extractable Pb content, unsupported ^{210}Pb and Pb isotope ratios from a soil core from the Lake Redó catchment.

Measurements from the soil core are shown in Figure 2. The unsupported ^{210}Pb measurements showed that all Pb deposited during the last 100 years was in the upper 5 cm. However, the isotopic composition indicated three different origins of Pb (Figure 3): in the 0-5 cm layer, the isotopic ratios were similar to those of surface sediments; in the 5-20 cm layer the isotopic composition resembled that of the ancient peak; and between 20 and 30 cm it was similar to that of the Cierco galena. Similarly, the total extractable Pb concentration changed in these three layers: 100 mg kg^{-1} (0-5 cm), 30-50 mg kg^{-1} (5-20 cm), and 20 mg kg^{-1} (20-30 cm). There is discrepancy between the isotopic composition of Pb in the bottom soil layer and in the background sediment. There are two possible explanations for this: first, the bottom soil layer might contain Pb from another source in the catchment which has no effect in the sediments; second, some Pb might have migrated from the more polluted upper layers towards the bottom, changing the average ratios.

4. Discussion

The only reasonable explanation for the ancient Pb peak is mining and smelting. The Pb isotope composition of sediment samples supports this hypothesis. Table I and Figure 1 show the $^{206}Pb/^{207}Pb$ ratio in several samples along the sediment core. The background $^{206}Pb/^{207}Pb$ ratio, obtained from the 299 mm sample, was 1.224. This figure agrees closely with the mean isotopic composition of pre-industrial sediments over Europe ($^{206}Pb/^{207}Pb \approx 1.21$; Grousset *et al.* 1994). This ratio decreased slightly at 201 mm, coinciding with the first appearance of excess Pb, and fell dramatically at the ancient peak. The $^{206}Pb/^{207}Pb$ ratio of the modern peak was practically the same as the ancient one, and during the period between peaks this ratio increased but remained below the background value. The great similarity of the $^{206}Pb/^{207}Pb$ ratios in the two largest peaks strongly suggests a similar origin. The origin of Pb added to gasoline in Europe has been low radiogenic Precambrian ores, mostly the Broken Hill ores (Australia), with a $^{206}Pb/^{207}Pb$ ratio $\approx$ 1.04 (Grousset *et al.* 1994). The fact that the isotope ratio of both modern and ancient peak was so similar, and well above that of gasoline Pb indicates that the proportion of gasoline Pb in the modern peak must be relatively low (see the reconstruction of anthropogenic Pb isotope ratios below).

To compare the isotopic composition of sediments with the possible sources of pollutant Pb, Figure 3 shows the $^{206}Pb/^{207}Pb$ versus $^{208}Pb/^{206}Pb$ ratios of sediments, soil from the Lake Redó catchment area, and galenas from the Bossost-Liat (Cardellach *et al.* in press) and Cierco mines. The effect of Pb coming from the mines on the isotopic ratios of both ancient and modern peaks could be clearly seen: their composition is a mixture of background sediment Pb and galena Pb.

The final Pb isotope composition of a sediment sample is a result of the proportions in which both natural Pb and anthropogenic Pb are mixed. Equation (1) describes this:

$$\frac{(^{206}Pb/^{207}Pb)_N \times [Pb]_N + (^{206}Pb/^{207}Pb)_A \times [Pb]_A}{[Pb]_N + [Pb]_A} = (^{206}Pb/^{207}Pb)_A \quad (1)$$

where the subscript *N* denotes the natural fraction, *A* the anthropogenic fraction, and *T* the total. The same equation may be written for the $^{208}Pb/^{206}Pb$ ratio. In these equations the only unknown variables were the anthropogenic ratios. We assumed the natural fraction ratios to be those of the background sample, whereas the total ratios were those measured for each sample. The natural Pb concentration was estimated from the Pb/Al ratio, and the anthropogenic Pb was the difference between the measured values and the estimates of natural Pb, as discussed above. The equations were solved to find the anthropogenic Pb isotope ratios in our samples (Table III).

Despite the variable total extractable Pb concentrations, the ratios for the anthropogenic fraction were practically the same in both modern and ancient peaks, clearly indicating the effects of mining activities on the composition of pollutant Pb. The similarity between the estimated anthropogenic Pb isotope ratios is remarkable, because one would expect lower values in the surface peak owing to the widespread leaded gasoline contribution. Such coincidence indicated that the proportion of Pb from gasoline in the surface peak is low.

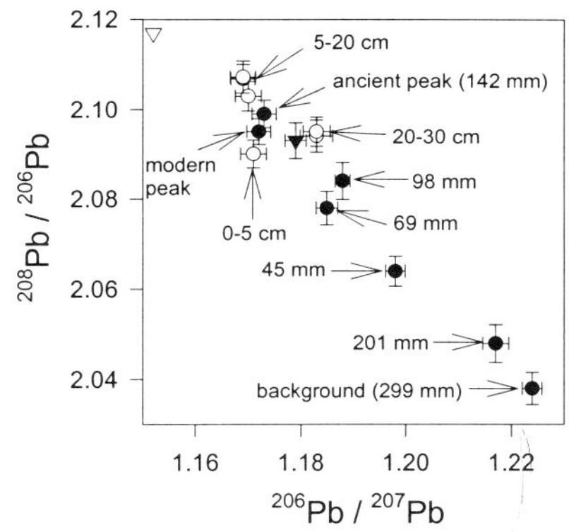

Fig. 3. ^{206}Pb/^{207}Pb plotted against ^{208}Pb/^{207}Pb for sediment, soil, and galena samples from Cierco and Bossost-Liat mines.

TABLE III
Reconstructed isotope ratios of anthropogenic Pb.

Depth (mm)	Calibrated date	^{206}Pb/^{207}Pb	^{208}Pb/^{206}Pb
3	1970 AD	1.166 ± 0.0028	2.102 ± 0.0037
45	1706 AD	1.182 ± 0.0040	2.080 ± 0.0073
69	1451 AD	1.165 ± 0.0041	2.099 ± 0.0074
98	1143 AD	1.166 ± 0.0034	2.111 ± 0.0087
142	674 AD	1.169 ± 0.0026	2.103 ± 0.0035
201	659 BC	1.197 ± 0.0140	2.077 ± 0.0251

The samples from 69 and 98 mm showed an anthropogenic Pb composition identical to that of the mining peaks. If it indicates mining operations, then higher Pb concentrations in the sediment would be expected. Another possible explanation is the release of Pb from polluted soils over a long period after the massive pollution. This aspect is currently being investigated. The change in the isotopic ratio of the sample from 45 mm indicates, following the above reasoning, either the end of pre-industrial mining activities or the exhaustion of at least the most mobile fraction of pollutant Pb stored in the catchment area.

The question of who was operating the mines in the Pyrenees in the c. 467 - 1105 AD period remains open. Roman remains exist in this area. No mining is historically or archaeologically documented, although it could well have taken place and so might be the cause of the initial rise of Pb concentration. But the largest Pb peak occurred after Roman times, in a period in which emissions at a continental scale were at a minimum. The fact that pollutant Pb had the same isotopic signature as the Pb deposited from mining during this century in the nearby areas demonstrated that mining on a relatively large scale was taking place in the region in this period.

5. Conclusions

Our data clearly showed an increase of anthropogenic Pb released to the atmosphere starting around 670 BC, in good agreement with the accepted chronologies for Pb pollution at a large scale. A surface peak of Pb concentration appearing in the lake sediments showed a straight relation with the mining activities in the nearby areas. A deeper peak appeared with Pb levels higher than those in the modern one. This peak corresponded to about 658 AD. Coincident isotopic signatures revealed that Pb in both peaks must have the same origin. Mining and smelting at a relative large scale should have then taken place in the surrounding area during post-Roman times.

Acknowledgments

This study was funded by MEC AMB-93-0814-CO2-01 and EU ENV4-CT95-007 projects and CIRIT 1995 SGR 00531 grant. We thank to Mr. M. Castells for providing historical information and galena samples from the Cierco mines.

References

Ani-Ragolta, I.: 1996, *^{210}Pb en sediments: Estudi i aplicació d'alguns models de datació* Research Report, Autonomous University of Barcelona.

Appleby, P.G. and Oldfield, F.: 1978, *Catena* **5**, 1.

Cardellach, E., Canals, A. and Pujals, I.: in press, *Estudio Geológico.*

Catalan, J.: 1989, *Water Resourc. Res.* **25**, 519.

Catalan, J.: 1992, *Can. J. Fish. Aquat. Sci.* **49**, 945.

Davis, A. and Galloway, J.N.: 1993, *Appl. Geochem.* **8**, 51.

Engstrom, D.R. and Wright Jr., H.E.: 1984, 'Chemical Stratigraphy of Lake Sediments as a Record of Environmental Change', in E.Y. Haworth and J.W.G. Lund (eds.), *Lake Sediments and Environmental History*, University of Minnesota Press, Minneapolis, pp. 11-67

Erel, Y. and Patterson C.C.: 1994, *Geochim. Cosmochim. Acta* **58**, 3289.

Gobeil, C., Johnson, W.K., Macdonald, R.W., and Wong, C.S.: 1995, *Environ. Sci. Technol.* **29**, 193.

Grousset, F.E., Quétel,C.R., Thomas, B., Buat-Ménard, P., Donard, O.F.X. and Bucher, A.: 1994, *Environ. Sci. Technol.* **28**, 1605.
Hong, S., Candelone, J.P., Patterson, C.C. and Boutron, C.F.: 1994, *Science* **265**, 1841.
Johnson, C.A., Cardellach, E., Tritlla, J. and Hanan, B.B.: 1996, *Econ. Geol.* **91**, 497.
Kheboian, C. and Bauer, C.F.: 1987, *Anal. Chem.* **59**, 1417 .
Moor, H. C., Schaller, T. and Sturm, M.: 1996, *Environ. Sci. Technol.* **30**, 2928.
Nriagu, J.O.: 1996, *Science* **272**, 223.
Petit, D., Mennessier, J.P., and Lamberts, L.: 1984, *Atmos. Environ.* **18**, 1189.
Renberg, I., Persson, M.W. and Emteryd, O.: 1994, *Nature* **368**, 323.
Ritson, P. I., Esser, B.K., Niemeyer, S. and Flegal, A.R.: 1994, *Geochim. Cosmochim. Acta* **58**, 3297.
Sanchez-Cabeza, J.A., Masqué, P. and Ani-Ragolta, I.: in press, *J. Radioanal. Nuc. Chem.* **227**
Sanchez-Cabeza, J.A., Masqué, P., Schell, W.R., Palanques, A., Valiente, M., Palet, C., Pérez-Obiol, R. and Pantaleón Cano, J.: 1993, 'A Record of Anthropogenic Environmental Impact in the Continental Shelf North of Barcelona City', in *Isotope Techniques in the Study of Past and Current Environmental Changes in the Hydrosphere and the Atmosphere*, Vienna, IAEA-SM-329/18, pp. 175-178.
Settle, D.M., and Patterson, C.C.: 1980, *Science* **207**, 1167.
Tessier, A., Campbell, P.G.C. and Bisson, M.: 1979, *Anal. Chem.* **51**, 844.
Tobin, M.J. and Schell, W.R.: 1988, 'Recent Developments in Sedimentation Modelling and the Statistical Reliability of the ^{210}Pb Dating Method.' Research Report McMaster University, Ontario, Canada, 15 pp.
Vogel, J.C., Fuls, A., Visser, E., and Becker, B.: 1993, *Radiocarbon* **35,** 73.

CRITICAL LOADS OF TRACE METALS IN SOILS: A METHOD OF CALCULATION

T. PAČES

Czech Geological Survey, Klárov 3, 118 21 Praha, Czech Republic

Abstract. Critical load of trace metals in soils is a function of biological uptake, leaching by percolating water, input of the trace metals due to bedrock weathering, and the norms set to protect soils from metal pollution. The exceedance of the critical load is a difference between the calculated load and the measured input of the metals by atmospheric deposition and by application of agrochemicals. A critical time is the time when the concentration of a trace metal in soil, which is the result of all inputs and outputs of the metal in the soil, reaches the value of the norm. Data on biologically important trace metals in a small agricultural catchment in the Czech Republic indicate that the soil concentrations of As, Cd and Pb will reach the norms set for the soils after 4.5, 61, and 980 years, respectively. The present mass balances of Cu and Zn in the soils indicate that their steady-state concentrations will be below the norms so that the atmospheric and agricultural inputs will never overshoot the calculated critical load.

Keywords: critical loads, critical times, Czech Republic, heavy metals, predictive model

1. Introduction

Terrestrial ecosystems may react adversely to high anthropogenic inputs of metals. The aim of this communication is to present a procedure for the calculation of critical loads of trace metals such as As, Cd, Cu, Pb and Zn in agricultural soils and forest soils. The term "critical load" was developed to set limits on acidic emissions in European forest ecosystems (Nilsson and Grennfelt, 1988). The original definition of critical load can be modified to be applicable to trace metals: "The critical load of a trace metal is a calculated atmospheric and agricultural input to a terrestrial ecosystem that will increase the concentration of the metal in soil to a steady state level not yet harmful to present organisms." When the real atmospheric and agricultural input is higher than the calculated critical load, the concentration of the metal in soil can reach a toxic level with respect to organisms of the ecosystem. The critical load values can serve as an indicator of atmospheric and agricultural pollution in regions where the high input of trace metals is of ecological concern and where a legislation should limit this type of pollution. The first evaluation of critical loads of heavy metals was made in the Netherlands (de Vries and Bakker, 1996). Maps of critical loads of Cd, Cu, Pb and Zn were prepared for the Czech Republic by Vrubel *et al.* (1996) using the method presented in this paper.

Trace metals in soils originate from chemical weathering of minerals, atmospheric deposition, dispersed inputs of agricultural chemicals and from industrial and domestic point sources of pollution. They are removed from soils by plant uptake and by leaching and runoff. We will exclude the domestic and industrial point sources from the calculation because the "critical load" is a concept useful to characterize the environmental impact of dispersed type of pollution.

The soil budget of a trace metal is illustrated in Figure 1. The mass balance equation (1) of the system expresses the change in the soil pool of the metal, P, over time t:

Water, Air, and Soil Pollution **105:** 451–458, 1998.

$$F_{atm} + F_{agr} + F_{wea} - F_{up} - F_{run} = dP/dt \qquad (1)$$

where F_x is flux (g ha^{-1} yr^{-1}) of the metal by a transport mechanism x. The fluxes considered in the model are: F_{atm} - input of metal by atmospheric deposition; F_{agr} - input of metal by application of agrochemicals; F_{wea} - input of metal by chemical weathering of bedrock; F_{up} - output of metal by biological uptake; F_{run} - output of metal by runoff of water. The pool of the metal is proportional to biologically available concentration in soil, c_s:

$$P = c_s \times \sigma \times h \times (1-p) \qquad (2)$$

where s is density, h is thickness, and p is porosity of biologically active soil.

The biologically available part of a metal in soil is defined by a chosen analytical method. Soil is leached with an extract solution (Jones, 1990; Zbíral, 1995) and the content of the metal in the extract is considered to be biologically available. Commonly used extract solutions include 2 M HNO_3, a mixture of CH_3COOH, NH_4Cl and NH_4F (Mehlich, 1978), or a mixture of CH_3COOH and HNO_3 with NH_4NO_3 and EDTA (Mehlich, 1984); other extract solutions are described by Morgan and Wolf (Wolf, 1982) and others (Lindsay and Norvell, 1978).

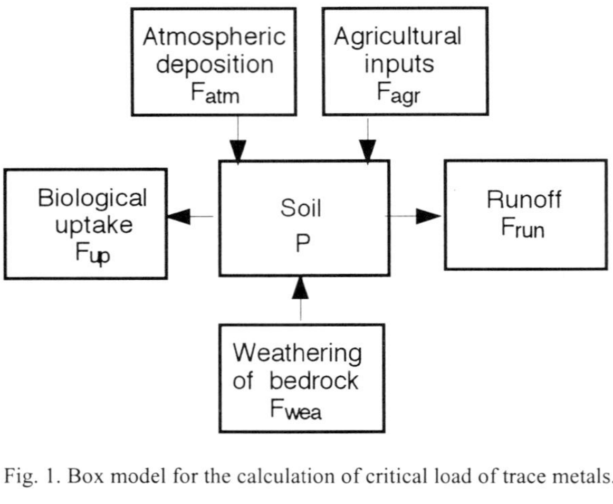

Fig. 1. Box model for the calculation of critical load of trace metals.

2. Model and Definitions

The pool of the biologically available trace metal in soil is a function of all inputs, outputs, sources and sinks in the system. Therefore, it is a function of time. Under pristine environmental conditions, the pool reaches a steady-state favorable for a given ecosystem. Due to the anthropogenically-induced acceleration of the input of trace metals, the pool of the metals becomes larger and the concentrations of the metals will increase.

The simple dynamic model assumes that the annual inputs are independent of the pool, while the outputs are proportional to the size of the pool.

$$F_{up} = k_{up} \times P \tag{3}$$
$$F_{run} = k_{run} \times P \tag{4}$$

where k_{up} and k_{run} are the first-order rate constants (yr^{-1}). After substitution of the first-order relationships to the mass balance equation (1), the change in the pool with time is

$$\frac{dP}{dt} = F_{atm} + F_{agr} + F_{wea} - (k_{up} + k_{run}) \times P \tag{5}$$

The critical load, $(F_{atm} + F_{agr})_{critical}$, is defined as an input reaching a steady-state in which the content of the trace metal is not yet harmful to organisms.

$$(F_{atm} + F_{agr})_{critical} = P_{critical} \times (k_{up} + k_{run}) - F_{wea} \tag{6}$$

where $P_{critical}$ is the still ecologically acceptable pool of the metal defined as

$$P_{critical} = c_{critical} \times \sigma \times h \times (1-p) \tag{7}$$

The critical concentration, $c_{critical}$, can be defined in two ways: the ecologically favorable steady-state concentration established under pristine conditions, or the concentration corresponding to an experimentally determined concentration not yet harmful to the ecosystem. The first definition is not practical because we rarely know the pristine fluxes needed to calculate the pristine steady-state concentration. A more practical definition is based on hygienic norms set for the concentrations of metals in soils. A state norm issued in the Czech Republic to protect soils requires determination of metal in the extract with 2 M HNO_3 solution (Regulation, 1994). The norms, c_n, are given in Table I. We define the critical concentration as being equal to the norm

$$c_{critical} = c_n \tag{8}$$

The exceedance of the critical load, EX, is the difference between the real input from atmosphere and from agricultural chemicals $(F_{atm} + F_{agr})$ and their calculated critical values:

$$EX = F_{atm} + F_{agr} - (F_{atm} + F_{agr})_{critical} \tag{9}$$

An ecologically, as well as politically, relevant question is how long it will take to reach the critical concentrations under the present rate of atmospheric deposition and agricultural application of a given metal. Substituting equation (2) to equation (5) and integrating from the present concentration in soil ($c_{s,present}$) at time t=0 to c_s at time t

$$c_s = \frac{F}{\vartheta \times k} (1-e^{-kt}) + c_{s,present} \times e^{-kt} \tag{10}$$

where

$\vartheta = \sigma \times h \times (1\text{-}p)$ is a mass to area conversion factor,

$F = F_{atm} + F_{agr} + F_{wea}$ is the total input, and

$k = k_{up} + k_{run}$ is the combined rate constant of the output.

The critical time, $t_{critical}$, when the soil reaches the critical concentration defined by the norm, c_n, is

$$t_{critical} = -\frac{1}{k} \times \frac{F - \vartheta \times k \times c_n}{F - \vartheta \times k \times c_{s,present}} \tag{11}$$

3. Application of the Model

Data representing agricultural land in the Bohemian-Moravian Upland in the Czech Republic are used to calculate the critical load and critical time for As, Cd, Cu, Pb and Zn. The characteristics of bedrock, soil and hydrology are summarized in Table I. The fluxes and related constants of the model are in Table II. The result of the calculation is in Table III. The atmospheric and agricultural inputs, biological uptake, and runoff were determined by monitoring the Vocadlo small catchment. The catchment and the results of the monitoring and mass balance evaluation of major elements are described in Pačes (1985, 1991). The data on trace metals were obtained by more recent monitoring in 1996.

The input of trace metals by weathering, F_{wea}, is calculated using the weathering rate of the bedrock determined by monitoring of mass balance in the small catchment (Pačes 1985, 1986). F_{wea} of the trace metals is assumed to be proportional to the total (chemical + mechanical) weathering rate of bedrock.

$$F_{wea} = c_r \times R \times \frac{W}{\Sigma A} \tag{12}$$

where

c_r is the concentration of the trace metal in bedrock,

W is the weathering rate of bedrock expressed as the net removal of base cations Na^+, K^+, Ca^{2+} and Mg^{2+} (eq ha^{-1} yr^{-1}),

R is the estimated ratio of the release of the trace metal by chemical weathering to its release by total weathering, and

ΣA is the sum of base oxides (Na_2O, K_2O, CaO, MgO) in bedrock (eq kg^{-1}).

The exceedances of atmospheric and agricultural inputs of As, Cd and Pb are positive values while the exceedances of Cu and Zn are negative values (Table III), indicating that Cu and Zn should never reach their critical values in soil if the present inputs and outputs last. The critical concentrations of As, Cd, and Pb in the soils will be reached after 45, 61 and 980 years, respectively. The meaning of the positive and negative exceedance values is illustrated in Figures 2 and 3. The plots illustrate that the calculated critical loads and their exceedances depend on the choice of the critical concentrations of the trace metals.

On the other hand, the time function does not depend on the c_n values. The progress of soil concentrations depends on our estimates of the atmospheric, agricultural, hydrological and geochemical fluxes in the mass balance equation of the model and on physical and chemical properties of the soils.

TABLE I
Field data for the calculation of critical load and its exceedance (source of data: Beneš 1993, 1994; Pačes 1975, 1985, 1991; Regulation, 1994)

Bedrock - gneiss						
W - weathering	eq ha^{-1} yr^{-1}	1,070				
R - Chemical/total weathering		0.52				
% by weight		Na_2O	K_2O	CaO	MgO	
		1.8	1.67	0.7	0.37	
Soil - Cambisol:						
p - porosity	%	40				
σ - density	kg m^{-3}	1,520				
h - thickness	m	0.3				
Water-runoff						
Q	L sec^{-1} km^{-2}	5.4				
		As	Cd	Cu	Pb	Zn
c_r - in rock	mg kg^{-1}	2.0	0.13	10	15	40
c_s - in dry soil	mg kg^{-1}	3.1	0.17	7.5	27.1	27.1
c_w - in runoff	μg L^{-1}	0.46	0.07	0.7	0.39	13
c_n - norm	mg kg^{-1}	4.5	0.40	50	80	100

TABLE II
Average characteristics of the biogeochemical mass balance of five trace elements in agricultural catchments with crystalline bedrock in the Czech Republic (F_{atm}, F_{agr} and F_{up} - Beneš, 1994).

		As	Cd	Cu	Pb	Zn
F_{atm}	g ha^{-1} yr^{-1}	83	9.3	106	181	473
F_{agr}	g ha^{-1} yr^{-1}	13.5	5.6	73	14.8	387
F_{wea}	g ha^{-1} yr^{-1}	0.81	0.053	4.07	6.1	16.3
F_{up}	g ha^{-1} yr^{-1}	9.4	2.6	59	26	249
F_{run}	g ha^{-1} yr^{-1}	0.79	0.12	1.2	0.67	22
k_{up}	yr^{-1}	1.1×10^{-3}	5.6×10^{-3}	2.9×10^{-3}	3.5×10^{-4}	3.36×10^{-3}
k_{run}	yr^{-1}	9.3×10^{-5}	2.6×10^{-4}	5.8×10^{-5}	9.0×10^{-6}	3.0×10^{-4}

TABLE III
Calculated critical load $(F_{atm}+F_{agr})_{crit}$, exceedance (EX),and critical times ($t_{critical}$) for an agricultural catchment in the Czech Republic.

		As	Cd	Cu	Pb	Zn
$(F_{atm}+F_{agr})_{crit}$	g ha^{-1} yr^{-1}	14	6.4	397	73	985
EX = $(F_{atm}+F_{agr})$ - $(F_{atm}+F_{agr})_{crit}$	g ha^{-1} yr^{-1}	83	8.6	-218	123	-125
$t_{critical}$	yr	45	61		980	

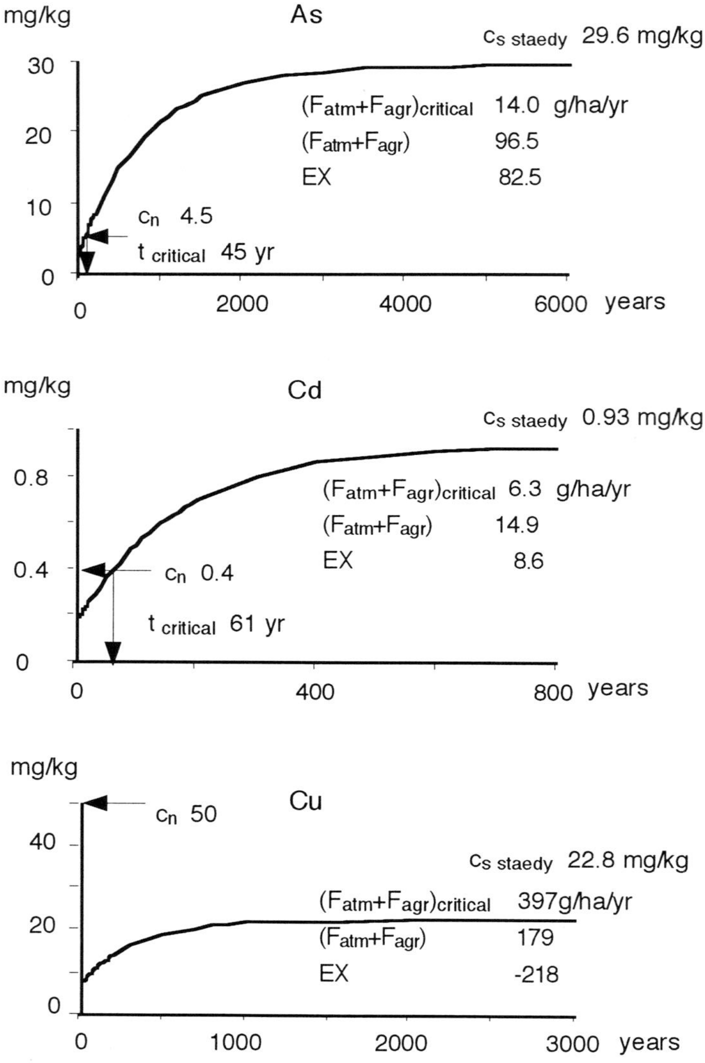

Fig. 2. Time function of the biologically available concentration of As, Cd and Cu in soil as a result of a simple mass balance calculation.

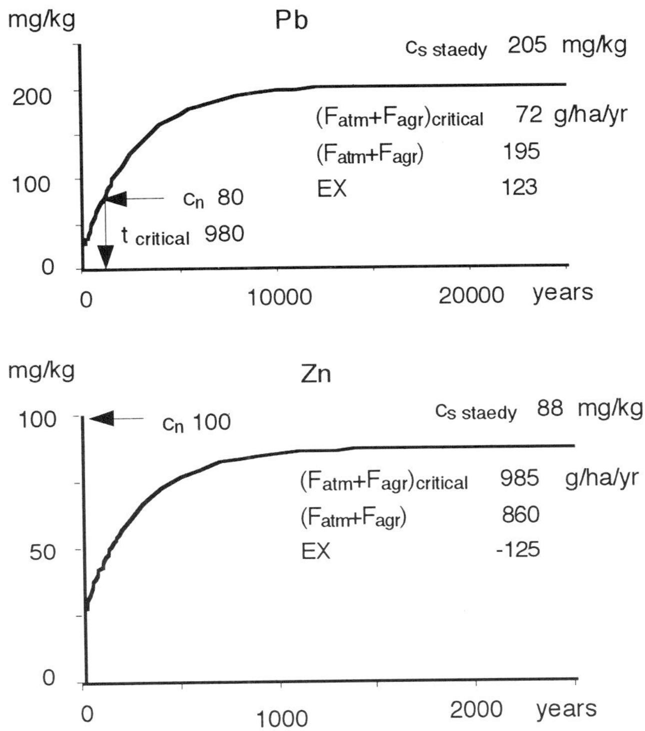

Fig. 3. Time function of the biologically available concentration of Pb and Zn in soil as a result of a simple mass balance calculation.

4. Discussion

The model enables us to calculation the critical time when a metal in soil will reach its environmentally critical concentration. The method of calculation is based on a simple hypothesis that the rates of runoff and biological uptake of the metal are proportional to the pool of the biologically available metal in soil. The first-order kinetics is not a realistic model of the complex soil-water-plant system. Therefore, the calculated critical loads and their exceedances are only approximate indicators of the behavior of trace metals in soil systems. They provide, however, a warning signal to initiate an environmental evaluation of possible pollution hazards.

The data used for the testing of the model were obtained in a small agricultural catchment representing typical Czech agricultural countryside. The results indicate that the present agricultural and atmospheric inputs of As, Cd and Pb are higher than their critical loads. The critical concentrations in soil will be reached after 45, 61 and 980 years, respectively. If the present inputs of Cu and Zn do not increase, then their atmospheric and agricultural inputs will never overcome their critical values and their steady state concentrations will never reach the critical concentrations. The critical loads, their exceedances, and the critical times depend on the choice of the hygienic norms. Therefore, the reliability of the norms is of utmost significance for the critical load evaluation. Critical time is an important parameter of the critical load evaluation because it indicates which of the trace metals may be the most acute threat to the soils.

This model of the behavior of trace metals in soils and the idea of critical load can be rightly criticized for its excessive simplicity. The real processes of fixation of trace metals in soils, their chemical speciation and role in plant physiology are complex (Bernquist 1986a,b, 1987) and require detailed data to clarify the threat of heavy metal pollution to crops and forests. The simple model perhaps can be applied on a regional or even continental scale to yield comparative indicators of metal pollution in areas with limited data.

References

Beneš, S.: 1993, *Obsahy a bilance prvků ve sférách Životního prostředí, Part. I*, Ministry of Agriculture of the Czech Republic, Praha, (in Czech).

Beneš, S.: 1994, *Obsahy a bilance prvků ve sférách Životního prostředí, Part. II*, Ministry of Agriculture of the Czech Republic, Praha, (in Czech).

Bernquist, B.: 1986a, *Metal Fluxes in Spruce and Beech Forest Ecosystems of South Sweden,* Ph.D. Dissertation, Department of Ecology, University of Lund, Lund.

Bernquist, B.: 1986 b, *Water, Air, Soil Pollut.* **31**, 901.

Bernquist, B.: 1987, *Water, Air, Soil Pollut.* **33**, 131.

de Vries, W. and Bakker, D.J.: 1996, *Manual for Calculating Critical Loads of Heavy Metals for Soils and Surface Waters, Report 114*, DLO Winand Staring Centre, Wageningen, TNO Institute of Environmental Sciences, Delf.

Jones, J.B.: 1990, *Comm. Soil Sci. Plant Anal.* **21**, 1091.

Lindsay, W.L. and Norvell, W.A.: 1978, *Soil Sci. Soc. Amer. J.* **42**, 421.

Mehlich, A:. 1978, *Comm. Soil Sci. Plant Anal.* **9**, 477.

Mehlich, A.: 1984, *Soil Sci. Plant Anal.* **15**, 1409.

Nilsson, J. and Grennfelt, P. (eds.): 1988, *Critical Loads for Sulfur and Nitrogen*, Report from a Workshop at Skokloster, Sweden, March 1988, Miljo rapport 1988 15, Nordic Council of Ministers, Copenhagen.

Pačes, T., Moldan, B., Drábek, K., and Zukriegelová, M.: 1975, *Geochemie prostfiedí modelového povodí Trnávky: Metodika a dílčí výsledky*, Archive of the Czech Geological Survey, Praha, (in Czech).

Pačes, T.: 1985, *Nature*, **315**, 31.

Pačes, T.: 1986, *J. Geol. Soc. of London*, **143,** 673.

Pačes, T.: 1991, 'Changes in Rates of Weathering and Erosion Induced by Acid Emissions and Agriculture in Central Europe', in F.M. Brower, A.J. Thomas and M.J. Chadwick (eds), *Land Use Changes in Europe*, Kluwer Academic Publishers, Dordrecht, pp. 317-323.

Regulation: 1994, *Regulation of Ministry of Environment for Agricultural Soil Protection, Act No. 13/1994*, Ministry of Environment of the Czech Republic, Prague.

Vrubel, J., Pačes, T., Čejková, G., Kantor, Č., Majer V., Sáňka, M., Trojáček, P. and Zapletal, M.: 1996, *Critical Loads of Heavy Metals for Soils in Czech Republic - Draft.* Ekotoxa, Opava and Czech Geological Survey, Prague.

Wolf, B.: 1982, *Comm. Soil Sci. Plant Anal.* **13**, 1005.

Zbíral, J.: 1995, *Analýza půd I, jednotné pracovní postupy*, State Institute for Agriculture Supervision and Testing, Brno (in Czech).

QUANTIFYING FINE-SCALE VARIABILITY IN POLLUTANT DEPOSITION IN COMPLEX TERRAIN USING ^{210}Pb INVENTORIES IN SOIL

D. FOWLER[1], R.I. SMITH[1], I.D. LEITH[1], A. CROSSLEY[1], R.W. MOURNE[2], D.W. BRANFORD[3] and M. MOGHADDAM[3]

[1]*Institute of Terrestrial Ecology, Bush Estate, Penicuik, Midlothian EH26 0QB, United Kingdom*
[2]*Faculty of the Built Environment, University of the West of England, Bristol, United Kingdom*
[3]*Department of Physics, University of Edinburgh, Edinburgh, United Kingdom*

Abstract. The accumulation of ^{210}Pb in organic material within the surface (0 to 20 cm depth) horizons of soil is used to quantify local variability in the atmospheric inputs through wet deposition, cloud droplet deposition and dry deposition of aerosols. The method has been applied to quantify the long-term (~50 yr) average enhancement in deposition as a consequence of orographic effects on a 800 m mountain in southwest Scotland. The ^{210}Pb inventory increases by a factor of 2.5 up the hillslope and is comparable to the modeled increase in wet deposition of major ions, and larger than the increase in rainfall with altitude by a factor of two. A second study site examined the increase in deposition beneath a Norway spruce canopy relative to open grassland at an elevation of 450 m in the Scottish Borders. The inventory of ^{210}Pb under the forest canopy exceeded that in the grassland by approximately 35 %, in good agreement with deposition estimates obtained from a continuous record of cloud frequency and meteorological variables.

Keywords: aerosol deposition, atmospheric deposition, orographic effects, ^{210}Pb, pollutant deposition, Scotland

1. Introduction

The inputs of pollutants to terrestrial surfaces have become the focus of interest in a range of both local and regional environmental issues in the last three decades. In the case of acid deposition, some of the most acid sensitive ecosystems occur in regions remote from major sources of the primary pollutants (SO_2, NO_x and NH_3) and in complex terrain (Grennfelt *et al.*, 1995). Similarly, the recent interest in deposited N and its effects on ecosystem productivity and species composition (Pitcairn *et al.*, 1995) extends over regional scales, and the ecosystems that have developed on very nutritionally poor soils are among those most sensitive to atmospheric inputs of N (Heil and Diemont, 1983). The pollutant deposition maps for Europe and North America generally are provided by low density monitoring networks of precipitation chemistry collectors and estimates of dry deposition largely from models. Thus, the quality of deposition data is often a major limitation in interpreting the cause of observed changes in the vegetation or soils.

The problems of quantifying deposition in complex terrain are exaggerated by orographic and airflow effects, which create large local variability in deposition (Choularton *et al.*, 1988). Some of the orographic effects such as seeder-feeder scavenging are well understood and can be simulated using airflow and precipitation scavenging models (Inglis *et al.*, 1995). However even here, the simple ridge or isolated mountain is much easier to simulate than a mountainous region, and in the latter the inputs are difficult to simulate with models and even harder to validate with field measurements.

The scale of deposition mapping ranges from the coarse output of long-range transport models such as the EMEP model, which has a grid size of 150 km × 150 km and was

Water, Air, and Soil Pollution **105:** 459–470, 1998.

designed to quantify the transboundary exchange of pollutants, to much finer resolution models that are nested within long-range transport schemes and provide estimates of deposition at 5 km × 5 km scales (Erisman *et al.*, 1993). In field measurement programs, site-specific data also are used to overcome the shortcomings of national or regional maps, but even here the measurements are subject to large uncertainty. In the case of forests for example, inputs of pollutants as aerosols, gases or cloud droplets are difficult to measure directly except by micrometeorological methods.

In this paper, we report measurements of the effects of land use and orography on the spatial variability in long-term average deposition of aerosols, and cloud and wet deposition using the naturally occurring isotope ^{210}Pb. The measurements are compared with direct measurements at upland sites in the United Kingdom and current national maps. The effects of forests on the long-term inputs also are measured and compared with models and other measurement methods.

2. Methods

2.1. The ^{210}Pb Approach

Several of the isotopes in the decay series from ^{238}U to stable ^{206}Pb have been applied in environmental research. The intermediate isotopes include gaseous ^{222}Rn, which is emitted from soil and has been used widely to estimate rates of soil-air transfer (e.g., Dorr and Munnich, 1990). The ^{210}Pb isotope formed by radioactive decay of ^{222}Rn through a series of short-lived intermediates has been widely applied as a tracer of atmospheric transfer (Graustein and Turekian, 1996). More commonly, the ^{210}Pb isotope is used to provide a date record, for example in lake sediments (Appleby and Oldfield, 1978). The application of the ^{210}Pb isotope in the study reported here does not require stratigraphic dating, as the total inventory (less the supported fraction) is used as a measure of total aerosol deposition (dry plus wet and occult deposition). The primary objective of the ^{210}Pb inventory method used here is to quantify the effects of orography and/or land use on long-term (> 10 year) inputs of pollutants in precipitation, cloud droplets or as aerosol. The method is not applicable for the dry deposition of gases. The underlying processes that make ^{210}Pb appropriate as a surrogate for the other deposited (and much more mobile) species is that the atmospheric ^{210}Pb essentially labels the atmospheric aerosols containing the other ions (notably SO_4^{2-}, NO_3^-, and NH_4^+), and when deposited onto organic soils, ^{210}Pb is readily sequestered by organic matter and accumulates in the surface organic horizons of the soil.

The origin of ^{210}Pb in the atmosphere is through radioactive decay from ^{238}U. The parent ^{238}U in soil decays through a sequence of relatively long-lived daughters to ^{226}Ra. The radioactive decay of ^{226}Ra creates gaseous ^{222}Rn, which readily diffuses from soil into the atmosphere. The ^{222}Rn,with a half life of 3.8 da, decays through a series of short lived Po daughters (Figure 1), initially to ^{210}Pb, which has a half life of 22.3 yr. The radioactive decay of gaseous ^{222}Rn creates particulate ^{210}Pb, which rapidly becomes attached to aerosols in the atmosphere. The aerosols 'tagged' in this way with ^{210}Pb include the common pollutants $(NH_4)_2SO_4$, NH_4HSO_4, NH_4NO_3 as well as other, mainly submicron-sized aerosols. These aerosols are readily activated into cloud droplets and removed from the atmosphere primarily by precipitation scavenging (Hales *et al.*, 1978). The same aerosols once incorporated in cloud or fog droplets also are efficiently captured and deposited onto vegetation. The deposition of cloud and fog droplets commonly and misleadingly referred to as occult deposition is not routinely measured in monitoring networks, but may contribute a significant proportion of the annual input for upland forests (Fowler *et al.*, 1989).

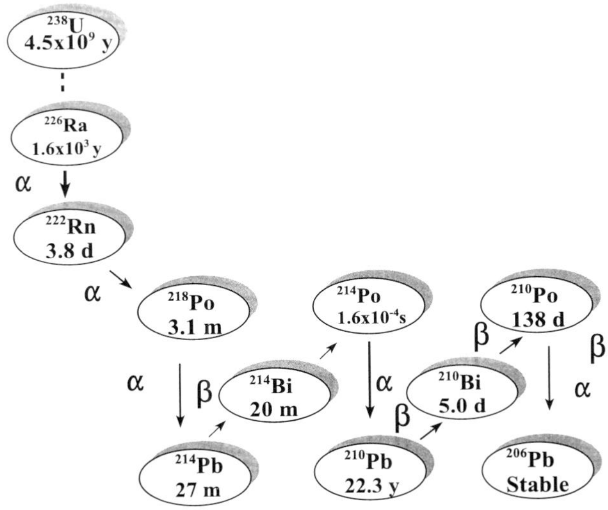

Fig. 1. The radioactive decay series leading to ^{222}Rn and ^{210}Pb.

The ^{210}Pb following deposition to terrestrial surfaces becomes readily attached to organic matter in soils, primarily in the surface layer as shown by Lewis (1972). In the absence of disturbance of the soil by cultivation or other physical disruption, the ^{210}Pb inventory in soil eventually reaches an equilibrium between the atmospheric input and radioactive decay. The only source of ^{210}Pb in addition to that from the atmosphere is the direct supply by radioactive decay from the parent ^{226}Ra within the soil without ever leaving the surface; this ^{210}Pb is known as the supported fraction. This quantity may be deduced directly by measuring the ^{226}Ra and ^{214}Pb concentration profiles in the same soil samples as the ^{210}Pb.

When a steady state between atmospheric supply and radioactive decay exists, the flux of ^{210}Pb from the atmosphere (Bg m^{-2} yr^{-1}) may be obtained as the product of the decay constant (λ; yr^{-1}) for ^{210}Pb and the inventory of ^{210}Pb in the soil profile (I; Bg m^{-2}), corrected for the supported fraction:

$$^{210}\text{Pb flux} = \lambda \times I$$

The method has been developed and applied by Graustein and Turekian (1983, 1989) to quantify the deposition of atmospheric aerosols to terrestrial surfaces. More recently, in a related study to the one reported here, Fowler *et al.* (1995) showed that this method quantitatively reproduced the increase in wet deposition observed at Great Dun Fell in northern England.

The absolute value for the ^{210}Pb inventory is influenced by larger-scale processes and by the sources of air scavenged by precipitation, as continental air is richer in ^{210}Pb than oceanic air. However, at a local scale, and to some extent over small regions (100 km × 100 km), the long-term average ^{210}Pb concentration in the atmosphere may be considered reasonably constant. Thus, variability in the soil inventory of ^{210}Pb provides a direct measure of the local variability in deposition of aerosols (by wet, dry and cloud deposition).

2.2. Sites and Collection Methods

The data reported here were obtained from two sites, the Merrick, a small mountain rising to 844 m (above sea level) from low ground approximately 7 km southwest at the summit at an elevation of 150 m. The mountain, on the southwest coast of Scotland is the first high ground on mainland Scotland encountered by westerly airflow from the Atlantic. The long-term average rainfall at the coast is approximately 1,000 mm, while at the base of the Merrick it is approximately 1,700 mm and at the summit 2,200 mm. Six sampling sites were used between the summit and an elevation of approximately 200 m asl. At each of the sampling sites, four replicate cores, each 7 cm in diameter and 30 cm deep, were taken from the hillside. The cores were subdivided on site to 0-1 cm, 1-5 cm, 5-10 cm, 10-15 cm, 15-20 cm and 20-30 cm horizons for analysis.

The second site, at Dunslair Heights is located in the Scottish Borders, 5 km north of Peebles at an altitude of 450 m. The measurements were made over a short transect (~100 m) across a boundary between open, grass-covered moorland and a conifer forest (*Picea abies*) with an age of approximately 50 years. At this site, four sampling locations were used, at each of which five replicate soil cores were taken to a depth of 20 cm using a 7 cm diameter corer.

Soil and vegetation samples were weighed, then dried at 80 °C for 24 hr. Samples were re-weighed then ground following removal of stones larger than 2 mm. The ground samples were then placed in standard holders for the detection of gamma-rays from the ^{210}Pb. The counting periods for data from the 46 keV gamma line were generally between 1 and 2 days. Calibration of the system to obtain absolute activities of each sample was achieved using standards prepared by the Institute of Terrestrial Ecology at Merlewood, UK. Correction for gamma ray self absorption was made using the method described by Cutshall *et al.* (1983). The supported fraction of ^{210}Pb from within soil (rather than atmospheric) sources was estimated from the 351.9 keV gamma line of ^{214}Pb, assuming the ^{210}Pb to be in an equilibrium with ^{214}Pb. Only samples from deeper soil horizons (i.e. below 20 cm) generally contained significant supported ^{210}Pb.

In addition to the ^{210}Pb measurements, we report on a set of direct measurements of cloud composition and meteorological measurements from Dunslair Heights. The site and methods are described by Crossley *et al.* (1993) and include continuous monitoring of cloud presence, samples of which are collected using a passive collector. The site also provides a continuous record of cloud, liquid water content, wind speed, precipitation rate, and air temperature. The cloud water samples are bulked for weekly analysis of all major anions and cations in solution. These data have been used in this study to calculate the cloud deposition rate onto the forest 150 m below and to the west of the cloud monitoring station.

3. Results and Discussion

3.1. Effects of Orography on ^{210}Pb Deposition

The profile of ^{210}Pb deposition at the Merrick is shown in Figure 2, along with the location of the sampling sites. The ^{210}Pb inventory increases from approximately 1,300 Bq m^{-2} on the lower slopes of the mountain to values close to 4000 Bq m^{-2} close to the summit, although considerable variability in the data is evident from the error bars. The uncertainty on Figure 2 is due mainly to within-site variability in the ^{210}Pb inventory between the four replicate soil samples at each site. The contribution to these uncertainties from gamma counting consistently varies between 7 % at the top of the profile to 17 % at the bottom. The cause of within-site variability is not known, but each of the four replicate samples was obtained within an overall site radius of 5 m. It is probable that microtopographic features and vegetation, which influence the drainage properties of the site, and the fraction of precipitation that drains vertically into the soil are the primary cause. No significant amount of ^{210}Pb was found below 10 cm at any of the sampling sites (Figure 3); the bulk of the ^{210}Pb is concentrated in the surface organic horizon.

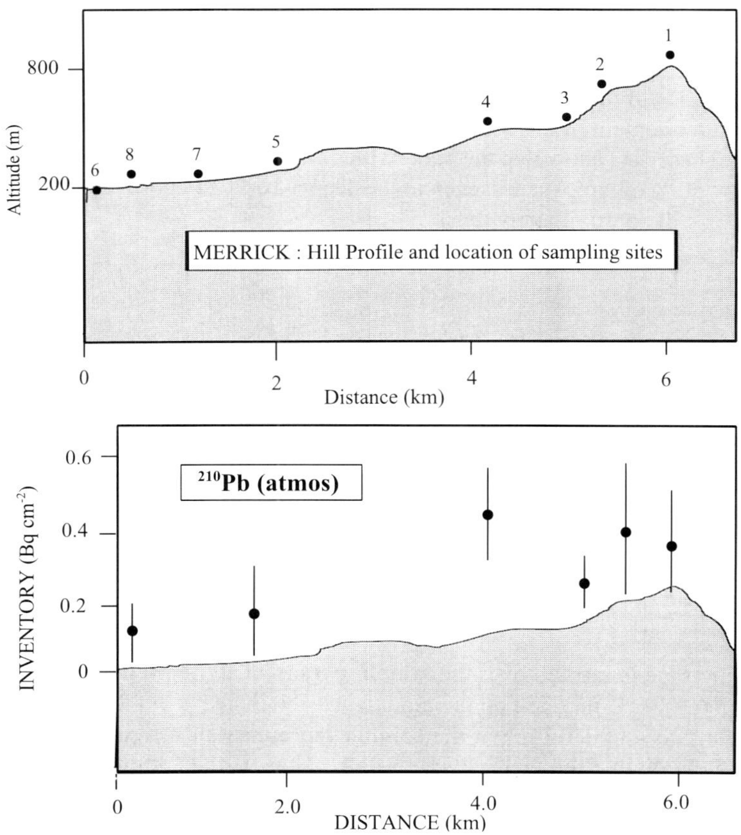

Fig. 2a. The sampling positions for ^{210}Pb inventories and altitudes on the western slopes of the Merrick.

Fig. 2b. The ^{210}Pb inventories and associated uncertainties (indicated as the range within the four replicate samples) in soil along a height transect of the western slopes of the Merrick.

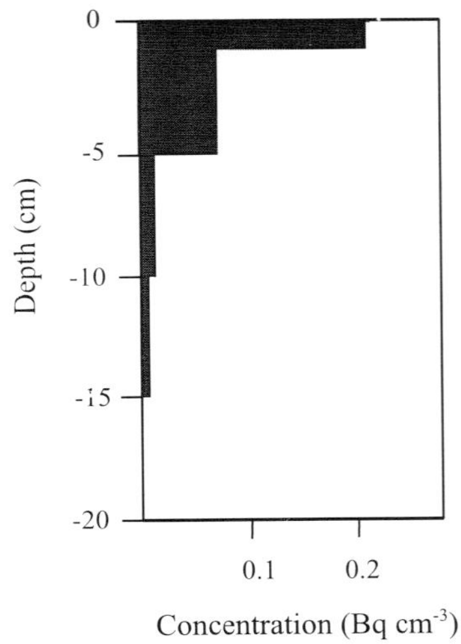

Fig. 3. Depth distribution of ^{210}Pb in soil at the Merrick site (site at 200 m).

Subsamples of the cores were subjected to loss on ignition analyses to quantify the organic matter content of the different horizons of the soil profile at sites on the Merrick (Table I). These data show that the bulk of the ^{210}Pb is contained in the horizon consisting of typically 50 % organic matter; even in the deeper 10-20 cm horizon, the organic matter content was still approximately 20%.

TABLE I
Loss of ignition; samples from the Merrick site (site at 200 m).

Depth interval	Loss on ignition (%)
0-1 cm	83
1-5 cm	26
5-7.5 cm	12
7.5-10 cm	11
10-15 cm	10

The increase in rainfall over the sampling transect at the Merrick is from 1,700 mm to 2,000 mm, a 30 % increase up the mountain. The increase in the ^{210}Pb inventory from 1,300 Bq m^{-2} to 4,000 Bq m^{-2} at the summit represents an increase by a factor of 2.5, approximately double the increase in rainfall. Thus, if the additional deposition at the higher elevation results primarily from seeder-feeder scavenging of the orographic cloud, as has been shown to explain the larger precipitation amounts *and* larger concentrations in precipitation at Great Dun Fell and elsewhere (Fowler *et al.*, 1987; Dore *et al.*, 1989).

It is possible to deduce the concentration in the feeder cloud required to account for the enhanced deposition at higher elevations. In this case, the concentration of ^{210}Pb in the feeder rain must be approximately a factor of two larger than that in the seeder rain. This finding is entirely consistent with the orographic enhancement of SO_4^{2-}, NO_3^- and NH_4^+ deposition at Great Dun Fell in the studies by Choularton *et al.* (1987).

The data are summarized in diagrammatic form in Figure 4, which shows the change in precipitation and ^{210}Pb inventory with altitude on the windward (western) slopes of the Merrick. Also shown is the increase in S deposition predicted from the simple model of orographic enhancement in wet deposition that currently is used for the UK wet deposition mapping. In the country-wide mapping, the coastal rainfall is used to characterize the seeder precipitation amount. The larger values inland are assumed to arise through seeder-feeder processes, as has been demonstrated at several sites in the uplands of western Britain. The predicted increase in wet deposition of SO_4^{2-}, based on studies from similar topography, is very similar to the observed increase in the ^{210}Pb inventory at the Merrick. It is difficult in the absence of more detailed studies to attempt to interpret the cause of the detailed variability in the ^{210}Pb profile at the Merrick other than the general upward trend, and the uncertainty in the data obscure any real differences between inventories at individual sampling locations. However, it is probable that there are local variations caused by smaller-scale topographic features on the hillside. The source of the additional ^{210}Pb at the summit must contain contributions from cloud droplet and aerosol deposition. However, the aerodynamic roughness length of the surface is likely to be no larger than 0.05 m to 0.10 m given that the surface was short grass (height $<$ 30 cm) and the surface was uniform. Taking a cloud frequency of 30 % at the summit, a mean wind velocity of 5 m sec^{-1}, and a cloud liquid water content of 100 mg m^{-3}, it may be shown that cloud droplet deposition would contribute only an additional 50 mm precipitation equivalent at the summit, 2 % of wet deposition, and that aerosol deposition is unlikely to exceed 5 % of the annual input flux of ^{210}Pb assuming a deposition velocity of 0.5 mm sec^{-1} and an air concentration of 0.01 Bq m^{-3}. It appears from Figure 4 that the majority of the orographic enhancement takes place at altitudes in excess of 300 m, although from only six sampling sites it is difficult to be more precise.

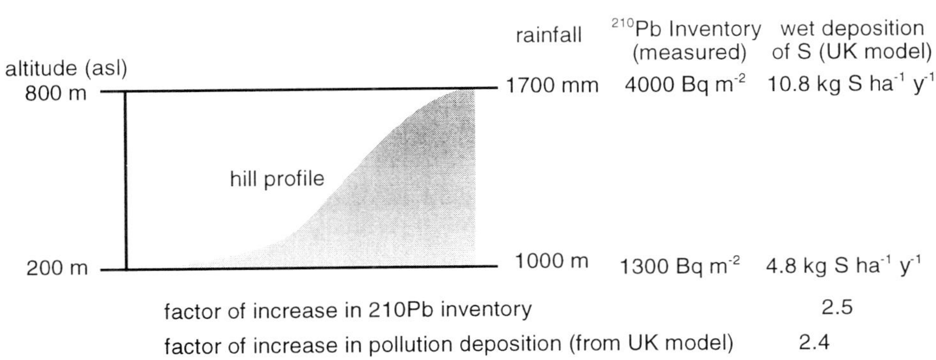

Fig. 4. Composition of measured and modeled deposition at the Merrick.

3.2. Effects of Forests on Deposition of Cloudwater and Aerosols

The four sampling locations provide estimates of the ^{210}Pb inventory in the grassland upwind of the forest edge in a westerly wind, which is the prevailing direction. The depth profile of ^{210}Pb in the soil samples from Dunslair Heights (Figure 5), like that at the Merrick (Figure 3), shows that the ^{210}Pb is concentrated in the surface horizons and that the contribution from horizons below 10 cm is small. Furthermore, the very different litter layer beneath the dense spruce canopy does not greatly modify the soil ^{210}Pb concentration profile (Figure 5) with more than 70 % of the ^{210}Pb inventory in the top 10 cm. The profiles of ^{210}Pb with depth are subject to errors in gamma counting (± 3 % at the surface; ± 8 % for the lower sections of the profile) and errors related to field sampling (± 10 %, mainly due to uncertainty in the area and depth measurements in field soils). Thus, the overall uncertainty increases from ± 10 % at the surface to ± 13 % for the lower sections of the profile. It appears, therefore, that the ^{210}Pb is effectively immobilized in the surface 10 cm. The three points of the transect inside the canopy show substantially larger inventories, respectively 34 %, 40 % and 38 % larger than in the open moorland. The precipitation inputs may be assumed to be the same for all sites (Figure 6). The increase may therefore be assumed to result from cloud droplet and aerosol capture by the trees in excess of that by the grassland.

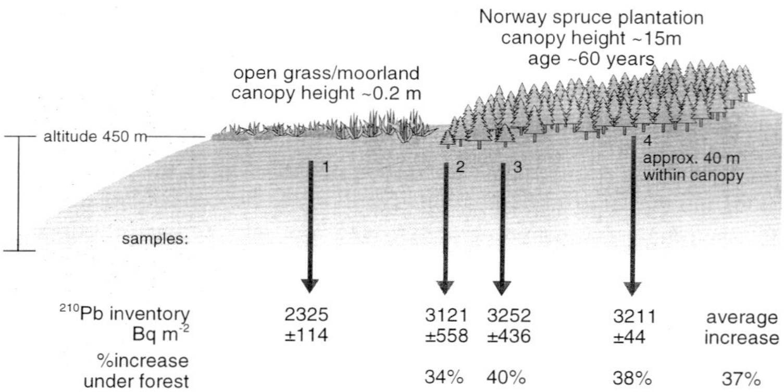

Fig. 5. Vertical profile of ^{210}Pb in soil at Dunslair Heights under grassland and forest at 450 m (asl).

The amount of the enhancement in ^{210}Pb also may be compared with independent estimates of cloud droplet and aerosol deposition. The summit of Dunslair Heights is equipped with continuous monitoring of cloud, wet deposition and standard meteorological variables (Crossley *et al.*, 1993). These data provide a direct estimate of cloud frequency and liquid water content which with wind velocity and roughness length for the forest and grassland may be used to calculate cloud droplet deposition. The procedure assumes first that rates of cloud droplet deposition may be approximated by those for momentum (Gallagher *et al.*, 1992) and uses the wind profile equation:

$$U_{(z)} = \frac{U_*}{k} \ln(z - d / zo)$$

in which $U_{(z)}$ is wind speed at reference height (z), k is von Karman's constant, and d is the zero plane displacement, assumed to be 63 % of canopy height (Monteith and Unsworth, 1987), from which U_* may be used to calculate the deposition velocity for cloud droplets $V_{d(z)}$ as $V_{d(z)} = U_*^2/U_{(z)}$ at reference height (z) above the canopy. The only additional requirement to estimate directly the cloud droplet deposition on the canopy is the enhancement in concentration of ^{210}Pb in hill cloud over that in rain. This has been estimated to be a factor of 2.2 larger than that in precipitation by Mourne (1993) from the measurements at Great Dun Fell. Table II summarizes these data in which cloud droplet deposition is calculated for both forest and moorland. The results show that the forest canopy receives 37 % more ^{210}Pb than the open moorland from cloud droplet deposition.

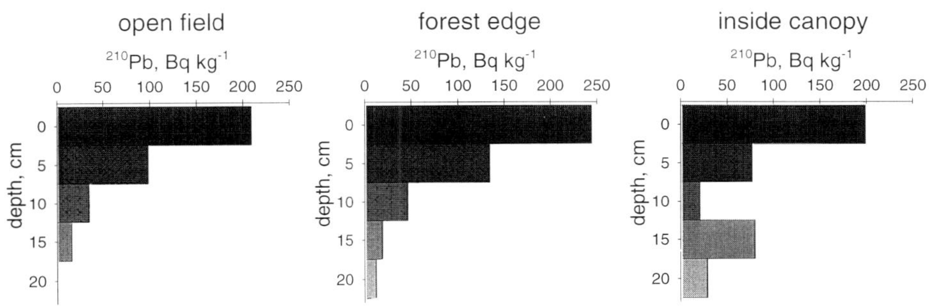

Fig. 6. Dunslair Heights ^{210}Pb inventories under grass and forest.

TABLE II
Cloud deposition at Dunslair Heights from direct measurements, altitude 600 m - measured cloud frequency, 1,257 hr yr^{-1} (14 %); inter-annual variability ± 20 % (Crossley *et al.*, 1988).

	Forest	Moorland
Canopy height	15 m	0.2 m
Measured precipitation	1,213 mm	1,213 mm
Wind speed at canopy height	5 m sec^{-1}	5 m sec^{-1}
Cloud deposition	181 mm	14 mm
Enhancement of ^{210}Pb in cloud: 2.2 (Mourne, 1993)		
Cloud deposition of ^{210}Pb expressed in mm precipitation equivalents	398 mm	31 mm
Total deposition	1,611 mm	1,244 mm

3.3. Cloud Deposition at Dunslair Heights

The measurement of ^{210}Pb inventories in open grassland and under the conifer forest at Dunslair Heights show the magnitude of the enhancement in deposition by the forest. The increase in the ^{210}Pb inventory of 37 % is entirely consistent with the direct estimate of cloud deposition from the cloud monitoring data. However, in the case of the directly measured ^{210}Pb, aerosol deposition is included. The monitoring station at Dunslair Heights does not monitor ^{210}Pb in aerosol and cloud, and the concentrations in air cannot be estimated with sufficient precision to justify speculating on the magnitude of this term. It seems probable, therefore, that the close match between forest enhancement deposition quantified by direct cloud measurements and by the ^{210}Pb inventory is fortuitous. The underlying cause of the close agreement is most likely that the cloud frequency at Dunslair Heights summit (602 m asl) overestimates the actual exposure of the forest at 450 m asl. This would lead to cloud droplet deposition on the forest being overestimated from the cloud data; the aerosol deposition component is of a similar magnitude to the overestimate in cloud deposition. These data, however, provide considerable support for the procedures that are used to calculate land-use-specific inputs of pollutants by cloud droplet deposition over regional scales (CLAG, 1997). They also demonstrate the importance of land use in input to catchments, especially in upland regions.

The UK modeling procedures may be compared directly with the measured forest enhancement using ^{210}Pb inventories and the cloud frequency measurements. The modeling is used to provide national maps at 20 km × 20 km resolution, and may be applied at a specific site using the national data including some key site characteristics (altitude and land use). The input data used include wet deposition from national pollutant concentration, wind speed and 5 km × 5 km precipitation fields and the simplified seeder-feeder scavenging scheme. For forest, a canopy height of 10 m is assumed, while for moorland 0.5 m is assumed. Cloud cover is calculated from the method of Weston (1993) and cloud droplets are assumed to deposit at the same rate as momentum (i.e., that $V_{d(z)}$ cloud droplets $\approx U_*^2/U_{(z)}$). The 20 km × 20 km grid resolution provides the average height in the grid square (in this case 341 m). By contrast, the site-specific model applies the same processes, but for the heights and meteorological conditions of the site. The results of this exercise are summarized in Table III. These yield larger forest enhancement by the site-specific method largely because the cloud frequency is larger for the actual height of the forest at (450 m) rather than the 20 km × 20 km grid square average of 341m.

These two modeling approaches may now be contrasted with the direct measurements by ^{210}Pb inventories and cloud monitoring. In the measurements, the striking agreement in the magnitude of the forest enhancement that has been discussed earlier is closest to the UK model estimate of between 35 % and 40 %. The larger effect produced by the site-specific model probably arises through uncertainty in the cloud frequency, which is estimated for the site from the UK cloud base statistics that are provided mainly by airports and are not, therefore, representative of the upland areas. Improving the accuracy of the cloud frequency estimates by including data sets from upland areas, including Dunslair Heights, Great Dun Fell and Holm Moss, would improve the estimates of cloud frequency in the UK uplands.

TABLE III

Cloud deposition at Dunslair Heights from: the UK model at 20 km × 20 km, altitude 341 m (cloud frequency from interpolated 3D surface; 505 hr yr^{-1}; Weston, 1992); and from a model specific to measurement site (cloud frequency from interpolated 3D surface; 760 hr yr^{-1}).

	From the UK model		From site-specific model	
	Forest	Moorland	Forest	Moorland
Canopy height	10 m	0.5 m	10 m	0.5 m
Wind speed at canopy height	6.3 m sec^{-1}	5.6 m sec^{-1}	5 m sec^{-1}	5 m sec^{-1}
Total S deposition	9.3 kg ha^{-1} yr^{-1}	6.7 kg ha^{-1} yr^{-1}	11.9 kg ha^{-1} yr^{-1}	7.2 kg ha^{-1} yr^{-1}
	Enhancement 39%		Enhancement 65%	

The potential of ^{210}Pb inventories to be used to validate our understanding of deposition in complex terrain has now been demonstrated in a range of environments. The increases in ^{210}Pb inventory due to orographic effects at the Merrick and to cloud droplet capture by aerodynamically rough surfaces at Dunslair Heights also is consistent with the work by Graustein and Turekian (1989). The advance reported here is that the enhancement in deposition at the forest site implied by the ^{210}Pb measurement is supported by entirely independent measurements of hill cloud frequency and the associated micrometeorology, which enabled the cloud droplet flux to be calculated directly. These comparisons support the current understanding of precipitation and cloud droplet scavenging processes. They also could be applied to study aerosol deposition in aerodynamically rough surfaces, especially forests. The question of aerosol deposition on forests remains an important issue in which there is considerable uncertainty. Wind tunnel studies support small rates of submission aerosol deposition on vegetation and forests, while field measurements suggest much larger deposition velocities, in the range 1 cm sec^{-1} for forests (Draaijers *et al.*, 1992; Erisman and Draaijers, 1995). The application of the depth profile of ^{210}Pb in the soil samples from Dunslair Heights, like that at the Merrick, shows that the ^{210}Pb is concentrated in the surface horizons; the contribution from horizons below 10 cm is small. Furthermore, the very different litter layer beneath the dense spruce canopy does not significantly modify the soil ^{210}Pb concentration profile, with more than 70 % of the ^{210}Pb inventory in the top 10 cm. It appears therefore that the ^{210}Pb is effectively immobilized in the surface 10 cm.

Acknowledgments

The authors gratefully acknowledge the support for this work from the UK Department of the Environment under contract EPG 1/3/31.

References

Appleby, P.G. and Oldfield, F.: 1978, *Catena* **5**, 1.
Choularton, T.W., Gay, M.J., Jones, A., Fowler, D., Cape, J.N. and Leith, I.D.: 1988, *Atmos. Environ.* **22**, 1363.

CLAG:1997, *Deposition Fluxes of Acidifying Compounds in the United Kingdom. Subgroup Report on Deposition Fluxes*, Department of the Environment.
Crossley, A., Wilson, D.B. and Milne, R.: 1992, *Environ. Pollut.* **75**, 81.
Cutshall, N.H., Larsen, I.L. and Olsen, C.R.: 1983, *Nuc. Instruments Meth.* **206**, 309.
Dore, A.J., Choularton, T.W., Fowler, D. and Storeton-West, R.L.: 1990, *Quart. J. Royal Meteorol. Soc.* **116**, 1193.
Dorr, H. and Munnich, K.: 1990, *Tellus* **42B**, 20.
Draaijers, G.P.J., Van Ek, R. and Meijers, R.: 1992, *Environmental Pollution* **75**, 243.
Erisman, J.W., Van Jaarsveld, H., Van Pul, A., Fowler, D., Smith, R. and Lovblad, G.: 1993, 'Comparison between Small-scale and Long-range Transport Modeling', in G, Lovblad, J.W. Erisman and D. Fowler (eds.), *Models and Methods for the Quantification of Atmospheric Input to Ecosystems*, Nordic Council of Ministers, Copenhagen, pp. 185-198.
Erisman, J.W. and Draaijers, G.P.J.: 1995, *Studies in Environmental Science*, **63**.
Fowler, D., Cape, J.N. and Unsworth, M.H.: 1989, *Philos. Trans. Royal Soc. Lond.* **B324**, 247.
Fowler, D.: 1987, 'The Transfer of Air Pollutants to the Ground by Wet and Dry Deposition', in S. Sandroni (ed.), *Regional and Long-range Transport of Air Pollution*, Elsevier, Amsterdam, pp. 95-127.
Graustein, W.C. and Turekian, K.K.: 1996, *Geophys. Res. Lett.* **23**, 539.
Grennfelt, P., Rodhe, H., Thörnelöf, E. and Wisniewski, J. (eds.): 1995, *Acid Reign '95?. Proceedings from the 5th International Conference on Acidic Deposition, Göteborg, Sweden*, Kluwer Academic Publishers, Dordrecht.
Hales, J.M.: 1978, 'Wet removal of Sulphur Compounds from the Atmosphere', in R.B. Husar, J.P. Lodge and D.J. Moore (eds.), *Sulphur in the Atmosphere*, Pergamon Press, pp. 389-401.
Heil, G.W. and Diemont, W.M.: 1983, *Vegetatio* **53**, 113.
Inglis, D.W.F., Choularton, T.W., Wicks, A.J., Fowler, D., Leith, I.D., Werkman, B. and Binnie, J.: 1995, *Water, Air, Soil Pollut.* **85**, 2119.
Lewis, D.M.: 1977, *Geochim. Cosmochim. Acta* **41**, 1557.
Monteith, J.L. and Unsworth, M.H.: 1990, *Principles of Environmental Physics*, Edward Arnold.
Mourne, R.W.: 1993, *Lead-210 as a Tracer for Acidic Deposition in Areas of Complex Topography*, Ph.D. Thesis, Edinburgh University, unpublished.
Pitcairn, C.E.R., Fowler, D. and Grace, J.: 1995, *Environ. Pollut.* **88**, 193.
Turekian, K.K., Benninger, L.K. and Dion, E.P. 1983. *J. Geophys. Res.* **88(C9)**, 5411.
Turekian, K.K., Graustein, W.C. and Cochran, J.K.: 1989, *Chem. Oceanogr.* **10**, 51.
Weston, K.J.: 1992, *Meteor. Mag.* **121**, 108.

FLUX OF BENZO(a)PYRENE TO THE GROUND SURFACE AND ITS DISTRIBUTION IN THE ECOSYSTEM

A. MILUKAITĖ

Institute of Physics, A.Goštauto 12, 2600 Vilnius, Lithuania

Abstract. Benzo(a)pyrene (BP) has been investigated in bulk atmospheric deposition, moss, needles of pine and some species of vascular plants. At two remote Lithuanian sites, for 1990-1995 the flux of benzo(a)pyrene from the atmosphere to the ground surface varied between 0.3 to 4.8 μg^{-2} mo^{-1}. Consequently the territory of Lithuania (65,000 km^2) yearly was exposed to 624-2574 kg of carcinogen. The distribution of BP in soil and various vascular plant tissues (*Trifolium tepens, Elitrygea repens, Thymus serpyllum)* indicates that benzo(a)pyrene is assimilated by flora. The concentration of BP is different in various organs of vascular plants and mostly depends on the degree of soil pollution. More than 300 samples of moss, mostly *Hylocomium splendens* and *Pleurozium schreberi* were analyzed for BP. From 3.1 to 896.0 $\mu g\ kg^{-1}$ of BP were measured in the moss samples. The flux of BP to the ground surface correlates well with its concentration in moss. A map of BP flux across Lithuania was created.

Keywords: benzo(a)pyrene deposition, Lithuania, monitoring, moss chemistry, plant tissue chemistry

1. Introduction

Under the influence of polycyclic aromatic hydrocarbon (PAH) fluxes from the atmosphere to the ground surface, various objects of the biosphere are exposed to mutagenic and carcinogenic effects (Griciūtė, 1979, Lee *et al.*, 1981). In comparison with investigations of PAH concentrations in atmospheric air, the fluxes of PAHs to the ground surface and the mechanism of PAH uptake by vegetation have received little study. The first investigations on PAH fluxes were focused on the input of aerosol organic substances to water basins and their incorporation into sediments (Youngblood and Blumer, 1975; Laflamme and Hites, 1978). The background flux of benzo(a)pyrene (BP) at remote sites in the Great Lakes region of North America in 1978-1979 was between 0.2 and 2.0 ng cm^{-2} yr^{-1} (Gschwend and Hites, 1981). Benzo(a)pyrene flux in bulk precipitation in Germany during 1980-1981 ranged from 22 mg ha^{-1} yr^{-1} in the Harz mountains to 86 mg ha^{-1} yr^{-1} at remote sites in the northern part of the country (Matzner, 1984). The flux of 11 PAHs in the Swedish background station at Rorvik for several periods in 1988 was 0.2-0.4 mg m^{-2} yr^{-1}, while for several periods in 1990, the flux of these 11 PAH's compounds ranged from 0.2 to 2.0 μg m^{-2} da^{-1} (Brorstrom-Lunden *et al.* 1991, 1994). At this station, the flux of benzo(a)pyrene in 1990 ranged from 0.005 to 0.070 μg m^{-2} da^{-1} (Brorstrom Lunden *et al.*, 1994). The monthly flux of benzo(a)pyrene was investigated at 24 remote and suburban locations of Lithuania from 1973 to 1977 (Milukaitė and Galvonaitė, 1991). Benzo(a)pyrene deposition to the ground surface ranged from 0.4 to 5.1 μg m^{-2} mo^{-1} (median of 1.1 μg m^{-2} mo^{-1}) and was influenced by distance from the sources and by meteorological conditions. About 860 kg yr^{-1} of benzo(a)pyrene was deposited to the surface across all of Lithuania over this period.

It is thought that PAHs, like many insoluble organic substances, have a minor influence on vegetation after deposition to the soil. However, traces of carcinogenic PAHs have been detected in vegetables (Grimmer, 1979; Yanysheva *et al.*, 1985). High concentrations of

Water, Air, and Soil Pollution **105:** 471–480, 1998.

benzo(a)pyrene, as well as of metals and other particulate pollutants, have been quantified in mosses. Thirteen organic and inorganic substances were quantified in the epiphytic moss *Hypnum cupressiforme* in Germany along a profile from the Baltic Sea to the Alps; average concentration of benzo(a)pyrene in moss from 37 sites ranged from 20 to 170 $\mu g\ kg^{-1}$ (Thomas, 1980). Later a model for the accumulation of some PAHs in *Hypnum cupressiforme* was developed and the use of moss for the environmental monitoring was confirmed (Thomas, 1984). Some investigations have been carried out on using leaves and conifer needles for the monitoring of some PAHs in the environment (Matzner, 1984; Milukaitė, 1986). However, systematic observations of PAH's (partly benzo(a)pyrene) fate and influence on ecosystems are absent.

The main objective of this paper is to show the pathway of benzo(a)pyrene migration from bulk deposition to soil and vegetation, with special emphasis on the forest ecosystem. Each PAH is different in its physical and chemical properties, and consequently, the fate and influence of each PAH on various components of the ecosystem may be different. We chose only benzo(a)pyrene, which in the environment mainly exist as particulates, and is the strongest carcinogen in PAH group (McVeety and Hites, 1988).

2. Methods

2.1. Sampling

To quantify BP flux from the atmosphere to the ground surface, bulk precipitation (wet and dry fallout) was collected at two sites in Lithuania. The first sampling site was located on the Curonian spit in the Baltic Sea (background environmental research station Preila) and the second site was in a suburb of Vilnius (at the Institute of Physics). Bulk precipitation was collected in polyethylene funnels with an area of 0.025 m^2 placed 2 m above the ground surface. At the first site, bulk precipitation was collected for a duration of 4-6 weeks from 1990-1995. At the second site, samples were collected monthly from 1993-1995.

More than 300 moss samples were gathered at wooded sites within a 16 x 16 km network across Lithuania from 1993-1995. The most common species among samples were *Hylocomium splendens* and *Pleurozium schreberi*. Retention of benzo(a)pyrene by these moss species was determined to be very similar (Milukaitė, 1996).

Samples of savory (*Thymus serpyllum*) were collected in the woods of the southern part of Lithuania. White clover (*Trifolium repens*), couch-grass (*Elitrygea repens*) and bent-grass (*Agrostis vulgaris*) were gathered in meadows near roads. Pine bark, wood and needles were collected from s single tree (*Pinus silvestris*) growing at 20 m from highway Vilnius-Kaunas. More than 20 separate samples of needles were gathered at various distances from the same highway.

Soil samples were gathered from the upper layer of ground (0-5 cm in depth) mostly at the same sites that were used for vegetation sampling. For the BP studies in the soil horizons, samples were gathered at 20 m from highway Vilnius-Kaunas.

2.2. Chemical and Instrumental Analysis of BP

Bulk precipitation was separated into particulate and liquid phases by filtration through a "Filtrak" filter for finest deposits; BP was determined in both phases. Filters with solid residue were dried under room conditions (20 °C). Twenty-five mL of n-hexane were added and left for 12-15 hr to soak. Extraction of BP in the solid-liquid system was carried out by placing a vessel in a vibrating apparatus at 8Hz for 1 hr. After evaporation to 5 mL, the extract was fractionated on a thin layer of Al_2O_3 with a solvent mixture of hexane:diethyl ether in the ratio 10:1. The BP fraction was eluted by 50 mL of n-hexane and evaporated to 5 mL. Quantitative analysis of BP was performed by the spectro-fluorescence method in liquid nitrogen at 77 °K. Excitation of BP was performed at λ = 298 nm and fluorescence was recorded at λ = 403 nm. The fluorescence analysis was performed by a diffraction spectrometer (DFS-12 "LOMO"). The sensitivity of the method is 0.01 ng mL^{-1} of chromatographic eluate (Milukaitė, 1986). The bulk precipitation filtrate was divided into portions of 0.5 L or less and extracted separately three times with 20 mL of n-hexane. The extracts were pooled, and after evaporation to 1 mL were analyzed as described above.

Five g of dried (at room temperature, 20 °C), homogenized and ground moss, other vegetation, or soil was mixed with 25 mL of n-hexane. After extraction, the analysis was analogous to that of the solid phase of the bulk precipitation.

3. Results and Discussion

The monthly flux of benzo(a)pyrene at the background station (Preila) ranged from 0.3 to 3.8 µg m^{-2} mo^{-1} over the past six years (Table I). The flux of BP was higher at a suburban site in the eastern part of Lithuania, where it varied from 0.6 to 4.8 µg m^{-2} mo^{-1}. The average annual flux of BP generally increased from year to year, except for 1993, when it was lowest at both sites. Across all of Lithuania, the total annual flux of benzo(a)pyrene ranged from 624 kg in 1993 to 2574 kg in 1995. The general sources of BP are associated with fuel combustion, and the use of organic fuel was calculated to be at a minimum in 1993 (Žukauskas *et al.*, 1997). In 1995, the use of oil products increased and the increasing influence of traffic has been noticeable in Lithuania recently.

Benzo(a)pyrene that is deposited to the ground surface is adsorbed by soil particles. Readman *et al.* (1984) determined that BP bound to lake sediments is inert to leaching, microbiological degradation or photodegradation. To assess the fate of BP in soil, the concentration of BP was determined in six horizons of a loam soil (Figure 1). The loam soil was chosen because about 50 % of Lithuanian soils are loams. Most of the BP deposited to the ground surface is accumulated in the upper layer (0-5 cm) of loam soil, where its concentration was 10 times higher than in deeper layers (10-20 cm). This depth distribution confirms that the adsorption of BP by surface soil particles is quite strong, whereas the BP washout rate from the soil is rather slow.

To study BP uptake from the soil by vegetation, the concentration of BP was quantfied in various organs of plants mostly distributed in meadow and forest biocenoses in Lithuania (Table II). The soil samples were collected from the same sites as the vegetation samples. The highest concentration of BP was observed in the well-developed roots of *Agrostis*

TABLE I

Flux (F) of benzo(a)pyrene with bulk deposition at remote sites of Lithuania ($\mu g\ m^{-2}\ mo^{-1}$).

Year	Preila (background station)				Suburb of Vilnius			
	F_{min}	F_{max}	$F_{avg} \pm SD$	Load ($kg\ yr^{-1}$)	F_{min}	F_{max}	$F_{avg} \pm S\ D$	Load ($kg\ yr^{-1}$)
1990	0.3	2.5	1.0 ± 0.7	780				
1991	05	3.6	1.7 ± 1.0					
1992	0.4	3.8	1.9 ± 1.1	1482				
1993	0.4	1.6	0.8 ± 0.5	624	0.6	2.5	1.2 ± 0.6	936
1994	1.0	3.0	2.1 ± 0.8	1638	1.4	3.8	2.2 ± 0.9	1716
1995	0.5	2.8	2.5 ± 0.4	1950	1.7	4.8	3.3 ± 1.1	2574

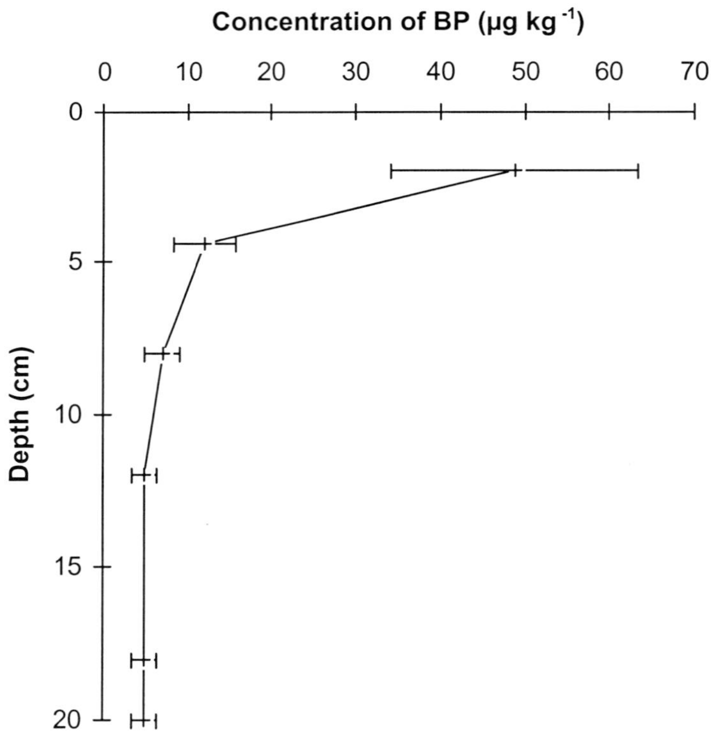

Fig. 1. Concentrations of BP in various horizons of a loam soil (n=4).

vulgaris and *Elitrygea repens*, whereas the concentration of BP in the leaves of these plants was low. The partition coefficient of BP (K_p) among the roots and soil is 10-25 times higher than that of leaves and soil, suggesting that the main pathway of BP uptake by these plants is through the roots. By contrast, the concentration of BP in *Thymus serpyllum* flowers, leaves and roots was very similar. Flowers of savory have relatively high concentrations of etheric oils (Purvinas and Skirgailaitė, 1975), which may influence the solubility of deposited BP and the capture of particulate BP from the atmosphere. Thus, *Thymus serpyllum* may take up BP both by roots and directly from the atmosphere.

Concentrations of BP in *Trifolium repens* and *Elitrygea repens* leaves as a function of soil pollution are presented in Figure 2. High concentrations of BP were observed in the leaves of *Trifolium repens*. Plants of different species may assimilate BP from the soil in different efficiencies. Leaves of *Trifolium repens* might take up considerable BP when soil BP concentrations are greater than 30 μg kg^{-1}, while at these concentrations *Elitrygea repens* might assimilate BP from the soil more slowly. A clear increase in BP concentration in *Elitrygea repens* is evident when soil concentrations are over 100 μg kg^{-1}. A background concentration of BP (5 μg kg^{-1}) was observed in both species at soil BP concentrations of 20 μg kg^{-1} (the maximum allowed soil concentration Griciūtė, 1979).

TABLE II
Distribution of BP in soil and plants

Biocenosis	Plant tissues	Plant BP (μg kg^{-1})	Soil type	Soil BP (μg kg^{-1})	K_p
Meadows					
Trifolium repens	Leaves & stems	12.0 ± 8.7	Loam	20.7±15	0.6
Elitrygea repens	Leaves	6.4 ± 1.3	Loam	34.0 ± 8.9	0.2
	Roots	86.8 ± 35.2			1.0
Agrostis vulgaris	Leaves	4.5 ± 0.7	Sand	75.0 ± 20.0	0.06
	Roots	116 ± 52			1.5
Thymus serpyllum	Flowers & leaves	13.5 ± 33.0	Sand	9.5 ± 2.8	1.4
	Roots	15.8 ± 5.2			1.7
Forests					
Pleurozium schreberi		35.7 ± 17.2	Loam	9.0 ± 4.6	4.0
Pinus sylvestris	Wood	7.1			
	Bark	87.0			
	Needles	47.5			

Translocation of BP from the soil through the roots appears to be a major pathway of BP appearance in the leaves of vegetation. A different mechanism of BP uptake from the environment may occur in forests. High concentrations of BP were determined in various organs of Scots pine (*Pinus silvestris*) growing at 20 m from the highway with traffic intensity of about 1,000 vehicles per hour. A BP concentration of 87 μg kg^{-1} of BP was measured in pine bark; this concentration is about 2 times higher than in the needles and

10 times higher than in the wood. Different concentrations of BP in various organs of pine may be explained by different contents of lipids in various organs of pine, rather than by the uptake of BP from the soil to the needles through the roots. In our opinion, particulate pollutants in conjunction with lipids may not be removed by wind or rain. Analysis of BP in 22 samples of pine needles gathered at a different distances from the road showed that BP concentration varies over a wide range. A concentration of 24.3 μg kg^{-1} BP was observed in needles at a distance of 200 m from the road and between 103 and 280 μg kg^{-1} closer to the road (up to 10 m). Hence, pine needles can accumulate not only such pollutants as S and heavy metals, but BP as well. Each class of pollutants deposited on the

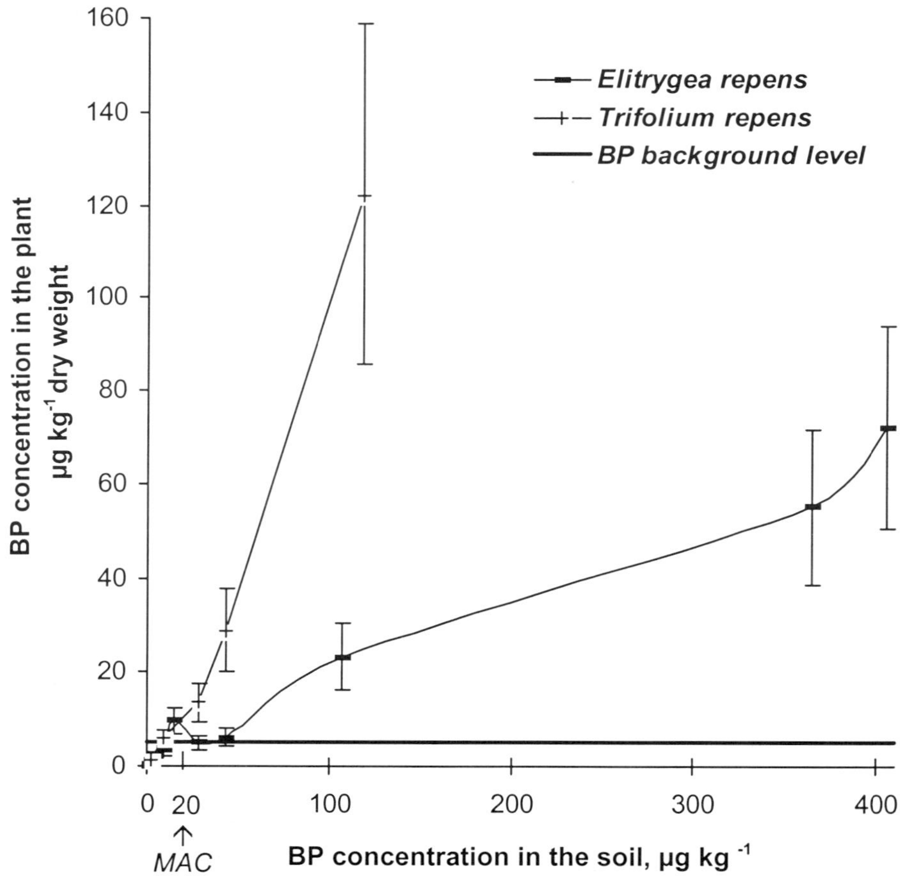

Fig.2. Accumulation of BP by *Trifolium repens* and *Elitrygea repens* leaves (n=3).

surface of needles may affect them, while the mechanism of action between conifer needle waxes and pollutants has yet to be determined. Our investigation showed that needles of pine might be used as an indicator of environmental pollution.

Low concentration of BP in the forest soil and a high partition coefficient between moss and soil (K_p=4) suggest that moss may accumulate BP directly from the atmosphere (Table II). The long residence time (3-5 years) allows moss to accumulate high quantities of BP. Within the analysis of more than 300 samples of moss, mainly *Pleurozium schreberi* and *Hylocomium splendens,* we determined that the concentration of BP in the moss samples varied from 3.1 μg kg^{-1} in the region of Panevėžys to 896 μg kg^{-1} in the Lazdijai region at the custom house near the Polish border. Furthermore, BP concentrations between 10-50 μg kg^{-1} were observed in more than 50 % of all moss samples. The average BP concentration among the samples was 54.7 μg kg^{-1} of moss (Milukaitė, 1996). According to physiological properties of moss, the concentration of BP in moss tissue should be related to BP flux from the atmosphere during the warm season of year (May-October), although such a relationship would depend on the stability of BP in moss tissue over the duration of observation (in our case, 3 years).

Table III
Partition of BP (%) among the parts of moss

Species of Moss				
Pleurozium schreberi (n=16)		*Hylocomium splendens* (n=7)		
brown	green	brown	green	light-green
52.2 ± 18.1	47.8 ± 20.0	37.7 ± 8.3	40.0 ± 18.0	22.3 ± 15.2

To determine BP stability in moss, BP concentration in the various parts of moss was measured. *Pleurozium schreberi*, 3 year of age (1991-1993) was divided into two parts: lower-brown part and upper-green part. The samples of *Hylocomium splendens* were divided into three parts: lower-brown, middle-green, upper-light green (recent growth). Every part of moss is related to a definite time period: 1.5 years for *Pleurozium schreberi* and 1 year for *Hylocomium splendens*. Partitioning of BP among the various parts of moss is presented in Table III. The quantity of accumulated BP in moss was comparatively stable from year to year for the duration of vegetation period. The lower quantity of BP in the *Hylocomium splendens* of the third year of growth (light green part) may be justified by the shorter period of exposure in the last year (samples were gathered in June-August of 1993). Furthermore, the lowest flux of BP was observed in 1993 (Table I).

On the basis of these findings, I propose predicting the BP flux to the ground surface using the concentration of BP in the moss in the following equation:

$$F = \frac{C \bullet W}{T} \qquad (1)$$

where, F is flux of BP ($\mu g\ m^{-2}\ mo^{-1}$);
C is BP concentration in moss ($\mu g\ kg^{-1}$);
W is weight of 1 m^2 moss ($kg\ m^{-2}$);
T is period of moss exposure to the atmosphere.

It was determined that 1 m^2 of *Pleurozium schreberi* weighs approximately 0.323±0.070 kg. Moss was exposed to the atmosphere for about18 months (May-October) over the 3 year period; for the rest of the year, moss in Lithuania was under snow cover. The mechanism of BP uptake from melting snow by moss is unknown and in our reasoning is unimportant. It follows that:

$$F = \frac{C \bullet 0.323}{18} \approx 0.02 \bullet C \qquad (2)$$

$$F \approx 0.02 \bullet C$$

The geographical distribution of BP flux in Lithuania, calculated from equation (2), is presented in Figure 3. The map of benzo(a)pyrene flux distribution was created by geostatistical gridding method "Kriging" with a linear model. The flux of BP over varies across Lithuania. A flux of BP of up to 2.0 $\mu g\ m^{-2}\ mo^{-1}$ occur over more than 50 % of the Lithuanian territory; a flux greater than 2.0 $\mu g\ m^{-2} mo^{-1}$ was observed near large towns (Vilnius, Kaunas, Šiauliai), the seaport of Klaipėda, and the regions of Kėdainiai and Jonava, where the fertilizer industry is developed. A high flux of BP was observed between the two highways that connect Klaipëda and Kaunas. The high flux of BP outward from Klaipëda may be related to using coal as a fuel for domestic heating in this region. Because the contours of the map were produced from observation data spaced on a 16x16 km grid across Lithuania, the determined fluxes of BP in the country mainly reflect local sources.

Applying the average BP concentration in moss (54.7 $\mu g\ kg^{-1}$) in equation (2) produces an estimated average flux of BP from the atmosphere to the ground surface during the warm season (of 1991-1993) of 1.1 $\mu g\ m^{-2}\ mo^{-1}$; the whole of the Lithuanian territory over this period would be exposed to 858 kg of BP. The calculated flux is comparable with summer bulk deposition values of BP in 1991-1993, measured at the Preila background station (0.4 - 1.5 $\mu g\ m^{-2} mo^{-1}$). We must take in consideration that the flux of BP in Lithuania during the summer months (May-October) is more than 2 times lower than during the rest of the year (Milukaitė and Mikelinskienė, 1997). Consequently, the yearly load of BP will be 1.5 times higher than that calculated for the warm season of the year only, or 1,287 kg. This calculated value of BP load is in accordance with the experimentally determined yearly load in 1991-1993 (Table I). Our investigation of BP concentration in moss confirms that moss may be used for quantitative evaluation of BP flux to the ground surface.

4. Conclusions

The results of this investigation show that: 1) the flux of benzo(a)pyrene in Lithuania varied from year to year, exposing ecosystems to various loads of carcinogen; 2) the uptake of benzo(a)pyrene by vegetation mostly is by the soil-roots-leaves and is related to BP concentrations in soil; 3) mosses accumulate BP and may be used for quantitative evaluation of benzo(a)pyrene deposition during the warm season of the year.

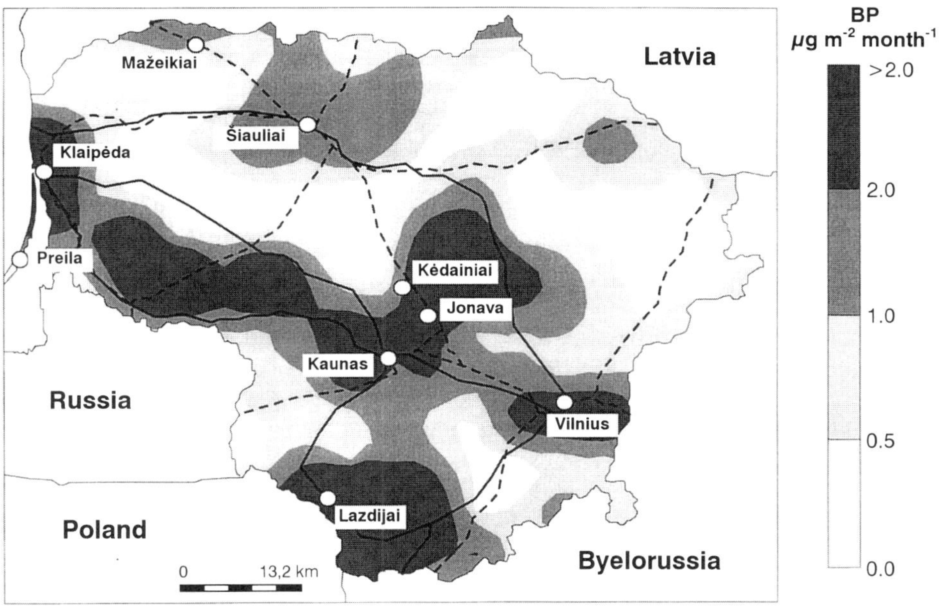

Fig. 3. Estimated distribution of BP flux in Lithuania in 1991-1993.

Acknowledgements

The author thanks the staff of Lithuanian Forestry Research Institute for the moss sampling, the senior scientist of the Botany Institute, Hab. Dr. R. Dušauskienė-Duž for help with consultations in plant physiology and assistant V. Juozefaitė for technical help.

References

Brorstrom-Lunden, E. and Lovblad, G.: 1991, *Atmos. Environ.* **25A,** 2251.

Brorstrom-Lunden, E., Lindskog, A. and Mowrer, J.: 1994, *Atmos. Environ.* **28**, 3605.

Griciūtė, L.: 1979, 'Carcinogenity of Polycyclic Aromatic Hydrocarbons', in H. Egan (ed.), *Environmental Carcinogens - Selected Methods of Analysis*, IARC, Lyon, pp. 3-15.

Grimmer, G.: 1979, 'Distribution of Polycyclic Aromatic Hydrocarbons in Environmental Samples',in H. Egan (ed.), *Environmental Carcinogens - Selected Methods of Analysis*, IARC, Lyon, pp. 55-65.

Gschwend , P.M. and Hites, R.A.: 1981, *Geochim. Cosmochim. Acta*, **45**, 2359.

Laflamme, R.E. and Hites, R.A.: 1978, *Geochim. Cosmoshim. Acta*, **42**, 289.

Lee, M., Novatny., and Bartle, K.: 1981, *Analytical Chemistry of Polycyclic Aromatic Compounds*, Academic Press, New York.

Matzner, E.: 1984, *Water, Air, Soil Pollut.*, **21**, 425.

McVeety, B. and Hites, R.A.: 1988, *Atmos. Environ.*, **22**, 511.

Milukaitė, A. and Galvonaitė, A.: 1991, *Atmos. Phys.*, Vilnius, **15**, 89. (in Russian with English summary).

Milukaitė, A.: 1986, 'Fine-structural Fluorimetry', in *Unified Methods of the Environment Background Monitoring*, Moscow, pp.46-53 (in Russian).

Milukaitė, A., Masiliūnas, L., Eitmanavičienė, N.: 1986, 'Benzo(a)pyrene in the Roadside Ecosystems', in *The Geographical Yearbook XXII-XXIII*, Vilnius, pp.191-196.

Milukaitė, A. and Mikelinskienė, A.: 1997, *J. Ecol. Chem.,* submitted.

Purvinas, E. and Skirgailaitė, V.: 1975, *Botany*, Vilnius.

Readman, J.W., Mantoura, R.F. and Rhead, M.M.: 1984, *J. Anal. Chem.*, **319**, 126.

Thomas, W. and Hermann, R.: 1980, *Staub-Reinhalt, Luft*, **40**, 440.

Thomas, W.: 1984, *Water, Air, Soil Pollut.*, **22**, 351.

Yanysheva, N., Kireeva, I., Chernichenko, I., Balenko, N. and Pavlova, N.: 1985, *Hygienic Problems of Environmental Protection from the Pollution by Carcinogens*, Kiev, pp.48-54 (in Russian).

Youngblood, W. and Blumer, M.: 1975, *Geochim. Cosmochim. Acta,* **38**, 1303.

Žukauskas, A., Jaskelevičius, B., Šiugžda, J. and Stumbras, A.: 1997, 'Anthropogenic pollution of the atmosphere in Lithuania', in L. Kairiūkštis (ed.), *Ecological Sustainability of Lithuania*, Vilnius, pp. 11-15.

AIRBORNE CONTAMINATION OF IMMATURE SOIL (LUSATIAN MINING DISTRICT) BY LIGNITE-DERIVED MATERIALS: ITS DETECTION AND CONTRIBUTION TO THE SOIL ORGANIC MATTER BUDGET

C. RUMPEL[1], H. KNICKER[2], I. KÖGEL-KNABNER[2] and R.F. HÜTTL[1]

[1]*Department of Soil Protection and Recultivation, Brandenburg Technical University of Cottbus, D-03013 Cottbus, Germany*
[2]*Lehrstuhl für Bodenkunde, Technische Universität München, D-85350 Freising Weihenstephan, Germany*

Abstract. Large areas in eastern Germany have been subjected to substantial airborne contamination by fly ash, soot and lignite dust. The objective of the study was to detect the input of lignite-derived airborne contamination into forest soils and to examine the chemical and structural characteristics of the soil organic matter, consisting of natural humic material and lignite-derived carbon in reforested immature mine soils. The mine soil developed on sandy overburden material that was excavated in open-cast lignite mines and had been relocated and deposited at a spoil bank. Samples were taken from the forest floor (L, Oh), the humic surface horizon (Ai), and the parent substrate (Cv) of an immature mine soil under a 25-year-old red oak (*Quercus rubra*), situated close to a briquette factory. The conceptual approach includes analyses of bulk soil as well as particle-size fractions for C and N contents, magnetic susceptibility, radiocarbon age and chemical structure by using ^{13}C CPMAS NMR spectroscopy. High magnetic susceptibility of the Oh and Ai horizon is the result of airborne contamination by lignite-derived ash. Fly ash contamination consisting of ferrimagnetic minerals contributes mainly to the < 20 μm fractions. In the Oh and Ai horizon, 44 % and 46 % of the C was found to be of anthropogenic origin. Structural information indicates that lignite-derived dust and/or soot are present in the coarse particle size fractions (6.3-200 μm). Anthropogenic C increased the C content as well as the contribution of alkyl and aromatic C species in the organic matter.

Keywords: ^{13}C CPMAS NMR, Germany, lignite dust, soil chemistry, soil contamination, soil organic matter

1. Introduction

In the eastern part of Germany, lignite has been the most important energy source for more than a century, producing high amounts of emissions (soot, fly ash, and lignite dust) when burned in power plants or briquette-producing factories. Consequently, large areas have been subject to airborne contamination with combustion products of lignite (Heinsdorf, 1993) and lignite dust. Airborne fly ash, soot, and lignite dust contain substantial amounts of C (up to 250 g/kg). Due to low decomposition rates (Robertson and Morgan, 1995), C contamination may accumulate in soil and may have an impact on the quantity and composition of soil organic matter (SOM) (Schmidt *et al.*, 1996).

Large amounts of airborne lignite-derived materials have accumulated at recultivated sites that often are located close to briquette-producing factories. The immature mine soils were subject to the long-term atmospheric deposition of lignite-derived organic and inorganic material.

The objective of this study was to detect the input of lignite-derived airborne contamination in a reforested mine soil and to examine the chemical and structural characteristics of the bulk organic material composed of a mixture of airborne lignite-derived materials and natural humic matter. The conceptual approach of the study includes the analyses of bulk soil as well as particle size fractions for C and N contents, magnetic susceptibility,

Water, Air, and Soil Pollution **105:** 481–492, 1998.

and chemical structure of the organic material by using ^{13}C-NMR spectroscopy. Radiocarbon dating was employed to quantify the amount of C derived from contamination.

2. Material and Methods

2.1. Study Site

An immature soil profile developing from forest-remediated Tertiary sandy overburden material, which was excavated in an open-cast lignite mine and which had been deposited after relocation, has been studied. The site is located in the Lusatian mining district (East Germany). The sampled soil profile occurs close to two briquette-producing factories. One is located 1,250 m west of the study site and the other 3,000 m to the northwest. The parent substrate (Cv) consists of Tertiary sand free of organic C and was forested 25 years ago with red oak (*Quercus rubra*). The sampled soil is in the initial state of soil development. Horizons were assigned according to the German Soil Classification (AG Boden, 1994). Two horizons of the forest floor were identified. The L-horizon consists of slightly decomposed plant material. Underneath, a 2 cm thick humified Oh horizon consists of an intimate mixture of decomposed plant material, lignite combustion products and lignite dust. During soil development, accumulation and incorporation of organic matter led to the formation of a dark-colored surface horizon (Ai). Samples were taken from the L, Oh, Ai, and Cv horizons of the soil profile. Airborne lignite combustion products and lignite dust accumulate in many places in the vicinity of the briquette-producing plant. The airborne, lignite-derived products were sampled at an afforested area exposed to the emission source, where this material formed a surface deposit up to 60 cm thick. During sampling, care was taken not to include plant litter or humic material.

2.2. Particle Size Separation

After air-drying the soil samples, the fraction > 2 mm was removed by dry sieving. Particle size separation of the Ai horizon was carried out by a combination of wet sieving and sedimentation following ultrasonic treatment (Christensen, 1985; 1992). Ultrasonic energy was 150 J mL^{-1} and had been calibrated by the procedure suggested by Schmidt *et al.* (1997). By wet sieving, the soil suspension (1:3 w/v) was separated into three sand fractions (2000-630 μm, 630-200 μm and 200-63 μm). Using sedimentation cylinders three silt fractions (63-20 μm, 20-6.3 μm and 6.3-2 μm) and a clay fraction (< 2 μm) were recovered and freeze-dried (Schmidt *et al.*, 1997).

2.3. pH, Magnetic Susceptibility, and Elemental Composition

The pH (H_2O) was measured with a glass electrode in the supernatant of a 1/2.5 (w/v) soil/water suspension. Magnetic susceptibility was determined with a Forgenta ferromagnetic analyzer FMA 5000. For calibration, a Kappa-Bridge KY-2 was used. Carbon and N were measured with a Leco CNH 1000 analyzer.

2.4. RADIOCARBON DATING

Radiocarbon dating (^{14}C) was obtained using the conventional macro-technique of liquid scintillation as described by Becker-Heidmann *et al.* (1988). A soil sample containing 6 g of carbon was burned and the carbon was subjected to benzol synthesis. After 6 weeks the ^{14}C activity of the benzol was recorded with a scintillation spectrometer (Packard Tri Carb Model 3320). ^{14}C data were corrected for isotopic effects. The amount of lignite C (x) present in the sample was calculated using the following relationship (Becker-Heidmann, 1992):

$$\mathbf{f}(x) = 1 - \exp(-(t_r - t) \times \ln 2/T), \qquad (1)$$

In this equation t_r is the ^{14}C age (Stuiver and Polach, 1977), t the ^{14}C age of soil, and T the Libby half-life of ^{14}C, i.e., 5,568 years (Libby, 1969). Assuming that the sample contains a mixture of recently formed soil organic matter derived from plant litter, and dead C (C without ^{14}C activity) from lignite, t equals 0. The maximum amount of lignite-derived C present in the sample is obtained as a percentage of the total organic C.

2.5. ^{13}C CPMAS NMR ANALYSIS

The solid-state ^{13}C NMR spectra were obtained on a Bruker MSL-100 NMR spectrometer at a frequency of 25.2 Hz. Cross-polarization with magic angle spinning (CPMAS) was applied (Schaefer and Stejskal, 1976) at 4-5 kHz. The ^{13}C chemical shifts were referenced to tetramethylsilane. A contact time of 1 msec was used. Pulse delays between 250 and 350 msec were chosen. To increase NMR sensitivity, NMR-analyses were carried out after treatment with 10 % HF described by Schmidt *et al.* (1997), who found that HF treatment of mineral soils leads to an enrichment of the organic material by the efficient removal of mineral matter including paramagnetic compounds. For quantification, the solid-state ^{13}C NMR spectra were integrated using the integration routine of the spectrometer. The chemical shift regions 0-45 ppm, 45-110 ppm, 110-160 ppm and 160-220 ppm were assigned to alkyl-C, O-alkyl-C, aromatic C, and carboxyl-C, respectively (Wilson, 1987).

3. Results and Discussion

3.1. MAGNETIC SUSCEPTIBILITY OF THE BULK MATERIAL AND PARTICLE SIZE FRACTIONS

Table I lists the magnetic susceptibility of the bulk soil samples and the forest floor. The magnetic susceptibility of the reference, lignite-derived material was as high as $281´10^{-8}$ m^3 kg^{-1}. The strong magnetic properties of the lignite-derived material is most probably due to particles of magnetite that are formed during the combustion of pyrite (Morris *et al.*, 1994). While the parent substrate (Cv) and the plant litter (L) are virtually free of magnetic minerals ($15´10^{-8}$ m^3 kg^{-1}), magnetic susceptibility values as high as $173´10^{-8}$ m^3 kg^{-1} in the forest floor (Oh) and $48´10^{-8}$ m^3 kg^{-1} in the Ai horizon have been recorded. Fine *et al.* (1993) explained enhanced magnetic susceptibility of mature soils by the formation of ferrimagnetic material during very long pedogenesis.

Compared with mature soils, the soil profile described in this study is not older than 25 years. This age refers to the time of stripping-off, relocation and redeposition of parent substrate. Thus, considering the very short time of pedogenesis, the high magnetic susceptibility of the sample can be better explained by the deposition of atmospheric particulate material containing fly ash produced by coal-fired power plants (Locke and Bertine, 1986; Stryszcz, 1994; Dearing *et al.,* 1995).

TABLE I

Chemical characteristics of the forest floor, mineral soil and reference, lignite-derived material.

Sample German classification	Depth (cm)	pH H_2O	Magnetic susceptibility (10^{-8} m^3 kg^{-1} soil)	C (g kg^{-1} soil)	N (g kg^{-1} soil)	C/N (wt /wt)
L		4.6	15	425	8.4	51
Oh	2	6.3	173	173	5.2	33
Ai	0-2	6.0	48	28	0.8	31
Cv	10	7.0	15	3	<0.1	>100
Reference lignite-derived material		6.3	281	325	3.9	81

The magnetic susceptibility values of individual size fractions of the Ai horizon range from 14 to 425´10^{-8} m^3 kg^{-1} (Table II). The maximum value, which was found for the 2-63 µm fraction, probably indicates that the mean size of the magnetic particles lies in this range.

3.2. Carbon and Nitrogen Content of Bulk Soil and Particle Size Fractions

In the forest floor a rapid decrease of organic carbon (OC) was found from 425 g OC kg^{-1} soil in the litter layer (L) to 173 g OC kg^{-1} soil in the humified O horizon (Oh) (Table I); 28 g OC kg^{-1} soil was found in the Ai horizon, and the parent substrate (Cv) was free of organic C. High C content and a C/N ratio of 51 in the litter layer indicate an input of plant material consisting of cellulose and lignin. In the course of decomposition and mineralization of plant litter, the C/N ratio decreases rapidly because of CO_2 liberation during microbial processes (Table I). The decrease of the C content in the soil profile with increasing depth and the observation that decomposition of the litter layer takes place within a one year interval suggest that rapid mineralization occurs at the study site. The C/N ratio of completely humified material is expected to be in the range of 10 to 38. Ratios between 22 and 25 were measured for comparable matured non-contaminated soils (Zezschwitz, 1980). The reference airborne lignite-derived material contains 325 g OC kg^{-1} and has a C/N ratio of 81 (Table I). The C/N ratio of 33 and 31 found in the Oh and Ai horizons of the studied soil profile probably indicates both the effect of humification of cellulose and lignin derived material and the addition of material with a high C/N ratio, i.e., carbonaceous particles of industrial origin. These data are in agreement with Schmidt *et al.* (1996), who observed elevated C content and C/N ratios of about 40 for a soil sample contaminated with brown coal dust. Our results suggest that, apart from mineral particles of fly ash, input of contaminating carbonaceous particles derived from deposition of lignite dust and/or soot occurs at the study site. In Table II, the results obtained from C and N analyses of the particle size fractions of the Ai horizon are listed.

TABLE II

Magnetic susceptibility, C and N content and distribution between particle size fractions of the Ai horizon.

Particle size	Weight	Magnetic susceptibility	OC		N		C/N
(μm)	(%)	(10^{-8} m^3 kg^{-1} soil)	(g kg^{-1} soil)	(% of total soil OC)	(g kg^{-1} soil)	(% of total soil N)	(wt/wt)
630-2000	8	14	6	2	<0.1	0	>100
200-630	36	17	8	10	<0.1	0	>100
63-200	37	35	21	26	<0.1	0	>100
20-63	8	110	40	11	1.0	11	40
6.3-20	5	250	185	29	5.0	32	37
2-6.3	3	425	130	12	6.0	21	22
<2	4	206	74	11	6.0	36	12

The highest C content (185 g OC kg^{-1} soil) of the Ai horizon was found in the 6.3-20 μm fraction. This corresponds to 29 % of the total soil C. In comparison with the highest N content in the smallest particle size fractions (5-6 g kg^{-1} in the < 6.3 μm fraction), the organic C shows a bimodal distribution with maxima in the 63-200 μm and 6.3-20 μm fractions. The C/N ratios decrease gradually from coarse (> 100 in fractions > 63 μm) to fine fractions (12 in fractions < 2 μm). The distribution pattern of C contrasts to that of natural soil where the highest C contents are found in the < 6.3 μm fraction (Christensen, 1996). This can be explained by an increasing degree of decomposition with decreasing particle size, the C/N ratio of the clay fraction being related to organic matter derived from microorganisms (Christensen, 1992). However, in the studied profile, most C is located in the coarse-size fractions and is associated with C/N ratios higher than those found by other authors for natural soils (e.g., Anderson *et al.*, 1981). An explanation for this may be the accumulation of lignite-derived material. Similar results were obtained in a recent study by Schmidt *et al.* (1996). Results from this study suggest that the airborne C-rich particles of brown coal dust accumulate mainly in the coarse to medium sized fractions of soil.

3.3. Radiocarbon Dating

^{14}C ages of 4,680 ± 60 and 5,080 ± 60 were recorded for the forest floor and the Ai horizon of the soil profile (Table III). According to Becker-Heidmann *et al.* (1988), a radiocarbon age of more than 5,000 years B.P. is very uncommon. Very high radiocarbon age is related to the material that consists of dead C, i.e., lignite. The ^{14}C age of a soil contaminated with lignite, which has lost its ^{14}C activity, can be used to determine the amount of contamination (Schmidt *et al.*, 1996; Rumpel *et al.*, 1996). Assuming that the high ^{14}C ages of the Oh and Ai horizon are related to airborne contamination, and using equation 1, 44 % and 46 % of the total OC are found to be derived from lignite.

TABLE III
Radiocarbon age and calculated lignite content of the horizons subject to airborne contamination.

Soil Horizon	Radiocarbon Age (years B.P.)	Lignite Content (% of total C)	Lignite Content (g OC kg^{-1} soil)
Oh	4,680 ± 60	44	76
Ai	5,080 ± 60	46	13

Normalized for the OC content of the two horizons (Table I), the data suggest that the 97 g $OCkg^{-1}$ soil and 15 g OC kg^{-1} soil accumulated in the Oh and Ai horizon is related to recent organic material. The anthropogenic material (fossil organic matter) represents 76 g $OCkg^{-1}$ soil of the Oh horizon and 13 g OC kg^{-1} soil of the Ai horizon. Due to constant input and slow decomposition rates (Shenour, 1966), the anthropogenic material might have been accumulating over the last 25 years. Our data show that this represents a considerable portion of the total organic fraction of the studied soils.

3.4. ^{13}C CPMAS NMR SPECTROSCOPY

Results of ^{13}C NMR spectroscopy (Figures 1 and 2) can elucidate the influence of lignite-derived material on the chemical composition of the soil organic matter by identifying structural characteristics of those products in the spectra of soil organic matter. The spectrum of the Oh horizon (Figure 1) shows the typical pattern of non-contaminated forest floors containing a large amount of plant litter (Kögel-Knabner, 1993; Baldock and Preston, 1995). Approximately 48 % of the total signal intensity of this spectrum can be found in the chemical shift region of O-alkyl carbon (45 to 110 ppm, Table IV), which is most probably due to polysaccharides of plant or microbial origin. Carbon derived from the aromatic part of lignin contributes to the signal intensity in the chemical shift region of aromatic C from 110 to 160 ppm. This region accounts for 26 % of the total relative signal intensity of the spectrum. The signals between 220 and 160 ppm (carboxylic, carbonyl and amide C) and between 45 and 0 ppm (alkyl C), contribute with approx. 7 % and 19 % to the total relative signal intensity of the spectrum, respectively. Compared to the Ai horizon (Figure 1), a decrease in signal intensity in the chemical shift region of O-alkyl C from 48 % to 41 % is observed. It occurs simultaneously with an increase in relative signal intensity in all other regions and can be explained by the preferred degradation of carbohydrates and the accumulation of more refractory organic material (Knicker and Lÿdemann, 1995). A similar decrease in relative intensities of the O-alkyl C was recorded for a non-contaminated forest soil by Preston *et al.* (1994).

The NMR spectrum of reference, lignite-derived material (Figure 1) consists mainly of aromatic and aliphatic C species (chemical shift regions 110-160 and 0-45 ppm). The ratio of signal intensities (alkyl C + aromatic C)/(O-alkyl C + carboxyl C) can be used as an indicator of contamination with lignite-derived C (Schmidt *et al.*, 1996). For plant material, this ratio was found to be in the range of 0.48 and 0.52 (unpublished data and data of Preston *et al.*, 1994). In plant litter, polysaccharide predominate and the contribution of the aromatic fraction was found to be small (Preston *et al.*, 1994). For the reference, lignite-derived material, the ratio (alkyl C + aromatic C)/(O-alkyl C + carboxyl C) is 1.3. Ratios of 0.8 and 1.0 for the Oh and Ai horizon of the investigated site indicate that part of the organic matter is derived from airborne lignite-derived contamination and increases the aromatic and the aliphatic carbon species of the organic matter.

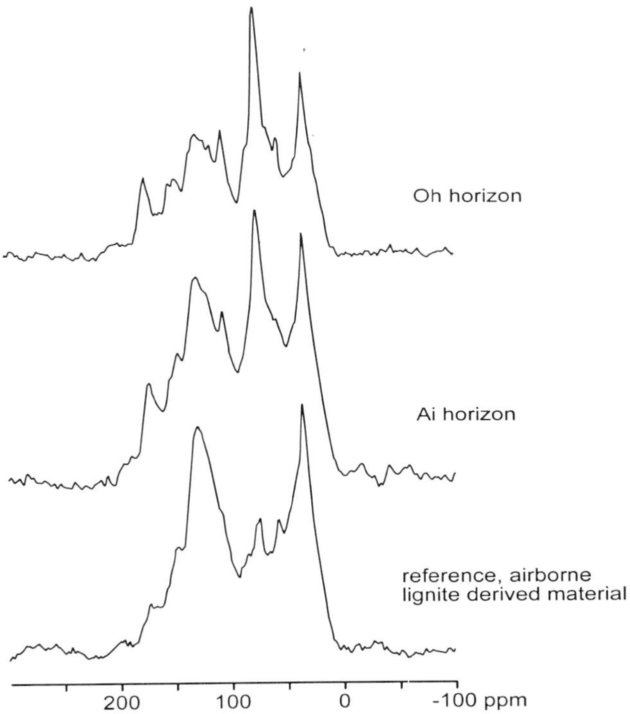

Fig. 1: ^{13}C CPMAS NMR spectra of the forest floor and Ai horizon of the studied soil profile and reference, lignite-derived material.

The amount of lignite-derived C present in the Oh and Ai horizon was determined from radiocarbon data using equation 1 and the total C content of the two horizons. Seventy-six and 13 g OC kg^{-1} soil were found to be derived from lignite material in the Oh and Ai horizon (Table III). By multiplication of the lignite OC content of the horizons with the relative intensities of the ^{13}C CPMAS NMR spectrum of reference, lignite-derived material (Table IV), it is possible to calculate the amount (g OC kg^{-1} soil) of alkyl, O-alkyl, aromatic and carboxyl C species in the Oh and Ai horizon related to lignite material (Table V). The amount of C related to airborne lignite-derived material contributing to each chemical shift region can be subtracted from the total amount of C (g OC kg^{-1} soil) present in the chemical shift regions of the original NMR spectrum, which was calculated by multiplication of the relative intensities of the spectrum with the total C content of the soil horizon.

TABLE IV
Relative intensities (%) of the ^{13}C CPMAS NMR spectra.

	Alkyl C 0-45 ppm	O-alkyl C 45-110 ppm	Aromatic C 110-160 ppm	Carboxyl C 160-210 ppm	(Alkyl + aromatic)/ (O-alkyl + carboxyl)
			Bulk soil		
Oh	19	48	26	7	0.8
Ai	21	41	29	9	1.0
Reference, lignite-derived material	23	34	34	9	1.3
			Particle size fractions		
630-200 μm	17	51	26	6	0.7
200-63 μm	18	42	33	7	1.0
63-20 μm	23	35	34	8	1.3
20-6.3 μm	22	39	30	9	1.1
6.3-2 μm	25	43	24	8	0.9
< 2 μm	28	55	11	6	0.6

The percentage of each chemical shift region without lignite-derived contamination obtained in this way represents the relative distribution of C species of the SOM which was formed during the decomposition of plant litter (Table V). After correction, the relative signal intensities in the O-alkyl region originating from decomposing plant litter increased from 41 to 47 % in the Ai horizon and from 48 to 58 % in the Oh horizon. Compared to the uncorrected spectra, the relative signal intensity in the aromatic region of the corrected spectra decreased by 2-7 % (Tables IV and V). Without the correction for the contamination, the relative intensities recorded in the four chemical shift regions of the spectra of the Oh and Ai horizon are similar. When the C species related to contamination are subtracted, clear changes due to decomposition can be observed in the profile. The alkyl and aromatic fraction increase by 3 % and 8 % from the Oh to Ai horizon, while the O-alkyl fraction decreases by 11 %. These observations are consistent with the findings of other authors (e.g. Kögel *et al.*, 1988; Zech *et al.*, 1985). The ratio of signal intensities (aromatic C + alkyl C)/(O-alkyl C + carboxyl C) was calculated to be 0.5 for the Oh horizon and 0.9 for the Ai horizon after subtraction of C species related to airborne lignite-derived material. After subtraction, the two horizons show characteristics for organic matter derived from the decomposition of plant material. Our data indicate that humification follows the same pattern as observed for natural soils and that lignite-derived contamination masks changes occurring during decomposition.

The ^{13}C NMR spectra of the particle size fractions are shown in Figure 2. Aromatic and aliphatic C species characteristic for airborne lignite-derived contamination have the greatest impact on the 6.3-200 μm fraction, leading to ratios of signal intensity (alkyl C + aromatic C)/(O-alkyl C + carboxyl C) similar to those of lignite-derived material (1.3). The greatest contribution of lignite-derived C species was noted in the 20-63 μm fraction. The NMR spectra of the 6.3-20 and 2-6.3 μm fraction show more pronounced signals in the chemical shift region of carbohydrates (45-110 ppm), indicating a higher contribution of plant-derived material.

TABLE V

Distribution of C species obtained from ^{13}C CPMAS NMR spectra of the organic matter in bulk soil, the C fraction from lignite-derived contamination and the resulting SOM after normalizing for the amounts of lignite-derived contamination.

Horizon	OC	Alkyl C	O-alkyl C	Aromatic C	Carboxyl C	(Alkyl C + aromatic C)/ (O-alkyl C + carboxyl C)
Oh				*g kg^{-1} soil*		
Bulk soil	173	33	83	45	12	0.8
Reference,lignite-derived material	76	17	26	26	7	1.3
			Relative intensities after subtraction (%)			
		17	58	19	6	0.5
Ai				*g kg^{-1} soil*		
Bulk soil	28	6	11	8	2	1.0
Reference, lignite-derived material	13	3	4	4	1	1.3
			Relative intensities after subtraction (%)			
		20	47	27	6	0.9

These observations are confirmed by the ratios of signal intensities (alkyl C + aromatic C)/(O-alkyl C+ carboxyl C) of the two fractions (1.0, 0.9). In the 630-200 μm fraction, a low ratio of signal intensity of 0.7 indicates a higher contribution of plant material. The clay fraction (< 2 μm) mainly consists of alkyl C and polysaccharides, most probably of microbial origin (Baldock *et al.*, 1992). The ratio of signal intensities (alkyl C + aromatic C)/(O-alkyl C + carboxyl C) is 0.6. Similar results were obtained for natural soils (Guggenberger *et al.*, 1995), indicating that the organic matter of the medium sand and clay fraction is relatively unaffected by lignin-derived C input. ^{13}C NMR data for the particle size fraction of the mine soil confirm the results obtained by the C and N analysis and are in agreement with the data reported by Schmidt *et al.* (1996). From these results, it can be concluded that the impact of airborne lignite-dust and/or soot on the composition of organic matter is greatest in the 63-6.3 μm fraction and that this contamination increases the contribution of aromatic and aliphatic C species.

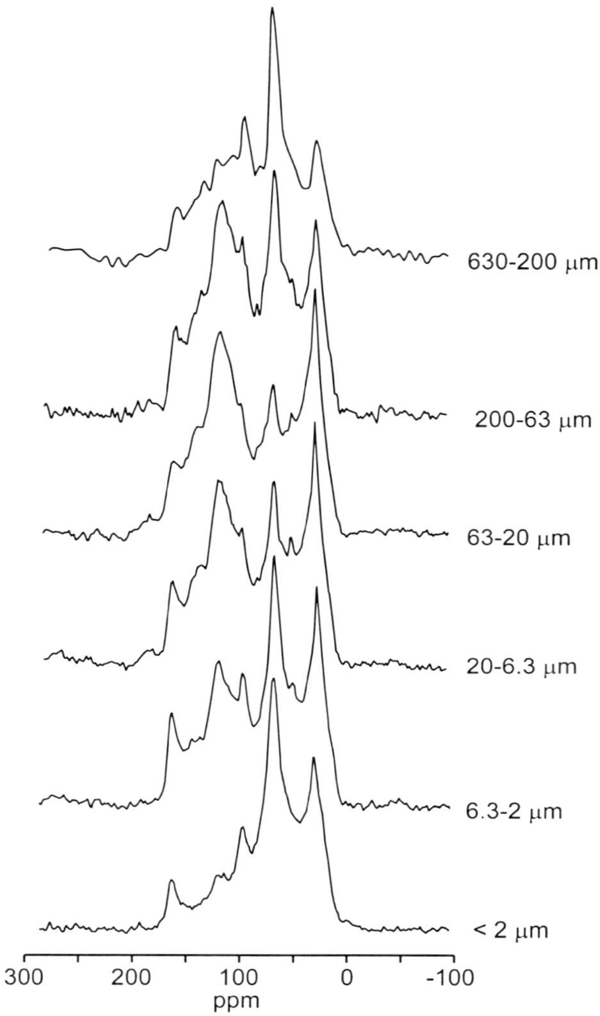

Fig. 2: ^{13}C CPMAS NMR spectra of the particle size fractions of the Ai horizon.

4. Conclusion

Contamination of a mine soil by inorganic particles of fly ash, as well as carbonaceous particles of lignite dust and soot, was detected using magnetic susceptibility measurements, C and N analysis, radiocarbon dating and ^{13}C CPMAS NMR spectroscopy. Inorganic fly ash contamination consisting of ferrimagnetic minerals is predominantly present in the 2-6.3 µm particle size fractions. Lignite derived dust and soot are found mainly in the coarse particle fractions (6.3-63 μm). The dust and soot increase the C content in soil and change the ratio of signal intensities (aromatic C + aliphatic C) to (O-alkyl C + carboxyl C) of bulk organic C.

Acknowledgments

The Deutsche Forschungsgemeinschaft is acknowledged for financial support. Dr. Stanjek (Technical University of Munich) is acknowledged for help with the calibration of the magnetic susceptibility data and for his comments on the manuscript. We are grateful to Dr. Becker-Heidmann (University of Hamburg) for the radiocarbon dating. We thank Prof. Lüdemann (University of Regensburg) for providing the NMR facilities. Mrs. Schubauer (Technical University of Munich) is acknowledged for her help with the preparation of the figures.

References

AG Boden: 1994, *Bodenkundliche Kartieranleitung*, Schweizbart'sche Verlagsbuchhandlung, Stuttgart.

Anderson, D.W., Saggar, S., Bettany, J.R. and Stewart, J.W.B.: 1981, *Soil Sci. Soc. Am. J.* **45**, 767.

Baldock, J.A., Oades, J.M., Waters A.G., Peng X., Vasallo, A.M. and Wilson, M.A.: 1992, *Biogeochemistry* **16**, 1.

Baldock, J.A. and Preston, C.M.: 1995, *Carbon Forms and Functions in Forest Soils*, Soil Science Society of America, Madison, pp. 89-118.

Becker-Heidmann, P. and Scharpenseel, H.-W.: 1992, *Radiocarbon* **34**, 535.

Becker-Heidmann P., Liang-wu, L. and Scharpenseel, H.W.: 1988, *Z. Pflanzenernähr. Bodenk.* **15**, 37.

Christensen, B.T.: 1985, *Acta Agric. Scand.* **35**, 175.

Christensen, B.T.: 1992, *Adv. Soil Sci.* **20**, 1.

Christensen, B.T.: 1996, *Adv. Soil Sci.* **22**, 97.

Dearing, J.A., Lees, J.A. and White, C.: 1995, *Geoderma* **68**, 309.

Fine, P., Singer, M.J., Verosub, K.L. and TenPasm J.: 1993, *Soil Sci. Am. J.* **57**, 1537.

Knicker, H. and Lüdemann, H.-D.: 1995, *Org. Geochem.* 23, 329.

Guggenberger, G., Zech, W., Haumaier, L. and Christensen, B.T.: 1995, *Europ. J. Soil Sci.* **46**, 147.

Kögel I., Hempfling R., Zech W., Hatcher P.G. and Schulten H.-R.: 1988, *Soil Sci.* **146**, 124.

Kögel-Knabner, I.: 1993, *Soil Biochemistry, Volume 8,* Marcel Dekker, pp. 101-137.

Libby, W. F.: 1969, *BI-Taschenbuch.* Berlin.

Heinsdorf, D.: 1993, *Water, Air, Soil Pollut.* **69**, 21.

Locke, G. and Bertine, K.K.: 1986, *Appl. Geochem.*, **1**, 345.

Morris, W.A., Versteeg, J.K., Marvin, C.H., McCarry, B.E. and Rukavina, N.A.: 1994, *Sci. Tot. Environ.* **152**, 153.

Preston, C.M., Hempfling, R., Schulten, H.-R., Schnitzer, M., Trofymow, J.A. and Axelson, D.E.: 1994, *Plant and Soil* **158**, 62.

Robertson, F.A. and Morgan, W.C.: 1995, *Aust. J. Soil Res.* **33**, 511.

Rumpel, C, Kögel-Knabner, I., Becker-Heidmann, P. and Hüttl, R.F.: 1996, *Proc. Int. Symp. Geochemistry of the Earth's Surface*, pp. 461-466.

Schaefer J. and Stejskal E.O.: 1976, *J. Amer. Chem. Soc.*, **98**, 1031.

Schmidt, M.W.I., Knicker H, Hatcher P.G. and Kögel-Knabner, I.: 1996, *Org. Geochem.* **25**, 29.

Schmidt, M.W.I., Knicker, H., Hatcher, P.G. and Kögel-Knabner, I.: 1997, *Europ. J. Soil Sci.*, **48**, 319.

Schmidt, M.W.I., Rumpel, C. and Kögel-Knabner, I.: 1997, *Europ. J. Soil Sci.*, submitted.

Shenour, E.: 1966, *Science* **155**, 991.

Stryszcz, Z., Magiera, T. and Bzowski, Z.: 1994, *Roczinki Gleboznawcze (Soil Science Annual) Suppl T.* **44**, 85.

Stuiver, M. and Polach, H.A.: 1977, *Radiocarbon* **19**, 355.

Wilson, M.A.: 1987, *NMR-Techniques and Application in Geochemistry and Soil Chemistry*, Pergamon Press, Oxford.

Zech, W., Kögel, I., Zucker, A. and Alt, H.: 1985, *Z. Pflanzenernähr. Bodenk.* **148**, 481.

Zezschwitz, E.: 1980, *Z. Pflanzenernähr. Bodenk.* **143**, 692.

EFFECTS OF EXOTIC PLANT SPECIES ON SOIL PROPERTIES IN HARDWOOD FORESTS OF NEW JERSEY

P.S. KOURTEV, J.G. EHRENFELD, and W.Z. HUANG

DEENR, Cook College, Rutgers University, New Brunswick, NJ 08903, USA

Abstract. Two exotic plant species, *Berberis thunbergii* and *Microstegium vimineum*, recently have invaded deciduous hardwood forests in the Northeast. We examined changes in soil properties that may be associated with this invasion in three parks in northern New Jersey. In each park, we collected soil and vegetation data along transects that were established across heavily infested areas and extended into uninvaded forest. The data were analyzed statistically by ANOVA and Canonical Correspondence Analysis (CCA). Significant differences were found between invaded and uninvaded plots in both soil and vegetation characteristics. Invaded areas have fewer oaks (*Quercus* spp.) in the canopy, and lack the native understory shrubs (*Vaccinium* spp.). The pH of soils in the invaded areas is significantly higher than in the uninvaded areas, and the litter and organic horizons are thinner. The data cannot show that the exotic species have caused these changes. However, the occurrence of contrasting soils in adjacent areas of native vegetation, with no evidence of differences in land-use history between areas, suggests that such a cause-and-effect relationship exists. We propose a feedback loop involving the exotic plants, and the presence of earthworms to explain these dramatic soil differences.

Keywords: *Berberis thunbergii*, invasive species, *Microstegium vimineum*, New Jersey, soil chemistry

1. Introduction

Invasions of exotic species have become a pervasive problem throughout the world (Kareiva, 1996; Vitousek *et al.*, 1996), and have occurred in most types of ecosystems in North America (Mooney and Drake, 1986). Most of these invasions have been by herbaceous species, which often invade herb-dominated communities such as old-field meadows, annual and perennial grasslands, and marshes. Prominent examples include the invasion of *Bromus tectorum* into the desert shrublands of the Great Basin (Mack, 1986), the invasion of *Lythrum salicaria* into marshes throughout the continent (Mitsch and Gosselink, 1993), and the replacement of perennial grasses by annuals in California (Mooney *et al.*, 1986). Invasions into closed-canopy forest are less common (Baker, 1986; Webb and Kaunziger, 1993; Robertson *et al.*, 1994).

The deciduous forests of the northeastern United States currently are experiencing a recent and rapid invasion by two exotic species. One, *Berberis thunbergii* DC. (common name, Japanese barberry) is a thorny, berry-bearing shrub that forms dense thickets. It has been widely introduced as an ornamental landscape plant, but has spread into otherwise undisturbed forests throughout the region. Ehrenfeld (1997) found that it is present in virtually all publicly-owned parks, forests, and nature preserves in New Jersey and surrounding areas. The other species, *Microstegium vimineum* (Trin.) A. Camus (common name, Japanese wiregrass), is a C4 grass, also originally native to Japan and eastern China, which prospers under light conditions ranging from full sunlight to 20 % of full sunlight (Barden, 1991; Hunt and Zaremba, 1992). It forms dense lawns on the forest floor. Both species occur on soils of widely varying moisture status (hydric to xeric). The two species often occur together, with *M. vimineum* carpeting the ground between *Berberis* shrubs.

Water, Air, and Soil Pollution **105:** 493–501, 1998.

Examination of herbarium records (Redman, 1995; Ehrenfeld, 1997) demonstrates that both species have become widespread only in the last 20-30 years.

We hypothesized that the substitution of these populations of mixed exotics for the native shrub understory of the deciduous forests would be associated with marked differences in the nature of the soils. Characteristics of the exotics that could promote changes in soil quality include a high density of very fine roots in the surface soils (as opposed to the presence of lower densities of coarser fine roots from the native woody shrubs and trees) and litter with higher nitrogen (N) and lower lignin content than the native shrubs or trees. Both characteristics would tend to promote more rapid nutrient cycling and a larger fraction of net N mineralization resulting from nitrification.

To determine whether soil characteristics differ between exotics-infested and exotics-free areas in mature closed-canopy forests, we sampled vegetation and soils on transects extending from dense exotic populations into undisturbed forest. Transects were established in three separate areas in New Jersey, to test the generality of the associations between soil characteristics and vegetation.

2. Materials and Methods

2.1. Study Sites

We studied plant-soil associations in three areas in New Jersey located at the same latitude (41°15' N): Morristown National Historical Park (MOR), and Allamuchy State Park (ALL), located on the pre-Cambrian gneisses of the Highlands Province, and Worthington State Forest (WOR) located on Paleozoic sandstones of the Valley and Ridge Province (Wolfe, 1977). All three have pronounced topographic relief. Soils are similar among the sites: MOR - Parker and Edneyville soils (Typic Dystrochrept and Typic Hapludult), respectively; ALL - Rockaway soils (Typic Fragiudult); WOR - Oquaga and Steinsburg soils (both Typic Dystrochrepts). All three soils have a high content of rocks (10-30 % fragments >6 cm) and are reported to be acidic (pH 4.5 - 5.5) (Fletcher, 1979).

In studies conducted in one of the field sites (Morristown National Historical Park) in 1979 and in 1986 (Ehrenfeld, 1982; Ehrenfeld and Dibeler, 1990), *Microstegium* was absent, and only a few isolated individuals of *Berberis* were observed. Although no quantitative data on prior conditions or the timing of the invasion are available for the other two parks, observations by park personnel suggest that neither exotic species was present at noticeable densities 10 years ago.

2.2. Sampling Methodology

Within each park, we chose three sites in which the two exotic species occurred in very dense populations adjacent to uninvaded forest. We then established parallel transects that extended from invaded into uninvaded areas. In all sites, there was no evidence of differences in land-use history between invaded and uninvaded areas: trees had similar diameters, there was no evidence of former agricultural field boundaries or agricultural land-use within the past 50 or more years, and there was no other evidence of differential human or natural disturbance across the transects. Samples were taken at 50 m intervals along each transect, and the first point on each transect was established at a random number of meters from the beginning of each transect. Approximately equal numbers of data points were sampled in exotics-invaded and exotics-free areas in each park.

At each sample point we determined slope angle with a clinometer, recorded the slope aspect, and noted the geomorphic position of the point (plateau at the bottom/top of a hill, slope or crest). The canopy was measured using a basal area gauge, with trees tallied by species. To measure the shrub stratum, a 5 m x 5 m plot was established with the sample point in the center. Counts of tall shrub stems and tree saplings (>2 m) by species were made in these plots, and one randomly selected 2.5 x 2.5 m quarter of the plot was used to count small shrub stems. The number of seedlings of all tree species also was recorded in the quarter-plot. Herbaceous cover, including that of *M. vimineum*, was recorded in cover classes (0-20, 20-40, 40-60, 60-80, 80-100 %) in four 1 m^2 quadrats placed in the corners of the 5 x 5 plot. The same method was used for estimating the cover by ferns, herbs and native grasses.

At each sample point we estimated the amount of large dead woody debris and the amount of exposed boulders by recording the number of large rocks and dead logs (>10 cm in diameter) intercepted by the line delimiting the 5 m x 5 m plot. At eight points along the plot perimeter we measured the soil depth, the thickness of the litter layer on the forest floor and the thickness of the organic horizon in the soil. From each sample point, we took eight soil samples (2.5 cm diameter cores) of the surface (0-10 cm) soil. In the laboratory, the composite soil samples were air-dried and analyzed for soil pH, soil organic matter (SOM %, as determined by loss on ignition after 8 hr at 500 °C) and total Kjeldahl N (TKN %, as determined by block digestion with conc. H_2SO_4 and 30 % H_2O_2 (Lowther, 1980) and Alpkem Autoanalyzer (method # A303-S071-00)).

2.3. Statistical Analysis

Although the parks are similar in their soil and vegetation characteristics, they constitute three separate and distinct sites. For that reason we did not group them; each park was analyzed separately. However, we considered results for a certain variable to be important only if they were significant in all the three parks and showed similar tendencies.

The data were statistically analyzed in two ways. First, we compared all environmental and vegetation variables between exotic-infested and exotics-free plots using ANOVA (StatView®, Abacus Concepts). We also performed canonical correspondence analysis (CCA), using CANOCO® version 3.1 (ter Braak, 1988). CCA was performed for each separate park in four ways. Every analysis included all the environmental variables, but we used different vegetation data - trees only, shrubs only, seedlings only and all vegetation data. The best correlation coefficients between the environmental axes and the species axes were obtained when all vegetation data were used and these are the results we report here.

3. Results

3.1. Vegetation

The canopy in all sites was composed of a mixture of oaks (*Quercus rubra*, *Q. velutina*, *Q. prinus* and *Q. alba*), beech (*Fagus grandifolia*), yellow-poplar (*Liriodendron tulipifera*), black birch (*Betula lenta*), ash (*Fraxinus americana*), maples (*Acer rubrum* and *A. saccharum*), and 12 other comparatively minor species. Total basal area differed between parks, but the invaded and uninvaded areas in each park were not significantly different (MOR: 15.60 vs. 14.25 m^2 ha^{-1}; ALL: 6.21 vs. 6.56 m^2 ha^{-1}; WOR: 9.15 vs. 10.69 m^2 ha^{-1}), nor was species richness of canopy trees per sample point different between invaded and uninvaded areas. However, the relative abundance of oaks and several other species differed significantly between invaded and uninvaded areas in each park (Table I). *Quercus* spp. tended to be significantly more abundant in the uninvaded areas (55 %, 73 % and 43 % of total basal area in uninvaded portions of MOR, ALL and WOR, respectively, vs. 23 %, 49 % and 41 % of total basal area, in the respective invaded areas), whereas *L. tulipifera* and *F. americana* were more abundant in the invaded areas (51 %, 5 % and 10 % of total basal area in the invaded sections of MOR, ALL and WOR, respectively). *B. lenta*, although abundant in all parks (12 %, 15 % and 44 % respectively of total basal area), did not differ significantly in basal area between invaded and uninvaded areas.

Shrub and vine species richness was low; only 13 species were observed. In uninvaded areas, the shrub layer was dominated by *Vaccinium stamineum* and *V. pallidum*, in the invaded areas, these species were significantly less abundant. In addition to the dense presence of *Berberis*, several other weedy and exotic shrubs and vines were found in the invaded areas. In MOR, heavy deer browse has eliminated most shrubs other than *Berberis*, which is defended by high concentrations of alkaloids (Weber *et al.*, 1989).

The mean *Microstegium* cover class was 3.6 ± 0.26 (corresponding to approximately 55 % cover) in MOR, and 1.6 ± 0.15 (corresponding to approximately 31 % cover) in WOR. It was absent from ALL. Herbaceous cover from ferns, sedges and native grasses was low in all parks, and did not differ among sample point types. *Microstegium* occured most commonly intermixed with *Berberis* shrubs, such that most invaded areas contained both species (coefficients of association between *Berberis* and *Microstegium* occurrence: MOR χ^2 = 20.936, WOR χ^2 = 20.795, $p < 0.0001$ for both; contingency table, Abacus Concepts, 1996).

3.2. Soil Characteristics

Several soil characteristics showed significant differences between invaded and uninvaded plots. Most notably, soil pH was significantly higher under the exotics in all three parks (n=58, $p < 0.01$ for MOR; n = 68, $p < 0.01$ for ALL; n = 67, $p < 0.01$ for WOR). Soils under the exotics were more variable in pH, as all sites had similar minimum observed values (Table II), but in the invaded areas, high pH values often were observed. In ALL, the maximum pH observed in an invaded area (a dense *Berberis* thicket surrounded by uninvaded forest) was 2.5 units higher than the maximum observed in the undisturbed forest (Table I); in the other sites, the differences in the maxima were approximately 1 unit.

TABLE I

Basal area of dominant tree and shrub species (mean ± standard error). Differences between invaded and uninvaded areas determined by analysis of variance (* $p < 0.05$; ** $p < 0.01$; *** $p < 0.001$).

	MOR		ALL		WOR	
Species	Invaded	Uninvaded	Invaded	Uninvaded	Invaded	Uninvaded
Quercus velutina	0.52** ± 0.20	2.76 ± 0.56	0.51 ± 0.14	0.33 ± 0.12	0.12 ± 0.06	0.28 ± 0.22
Q. rubra	1.07* ± 0.29	0.30 ± 0.12	1.10** ± 0.21	2.02 ± 0.32	0.83 ± 0.20	1.47 ± 0.51
Q. prinus	1.15** ± 0.56	4.12 ± 0.88	0.30** ± 0.09	2.01 ± 0.47	0.58** ± 0.15	2.66 ± 1.05
Q. alba	0.82 ± 0.26	0.60 ± 0.21	1.12 ± 0.24	0.46 ± 0.18	0.22 ± 0.07	0.40 ± 0.42
Liriodendron tulipifera	6.78** ± 1.05	2.13 ± 0.48	0.12 ± 0.07	0.04 ± 0.04	0.24 ± 0.09	0.09 ± 0.06
Fagus grandifolia	0.41 ± 0.13	0.55 ± 0.48	0.00 ± 0.00	0.00 ± 0.00	0.07 ± 0.05	0.00 ± 0.00
Fraxinus americana	1.15** ± 0.03	0.00 ± 0.00	0.20 ± 0.07	0.03 ± 0.03	0.66 ± 0.05	0.17 ± 0.10
Vacinium stamineum	0.00 ± 0.00	0.00 ± 0.00	1.50** ± 0.41	4.60 ± 1.00	1.40** ± 0.81	5.90 ± 1.80
V. pallidum	0.01 ± 0.00	0.00 ± 0.00	1.48* ± 0.44	4.50 ± 0.95	0.21* ± 0.11	3.43 ± 2.20
Berberis thunbergii	3.07*** ± 0.73	0.00 ± 0.00	3.71*** ± 0.80	0.00 ± 0.00	4.80*** ± 0.69	0.00 ± 0.00

TABLE II

Ranges of mineral soil pH in invaded and uninvaded sample points of three study areas.

	Invaded			Univaded		
Site	Minimum	Mean	Maximum	Minimum	Mean	Maximum
MOR	4.23	5.06	6.69	4.23	4.93	5.88
ALL	3.92	3.92	7.66	3.98	4.42	5.08
WOR	4.11	4.11	5.63	4.67	4.61	4.99

The presence of exotic species was associated with a thinner litter layer at two of the three study sites (Figure 1). In many of the invaded areas, the litter layer was com-pletely missing. A similar pattern was found for the thickness of the organic horizon (Oe + Oa), but the differences were significant at all three parks ($p < 0.01$ for ALL; $p < 0.01$ for MOR; $p < 0.01$ for WOR; sample sizes as for pH). The association between these alterations in soil characteristics and the exotic plants is further illustrated by the significant regressions with *Berberis* density (Table III). Other soil characteristics (slope angle, rockiness, logs, soil depth, SOM %, TKN %) showed no significant differences between parks, except for a significantly higher total N in soil from invaded plots in WOR ($n = 67$, $p = 0.05$).

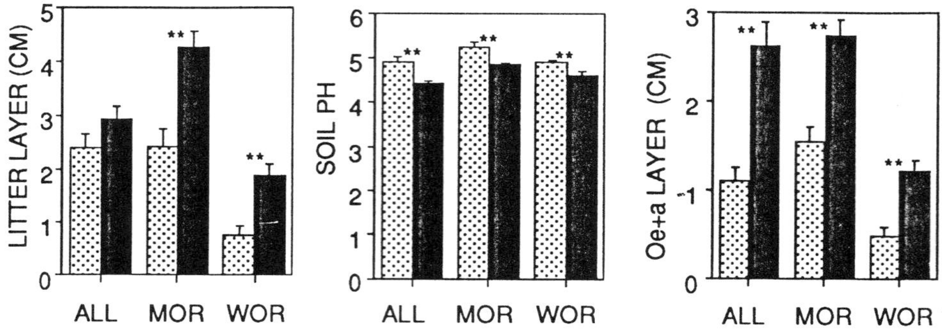

Fig. 1. Soil properties of invaded (stippled bars) and uninvaded (solid bars) areas of the three study sites. Statistical differences determined by analysis of variance (* $p < 0.05$; ** $p < 0.01$; *** $p < 0.001$).

TABLE III
Regression coefficients (R) for the environmental factors and barberry densities (*$p < 0.05$; **$p < 0.01$; ***$p < 0.001$).

	MOR	ALL	WOR
Litter layer thickness	-0.527***	n.s.	-0.308*
Organic horizon thickness	-0.627***	-0.427***	-0.425***
pH	0.503**	0.266*	0.350**
Quercus spp. basal area	-0.369**	-0.291*	-0.253*
L. tulipifera basal area	0.637***	n.s.	n.s.
Vaccinium spp. Basal area	n.s.	-0.298*	-0.315**

3.3. Canonical Correspondence Analysis

The canonical axes extracted by CCA in the analysis of each park were highly correlated with the species axes (Table IV). In the three parks, the first two canonical axes were composed of different environmental variables, but several similar trends, which completely agree with the results from the ANOVA, are evident (Figure 2 gives an example). The angle between the arrows representing environmental factors and species scores on the ordination plots are indicative of the correlation between them (ter Braak, 1988): small, acute angles indicate a high positive correlation, while large obtuse angles indicate a strong negative correlation. The following two trends are evident for all three parks: 1) a high positive correlation between *Berberis* or *Microstegium* and soil pH and between native blueberries and oak trees and the thickness of litter layer and organic horizon in the soil; 2) a strong negative correlation between *Berberis* and *Microstegium* and the native shrubs and oaks and between *Berberis* and *Microstegium* and the thickness of litter layer and the soil organic horizon.

TABLE IV

Characteristics of the CCA axes (corr - correlation coefficient between a species axis and the corresponding environmental axis, λ - eigenvalues).

	Axis1		Axis2	
Site	corr.	λ	corr.	λ
MOR	0.818	0.52	0.768	0.18
ALL	0.835	0.45	0.701	0.16
WOR	0.809	0.43	0.631	0.18

4. Discussion

The association of the two exotic species with thinner litter and the O_{e+a} horizons, and higher mineral soil pH is clear from both the analyses of individual variates and from the CCAs. The significant correlations between these variables and the density of *Berberis* (Table III) strongly support the association of these changes with the invasions. A distinct organic horizon and acidic mineral soil (pH ~ 4.5) are the expected characteristics for these oak-dominated forests, as is reflected in the descriptions published in 1979, prior to the spread of the exotics, and are the characteristics found throughout the region in this vegetation type (Fletcher, 1979). The alteration in soil properties and reaction in the relatively short time since the invasion occurred is thus striking.

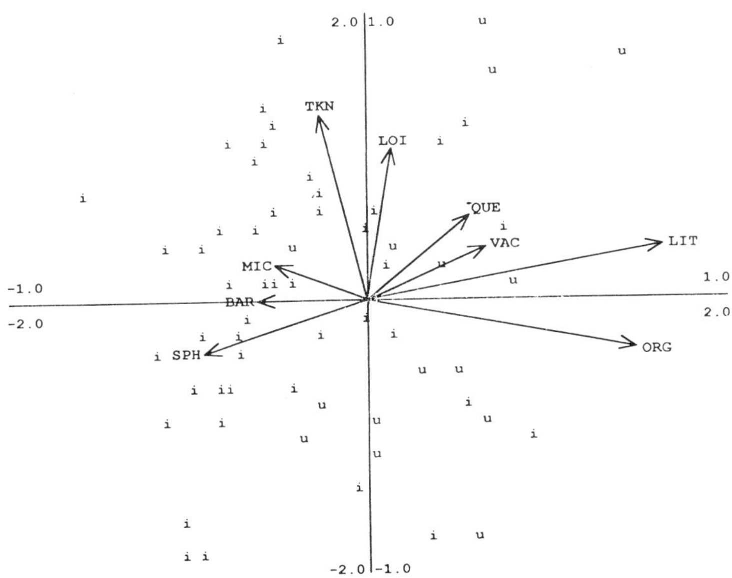

Fig. 2. Example of a CCA analysis in WOR; points marked by "u" refer to uninvaded areas; points marked by "i" refer to invaded areas. Abbreviations: BAR - barberry, LIT - thickness of litter layer, LOI - organic matter in soil, MIC - *Microstegium*, ORG - thickness of organic horizon, QUE - *Quercus prinus*, SPH - soil pH, TKN - TKN, VAC - *Vaccinium stamineum*; Scales : -1.0 to 1.0 for environmental factors, -2.0 to 2.0 for species and sample points.

While these data clearly demonstrate an association between the presence of the two exotic plant species and different soil structure and chemistry, they cannot resolve the question of whether the soil changes preceded the exotic invasions (and indeed promoted their spread) or whether the exotics have caused the changes. The study areas in each park were carefully selected to avoid any obvious differences in land-use history in the invaded and uninvaded areas. While total basal area of oaks was significantly higher in the uninvaded than the invaded stands, there was still substantial oak presence in the invaded areas (23-49 % of total basal area). Tree species with less acidic litter (Coté and Fules, 1994), such as *Fraxinus*, constituted at most 10 % of the total basal area. Thus, the differences in canopy composition are unlikely to have led to differences in soil qualities.

While conducting the sampling described above, we observed a high density of earthworms in the invaded soils, while earthworms were present but rare in the uninvaded soils. Earthworms are well known to incorporate litter material into the mineral soil (Edwards and Bohlen, 1996; Coleman and Crossley, 1996); an increased abundance of earthworms in the invaded areas is likely to provide an explanation for the decrease in litter and organic horizon thickness. There is less evidence for the role of earthworms in altering soil pH. There are some reports of increased P, K and Ca availability in earthworm casts and earthworm-worked soils (Atlavinyte and Vanagas, 1973; Blair *et al.*, 1995; Edwards and Bohlen, 1996). Casts may have a pH higher than the bulk soil (Edwards and Lofty, 1977; Edwards and Bohlen, 1996). However, there have been relatively few studies that address the potential effects of earthworms on the pH of the bulk soil over time.

We speculate that the exotic plants may stimulate localized increases in earthworm populations, and these localized increases may then cause localized increases in pH through a reduction in organic acid inputs as the litter and organic horizons are incorporated, and through a net upward movement of base-rich soil material in casts of endogeic species. However, it is not clear why or how earthworms become abundant in areas of high exotic plant density. The paucity of earthworms in the noninvaded soils suggests that the plant invasion may provide a environment that is more conducive to earthworm growth than the uninvaded soils. One possible mechanism for this facilitation could be increased pH in the rhizospheres of the exotics, caused by preferential uptake of NO_3^- over NH_4^+ (Haynes 1986). Both species of exotics have abundant shallow, very fine roots and are endomycorrhizal (unpubl. observation by JE). These characteristics, together with the rapid growth of exotics, suggests that they may preferentially utilize NO_3^- as a source. If so, NO_3^- uptake could account for localized improvement of the soil environment for earthworm activity; increased earthworms would then expand the zone of soil with increased pH. If localized increase in pH then stimulated increased nitrification, a positive feedback loop would be established in which the exotic plants and earthworms mutually facilitate each other, driving the rapid and extensive changes in soil qualities that we have documented here. This hypothesized feedback loop is currently under investigation in our laboratory.

Acknowledgments

This work was supported by USDA National Research Initiative, Competitive Grants Program (95-37101-1701). We thank A. Cantelmo, Jr., J. Lambe, M. Pekocz, M. Kourteva, and P. Yakovleva for assistance in the field work.

References

Abacus Concepts : 1996, *Statview Reference*, Abacus Concepts, Inc, Berkeley, CA.
Atlavinyte, O. and Vanagas, J.: 1973, *Pedobiologia* **13**, 344.
Baker, H.G.: 1986, in H.A. Mooney and J.A. Drake (eds), *Ecology of Biological Invasions of North America and Hawaii*, Springer-Verlag, NY, pp. 44-57.
Barden, L.: 1991, *Amer. Midl. Nat.* **118**, 40.
Blair, J.M., Parmelee, R.W. and Lavelle, P,: 1995, in P.F. Hendrix (ed.), *Earthworm Ecology and Biogeography in North America*, CRC Press, Inc, pp. 127-158
Coleman, D.C. and Crossley, D.C., Jr.: 1995, *Fundamentals of Soil Ecology*, Academic Press, San Diego, CA.
Coté, B. and Fyles, J. W.: 1994, *Can. J. For. Res.* **24**, 192
Dibeler, B.J. and Ehrenfeld, J.G.; 1990, *Bull. N.J. Acad. Sci.* **35**, 1.
Edwards, C.A. and Bohlen, P.J.: 1996, *Biology and Ecology of Earthworms*, Chapman & Hall, London.
Edwards, C.A. and Lofty, J. R.: 1977, *Biology of Earthworms*, Chapman and Hall, London.
Ehrenfeld, J.G.: 1982, *Bull. N.J. Acad. Sci.* **27**, 1.
Ehrenfeld, J.G., *Bull. Torrey Bot. Club* **124**, 210.
Fletcher, S.J.: 1975, *Soil Survey of Warren County, New Jersey.* U.S. Dept. Agric. Soil. Conserv. Service, Somerset, NJ.
Haynes, R.J.: 1986, *Mineral Nitrogen in the Plant-Soil System*, Academic Press, Orlando, Florida, 483 pp.
Hunt, D.M. and Zaremba, R.E.: 1992, *Rhodora* **94**, 167.
Kareiva, P.: 1996, *Ecology* **77**, 1651.
Lowther, J.R.: 1980, *Comm. Soil Sci. Plant Anal.* **11**, 175.
Mack, R.N.: 1986, in H.A. Mooney and J.A. Drake (eds.), *Ecology of Biological Invasions of North America and Hawaii*, Springer-Verlag, NY, pp. 191-213.
Mitsch, W. and Gosselink, J.G.: 1993, *Wetlands, 2nd ed.*, Van Nostrand Reinhold, New York.
Mooney, H.A., Hamburg, S.P. and Drake, J.A.: 1986, in H.A. Mooney and J.A. Drake, (eds.), *Ecology of Biological Invasions of North America and Hawaii*, Springer-Verlag, NY, pp. 250-269.
Mooney, H.A. and Drake, J.A., (eds): 1986, *Ecology of Biological Invasions of North America and Hawaii*, Springer-Verlag, NY, 321 pp.
Redman, D.E.: 1995, *Castanea* **60**, 270.
Robertson, D.G., Yurlina, M.E, and Handel, S.N.: 1994, *Bull. Torrey Bot. Club* **121**, 119.
ter Braak, C.J.F.: (1987-1992), *CANOCO - A FORTRAN Program for Canonical Community Ordination*, Microcomputer Power, Ithaca, NY.
Vitousek, P.M., D'Antonio, C.M., Loope, L.L. and Westbrooks, R.: 1996, *Amer. Scient.* **84**, 468.
Webb, S.L. and Kaunziger, C.K.: 1993, *Bull. Torrey Bot. Club* **120**, 343.
Weber, J.F., LeRay, A.M. and Bruneton, J.: 1989, *J. Nat. Products* **52**, 81.
Wolfe, P.E.: 1977, *The Geology and Landscapes of New Jersey*, Crane Russak, NY.

LIST OF REVIEWERS

JOHN ABER, *University of New Hampshire, Durham, New Hampshire, USA*
MARY BETH ADAMS, *U.S. Forest Service, Parsons, West Virginia, USA*
JOHN K. ADAMSON, *Institute of Terrestrial Ecology, Cumbria, United Kingdom*
LOTTA ANDERSSON, *Linköping University, Linköping, Sweden*
RICHARD S. ARTZ, *National Oceanic and Atmospheric Administration, Silver Spring, Maryland, USA*
ROGER BALES, *University of Arizona, Tucson, Arizona, USA*
JILL BARON, *Natural Resource Ecology Laboratory, Fort Collins, Colorado, USA*
DAN BERGGREN, *Swedish University of Argicultural Sciences, Uppsala, Sweden*
JORGEN BILLE-HANSEN, *Danish Forest and Landscape Research Institute, Hørsholm , Denmark*
MICHAEL BILLETT, *University of Aberdeen, Aberdeen, United Kingdom*
SIMON BOTTRELL, *University of Leeds, Leeds, United* Kingdom
VLADIMIR BOUŠKA, *Charles University, Prague, Czech Republic*
OWEN BRICKER, *U.S. Geological Survey, Reston, Virginia, USA*
BERT BRINKMAN, *Institute of Forestry and Nature Research, Den Burg, the Netherlands*
KEITH BULL, *Institute of Terrestrial Ecology, United Kingdom*
JIŘÍ ČERNÝ, *Czech Geological Survey, Prague, Czech Republic*
M. ROBBINS CHURCH, *U.S. Environmental Protection Agency, Corvallis, Oregon, USA*
R.M. COX, *Canadian Forest Service, Fredericton, New Brunswick, Canada*
ETIENNE DAMBRINE, *INRA, Seichamps, France*
A.W.H. DAMMAN, *University of Connecticut, Storrs, Connecticut, USA*
W. DEVRIES, *Winand Staring Center, Wageningen, The Netherlands*
DAVID R. DEWALLE, *Pennsylvania State University, University Park Pennsylvania, USA*
NANCY DISE, *The Open University, Milton Keynes, UK*
IAN DOUGLAS, *University of Manchester, Manchester, United Kingdom*
CHARLES T. DRISCOLL, *Syracuse University, Syracuse, New York, USA*
JEFF DUKES, *Stanford University, Stanford, California, USA*
BRIDGET EMMETT, *Institute of Terrestrial Ecology, Bangor, Gwynedd, United Kingdom*
YIGAL EREL, *The Hebrew University, Jerusalem, Israel*
KEITH ESHLEMAN, *Appalachian Environmental Laboratory, Frostburg, Maryland, USA*
MARK FENN, *Pacific Southwest Forest Range Experiment Station, Riverside, California, USA*
K.-H. FEGER, *Albert-Ludwig Universität, Freiburg, Germany*
IVAN FERNANDEZ, *University of Maine, Orono, Maine, USA*
THOMAS FISHER, *Horn Point Environmental Laboratory, Cambridge, Maryland, USA*
MARTIN FORSIUS, *Finnish Environment Institute, Helsinki, Finland*
MARIA-CRISTINA FORTI, *Instituto Nacional de Pesquisas Espacias, São José dos Campos, Brazil*
DAVID FOWLER, *Institute of Terrestrial Ecology, Midlothian, United Kingdom*
CHRIS FREEMAN, *University of Wales, Bangor, Wales, United* Kingdom
STEPHEN E. FROLKING, *University of New Hampshire, Durham, New Hampshire, USA*
GERHARD GEBAUER, *University of Bayreuth, Bayreuth, Germany*
ANNE E. GIBLIN, *The Ecosystems Center, Woods Hole, Massachusetts, USA*
P. GOOVAERTS, *Universite Catholique Louvain, Louvain-la-Neuve, Belgium*
NEVILLE GORHAM, *University of Minnesota, Minneapolis, Minnesota, USA*
PER GUNDERSEN, *Danish Forest and Landscape Research Institute, Hørsholm, Denmark*
JON PETTER GUSTAFFSON, *Royal Institute of Technology, Stockholm, Sweden*
JANE R. HALL, *Institute of Terrestrial Ecology, Huntingdon, United Kingdom*
RON HARRIMAN, *Freshwater Fisheries Laboratory, Perthshire, United Kingdom*
A.L. HEATHWAITE, *University of Sheffield, Sheffield, United Kingdom*
LARS O. HEDIN, *Cornell University, Ithaca, New York, USA*
ARNE HENRIKSEN, *Norwegian Institute for Water Research, Oslo, Norway*
ALAN T. HERLIHY, *Oregon State University, Corvallis, Oregon, USA*
MADS F. HOVMAND, *National Environmental Research Institute, Hørsholm, Denmark*

JAKUB HRUŠKA, *Czech Geological Survey, Prague, Czech Republic*
REINHARD HÜTTL, *Brandenburg Technical University of Cottbus, Cottbus, Germany*
MATTI JOHANSSON, *Finnish Environment Institute, Helsinki, Finland*
ARTHUR H. JOHNSON, *The University of Pennsylvania, Philadelphia, Pennsylvania, USA*
CHRISTOPHER E. JOHNSON, *Syracuse University, Syracuse, New York, USA*
DALE W. JOHNSON, *Desert Research Institute, University of Nevada, Reno, Nevada, USA*
CAROL A. JOHNSTON, *University of Minnesota, Duluth, Minnesota, USA*
JOHN R.M. KELSO, *Canada Department of Fisheries and Oceans, Sault Ste. Marie, Ontario, Canada*
MARTIN KERNAN, *Environmental Change Research Center, University College, London, United Kingdom*
MICHAEL KERTESZ, *Federal Technical Institute of Technology, ETH-Center, Zurich, Switzerland*
PIRKKO KORTELAINEN, *Finnish Environment Institute, Helsinki, Finland*
PAVEL KRÁM, *Czech Geological Survey, Prague, Czech Republic*
BOHDAN KŘÍBEK, *Czech Geological Survey, Prague, Czech Republic*
MICHAEL KROM, *University of Leeds, Leeds, United Kingdom*
HANS KROS, *Winand Staring Center, Wageningen, The Netherlands*
JANA KUBIZŇÁKOVÁ, *Czech Academy of Sciences, České Budějovice, Czech Republic*
KATE LAJTHA, *Oregon State University, Corvallis, Oregon, USA*
GREGORY B. LAWRENCE, *U.S. Geological Survey, Troy, New York, USA*
BRUCE LAZERTE, *Ontario Ministry of the Environment, Dorset, Ontario, Canada*
SEBESTIEN LEMONTAGNE, *University of Waterloo, USA*
AHTI LEPISTÖ, *Finnish Environment Institute, Helsinki, Finland*
STEPHEN LINDBERG, *Oak Ridge National Laboratory, Oak Ridge, Tennessee, USA*
GUN LÖVBLAD, *Swedish Environmental Research Institute, Göteborg, Sweden*
LARS LUNDIN, *Swedish University of Argicultural Sciences, Uppsala, Sweden*
PAMELA A. MATSON, *University of California, Berkeley, California, USA*
EGBERT MATZNER, *BITÖK, University of Bayreuth, Bayreuth, Germany*
MICHAEL MCCLAIN, *University of Washington, Seattle, Washington, USA*
KARL-MEIWES, *Lower Saxony Forest Experiment Station, Göttingen, Germany*
THOMAS MEIXNER, *University of Arizona, Tucson, Arizona, USA*
ERIC MILLER, *Dartmouth College, Hanover, New Hampshire, USA*
MYRON J. MITCHELL, *State University of New York, Syracuse, New York, USA*
FILIP MOLDAN, *Swedish Environmental Research Institute, Göteborg, Sweden*
ARVIN MOSIER, *USDA Agricultural Research Service, Ft. Collins, Colorado, USA*
FILIP MOLDAN, *Swedish Environmental Research Institute, Göteborg, Sweden*
DON MONTEITH, *Environmental Change Research Center, University College, London, United Kingdom*
STEPHEN S. MULKEY, *University of Missouri, St. Louis, Missouri, USA*
PETER MURDOCH, *U.S. Geological Survey, Troy, New York, USA*
COLIN NEAL, *Institute of Hydrology, Oxon, United Kingdom*
LARS-OWE NILSSON, *Swedish University of Agricultural Science, Uppsala, Sweden*
STEPHEN A. NORTON, *University of Maine, Orono, Maine, USA*
MARTIN NOVÁK, *Czech Geological Survey, Prague, Czech Republic*
ULF NYSTROM, *Göteborg University, Göteborg, Sweden*
FRANK OLDFIELD, *University of Liverpool, Liverpool, UK*
LARRY L, OLMSTEAD, *Duke Power Company, Huntersville, North Carolina, USA*
JOSEF PACYNA, *Norwegian Institute for Air Research, Kjeller, Norway*
LINDA H. PARDO, *USDA Forest Service, Burlington, Vermont, USA*
NORMAN E. PETERS, *U.S. Geological Survey, Atlanta, Georgia, USA*
M. POSCH, *Winand Staring Center, Wageningen, The Netherlands*
ROBERT G. QUALLS, *University of Nevada, Reno, Nevada, USA*
GERT JAN REINDS, *Winand Staring Center, Wageningen, The Netherlands*
WILLIAM A. REINERS, *University of Wyoming, Laramie, Wyoming, USA*
BRIAN REYNOLDS, *Institute of Terrestrial Ecology, Bangor, Gwynedd, United Kingdom*
WOLFGANG SCHAAF, *Brandenburg Technical University of Cottbus, Cottbus, Germany*
TIMOTHY SCHERBATSKOY, *University of Vermont, Burlington, Vermont, USA*
RAY SEMKIN, *University of Toronto, Toronto, Ontario, Canada*

-REY A. SIMMONS, *West Virginia Wesleyan College, Buckhannon, West Virginia, USA*
KO SOVERI, *Water and Environment Research Institute, Helsinki, Finland*
N STODDARD, *Dynamac Corporation, Corvallis, Oregon, USA*
BERT STOTTLEMYER, *National Biological Service, Fort Collins, Colorado, USA*
R SULOVSKÝ, *Masaryk University, Brno, Czech Republic*
RALD SVERDRUP, *Lund University, Lund, Sweden*
GAI TECLE, *Northern Arizona University, Flagstaff, Arizona, USA*
HAEL TOBIN, *University of Pittsburgh, Pittsburgh, Pennsylvania, USA*
N VALIELA, *Marine Biological Laboratory, Woods Hole, Massachusetts, USA*
O VAN BREEMEN, *Wageningen Agricultural University, Wageningen, The Netherlands*
. VELBEL, *Michgan State University, East Lansing, Michigan, USA*
EF VESELÝ, *Czech Geological Survey, Prague, Czech Republic*
ER M. VITOUSEK, *Stanford University, Stanford, California, USA*
HARD H. WARING, *Oregon State University, Corvallis, Oregon, USA*
RIBETH WATWOOD, *Idaho State University, Pocatello, Idaho, USA*
ELMAN WIEDER, *Villanova University, Villanova, Pennsylvania, USA*
RISTOPHER J. WILLIAMS, *Villanova University, Villanova, Pennsylvania, USA*
EPH B. YAVITT, *Cornell University, Ithaca, New York, USA*